FINITE PACKING AND COVERING

Finite arrangements of convex bodies were intensively investigated in the second half of the twentieth century. Connections were made to many other subjects, including crystallography, the local theory of Banach spaces, and combinatorial optimization. This book, the first one dedicated solely to the subject, provides an in-depth, state-of-the-art discussion of the theory of finite packings and coverings by convex bodies. It contains various new results and arguments, besides collecting those scattered throughout the literature, and provides a comprehensive treatment of problems whose interplay was not clearly understood before. To make the material more accessible, each chapter is essentially independent, and two-dimensional and higher dimensional arrangements are discussed separately. Arrangements of congruent convex bodies in Euclidean space are discussed, and the density of finite packing and covering by balls in Euclidean, spherical, and hyperbolic spaces is considered.

CAMBRIDGE TRACTS IN MATHEMATICS

General Editors

B. BOLLOBAS, W. FULTON, A. KATOK, F. KIRWIN,
P. SARNAK, B. SIMON

154   Finite Packing and Covering

Károly Böröczky Jr.

Alfréd Rényi Institute of Mathematics, Hungarian Academy of Sciences, Budapest

Finite Packing and Covering

CAMBRIDGE
UNIVERSITY PRESS

PUBLISHED BY THE PRESS SYNDICATE OF THE UNIVERSITY OF CAMBRIDGE
The Pitt Building, Trumpington Street, Cambridge, United Kingdom

CAMBRIDGE UNIVERSITY PRESS
The Edinburgh Building, Cambridge CB2 2RU, UK
40 West 20th Street, New York, NY 10011-4211, USA
477 Williamstown Road, Port Melbourne, VIC 3207, Australia
Ruiz de Alarcón 13, 28014 Madrid, Spain
Dock House, The Waterfront, Cape Town 8001, South Africa

http://www.cambridge.org

First published 2004

Printed in the United States of America

Typeface Times Roman 10.25/13 pt.   *System* LATEX 2$_\varepsilon$ [TB]

A catalog record for this book is available from the British Library.

Library of Congress Cataloging in Publication Data
Böröczky, K.
Finite packing and covering / Károly Böröczky Jr.
p.   cm. – (Cambridge tracts in mathematics ; 154)
Includes bibliographical references and index.
ISBN 0-521-80157-5
1. Combinatorial packing and covering.   I. Title.   II. Series.
QA166.7.B67   2004
511′.6 – dc22   2003065450

ISBN 0 521 80157 5 hardback

To Csilla and Csenge

Contents

Preface

The year 1964 witnessed the publication of two fundamental monographs about infinite packing and covering: László Fejes Tóth's *Regular Figures*, which focused on arrangements in surfaces of constant curvature, and Claude Ambrose Rogers's *Packing and Covering*, which discussed translates of a given convex body in higher dimensional Euclidean spaces. This is the finite counterpart of the story told in these works. I discuss arrangements of congruent convex bodies that either form a packing in a convex container or cover a convex shape. In the spherical and the hyperbolic space I only consider packings and coverings by balls. The most frequent quantity to be optimized is the density, which is the ratio of total volume of the congruent bodies over the volume either of the container or of the shape that is covered. In addition, extremal values of the surface area, mean width, or other fundamental quantities are also investigated in the Euclidean case. A fascinating feature of finite packings and coverings is that optimal arrangements are often related to interesting geometric shapes.

The main body of the book consists of two parts, followed by the Appendix, which discusses some important background information and prerequisites. Part 1 collects results about planar arrangements. The story starts with Farkas Bolyai and Axel Thue, who investigated specific finite packings of unit discs in the nineteenth century. After a few sporadic results, the theory of packings and coverings by copies of a convex domain started to flourish following the work of László Fejes Tóth. Here are some of the highlights: Given a convex domain K in $\mathbb{R}^2$, Hugo Hadwiger proposed to find the minimal number of non-overlapping translates of K touching K. The Hadwiger number of K turns out to be eight if K is a parallelogram, and six otherwise (see Theorem 2.9.1). For density-type problems, when K is centrally symmetric, the hexagon bound holds for packing a high number of congruent copies of K into any convex container (see Theorem 1.4.1) and for covering any convex shape by at least seven translates of K (see Theorem 2.8.1). In addition, the optimal packing of n unit discs has been determined for about 96% of all n (see Section 4.3).

Concerning packings and coverings by equal spherical discs, the optimal arrangements have been determined when the number of discs is small, and the corresponding triangle bound has been verified with respect to any convex domain (see Sections 4.4 and 5.4). Moreover, the Euclidean triangle bound has been verified for both packings or coverings of equal circular discs with respect to large circular discs in the hyperbolic plane (see Sections 4.6 and 5.5).

Part 2 considers problems in higher dimensional spaces. For arrangements in $\mathbb{R}^d$, an important inspiration was László Fejes Tóth's celebrated Sausage Conjecture; namely, if $d \geq 5$ and the convex hull of n nonoverlapping unit balls is of minimal volume then the centers are aligned, hence the convex hull is a "sausage". By now the conjecture has been verified for $d \geq 42$. Because the argument is very involved in lower dimensions, we present the proof only when the dimension is very high (see Section 8.3). Concerning mean projections, let K be a convex body. If a certain mean projection of the convex hull of n congruent copies of K is minimal then the convex hull is approximately some ball for large n (see Theorem 7.3.1). In addition, a convex compact set of maximal mean width that is covered by n congruent copies of K is close to some segment (see Theorem 7.4.1). Chapter 10 discusses the so-called parametric density of finite arrangements of translates of K, a notion due to Jörg M. Wills. For example, assuming that K is centrally symmetric and the parameter ϱ is not too small, the convex hull of the translates in the optimal lattice packing is essentially homothetic to a certain polytope, namely, to the so-called Wulff-shape, which is well known from crystallography.

Part 2 also discusses local translative arrangements in $\mathbb{R}^d$. The Hadwiger number is proved to be exponential in the dimension d where the optimal upper bound is $3^d - 1$, and the optimal lower bound is 12 in $\mathbb{R}^3$ (see Sections 9.6 and 9.7). In addition, a family of nonoverlapping translates of K that do not overlap K is called a translational cloud if every half line emanating from K intersects the interior of at least one translate. It has been verified that the minimal cardinality of clouds is exponential in d^2 (see Sections 9.12 to 9.14).

Chapter 6 of Part 2 is concerned with packing and covering by equal spherical balls on S^d for $d \geq 3$. For example, the vertices of any simplicial Euclidean regular $(d + 1)$-polytope determine an optimal ball packing on S^d. In addition, the optimal packings have been established in various cases using the so-called linear programming bound. For coverings, the optimal arrangements are known only in a very few cases, but the density estimates are very sound.

I do not discuss the arguments that use the linear programming bound in detail because related topics are thoroughly discussed in the recent monographs by J. H. Conway and N. J. A. Sloane [CoS1999], T. Ericson and V. Zinoviev [ErZ2001], and Ch. Zong [Zon1999a]. In addition to these results, many others are only surveyed at the ends of the chapters. The basic methods of proofs in this book are local volume estimates and the use of isoperimetric type inequalities. The theory of mixed volumes (the so-called Brunn–Minkowski theory) is an essential ingredient of various arguments, whereas many others are based on ingenious, rather elementary ideas. The results are not presented in their most refined form in many cases; the emphasis is rather on the underlying ideas.

I tried to keep the chapters as independent as possible; hence some ideas are repeated in a more general form in a subsequent chapter. Yet Chapter 3 and Chapter 10 rely heavily on Chapter 2 and Chapter 9, respectively. To avoid interrupting the flow of the presentation, the history of the topics is told only at the end of the sections, which is also the place to review related results.

Besides the monographs mentioned above, the interested reader may consult the forthcoming books by P. Brass, W. Moser, and J. Pach [BMP?], U. Betke, M. Henk, and J. M. Wills [BHW?], and P. Gruber [Gru?b] for in-depth discussion of many related topics.

Finally, I have to express my gratitude to many people whose help was essential in getting this project done. First, I thank my father, whose guidance from my early days led me into the Reign of Beauty and Deception called Geometry. The manuscript has been improved to a great extent by conversations with I. Bárány, U. Betke, A. Bezdek, K. Bezdek, G. Fejes Tóth, Z. Füredi, P. M. Gruber, M. Henk, A. Heppes, D. Ismailescu, W. Kuperberg, J. C. Lagarias, D. Larman, E. Makai, G. Moussong, J. Pach, T. Réti, A. Schürmann, R. Strausz, P. G. Szabó, I. Talata, T. Tarnai, Á. Tóth, G. Wintsche, and Ch. Zong. I learned the fundamental facts and methods concerning finite packings under the guidance of T. Bisztriczky and J. M. Wills. I would like to thank Béla Bollobás, Editor of the series Cambridge Tracts in Mathematics, for constant encouragement, and Roger Astley, Senior Editor at Cambridge University Press, for patiently answering all my inquiries. Finally, there are no words that express how much I owe to Csilla, who provides continuous inspiration in all aspects of my life.

Notation

$\#X$:   the cardinality of X if X is finite

$\operatorname{conv}X$:   the minimal convex set containing X

$V(\cdot)$:   volume (Lebesgue measure)

$\mathcal{H}^k(\cdot)$: k-dimensional Hausdorff measure

$C + K$:   the family of $x + y$, where $x \in C$ and $y \in K$

$C \ominus K$:   the family of points x satisfying $x + K \subset C$

$L^{\perp}$:   the orthogonal complement of the affine subspace L

$u^{\perp}$:   the family of vectors orthogonal to the nonzero vector u

$X|L$:   the orthogonal projection of a set X into L

X/L: the orthogonal projection of a set X into $L^{\perp}$

$\operatorname{aff}X$:   the minimal affine subspace containing a set X

$\chi_K(\cdot)$:   the characteristic function of a set K

∂K: the relative boundary of a compact convex set K

$h_K(\cdot)$:   the support function of a compact convex set K

$|K|$:   the relative content of a compact convex set K

$\dim K$:   the dimension of the affine hull of a compact convex set K

$\delta(K)$:   the packing density of a convex body K

$\delta_T(K)$:   the translative packing density of a convex body K

$\vartheta(K)$:   the covering density of a convex body K

$\vartheta_T(K)$:   the translative covering density of a convex body K

$\|\cdot\|_K$:   the norm with respect to the o-symmetric convex body K

$\langle \cdot, \cdot \rangle$:   scalar product

$\|\cdot\|$:   Euclidean norm

$\det \Lambda$:   the determinant of a lattice Λ

B^d: the Euclidean unit ball in $\mathbb{R}^d$

κ_d: $V(B^d)$

S^d: the d-sphere

H^d: hyperbolic d-space

$\mathbb{S}/\Lambda$:   the quotient of the space $\mathbb{S}$ with respect to a lattice Λ

$\lfloor x \rfloor$:   the largest integer that is at most x

$\lceil x \rceil$:   the smallest integer that is at least x

Part 1

Arrangements in Dimension Two

1

Congruent Domains in the Euclidean Plane

Let K be a convex domain. According to the classical result of L. Fejes Tóth [FTL1950], the density of a packing of congruent copies of K in a hexagon cannot be denser than the density of K inside the circumscribed hexagon with minimal area. Besides this statement, we verify that the same density estimate holds for any convex container provided the number of copies is high enough. In addition, we show that if K is a centrally symmetric domain then the inradius and circumradius of the optimal convex container cannot be too different. Following L. Fejes Tóth [FTL1950] in case of coverings, the analogous density estimate is verified under the "noncrossing" assumption, which essentially says that the boundaries of any two congruent copies intersect in two points. In case of both packings and coverings, congruent copies can be replaced by similar copies of not too different sizes. Finally, we verify the hexagon bound for coverings by congruent fat ellipses even without the noncrossing assumption, a result due to A. Heppes.

Concerning the perimeter, we show that the convex domain of minimal perimeter containing n nonoverlapping congruent copies of K gets arbitrarily close to being a circular disc for large n. However, if the perimeter of the compact convex set D covered by n congruent copies of K is maximal then D is close to being a segment for large n.

1.1. Periodic and Finite Arrangements

Let K be a convex domain. Given an arrangement of congruent copies of K that is periodic with respect to some lattice Λ (see Section A.13) and given m equivalence classes, it is natural to call $m \cdot A(K)/\det \Lambda$ the density of the arrangement. We define the *packing density* $\delta(K)$ to be the supremum of the densities of periodic packings of congruent copies of K and the *covering density* to be the infimum of the densities of periodic coverings by congruent copies of K. In addition, we define $\Delta(K) = A(K)/\delta(K)$ and $\Theta(K) = A(K)/\vartheta(K)$.

3

It is not hard to show that optimal clusters asymptotically provide the same densities as periodic arrangements (see Lemma 1.1.2). Our main result is that, in the planar case, finite packings are not denser (asymptotically) than periodic packings, and the analogous statement holds for coverings. We note that this is a planar phenomenon: Say, if $d \geq 3$ and K is a right cylinder whose base is a $(d-1)$-ball, then linear arrangements are of density one, whereas any periodic packing is of density at most δ' for some $\delta' < 1$, and any periodic covering is of density at least ϑ' for some $\vartheta' > 1$ (see Lemma 7.2.5).

Theorem 1.1.1. *Let K be a convex domain, and let n tend to infinity.*

> *(i) If D_n is a convex domain of minimal area containing n nonoverlapping congruent copies of K then $A(D_n) \sim n \cdot \Delta(K)$.*
>
> *(ii) If $\widetilde{D}_n$ is a convex domain of maximal area that can be covered by n congruent copies of K then $A(\widetilde{D}_n) \sim n \cdot \Theta(K)$.*

Since periodic arrangements correspond canonically to finite arrangements on tori (see Section A.13), $\Delta(K)$ is the infimum of $V(T)/m$ over all tori T and integers m such that there exists a packing of m embedded copies of K on T, and $\Theta(K)$ is the supremum of $V(T)/m$ over all tori T and integers m such that there exists a covering of T by m embedded copies of K. The first step towards verifying Theorem 1.1.1 is the case of clusters.

Lemma 1.1.2. *Given convex domains K and D with $r(D) > R(K)$, let N be the maximal number of nonoverlapping congruent copies of K inside D, and let M be minimal number of congruent copies of K that cover D. Then*

> *(i)* $\left(1 + \frac{R(K)}{r(D)}\right)^2 \cdot A(D) \geq N \cdot \Delta(K) \geq \left(1 - \frac{R(K)}{r(D)}\right)^2 \cdot A(D);$
>
> *(ii)* $\left(1 + \frac{R(K)}{r(D)}\right)^2 \cdot A(D) \geq M \cdot \Theta(K) \geq \left(1 - \frac{R(K)}{r(D)}\right)^2 \cdot A(D).$

Remark. Instead of the upper bound in (i), we actually prove the stronger estimate $A(D + R(K)\, B^2) \geq N \cdot \Delta(K)$.

Proof. We place K and D in a way that $K \subset R(K)B^2 \subset D$. In particular, assuming that K' is congruent to K, if the circumcentre c of K' lies outside $D + R(K)B^2$ then K' avoids D, and if $c \in (1 - R(K)/r(D))D$ then $K' \subset D$. Given a torus T, we write the same symbol to denote a convex domain in $\mathbb{R}^2$ and its embedded image on T.

We present the proof only for packings because the case of coverings is completely analogous. Let $T = \mathbb{R}^2/\Lambda$ be any torus satisfying that

$C = D + R(K) B^2$ embeds isometrically into T, and let $K_1, \ldots, K_m$ be the maximal number of nonoverlapping embedded copies of K on T. Writing x_i to denote the circumcentre of K_i, we have (see (A.50))

$$\int_T \#((C + x) \cap \{x_1, \ldots, x_m\}) \, dx = m \cdot A(C). \qquad (1.1)$$

Thus there exists a translate $C + x$ that contains at most $k \leq m \cdot A(C)/A(T)$ points out of $x_1, \ldots, x_m$, say, the points $x_{i_1}, \ldots, x_{i_k}$. After replacing $K_{i_1}, \ldots, K_{i_k}$ by the N nonoverlapping embedded copies of K contained in $x + D$, we obtain a packing of $m - k + N$ embedded copies of K on T. In particular, $N \leq k$ follows by the maximality of m. We conclude

$$A(D + R(K) B^2) \geq N \cdot \Delta(K),$$

which in turn yields the upper bound in (i).

Turning to (ii), we let $\lambda < 1$ satisfy $\lambda \cdot A(C)/\Delta(K) > \lceil A(C)/\Delta(K) \rceil - 1$. It follows by the definition of $\Delta(K)$ that there exist a torus $T = \mathbb{R}^2/\Lambda$ and m nonoverlapping embedded copies $K_1, \ldots, K_m$ of K on T satisfying $A(T) < \lambda^{-1} m \Delta(K)$, and D embeds isometrically into T. We define $C = (1 - R(K)/r(D))D$; hence, (1.1) yields that some translate $C + x$ contains at least $m \cdot A(C)/A(T)$ points out of the circumcentres of $K_1, \ldots, K_m$. We may assume that these points are the circumcentres of $K_1, \ldots, K_l$; therefore, $l \geq \lambda \cdot A(C)/\Delta(K)$ and $K_1, \ldots, K_l$ are contained in $D + x$. Thus $N \geq l \geq A(C)/\Delta(K)$ by the definition of λ, completing the proof of Lemma 1.1.2. $\qquad \square$

Proof of Theorem 1.1.1. We present the argument only for packings because just the obvious changes are needed for the case of coverings. In the following the implied constant in $O(\cdot)$ always depends only on K.

Theorem 1.1.1 for packings follows from the following statement: If $\varepsilon > 0$ is small, and $n > 1/\varepsilon^5$ then

$$A(D_n) = (1 + O(\varepsilon)) \cdot n\Delta(K). \qquad (1.2)$$

Dense clusters show (see Lemma 1.1.2) that

$$A(D_n) \leq (1 + O(\varepsilon)) \cdot n\Delta(K).$$

Therefore, it is sufficient to verify that

$$A(D_n) \geq (1 - O(\varepsilon)) \cdot n\Delta(K). \qquad (1.3)$$

If $r(D_n) > 1/\varepsilon$ then (1.3) follows from Lemma 1.1.2. Thus we assume that $r(D_n) \leq 1/\varepsilon$, a case that requires a more involved argument. We actually prove that there exists a rectangle R that contains certain N congruent copies of K, where

$$\frac{A(R)}{N} \leq (1 + O(\varepsilon)) \cdot \frac{A(D_n)}{n}. \tag{1.4}$$

Since the minimal width w of D_n is at most $3/\varepsilon$ according to the Steinhagen inequality (Theorem A.8.2), there exists a rectangle $\widetilde{R}$ such that its sides touch D_n, and two parallel sides of $\widetilde{R}$ are of length w. We say that these sides are vertical; hence, D_n has a vertical section of length w. Writing l to denote the length of the horizontal sides, we have $A(D_n) \geq wl/2$. For $k = \lceil 1/\varepsilon \rceil$, we decompose $\widetilde{R}$ into k^3 congruent rectangles $R_1, \ldots, R_{k^3}$ in this order, where the vertical sides of R_i are of length w and the horizontal sides are of length l/k^3.

Out of the circumcentres of the n nonoverlapping congruent copies of K that lie in D_n, let n_i be contained in R_i. Now the total area of $R_1, \ldots, R_{k^2+1}$ and of $R_{k^3-k^2}, \ldots, R_{k^3}$ is

$$\left(1 + \frac{1}{k^2}\right)\frac{2wl}{k} \leq (1 + O(\varepsilon))4\varepsilon A(D_n) \leq (1 + O(\varepsilon))4\Delta(K) \cdot \varepsilon n,$$

and hence $\sum_{i=k^2+1}^{k^3-k^2} n_i \geq (1 - O(\varepsilon))n$. In particular, there exists some index j such that $k^2 + 1 \leq j \leq k^3 - k^2$ and

$$\frac{A(R_j \cap D_n)}{n_j} \leq (1 + O(\varepsilon)) \cdot \frac{A(D_n)}{n}. \tag{1.5}$$

Let R' be the rectangle whose sides are vertical and horizontal, with each touching $R_j \cap D_n$. We write a to denote the common length of the vertical sides of R', which readily satisfies $a \geq 2r(K)$. Since $w/k^2 < 4\varepsilon$, we deduce that $R_j \cap D_n$ contains a rectangle whose horizontal side is of length l/k^3, and the vertical side is of length $a - 8\varepsilon$. In particular, $A(R')$ is at most $(1 + O(\varepsilon))A(R_j \cap D_n)$. Finally, the rectangle R whose horizontal sides are of length $l/k^3 + 2R(K)$ and vertical sides are of length a contains $N = n_j$ nonoverlapping congruent copies of K. Now $3l/\varepsilon \geq n A(K)$ yields $l/k^3 \geq 1/[4A(K)\varepsilon]$. Thus we conclude (1.4) by (1.5).

Since the arrangement in R induces a periodic packing of K, (1.4) readily yields (1.3) and hence Theorem 1.1.1 as well. □

Remark 1.1.3. Given a strictly convex domain K, if D_n is a convex domain with minimal area that contains n nonoverlapping congruent copies of K then $r(D_n)$ tends to infinity.

We sketch the argument for Remark 1.1.3: We suppose indirectly that there exists a subsequence of $\{r(D_n)\}$ that is bounded by some $\omega > 0$. For any $\varepsilon > 0$, the proof of Theorem 1.1.1 yields a parallel strip Σ_ε and a packing

Ξ of congruent copies of K inside Σ_ε such that the packing is periodic with respect to a vector parallel to Σ_ε, the width of Σ_ε is at most 3ω, and the density of the packing Ξ inside Σ_ε is at least $(1 - \varepsilon)\delta(K)$. We reflect this arrangement through one of the lines bounding Σ_ε and write Ξ′ to denote the image of Ξ. Because K is strictly convex, there exist positive v_1 and v_2 depending only on K with the following property: Translating the packing Ξ′ first parallel to Σ_ε by a vector of length v_1, then towards Σ_ε orthogonally by a vector of length v_2, we obtain an arrangement Ξ″ such that the union of Ξ and Ξ″ forms a packing. If $\varepsilon < v_2/(2\omega)$ then the union of Ξ and Ξ″ determines a periodic packing in the plane whose density is larger than $\delta(K)$. This contradiction verifies that $r(D_n)$ tends to infinity.

Open Problems.

(i) Let K be a convex domain that is not a parallelogram. We write D_n ($\widetilde{D}_n$) to denote a convex domain with minimal (maximal) area that contains n nonoverlapping congruent copies of K (i.e., is covered by n congruent copies of K). Is

$$r(D_n), r(\widetilde{D}_n) > c\sqrt{n}$$

for a suitable positive constant c depending on K? If the answer is yes then the ratio $R(D_n)/r(D_n)$ stays bounded as n tends to infinity, and a similar property holds for $\widetilde{D}_n$.

For packings, various partial results support an affirmative answer: The statement holds if K is centrally symmetric (see Corollary 1.4.3) or the packing is translative (see Theorem 2.4.1). Strengthening the method of Remark 1.1.3 yields that $r(D_n) > c\sqrt[3]{n}$ holds if K is any strictly convex domain. For coverings, the statement holds if K is a fat ellipse (see Theorem 1.7.1) or if K is centrally symmetric and only translative coverings are allowed (see Corollary 2.8.2).

(ii) Is $\vartheta(K) \leq 2\pi/\sqrt{27} = 1.2091\ldots$ for any convex domain K; namely, is the covering density maximal for circular discs (see Theorem 1.7.1)?

D. Ismailescu [Ism1998] proved $\vartheta(K) \leq 1.2281\ldots$ for any convex domain K. However, $\vartheta(K) \leq 2\pi/\sqrt{27}$ if K is centrally symmetric (see L. Fejes Tóth [FTL1972]).

(iii) Does it hold for any convex domain that there exist a periodic packing whose density is the packing density and a periodic covering whose density is the covering density? It is known that there exist no optimal lattice arrangement for the typical convex domain (see G. Fejes Tóth and T. Zamfirescu [FTZ1994] and G. Fejes Tóth [FTG1995a]).

Comments. The packing and covering densities were originally introduced in the framework of infinite packing and covering of the space (see G. Fejes Tóth and W. Kuperberg [FTK1993]). Readily, $\Theta(K) \leq A(K) \leq \Delta(K)$. W.M. Schmidt [Sch1961] proved that $\Theta(K) = A(K)$ or $\Delta(K) = A(K)$ if and only if some congruent copies of K tile the plane (see also Lemma 7.2.5).

According to the hexagon bound of L. Fejes Tóth [FTL1950] (see Theorem 1.3.1), $\Delta(K)$ is at least the minimal area of circumscribed hexagons for any convex domain K, where equality holds if K is centrally symmetric. Concerning absolute lower bounds on the packing density, G. Kuperberg and W. Kuperberg [KuK1990] verified that $\delta(K) > \sqrt{3}/2 = 0.8660\ldots$ holds for any convex domain K. In addition a beautiful little theorem of W. Kuperberg [Kup1987] states that $\delta(K)/\vartheta(K) \geq 3/4$, where equality holds for circular discs. It is probably surprising but the packing density, $\pi/\sqrt{12} = 0.9068\ldots$ of the unit disc is not minimal among centrally symmetric convex domains, which is shown say by the regular octagon. By rounding off the corners of the regular octagon, K. Reinhardt [Rei1934] and K. Mahler [Mah1947] proposed a possible minimal shape whose density is $0.9024\ldots$. P. Tammela [Tam1970] proved that $\delta(K) > 0.8926$ for any centrally symmetric convex domain K.

Concerning coverings, D. Ismailescu [Ism1998] proved $\vartheta(K) \leq 1.2281\ldots$ for any convex domain K. For very long the only convex domains with known covering densities were the tiles (when the covering density is one), and circular discs (when the covering density is $2\pi/\sqrt{27}$ according to R. Kershner [Ker1939]; see also Corollary 5.1.2). Recently A. Heppes [Hep2003] showed that the covering density of any "fat ellipse" (when the ratio of the smaller axis to the greater axis is at least 0.86) is $2\pi/\sqrt{27}$ (see also Theorem 1.7.1). A substantial improvement is due to G. Fejes Tóth [FTG?b]: On the one hand [FTG?b] generalized A. Heppes' theorem to ellipses when the ratio of the smaller axis to the greater axis is at least 0.741. On the other hand if K is a centrally symmetric convex domain and $r(K)/R(K) \geq 0.933$ then [FTG?b] proves that $\Theta(K)$ is the maximal area of polygons with at most six sides inscribed into K. Readily if K is either type of the convex domains considered in [FTG?b], and $C \subset K$ is a convex domain that contains a centrally symmetric hexagon of area $\Theta(K)$ then $\Theta(C) = \Theta(K)$.

1.2. The Hexagon Bound for Packings Inside an Octagon

Given a convex domain K, we write $H(K)$ to denote a circumscribed convex polygon with at most six sides of minimal area. The aim of this section is to verify the *hexagon bound* for packings of congruent copies of K

inside a hexagon; namely, the density is at most $A(K)/A(H(K))$. Later we will prove the hexagon bound with respect to any convex container (see Theorem 1.4.1).

Theorem 1.2.1. *If a polygon D of at most eight sides contains $n \geq 2$ congruent copies of a given convex domain K then*

$$A(D) \geq n \cdot A(H(K)).$$

The main idea of the proof Theorem 1.2.1 is to define a cell decomposition of D into convex cells in a way such that each cell contains exactly one of the congruent copies of K; hence, the average number of sides of the cells is at most six according to the Euler formula. Then we verify that the minimal areas of circumscribed k-gons are convex functions of k (see Corollary 1.2.4), and we deduce that the average area of a cell is at least $A(H(K))$. Unfortunately, we cannot proceed exactly like this because no suitable cell decomposition of D may exist. In spite of this we can still save the essential properties of a cell decomposition (see Lemma 1.2.2) and verify the hexagon bound. Lemma 1.2.2 is presented in a rather general setting because of later applications.

Lemma 1.2.2. *Let D be a convex domain that contains the nonoverlapping convex domains $K_1, \ldots, K_n$, $n \geq 2$. Then there exist nonoverlapping convex domains $\Pi_1, \ldots, \Pi_n \subset D$ satisfying the following properties:*

(i) $K_i \subset \Pi_i$.
(ii) $\Pi_1, \ldots, \Pi_n$ cover ∂D.
(iii) Π_i is bounded by $k_i \geq 2$ convex arcs that we call edges. The edges intersecting $\text{int } D$ are segments, and the rest of the edges are the maximal convex arcs of $\partial D \cap \Pi_i$.
(iv) The number b of edges contained in ∂D satisfy

$$\sum_{i=1}^{n}(6 - k_i) \geq b + 6.$$

In addition, if D is a polygon of at most eight sides and k_i^* denotes the number of sides of Π_i then $\sum_{i=1}^{n}(6 - k_i^*) \geq 0$.

Proof. Let $\Pi_1, \ldots, \Pi_n$ be nonoverlapping convex domains such that $K_i \subset \Pi_i \subset D$ and the total area covered by the convex domains $\Pi_1, \ldots, \Pi_n$ is maximal under these conditions. Since two nonoverlapping convex sets can be separated by a line, each Π_i is the intersection of a polygon P_i and D.

Now int $P_i \cap \partial D$ consists of finitely many convex arcs whose closures we call edges of Π_i. The rest of the edges of Π_i are the segments of the form $s \cap D$, where s is a side of P_i that intersects int D, and the vertices of Π_i are the endpoints of the edges.

It may happen that $\Pi_1, \ldots, \Pi_n$ do not cover D, and we call the closure of a connected component of int $D \backslash \cup_{i=1}^n \Pi_i$ a hole. Let Q be a hole. Then there exists an edge e_1 of some Π_{i_1} such that e_1 intersects int D, and $e_1 \cap \partial Q$ contains a segment s_1, where we assume that s_1 is a maximal segment in $e_1 \cap \partial Q$. Since Π_{i_1} cannot be extended because of the maximality of $\sum A(\Pi_j)$, one endpoint v_2 of s_1 is contained in the relative interior of e_1; hence, $v_2 \in \text{int} D$. Therefore, v_2 is the endpoint of an edge e_2 of some Π_{i_2} such that $e_2 \cap \partial Q$ contains a maximal segment s_2. Continuing this way we obtain that ∂Q is the union of segments $s_1, \ldots, s_k$ with the following properties (where $s_0 = s_k$): s_j is contained in an edge e_j of some Π_{i_j}, and $s_j \cap s_{j-1}$ is a common endpoint $v_j \in \text{int} D$ that is an endpoint of e_j and not of e_{j-1} for any $j = 1, \ldots, k$; moreover, different s_i and s_j do not intersect otherwise. We deduce that Q is a convex polygon and $Q \subset \text{int} D$.

Now (ii) readily follows; namely, $\Pi_1, \ldots, \Pi_n$ cover ∂D. Next, we construct a related cell decomposition Σ of D by cells $\widetilde{\Pi}_1, \ldots, \widetilde{\Pi}_n$. If there exists no hole then $\widetilde{\Pi}_i = \Pi_i$. Otherwise, let $\{Q_1, \ldots, Q_m\}$ be the set of holes, let $q_j \in \text{int} Q_j$, and we define $\widetilde{\Pi}_i$ to be the union of Π_i and all triangles of the form conv$\{q_j, s\}$ such that s is a side of Q_j and $s \subset \Pi_i$. In particular, the number of edges of Σ contained in $\widetilde{\Pi}_i$ is at least k_i; hence $\sum (6 - k_i) \geq b + 6$ is a consequence of Lemma A.5.9. If, in addition, D is a polygon of at most eight sides then $\sum k_i^* \leq 8 + \sum k_i$; thus $b \geq 2$ completes the proof of Lemma 1.2.2. $\square$

Given the convex domain K, let $t_K(m)$ denote the minimal area of a circumscribed polygon of at most m sides for any $m \geq 3$. Next, we show that $t_K(m)$ is a convex function of m, more precisely, that $t_K(m)$ is even strictly convex if K is strictly convex.

Lemma 1.2.3. *If K is a strictly convex domain and $m \geq 4$ then*

$$t_K(m - 1) + t_K(m + 1) > 2 t_K(m).$$

Proof. For any $m \geq 3$, we choose a circumscribed polygon Π_m of minimal area among the circumscribed polygons of at most m sides. Since K is strictly convex, Π_m is actually an m-gon, and each side of Π_m touches K at the midpoint of the side.

Let $m \geq 4$, and let $3 \leq k \leq l$ satisfy that $A(\Pi_k) + A(\Pi_l)$ is minimal under the condition $k + l = 2m$. We suppose that $k < m$ and seek a contradiction. The idea is to decrease the total area of Π_k and Π_l by interchanging certain sides. We write $p_1, \ldots, p_k$ and $q_1, \ldots, q_l$ to denote the midpoints of the sides of Π_k and Π_l, respectively, according to the clockwise orientation, and we write e_i and f_j to denote the side of Π_k and Π_l containing p_i and q_j, respectively. For $p, q \in \partial K$, let $[p, q)$ denote the semi open arc of ∂K, which starts at p and terminates at q according to the clockwise orientation, and the arc contains p and does not contain q. The k semi open convex arcs $[p_{i-1}, p_i)$ on ∂K (with $p_0 = p_k$) contain the $l \geq k + 2$ midpoints for Π_l, and hence either there exists $[p_{i-1}, p_i)$, which contains say q_1, q_2, q_3, or there exist two semi open arcs of the form $[p_{t-1}, p_t)$ such that each contains two midpoints from Π_l. In the first case, let Π'_{k+1} be obtained from Π_k by cutting off the vertex $e_{i-1} \cap e_i$ by aff f_2, and let Π''_{l-1} be obtained from Π_l by removing the side f_2, and hence aff $f_1 \cap$ aff f_3 is the new vertex of Π''_{l-1}. Then Π'_{k+1} and Π''_{l-1} have $k + 1$ and $l - 1$ sides, respectively, and $\Pi''_{l-1} \backslash \Pi_l$ is strictly contained in $\Pi'_{k+1} \backslash \Pi_k$. Therefore,

$$A(\Pi'_{k+1}) + A(\Pi''_{l-1}) < A(\Pi_k) + A(\Pi_l).$$

This is absurd, and hence we may assume that $q_1, q_2 \in [p_1, p_2)$ and $q_{j-1}, q_j \in [p_{i-1}, p_i)$ for $i \neq 2$. In this case let $\Pi'_{k'}$ be the circumscribed k'-gon defined by affine hulls of

$$e_1, f_2, \ldots, f_{j-1}, e_i, \ldots, e_k.$$

In addition, let $\Pi''_{l'}$ be the circumscribed l'-gon defined by affine hulls of

$$f_1, e_2, \ldots, e_{i-1}, f_j, \ldots, f_l;$$

thus $k' + l' = 2m$. When constructing $\Pi'_{k'}$ and $\Pi''_{l'}$, we remove the part of Π_k at the corner enclosed by e_1 and e_2 and cut off by $\partial \Pi_l$, and we add two nonoverlapping domains contained in this part (where one of the domains degenerates if $q_1 = p_1$). Because the situation is analogous at the corner of Π_k enclosed by e_{i-1} and e_i, we deduce that

$$A(\Pi'_{k'}) + A(\Pi''_{l'}) \leq A(\Pi_k) + A(\Pi_l).$$

The polygons were constructed in a way that $f_{j-2} \cap f_{j-1}$ is a common vertex for $\Pi'_{k'}$ and Π_l, whereas $f_j \cap f_{j-1}$ is a vertex for Π_l but not for $\Pi'_{k'}$, and hence q_{j-1} is not the midpoint of the side of $\Pi'_{k'}$ containing it. Therefore, there exists a circumscribed k'-gon whose area is less than $A(\Pi'_{k'})$, which contradicts the minimality of $A(\Pi_k) + A(\Pi_l)$. $\square$

Given any convex domain K, we define $t_K(2) = (3/2) t_K(3)$. It is not hard to see that $t_K(3) \leq 2 t_K(4)$; thus Lemma 1.2.3 and approximation yield that $t_K(m-1) + t_K(m+1) \geq 2 t_K(m)$ holds for any $m \geq 3$. Defining $t_K(s)$ to be linear for $s \in [m-1, m]$, we deduce the following:

Corollary 1.2.4 (Dowker theorem for circumscribed polygons). *Given a convex domain K, the function $t_K(s)$ can be extended to be a convex and decreasing function for any real $s \geq 2$.*

After all these preparations, the proof of the main result is rather simple.

Proof of Theorem 1.2.1. Given the packing of n congruent copies of K inside the convex polygon D of at most eight sides, we construct the system of cells according to Lemma 1.2.2. It follows by Lemma 1.2.2 (iv) that the average number of sides is at most six; hence, the Dowker theorem (Corollary 1.2.4) yields that the average area of the cells is at least $A(H(K))$. In turn, we conclude Theorem 1.2.1.   □

Comments. Theorem 1.2.1 is proved by L. Fejes Tóth [FTL1950], and Corollary 1.2.4 is due to C. H. Dowker [Dow1944].

1.3. The Hexagon Bound for Periodic Packings

In this section we estimate the packing density of a convex domain K. We recall that $H(K)$ denotes a circumscribed convex polygon with at most six sides of minimal area.

Theorem 1.3.1. *If K is a convex domain then $\Delta(K) \geq A(H(K))$. If in addition K is centrally symmetric then $\Delta(K) = A(H(K))$, and there exists a lattice packing of K that is a densest periodic packing.*

In light of the hexagon bound of Theorem 1.2.1, we only need to construct suitably efficient lattice packings of centrally symmetric domains. The first step is

Lemma 1.3.2. *If K is an o-symmetric convex domain then $H(K)$ can be chosen to be o-symmetric.*

Remark. There may exist circumscribed hexagons of minimal area that are not centrally symmetric. Say the regular octagon can be obtained from a square by cutting off four corners, and one circumscribed hexagon of minimal area is obtained by cutting off two neighbouring corners of the square.

Proof. We may assume that K is strictly convex, in which case we verify that any circumscribed hexagon H of minimal area is o-symmetric. The minimality of $A(H)$ yields that H has six sides that touch K in their midpoints. We write $p_1, \ldots, p_6$ to denote these midpoints in clockwise order and write l_i to denote the affine hull of the side touching at p_i. In addition, let p'_i and l'_i denote the reflected image of p_i and l_i, respectively, through o.

We suppose that p'_1, is not among $p_2, \ldots, p_6$ for some i, and we seek a contradiction. We may assume that the shorter arc $p_1 p_2$ of ∂K contains p'_{j-1} and p'_j for some $4 \le j \le 6$ and $p'_j \ne p_2$; hence, the shorter arc $p'_1 p'_2$ contains p_{j-1} and p_j, and $p_j \ne p'_2$. Now let P be the circumscribed $(2j - 4)$-gon determined by

$$l_2, \ldots, l_{j-1}, l'_2, \ldots, l'_{j-1},$$

and let Q be the circumscribed $(16 - 2j)$-gon determined by

$$l_j, \ldots, l_6, l_1, l'_j, \ldots, l'_6, l'_1.$$

Thus $A(P) + A(Q) \le 2A(H)$. Since p_1 is the midpoint of the side $l_1 \cap H$, it is not the midpoint of the side $l_1 \cap Q$; hence, Q is not a circumscribed $(16 - 2j)$-gon of minimal area. Therefore, the total area of P and of some circumscribed $(16 - 2j)$-gon is less than $2A(H)$, contradicting the Dowker theorem (Corollary 1.2.4). In turn, we conclude Lemma 1.3.2. □

Proof of Theorem 1.3.1. Let K be a convex domain. Applying Theorem 1.2.1 to large octagons leads to $\Delta(K) \ge A(H(K))$ (see Lemma 1.1.2). Next let $K = -K$; hence, we may assume that $H(K) = -H(K)$ (see Lemma 1.3.2). Writing $p_1, \ldots, p_6$ to denote the midpoints of the sides of $H(K)$ according to the clockwise orientation, we have $p_2 - p_1 = p_3$. Therefore, $2p_1$ and $2p_2$ generates a lattice Λ such that $\Lambda + H(K)$ is a tiling of the plane, completing the proof of Theorem 1.3.1. □

Comments. Theorem 1.3.1 is due to L. Fejes Tóth [FTL1950], and C. H. Dowker [Dow1944] proves Lemma 1.3.2 (actually for circumscribed $2m$-gons for any $m \ge 2$). According to G. Fejes Tóth and L. Fejes Tóth [FTF1973a], if a convex domain has k-fold rotational symmetry then there exists a circumscribed km-gon of minimal area that also has k-fold rotational symmetry. For a typical convex domain K, G. Fejes Tóth [FTG1995a] constructs periodic packings that are denser than any lattice packing, and his method yields that $\Delta(K) > A(H(K))$.

1.4. Packings Inside Any Convex Container

If K is a centrally symmetric convex domain then we consider packings of congruent copies of K inside any convex container. We recall that $H(K)$ denotes a circumscribed convex polygon with at most six sides of minimal area.

Theorem 1.4.1. *Given a centrally symmetric convex domain K, there exists $n(K)$ such that if a convex domain D contains $n \geq n(K)$ nonoverlapping congruent copies of K then*

$$A(D) \geq n \cdot A(H(K)).$$

Remarks. Lemma 1.1.2 and Theorem 1.3.1 show that the constant $A(H(K))$ is optimal. Moreover, the condition that n has to be large in Theorem 1.4.1 is optimal according to Example 1.4.2.

Proof. If K is a polygon of at most six sides then the hexagon bound readily holds. We verify that if K is not a parallelogram then there exist positive constants γ_1 and γ_2 depending on K such that

$$A(D) > n \cdot A(H(K)) + \gamma_1 P(D) - \gamma_2. \tag{1.6}$$

We write $K_1, \ldots, K_n$ to denote the nonoverlapping congruent copies of K in D, and we assume that D is the convex hull of these domains. Let $\Pi_1, \ldots, \Pi_n$ be the convex domains associated to $K_1, \ldots, K_n$ by Lemma 1.2.2.

In this proof σ always denote an edge of some Π_i that is contained in ∂D. Let $\sigma \subset \Pi_i$. We write x_i to denote the centre of a circular disc of radius $r(K)$ inscribed into K_i and write $C(\sigma, x_i)$ to denote the union of all segments connecting x_i to the points of σ. Since any line tangent to σ avoids K_i, we have

$$A(C(\sigma, x_i)) \geq \frac{r(K)}{2} \cdot |\sigma|, \tag{1.7}$$

where $|\cdot|$ stands for the arc length. Thus there exist positive constants λ and c_1 such that if $|\sigma| > \lambda$ then

$$A(C(\sigma, x_i)) \geq t_K(2) + c_1 |\sigma|. \tag{1.8}$$

In addition, we define the acute angle α by the formula $\sin 3\alpha = r(K)/\lambda$; hence, if the distance of x from K_i is at most λ for some $x \notin K_i$ then the angle of the two half lines emanating from x and tangent to K_i is at least 3α.

The following considerations are based on the total curvature $\beta(\sigma)$ of σ (see Section A.5.1). Let $|\sigma| \le \lambda$ and $\beta(\sigma) < \alpha$, and we write p and q to denote the endpoints of σ. Next we assign a compact convex set $\Omega(\sigma)$ to σ. If σ is a segment then we define $\Omega(\sigma)$ to be σ itself. If $\beta(\sigma) > 0$ then σ intersects K_i; thus, we may define $\Omega(\sigma)$ to be the triangle pqs, where the half line ps is tangent to σ, and the line sq contains the other edge of Π_i emanating from q. Since the angle of $\Omega(\sigma)$ at q is at most $\pi - 2\beta(\sigma)$ according to the definition of α, and at p is at most $\beta(\sigma)$, we deduce that

$$A(Q(\sigma)) \le \frac{1}{2} \lambda^2 \sin 2\beta(\sigma) \le c_2 \beta(\sigma). \tag{1.9}$$

Given any Π_i, adding all $\Omega(\sigma)$ with $\sigma \subset \Pi_i$ we obtain a k_i-gon; thus,

$$A(\Pi_i) \ge t_K(k_i) + \sum_{\substack{\sigma \subset \Pi_i \\ |\sigma| > \lambda}} c_1 |\sigma| - \sum_{\substack{\sigma \subset \Pi_i \\ |\sigma| \le \lambda \\ \beta(\sigma) < \alpha}} c_2 \beta(\sigma) - \sum_{\substack{\sigma \subset \Pi_i \\ |\sigma| \le \lambda \\ \beta(\sigma) \ge \alpha}} t_K(2). \tag{1.10}$$

Now the total curvature of ∂D is 2π, which in turn yields that

$$A(D) \ge \sum_{i=1}^{n} t_K(k_i) + \sum_{|\sigma| > \lambda} c_1 |\sigma| - c_2 2\pi - \frac{2\pi}{\alpha} t_K(2). \tag{1.11}$$

It follows by the convexity of $t_K(\cdot)$ (see Corollary 1.2.4) that

$$t_K(k_i) \ge A(H(K)) + (t_K(5) - t_K(6)) \cdot (6 - k_i), \tag{1.12}$$

where $t_K(5) - t_K(6) > 0$. We write b to denote the number of edges of $\Pi_1, \ldots, \Pi_n$ that are contained in ∂D; hence, Lemma 1.2.2 (iv) leads to

$$\sum_{i=1}^{n} t_K(k_i) \ge n \cdot A(H(K)) + (t_K(5) - t_K(6)) \cdot b$$

$$\ge n \cdot A(H(K)) + \sum_{|\sigma| \le \lambda} \frac{t_K(5) - t_K(6)}{\lambda} \cdot |\sigma|.$$

In turn, we conclude (1.6) by (1.11). Now $A(D) \ge nA(K)$ and the isoperimetric inequality Theorem A.5.7 yield that $P(D) > 2\sqrt{A(K)n/\pi}$; therefore, $A(D) > n \cdot A(H(K))$ holds for large n. $\qquad \square$

One may hope that the $n(K)$ in Theorem 1.4.1 can be chosen to be an absolute constant. We present now an example showing that this is not the case:

Example 1.4.2. Given any $c \in \mathbb{R}$, there exists a convex domain K such that any suitable $n(K)$ in Theorem 1.4.1 satisfies $n(K) \ge c$.

Let $\varepsilon > 0$ satisfy $(1/12) \ln(1/4\varepsilon) > c$. We fix a system of (x, y) coordinates, and for $p = (1, 0)$ and $q = (0, 1)$, we choose the points p_0 and q_0 in the positive corner and on the hyperbole of equation $x \cdot y = \varepsilon$ in a way that the lines pp_0 and qq_0 are tangent to the hyperbole. We write γ to denote the union of the segments pp_0 and qq_0 and the hyperbole arc between p_0 and q_0. If s is a segment of length $2(n - 1)$ parallel to the first coordinate axis then $s + K$ contains n nonoverlapping translates of K, and

$$A(s + K) = 4(n - 1) + A(K) = 4n - \varepsilon \cdot \ln \frac{1}{4\varepsilon} - 2\varepsilon.$$

However, the two coordinate axes and any tangent to the hyperbolic arc bound a triangle of area 2ε; hence, $A(H(K)) \geq 4 - 12\varepsilon$. Therefore, $n(K) \geq (1/12) \ln(1/4\varepsilon) > c$.

Next we investigate the structure of the optimal packing of a large number of copies.

Corollary 1.4.3. *Let K be a centrally symmetric convex domain that is not a parallelogram, and let D_n be a convex domain of minimal area that contains n nonoverlapping congruent copies of K. Then*

$$c_1 \sqrt{n} < r(D_n) \leq R(D_n) < c_2 \sqrt{n},$$

where $c_1, c_2 > 0$ depend only on K.

Proof. Since $H(K)$ can be assumed to be centrally symmetric according to Lemma 1.3.2, $(\sqrt{n} + 2)H(K)$ contains n nonoverlapping translates of $H(K)$; hence, (1.6) yields that $P(D_n) \leq c_2\sqrt{n}$. Therefore, readily, $R(D_n) \leq c_2\sqrt{n}$, and we deduce by $r(D_n) P(D_n) \geq A(D_n)$ (cf. (A.8)) that $r(D_n) \geq c_1\sqrt{n}$. $\qquad\square$

Comments. Theorem 1.4.1 and Corollary 1.4.3 are proved in K. Böröczky Jr. [Bör2003b].

1.5. Noncrossing Coverings

The covering density of a convex domain K is only known in a few cases, namely, if K is a tile (hence the density is one) or K is centrally symmetric and close to being a circular disc (see the Comments to Section 1.1). Therefore, we consider a restricted class of coverings: We say that two convex domains C_1 and C_2 are *noncrossing* if there exist complementary half planes l^- and

l^+ such that $l^- \cap C_1 \subset C_2$ and $l^+ \cap C_2 \subset C_1$. In addition, an arrangement of convex domains is called noncrossing if any pair is noncrossing. The typical example is a family of homothetic convex domains (see Lemma 2.1.1). We remark that if a sequence of n noncrossing convex domains tend to some n convex domains then the limiting convex domains are noncrossing as well.

Given any convex domain K, we write $\widetilde{H}(K)$ to denote an inscribed polygon of at most six sides whose area is maximal. We verify the *hexagon bound* for finite noncrossing coverings, namely, that the density is at least $A(K)/A(\widetilde{H}(K))$.

Theorem 1.5.1. *If a convex domain D is covered by n noncrossing congruent copies of a convex domain K then*

$$A(D) \le n\, A(\widetilde{H}(K)),$$

provided that either D is a polygon of at most eight sides or $n \ge \widetilde{n}(K)$, where $\widetilde{n}(K)$ depends only on K.

Remark. If K is centrally symmetric then the constant $A(\widetilde{H}(K))$ is optimal in Theorem 1.5.1. The reason is that $\widetilde{H}(K)$ can be assumed to be centrally symmetric according to Corollary 1.5.5; hence, there exists a lattice tiling by translates of $\widetilde{H}(K)$ (cf. Lemma 2.13).

It is widely believed that Theorem 1.5.1 holds without the assumption that the copies are noncrossing. This conjecture seems to be rather obvious to believe, yet some strange things do happen.

Example 1.5.2. There exist convex domains K and D such that D can be covered by two congruent copies of a given convex domain K, but this cannot be done in a noncrossing manner.

Let D be a unit square, and let K be the convex hull of the midpoints of the sides of D and two opposite vertices. Then K and a rotated image by $\pi/2$ cover D. However, it is not hard to see that two noncrossing copies of K cannot cover D.

The first step in proving Theorem 1.5.1 is to define a suitable cell decomposition into convex cells:

Lemma 1.5.3. *Let the convex domain D be covered by the pairwise noncrossing convex domains $K_1, \ldots, K_n$, $n \ge 2$, in a way that no $(n-1)$ copies* *out of* $K_1, \ldots, K_n$ *cover* D. *Then there exists a cell decomposition of* D *into cells* $\Pi_1, \ldots, \Pi_n \subset D$ *such that* $\Pi_i \subset K_i$ *for* $i = 1, \ldots, n$, *and*

$$\sum_{i=1}^{n}(6 - k_i) \geq b + 6,$$

where k_i *denotes the number of edges of* Π_i, *and* b *denotes the total number of edges contained in* ∂D. *In addition, if* D *is a polygon of at most eight sides and the number of sides of* Π_i *is* k_i^* *then* $\sum_{i=1}^{n}(6 - k_i^*) \geq 0$.

Proof. There exists a covering $\Pi_1, \ldots, \Pi_n$ of D by compact convex sets such that $\Pi_i \subset K_i$ for $i = 1, \ldots, n$, the sets $\Pi_1, \ldots, \Pi_n$ are pairwise noncrossing, and $\sum_i A(\Pi_i)$ is minimal under the previous two conditions. Now each Π_i is a convex domain because each K_i is needed to cover D.

We suppose that there exist some Π_i and Π_j, $i \neq j$, that overlap, and we seek a contradiction. The idea is that we can cut off a small part of certain Π_i in a way that the resulting family still forms a noncrossing cover of D. The difficulty is to ensure that the resulting family is pairwise noncrossing.

For $i < j$, let l_{ij} be a line witnessing that Π_i and Π_j are noncrossing, and let l_{ij}^+ and l_{ij}^- denote the half planes such that $l_{ij}^+ \cap \Pi_j$ contains $l_{ij}^+ \cap \Pi_i$ and $l_{ij}^- \cap \Pi_i$ contains $l_{ij}^- \cap \Pi_j$, respectively. We may assume that Π_2 overlaps l_{12}^-, and hence there exist a supporting line l to Π_2 that intersects Π_2 in a single point p, where $p \in \mathrm{int}\, l_{12}^-$.

After possibly renumbering the domains $\Pi_3, \ldots, \Pi_n$, we may assume the following property for some $2 \leq m \leq n$: If, for $i \geq 2$, Π_i contains p, and p has a neighbourhood U_i such that $\Pi_i \cap U_i \subset \Pi_2 \cap U_i$ then $i \leq m$, and $i > m$ otherwise. Thus there exists a half plane $\tilde{l}^+$ such that the line $\tilde{l}$ bounding $\tilde{l}^+$ is parallel to l, the interior of $\tilde{l}^+$ contains p, and $\tilde{l}^+ \cap \Pi_i \subset \tilde{l}^+ \cap \Pi_2$ holds for $i = 2, \ldots, m$. In addition, we may assume that if a domain Π_k or a line l_{ij} does not contain p then it does not intersect the cap $\tilde{l}^+ \cap \Pi_2$.

Let us define Π_i' to be the closure of $\Pi_i \backslash \tilde{l}^+$ if $i = 2, \ldots, m$, and $\Pi_i' = \Pi_i$ otherwise. Then $\Pi_1', \ldots, \Pi_n'$ cover D because $\tilde{l}^+ \cap \Pi_2$ lies in $l_{12}^- \cap \Pi_1$. It also follows that each Π_i' is a convex domain and $\Pi_i' \subset K_i$. Next, we claim that Π_i' and Π_j' do not cross for $i < j$. We may assume that $2 \leq i \leq m$, $j > m$, $p \in l_{ij}$, and $\Pi_j' = \Pi_j$ overlaps l_{ij}^- because otherwise l_{ij} witnesses that Π_i' and Π_j' are noncrossing. We may also assume that Π_i' and Π_j' overlap; hence, the half line $\tilde{l} \cap l_{ij}^-$ intersects $\partial \Pi_j$ in a point $p' \in \mathrm{int}\, l_{ij}^-$. If $q \neq p$ denotes the other endpoint of $l_{ij} \cap \Pi_j$ then the line $p'q$ witnesses that Π_i' and Π_j' are noncrossing, completing the proof of the claim. Since the total area of $\Pi_1', \ldots, \Pi_n'$ is less than the total area of $\Pi_1, \ldots, \Pi_n$, we have arrived at a contradiction. Therefore, $\Pi_1, \ldots, \Pi_n$ determine a cell decomposition of D, and Lemma A.5.9 yields the inequality of Lemma 1.5.3. $\square$

Next we show a concavity property of inscribed m-gons of maximal area.

Lemma 1.5.4. *Let K be a smooth convex domain, and let $\widetilde{\Pi}_m$ be an inscribed m-gon of maximal area. Then, for $m \geq 4$,*

(i) $A(\widetilde{\Pi}_{m-1}) + A(\widetilde{\Pi}_{m+1}) < 2\,A(\widetilde{\Pi}_m)$;
(ii) $\widetilde{\Pi}_m$ is o-symmetric if m is even and K is o-symmetric.

Proof. A simple property of $\widetilde{\Pi}_m$ we need is that the tangent at each vertex is parallel to the diagonal connecting the neighbouring vertices. It follows that the tangents to the endpoints of a side e intersect each other on the exterior side of e.

For $p, q \in \partial K$, let $[p, q)$ denote the arc of ∂K that starts at p and terminates at q according to the clockwise orientation where $[p, q)$ contains p and does not contain q. We write $\Omega_{p,q}$ to denote the area of the convex hull of the closed arc $[p, q]$. If Π is a polygon whose vertices $r_1, \ldots, r_n$ lie on ∂K in clockwise order then, setting $r_0 = r_n$, we have

$$A(\Pi) = A(K) - \sum_{i=1}^{n} \Omega_{r_{i-1}, r_i}.$$

Given $m \geq 4$, let $A(\widetilde{\Pi}_k) + A(\widetilde{\Pi}_l)$ be maximal under the conditions $k + l = 2m$ and $3 \leq k \leq m$. We suppose that $k < m$ and seek a contradiction. The idea is to partition the vertices of $\widetilde{\Pi}_k$ and $\widetilde{\Pi}_l$ into two suitable groups and in this way obtain two inscribed polygons whose total area is larger than $A(\widetilde{\Pi}_k) + A(\widetilde{\Pi}_l)$. We write $p_1, \ldots, p_k$ and $q_1, \ldots, q_l$ to denote the vertices of $\widetilde{\Pi}_k$ and $\widetilde{\Pi}_l$, respectively, according to the clockwise orientation. Since $l \geq k + 2$, we may assume that either $[p_1, p_2)$ contains q_1, q_2, q_3 or there exist two semi open arcs of the form $[p_{t-1}, p_t)$ such that each contains two vertices of $\widetilde{\Pi}_l$. In the first case, let $\widetilde{\Pi}_{k+1}$ be obtained from $\widetilde{\Pi}_k$ by adding the vertex q_2, and let $\widetilde{\Pi}''_{l-1}$ be obtained from $\widetilde{\Pi}_l$ by removing q_2 from the set of vertices. We claim that

$$A(\widetilde{\Pi}'_{k+1}) + A(\widetilde{\Pi}''_{l-1}) > A(\widetilde{\Pi}_k) + A(\widetilde{\Pi}_l), \tag{1.13}$$

which is equivalent to the inequality

$$-\Omega_{p_1, q_2} - \Omega_{q_2, p_2} - \Omega_{q_1, q_3} > -\Omega_{q_1, q_2} - \Omega_{q_2, q_3} - \Omega_{p_1, p_2}. \tag{1.14}$$

Since the tangents at p_1 and p_2 intersect on the exterior side of $p_1 p_2$, the sum of the areas of the triangles $p_1 q_1 q_2$ and $q_2 q_3 p_2$ is less than the area of the quadrilateral $p_1 q_1 q_3 p_2$ (here $p_1 = q_1$ is possible). In turn, we deduce (1.14); hence (1.13) holds as well. This is absurd; therefore, we may assume that $[p_1, p_2)$ contains q_1, q_2, and $[p_i, p_{i+1})$ contains q_j, q_{j+1} for some $2 \le i \le k - 1$ and $2 \le j \le l - 1$. Then let the inscribed k'-gon $\widetilde{\Pi}'_{k'}$ be the convex hull of $q_2, \ldots, q_j$ and $p_{i+1}, \ldots, p_k, p_1$, and let the inscribed l''-gon $\widetilde{\Pi}''_{l''}$ be the convex hull of $q_{j+1}, \ldots, q_l, q_1$ and $p_2, \ldots, p_i$. We observe $k' + l'' = 2m$ and prove the inequality

$$A(\widetilde{\Pi}'_{k'}) + A(\widetilde{\Pi}''_{l''}) \ge A(\widetilde{\Pi}_k) + A(\widetilde{\Pi}_l) \tag{1.15}$$

by verifying the pair of inequalities

$$-\Omega_{p_1,q_2} - \Omega_{q_1,p_2} \ge -\Omega_{p_1,p_2} - \Omega_{q_1,q_2}; \tag{1.16}$$

$$-\Omega_{p_i,q_{j+1}} - \Omega_{q_j,p_{i+1}} \ge -\Omega_{p_i,p_{i+1}} - \Omega_{q_j,q_{j+1}}. \tag{1.17}$$

To prove (1.16), we may assume $p_1 \ne q_1$. Since the tangents at p_1 and p_2 intersect on the exterior side of $p_1 p_2$, the lines $p_1 q_1$ and $p_2 q_2$ intersect on the exterior side of $q_1 q_2$; thus the area of the quadrilateral $p_1 q_1 q_2 p_2$ is larger than the sum of the areas of the triangles $p_1 q_1 p_2$ and $p_1 q_2 p_2$. We deduce (1.16), and also (1.17) by similar argument. It follows by (1.15) and by the maximality of $A(\widetilde{\Pi}_k) + A(\widetilde{\Pi}_l)$ that equality holds in (1.15); hence, $p_1 = q_1$ and $p_i = q_j$. Now the tangent l at p_1 is parallel to the line $p_k p_2$; thus, l is not parallel to the line $p_k q_2$. Therefore, $A(\widetilde{\Pi}_{k'}) > A(\widetilde{\Pi}'_{k'})$, which contradicts the maximality of $A(\widetilde{\Pi}_k) + A(\widetilde{\Pi}_l)$, and we conclude (i).

Turning to (ii), we suppose that $\widetilde{\Pi}_{2s}$, $s \ge 2$, is not symmetric through the centre of K, and we seek a contradiction. We write $p_1, \ldots, p_{2s}$ to denote the vertices of $\widetilde{\Pi}_{2s}$ according to the clockwise orientation and write p'_i to denote the image of p_i by reflecting through the centre of K. Since p'_j is not among $p_1, \ldots, p_{2s}$ for some j, we may assume that the arc $[p_1, p_2)$ contains p'_i and p'_{i+1} for some $3 \le i \le 2s - 2$; hence, the arc $[p'_1, p'_2)$ contains p_i and p_{i+1}. By considering the inscribed $2k$-gon with vertices $p_{i+1}, \ldots, p_{2s}, p_1$ and $p'_{i+1}, \ldots, p'_{2s}, p'_1$, where $k = 2s - i + 1$, and the inscribed $2l$-gon with vertices $p_2, \ldots, p_i$ and $p'_2, \ldots, p'_i$, where $l = i - 1$, the preceding argument yields $A(\widetilde{\Pi}_{2k}) + A(\widetilde{\Pi}_{2l}) > 2A(\widetilde{\Pi}_{2s})$. This contradicts (i), completing the proof of Lemma 1.5.4. $\qquad\square$

Given a convex domain K and $m \ge 3$, we write $\widetilde{t}_K(m)$ to denote the maximal area of an inscribed polygon of at most m sides. In particular,

$A(\widetilde{H}(K)) = \widetilde{t}_K(6)$. We define $\widetilde{t}_K(2) = 0$ because $\widetilde{t}_K(2) + \widetilde{t}_K(4) \leq 2\widetilde{t}_K(3)$ holds this way. Since any convex domain can be approximated by smooth convex domains, Lemma 1.5.4 yields the following:

Corollary 1.5.5 (Dowker theorem for inscribed polygons). *Given a convex domain K,*

(i) $\widetilde{t}_K(m)$ is a concave function of $m \geq 2$;
(ii) $\widetilde{H}(K)$ can be assumed to be o-symmetric if K is o-symmetric.

After all the hard work about related cell decompositions has been done, the proof of our main result is relatively straightforward.

Proof of Theorem 1.5.1. We may assume that K is not a polygon with at most six sides, and hence $\widetilde{t}_K(7) > \widetilde{t}_K(6)$. We may also assume that no $n - 1$ copies out of the n congruent copies cover D. Let $\Pi_1, \ldots, \Pi_n$ form the cell decomposition of Lemma 1.5.3. If D is a polygon of at most eight sides then the average number of sides of $\Pi_1, \ldots, \Pi_n$ is at most six; hence, Corollary 1.5.5 yields that the average area of the cells is at most $A(\widetilde{H}(K))$.

In general, we write Π_i' to denote the convex hull of the k_i vertices of Π, where Π_i' might be a segment. Then $A(\Pi_i') \leq \widetilde{t}_K(k_i)$ holds by definition; hence, Corollary 1.5.5 yields

$$A(\Pi_i') \leq \widetilde{t}_K(6) + (\widetilde{t}_K(7) - \widetilde{t}_K(6)) \cdot (k_i - 6).$$

We conclude by Lemma 1.5.3 (ii) that

$$\sum_{i=1}^{n} A(\Pi_i') \leq n \cdot A(\widetilde{H}(K)) - (\widetilde{t}_K(7) - \widetilde{t}_K(6)) \cdot (b + 6). \qquad (1.18)$$

Let σ be an edge of Σ that is contained in ∂D, and we write $\beta(\sigma)$ to denote the total curvature of σ (see Section A.5.1). If $\beta(\sigma) \leq 2\pi/3$ then the area of $\text{conv } \sigma$ is at most the area of the isosceles triangle whose equal sides enclose an angle of $\pi - \beta(\sigma)$, and the opposite side is of length $\text{diam } K$. However, if $\beta(\sigma) \geq 2\pi/3$ then we simply use the fact that σ is contained in a translate of K, and we deduce that

$$A(\text{conv } \sigma) \leq c \cdot \beta(\sigma) \quad \text{for } c = \max \left\{ \frac{3 A(K)}{2\pi}, \frac{\sqrt{27}}{8\pi} (\text{diam } K)^2 \right\}. \qquad (1.19)$$

Now the total curvature of ∂D is 2π and hence

$$A(D) \leq n \cdot A(\tilde{H}(K)) + c \cdot 2\pi - (\tilde{t}_K(7) - \tilde{t}_K(6)) \cdot (b + 6).$$

If $b \geq c \cdot 2\pi/[\tilde{t}_K(7) - \tilde{t}_K(6)]$ then readily $A(D) \leq nA(\tilde{H}(K))$, thus let $b < c \cdot 2\pi/[\tilde{t}_K(7) - \tilde{t}_K(6)]$. Since the diameter of D is at most $b \cdot \operatorname{diam} K$, D can be covered by some n_0 translates of $\tilde{H}(K)$, where n_0 depends on K. Therefore, we may choose $\tilde{n}(K) = n_0$, completing the proof of Theorem 1.5.1. $\qquad\qquad\square$

Open Problems.

(i) Prove that Theorem 1.5.1 holds without the assumption that the copies are noncrossing. It is known to hold when K is close to being a circular disc (see the Comments below).

(ii) If K is centrally symmetric then prove that the $\tilde{n}(K)$ in Theorem 1.5.1 can be chosen to be an absolute constant.

We note that, for translative coverings, $\tilde{n}(K) = 7$ works according to Theorem 2.8.1.

Comments. L. Fejes Tóth [FTL1950] verifies the hexagon bound for noncrossing coverings of a hexagon, and K. Böröczky, Jr. [Bör2003] proves Theorem 1.5.1. According to G. Fejes Tóth [FTG?b], Theorem 1.5.1 holds without the noncrossing assumption if K is close to being a circular disc. More precisely, it holds when either K is an ellipse such that the ratio of the smaller axis to the greater axis is at least 0.741 or some circular disc of radius R contains K in a way such that the concentric circular disc of radius $0.933R$ is contained in K.

Example 1.5.2 is due to A. Heppes (see [Hep2003]). According to G. Wegner [Weg1980], for arbitrary $n > 1$ there exist convex domains D and K such that D can be covered by n congruent copies of K, but it cannot be done in a pairwise noncrossing manner.

The fact that, for any finite noncrossing covering of a convex domain D, there exists an associated cell decomposition of D (see Lemma 1.5.3) is due to L. Fejes Tóth [FTL1950] but his argument seemed to have some gaps. A detailed proof is provided in R. P. Bambah and C. A. Rogers [BaR1952].

Corollary 1.5.5 is proved by C. H. Dowker [Dow1944]. In addition, G. Fejes Tóth and L. Fejes Tóth [FTF1973a] show that if a convex domain has k-fold rotational symmetry then there exists an inscribed km-gon of maximal area that also has k-fold rotational symmetry.

The constant $A(\tilde{H}(K))$ is not optimal in Theorem 1.5.1 for the typical convex domain K when n is large. This can be shown by extending the method of G. Fejes Tóth and T. Zamfirescu [FTZ1994].

1.6. Packing and Covering by Similar Convex Domains

Let us call the convex domains C and K *similar* if C is congruent to λK for some $\lambda \neq 0$. We have already seen that the hexagon bound holds for finite packings and for finite noncrossing coverings of congruent domains if the number of domains is high enough. In this section we verify the same estimates for the density even when the domains are allowed to be similar but their sizes are not too different. Naturally, the hexagon bound trivially holds if K is a polygon of at most six sides; hence, we may assume that this is not the case. Thus, $H(K)$ denotes a circumscribed hexagon of minimal area, and $\tilde{H}(K)$ denotes an inscribed hexagon of maximal area.

Theorem 1.6.1. *Let K be a convex domain that is not a polygon of at most six sides. If the convex domain D contains the nonoverlapping affine images $K_1, \ldots, K_n$ of K satisfying $A(K_i)/A(K_j) \geq [t_K(6) - t_K(7)]/[t_K(5) - t_K(6)]$ for $i, j = 1, \ldots, n$ then*

$$\frac{A(K_1) + \cdots + A(K_n)}{A(D)} \leq \frac{A(K)}{A(H(K))},$$

provided that either D is a polygon of at most eight sides or each K_i is similar to K and $n \geq n(K)$, where $n(K)$ depends only on K.

Remark. If K is a strictly convex then $[t_K(6) - t_K(7)]/[t_K(5) - t_K(6)] < 1$ according to Lemma 1.2.3.

Theorem 1.6.2. *Let K be a convex domain that is not a polygon of at most six sides. If the convex domain D is covered by the noncrossing affine images $K_1, \ldots, K_n$ of K satisfying $A(K_i)/A(K_j) \geq [\tilde{t}_K(7) - \tilde{t}_K(6)]/[\tilde{t}_K(6) - \tilde{t}_K(5)]$ for $i, j = 1, \ldots, n$ then*

$$\frac{A(K_1) + \cdots + A(K_n)}{A(D)} \geq \frac{A(K)}{A(\tilde{H}(K))},$$

provided that either D is a polygon of at most eight sides or each K_i is similar to K and $n \geq n(K)$, where $n(K)$ depends only on K.

Remark. If K is a smooth then $[\tilde{t}_K(7) - \tilde{t}_K(6)]/[\tilde{t}_K(6) - \tilde{t}_K(5)] < 1$ (see Lemma 1.5.4).

We only prove Theorem 1.6.1 because to prove Theorem 1.6.2, analogous changes are needed compared to the case of congruent domains.

Proof of Theorem 1.6.1. Let $\mu = [t_K(6) - t_K(7)]/[t_K(5) - t_K(6)]$; hence, we may assume that $A(K) = 1$ and

$$\mu \leq A(K_1), \ldots, A(K_n) \leq 1.$$

The core of the argument is the claim that if $m \geq 2$ then

$$t_K(m) \cdot A(K_i) \geq A(H(K)) \cdot A(K_i) + (t_K(6) - t_K(7)) \cdot (6 - m) \quad (1.20)$$

holds for $i = 1, \ldots, n$. The claim follows directly by the convexity of $t_K(\cdot)$ and $A(K_i) \leq 1$ if $m \geq 6$. However, if $m < 6$ then

$$t_K(m) \cdot A(K_i) \geq A(H(K)) \cdot A(K_i) + (t_K(5) - t_K(6)) A(K_i) \cdot (6 - m),$$

and thus $A(K_i) \geq [t_K(6) - t_K(7)]/[t_K(5) - t_K(6)]$ yields (1.20).

Let us construct the cells $\Pi_1, \ldots, \Pi_n$ given by Lemma 1.2.2. If D is a polygon of at most eight sides then the Π_is are actual polygons whose average number of sides is at most six; hence, the hexagon bound holds for any $n \geq 2$ according to (1.20). If D is any convex domain then we copy the proof of Theorem 1.5.1. The essential changes are that (1.12) is replaced by (1.20), and $r(K)$ in the formulae is replaced by $\sqrt{\mu}\, r(K)$, which is a lower bound for the inradius of any K_i. $\qquad\square$

Naturally, the density of a finite packing or noncrossing covering can be arbitrarily close to one if we allow similar copies of any sizes. Next we show that the conditions on the size of the affine images in Theorems 1.6.1 and 1.6.2 are close to being optimal if K is a circular disc.

Remark 1.6.3. According to Theorem 1.6.1, if $n \geq 2$ nonoverlapping ellipses of areas at most 1 and at least

$$\sqrt{\frac{6\tan\frac{\pi}{6} - 7\tan\frac{\pi}{7}}{5\tan\frac{\pi}{5} - 6\tan\frac{\pi}{6}}} = 0.5520\ldots$$

are contained in an octagon then the ratio of the total area of the ellipses to the area of the octagon is at most $\pi/\sqrt{12}$. However, we now construct packings of circular discs of area either 1 or 0.4169 inside large octagons such that the density is larger than $\pi/\sqrt{12}$.

Let $Trap(r)$ be the symmetric trapezoid whose circumradius is $1 + r$, with the shorter side out of the two parallel ones being $2r$ and the unit disc centred at the circumcentre of $Trap(r)$ touching the other two sides. We note that this unit disc lies in *Trap(r)*. Let us consider the periodic tiling of $\mathbb{R}^2$ by translates of *Trap(r)* and $-Trap(r)$ such that each translate of *Trap(r)* shares a common side with four translates of $-Trap(r)$, and vice versa. The circular discs of radius r centred at the vertices of the tiles together with the unit discs centred at the circumcentres of the tiles form a periodic packing whose density is $(\pi + \pi \cdot r^2)/A(Trap(r))$, which is greater than $\pi/\sqrt{12}$ for $r = 0.6457$. Therefore, the density of the circular discs lying in some large octagon is greater than $\pi/\sqrt{12}$.

Remark 1.6.4. According to Theorem 1.6.2, if $n \geq 2$ noncrossing ellipses of areas at most 1 and at least

$$\frac{7 \sin \frac{2\pi}{7} - 6 \sin \frac{2\pi}{6}}{6 \sin \frac{2\pi}{6} - 5 \sin \frac{2\pi}{5}} = 0.6275\ldots$$

cover an octagon then the ratio of the total area of the ellipses to the area of the octagon is at least $2\pi/\sqrt{27}$. However, we verify that large octagons can be covered by circular discs of area either 1 or 0.511 in a way that the density is larger than $2\pi/\sqrt{27}$.

Given $\pi/4 < \alpha < \pi/3$, let $R(\alpha) = -\cos 4\alpha/\cos \alpha$. Our basic building block is the symmetric trapezoid $\widetilde{Trap}(\alpha)$ whose circumradius is $R(\alpha)$, and the two nonparallel sides and the longer parallel side are equal to $2R(\alpha) \cdot \sin \alpha$. Let us consider the four circular discs of radius

$$r = \sin 4\alpha - \cos 4\alpha \cdot \tan \alpha$$

centred at the vertices and the unit disc centred at the circumcentre of $\widetilde{Trap}(\alpha)$. We have defined $\widetilde{Trap}(\alpha)$ in a way that the circular discs centred at the endpoints of the short side share a common point with the unit disc, and the unit disc intersects the long side at two points such that each has distance r from the nearest vertex. In particular, the five circular discs cover $\widetilde{Trap}(\alpha)$. Let us tile the plane by translates of $\widetilde{Trap}(\alpha)$ and $-\widetilde{Trap}(\alpha)$ in a way that each translate of $\widetilde{Trap}(\alpha)$ shares a common side with four translates of $-\widetilde{Trap}(\alpha)$, and vice versa. We place circular discs of radius r at the vertices of the tiling and place unit discs at the circumcentres of the tiles. The density of this periodic covering is $(\pi + \pi \cdot r^2)/A(\widetilde{Trap}(\alpha))$, which is less than $2\pi/\sqrt{27}$ when $r = 0.7149$. Therefore, the density of the circular discs covering some large octagon is less than $2\pi/\sqrt{27}$.

Comments. It was L. Fejes Tóth who observed around 1950 that packings of circular discs of slightly different sizes cannot be denser than $\pi/\sqrt{12}$. The qualitative bounds of Theorems 1.6.1 and 1.6.2 on the ratios of the areas are due to K. Böröczky (unpublished) and to G. Fejes Tóth [FTG1972] in the case when D is a hexagon and K is any convex domain. The arrangements in Remarks 1.6.3 and 1.6.4 were constructed by J. Molnár [Mol1959].

1.7. Coverings by Fat Ellipses

The advantage of the noncrossing assumption on coverings is that there exists a suitable cell decomposition in that case. Fortunately the methods leading to the hexagon bound for noncrossing coverings work even when the domains are not congruent, but are not too different affine images. Capitalizing on this fact, recently the hexagon bound has been proved for any coverings by congruent copies of a convex domain K provided that K is close to being a circular disc (see the Comments). We illustrate the results by the following statement:

Theorem 1.7.1. *Let K be an ellipse such that the larger axis is one, and the smaller axis is at least 0.86. If $n \geq 2$ congruent copies of K cover a convex domain D then*

$$A(D) \leq n \cdot \frac{\sqrt{27}}{2\pi} A(K),$$

provided either D is a polygon of at most eight sides or $n \geq 16{,}000$.

Remark. It follows that $\vartheta(K) = 2\pi / \sqrt{27}$.

Let the smaller and the larger axis of K be a and 1, respectively; hence,

$$a > \sqrt[3]{\frac{\widetilde{t}_K(7) - \widetilde{t}_K(6)}{\widetilde{t}_K(6) - \widetilde{t}_K(5)}} = 0.8561 \ldots \tag{1.21}$$

We observe that the area of K is $a\pi$, and the curvature at any point of ∂K is at least a. (One may use (A.6) to calculate the curvature because a part of the boundary is the graph of the function $a\sqrt{1 - t^2}$.)

The main idea of the proof is to replace the covering by congruent copies of K by a noncrossing covering of not larger density such that the new covering consists of congruent copies of K and circular discs of radius $1/a$. In this way we can apply the results about coverings by not too different noncrossing affine images.

Proposition 1.7.2. *If two congruent copies of K cross then their union can be covered by a circular disc of radius $1/a$.*

Proof. It is sufficient to prove that the centres of the two crossing congruent copies of K are at most $2(1/a - 1)$ apart. To verify this claim, we may assume that the two ellipses have a common supporting half plane at some point p. Let u_p be the endpoint of the common interior unit normal at p; hence, it is sufficient to prove that the distance of u_p from both centres is at most $1/a - 1$.

Concerning either of the ellipses, we may assume that the boundary of the ellipse is parameterized as $(\cos t, a \sin t)$, where $p = (\cos \theta, a \sin \theta)$. In particular,

$$u_p = \left(\cos \theta \cdot \left(1 - \frac{a}{\sqrt{a^2 \cos^2 \theta + \sin^2 \theta}} \right), \ \sin \theta \cdot \left(a - \frac{1}{\sqrt{a^2 \cos^2 \theta + \sin^2 \theta}} \right) \right).$$

Since both $\cos \theta$ and $\sin \theta$ are multiplied by a number of absolute value at most $1/a - 1$, we conclude Proposition 1.7.2. $\qquad \square$

Proof of Theorem 1.7.1. We write $K_1, \ldots, K_n$ to denote the congruent copies of K that cover the convex domain D. If K_i and K_j cross then we replace them by one circular disc of radius $1/a$ covering K_i and K_j. Continuing in this manner, we obtain $m \le n$ convex domains $C_1, \ldots, C_m$ covering D such that each is either congruent to K or a circular disc of radius $1/a$, and if C_i and C_j are congruent to K then they do not cross. Since $a^3 > 1/2$, it follows that the total area of $C_1, \ldots, C_m$ is at most the total area of $K_1, \ldots, K_n$.

Let us suppose that some C_i and C_j cross and seek a contradiction. Since two circular discs never cross, say C_i is a congruent to K, and C_j is a circular disc of radius $1/a$. Now the boundaries of C_i and C_j intersect in at least four points; hence, there exist two common boundary points p and q such that the shorter circular arc of radius $1/a$ connecting p and q is not contained in C_i. This fact contradicts Lemma A.5.5, and we deduce that $C_1, \ldots, C_m$ form a noncrossing family. In particular, if D is a polygon of at most eight sides then we conclude Theorem 1.7.1 by Theorem 1.6.2.

To calculate the threshold for n if D is any convex domain, we may assume that D is not covered by a proper subset of $\{C_1, \ldots, C_m\}$. Let $\Pi_1, \ldots, \Pi_m$ be the cells given by Lemma 1.5.3, and we write Π'_i to denote the convex hull of the vertices of Π_i. The concavity of $\widetilde{t}_K(\cdot)$ and the condition (1.21) yield that

$$A(\Pi'_i) \le \frac{\sqrt{27}}{2\pi} A(C_i) + \left(\frac{7}{2} \sin \frac{2\pi}{7} - 3 \sin \frac{2\pi}{6} \right) \cdot \frac{k_i - 6}{a^2},$$

where k_i denotes the number of edges of Π_i. In particular, we deduce by Lemma 1.5.3 (ii) that

$$\sum_{i=1}^{m} A(\Pi'_i) \leq \frac{\sqrt{27}}{2\pi} \sum_{i=1}^{m} A(C_i) - \left(\frac{7}{2} \sin \frac{2\pi}{7} - 3 \sin \frac{2\pi}{6}\right) \cdot \frac{b+6}{a^2},$$

where some b circular discs of radius $1/a$ cover ∂D; hence $b \geq a P(D)/(2\pi)$. According to the argument leading to (1.19), if σ is a boundary edge of some Π_i of total curvature $\beta(\sigma)$ then $A(\text{conv}\,\sigma) \leq 3\beta(\sigma)/(2a^2)$; therefore,

$$A(D) \leq n \cdot A(\widetilde{H}(K)) - \left(\frac{7}{2} \sin \frac{2\pi}{7} - 3 \sin \frac{2\pi}{6}\right) \cdot \left(\frac{P(D)}{2\pi a} + \frac{6}{a^2}\right) + \frac{3\pi}{a^2}.$$

$$(1.22)$$

If $P(D) \geq 456$ then (1.22) implies Theorem 1.7.1; thus, let $P(D) \leq 456$. Since $P(D + 2B^2) \leq 456 + 4\pi$, the isoperimetric inequality Theorem A.5.7 yields that $A(D + 2B^2) < (456 + 4\pi)^2/(4\pi)$. Now $\widetilde{H}(K)$ is an affine regular hexagon of area $A(K)\sqrt{27}/(2\pi) = a\sqrt{27}/2$, and its translates tile the plane. Since any tile intersecting D is contained in $D + 2B^2$, at most $16,000$ of the tiles cover D. In turn, we conclude Theorem 1.7.1. $\square$

Corollary 1.7.3. *If K is an ellipse as in Theorem 1.7.1, and D_n is a convex domain of maximal area that can be covered by n congruent copies of K, then $c_1\sqrt{n} < r(D_n) \leq R(D_n) < c_2\sqrt{n}$, where c_1 and c_2 are positive absolute constants.*

Proof. Since $(\sqrt{n} - 2)\widetilde{H}(K)$ is covered by n translates of $\widetilde{H}(K)$, its area is a lower bound for $A(D_n)$. It follows by (1.22) that $P(D_n) \leq c_2\sqrt{n}$ for some absolute constant c_2. Therefore, $R(D_n) \leq c_2\sqrt{n}$ and the formula $r(D_n) P(D_n) \geq A(D_n)$ (cf. (A.8)) yield $r(D_n) \geq c_1\sqrt{n}$. $\square$

Comments. Theorem 1.7.1 is due to A. Heppes [Hep2003]. It has been substantially improved by G. Fejes Tóth [FTG?b]. On the one hand [FTG?b] generalized Theorem 1.7.1 to ellipses when the ratio of the smaller axis to the greater axis is at least 0.741. On the other hand if K is a centrally symmetric convex domain and $r(K)/R(K) \geq 0.933$ then [FTG?b] proves all the analogues of the statements and corollaries of Theorem 1.7.1; for example, $\Theta(K) = A(\widetilde{H}(K))$.

1.8. Minimal Perimeter and Diameter for Packings

Given a convex domain K, we show that the asymptotic shape of a convex domain of minimal perimeter containing n nonoverlapping congruent copies of K is a circular disc:

Theorem 1.8.1. *Let K be a convex domain, and let D_n be a convex domain of minimal perimeter such that D_n contains n nonoverlapping congruent copies of K. Then*

$$2\sqrt{\pi \Delta(K)} \cdot \sqrt{n} - 2\pi R(K) \leq P(D_n) \leq 2\sqrt{\pi \Delta(K)} \cdot \sqrt{n} + 2\pi R(K),$$

and $\lim_{n \to \infty} D_n / \sqrt{n} = B$, where B is a circular disc of area $\Delta(K)$.

Remark. Let d_n be the minimal diameter of the union of n nonoverlapping congruent copies of K. Then the proof of Theorem 1.8.1 shows that $d_n \sim \sqrt{4\Delta(K)n/\pi}$ as n tends to infinity, and the convex hull in the optimal packing is close to some circular disc of area $n\Delta(K)$.

Proof. Let $\Delta = \Delta(K)$ and $R = R(K)$. Since $A(D_n + RB^2) \geq n\Delta$ according to Lemma 1.1.2, the isoperimetric inequality theorem (Theorem A.5.7) yields that $P(D_n + RB^2) \geq 2\sqrt{\pi \Delta n}$, and, in turn, the lower bound for $P(D_n)$ is a consequence of $P(D_n + RB^2) = P(D_n) + 2\pi R$. However, Lemma 1.1.2 also yields that the circular disc B_n of radius $\sqrt{\Delta/\pi}\sqrt{n} + R$ contains n nonoverlapping congruent copies of K; hence, the upper bound for $P(D_n)$ follows from $P(D_n) \leq P(B_n)$. Assuming that the origin is the circumcentre of D_n, a fixed circular disc contains $(1/\sqrt{n})(D_n + RB^2)$ for any n; thus, any subsequence contains a convergent subsequence according to the Blaschke selection theorem (Theorem A.3.1). If B is the limit of such a subsequence then its area is at least Δ, and its perimeter is $2\sqrt{\pi \Delta}$; therefore, the isoperimetric inequality theorem (Theorem A.5.7) yields that B is the circular disc of area $\Delta(K)$.

Concerning the minimal diameter, we note that $\operatorname{diam} D \geq P(D)/\pi$ for any convex domain D, which follows by considering $(D - D)/2$. $\qquad \square$

Comments. Theorem 1.8.1 was proved independently by K. Böröczky Jr. [Bör1994] and Ch. Zong [Zon1995a].

1.9. Covering the Maximal Perimeter

We show that a compact convex set of maximal perimeter that can be covered by n congruent copies of a given convex domain K is close to some segment for large n. It is equivalent to show that the inradius is small because if s is a diameter of a convex domain D then

$$D \subset s + 3r(D) \cdot B^2. \tag{1.23}$$

To verify (1.23), let h be the maximal distance of points of D from s. Then D contains a triangle T such that s is a longest side, and the distance of the opposite vertex from s is h. Therefore, (1.23) follows by $hs/2 = A(T) \le 3s \cdot r(T)/2$.

Theorem 1.9.1. *If K is a convex domain and D_n is a compact convex set of maximal perimeter whose boundary is covered by $n \ge 24$ congruent copies of K then*

(i) $r(D_n) \le \frac{144}{n} \cdot \operatorname{diam} K$;
(ii) $2n \cdot \operatorname{diam} K \ \le \ P(D_n) \ \le \ \left(2n + \frac{144}{n}\right) \cdot \operatorname{diam} K$.

Remark. The analogous statements hold if not only the boundary of D_n but also the whole set D_n is covered by the n congruent copies of K.

Proof. We assume that $\operatorname{diam} K = 1$, and write $K_1, \ldots, K_n$ to denote the congruent copies of K covering ∂D_n. Since any segment of length $2n$ can be covered by n congruent copies of K, we deduce

$$P(D_n) \ge 2n. \tag{1.24}$$

We divide the proof of Theorem 1.9.1 into two steps: First we show that $r(D_n) \le 1$ and then that D_n is really thin.

We suppose that $r(D_n) > 1$, and we seek a contradiction. We recall that $C = D_n \ominus (1/2)B^2$ is the family of points x such that $x + (1/2)B^2 \subset D_n$. In addition, let σ_i be the family of points ∂C that are the closest points of C to some point of $K_i \cap \partial D_n$. We observe that if $y, z \in K_i \cap \partial D_n$ then the segment yz does not intersect ∂C; hence, σ_i is a convex arc. It also follows that K_i contains a segment s_i that connects two points of $K_i \cap \partial D_n$, and σ_i is the family of points ∂C that are the closest points of C to some point of s_i. Since the closest point map decreases the distance (see (A.3)), we deduce that the length of σ_i is at most the length of s_i; hence, $P(C) \le n$. However,

D_n is a subset of a translate of $r(D)/(r(D) - 0.5) \cdot C$, which in turn yields $P(D_n) < 2n$. This contradiction with (1.24) proves that $r(D_n) \leq 1$.

Next we show that $r(D_n)$ is very small. Let s be a diameter of D_n, and let h be the maximal distance of the points of D_n from s; hence, $h \leq 3$ follows by (1.23). The crucial claim is that

$$\text{diam } D_n \leq \left(1 - \frac{h^2}{72}\right) \cdot n. \tag{1.25}$$

We may assume that $\text{diam } D_n \geq (1 - 3^2/72) \cdot n = 7n/8$; hence, s contains a segment s' of length at least $7n/12$ such that any line orthogonally intersecting s' intersects D_n in a segment of length at least $h/3$. Next we write σ and σ' to denote the two arcs of ∂D_n connecting the endpoints of s. If K_i intersects σ then let p_i and q_i be the furthest points of $\sigma \cap K_i$, and otherwise let $p_i = q_i$ be any point of σ. We define the points $p_i', q_i' \in \sigma'$ analogously, and we claim that if the projection of K_i into aff$s$ is contained in s' then

$$d(p_i, q_i) + d(p_i', q_i') \leq 2 - \frac{h^2}{18}. \tag{1.26}$$

We may assume that p_i, q_i, p_i', and q_i' are vertices of a quadrilateral Q. Since Q has an angle of at least $\pi/2$, and its diagonals are at most one, we deduce that either $p_i q_i$ or $p_i' q_i'$ is of length at most $\sqrt{1 - h^2/9}$. In turn, the claim (1.26) readily follows. Now the number of K_i whose projection into aff$s$ is contained in s' is at least $7n/12 - 2 \geq n/2$; therefore, $\text{diam} D_n$ is at most

$$\frac{1}{2} \left(\sum_{i=1}^{n} d(p_i, q_i) + d(p_i', q_i') \right) \leq \frac{n}{2} + \frac{n}{2} \cdot \left(1 - \frac{h^2}{36}\right) = \left(1 - \frac{h^2}{72}\right) \cdot n,$$

completing the proof of the claim (1.25). In particular, (1.25) yields

$$P(D_n) < 2 \cdot \text{diam } D_n + 4h \leq 2n - \frac{n \cdot h^2}{36} + 4h. \tag{1.27}$$

Optimizing the upper bound in h leads to Theorem 1.9.1 (ii), and $r(D_n) \leq 144/n$ follows by $r(D_n) \leq h$ and $P(D_n) \geq 2n$. □

Let us show that the estimates of Theorem 1.9.1 are essentially optimal.

Example 1.9.2. If K is the unit disc then the optimal D_n in Theorem 1.9.1 satisfies $r(D_n) = 1/n$ and $P(D_n) > 4n(1 + n/3)$.

If K is an isosceles triangle whose equal angles are $\pi/6$ then the optimal D_n in Theorem 1.9.1 is a segment of length $n \, \text{diam } K$ for large n.

The case of the unit disc follows by Theorem 5.7.1. If K is the isosceles triangle and the longest side is one then the inequality (1.27) in the preceding proof can be replaced by $P(D_n) \leq 2n - cn \cdot h + 4h$ for suitable positive absolute constant c. Therefore, $h = 0$ for large n.

1.10. Related Problems

Separable Packings

Let K be a convex domain. We write $\Pi(K)$ to denote K if K is a triangle, and a quadrilateral of minimal area containing K otherwise. Moreover, the congruent copies $K_1, \ldots, K_n$ of K are called *separable* if for any $i \neq j$, K_i and K_j can be separated by a line that does not intersect the interiors of $K_1, \ldots, K_n$. According to G. Fejes Tóth and L. Fejes Tóth [FTF1973b], if a quadrilateral Q contains n separable congruent copies of K then $A(Q) \geq nA(\Pi(K))$. If K is centrally symmetric then [FTF1973b] observes that $\Pi(K)$ can be chosen to be a parallelogram by a theorem of C.H. Dowker [Dow1944], and hence the coefficient $A(\Pi(K))$ is optimal for any n.

Snakes

Let K be a convex domain in $\mathbb{R}^2$. A *snake* for K is a family $K_1, \ldots, K_n$ of nonoverlapping congruent copies of K such that K_i and K_j touch if and only if $|i - j| = 1$. A snake is called limited if it is not the proper subset of any other snake. It follows by the ideas of K. Böröczky, Jr. and V. Soltan [BöS1995] that there exists a limited snake of cardinality ten for any convex domain K, and the number ten cannot be lowered if K is a circular disc. T. Bisztriczky and H. Harborth [BiH2001] assume that K is some regular polygon, and K_i and K_j share a common edge if $|i - j| = 1$. Under these conditions, [BiH2001] proves that the smallest cardinality of a limited snake is 20, 19 and 13 if K is a triangle, square and hexagon, respectively.

Clouds

Let K be a convex domain. A *cloud* for K is a packing $K_1, K_2, \ldots$ where each K_i is congruent with K and does not overlap K, and any ray emanating from K intersects the interior of some K_i. In other words, any light beam moving towards K from infinity is blocked by the interior of some K_i.

The problem of finding the minimal cardinality of a cloud appeared first in L. Fejes Tóth [FTL1959]. He posed the problem for unit balls in dimension three, and he only required that rays starting from the centre of the given ball are blocked. In our definition of a cloud, we required that open clouds block light rays from a closed shape. Some papers call these clouds deep clouds (see, e.g., I. Talata [Tal2000a]).

L. Szabó and Z. Ujváry–Menyhárt [SzU2002] verified for a planar convex domain K that the minimal cardinality of a cloud is 9 if K is a circular disc and at least 4 and at most 8 otherwise.

Newton Number

The *Newton number* $N(K)$ of a convex domain K is the maximal number of nonoverlapping copies of K that touch K. Readily $N(K) \geq 6$, with equality for circular discs. According to L. Fejes Tóth [FTL1967b],

$$N(K) \leq (4 + 2\pi) \cdot \frac{\text{diam } K}{w(K)} + 2 + \frac{w(K)}{\text{diam } K}$$

where $w(K)$ is the width of K (see Section A.8). It follows that $N(K) = 64$ if K is the isosceles triangle with $w(K)/\text{diam } K = \sin(\pi/9)$. In addition, the Newton numbers of various special convex domains have been determined: The regular n-gon P_n satisfies $N(P_3) = 12$, $N(P_4) = 8$, and $N(P_n) = 6$ for $n \geq 5$. This is proved for all regular polygons other than pentagons by K. Böröczky [Bör1971], and the case of regular pentagon is settled by J. Linhart [Lin1973]. Another interesting example is the isosceles triangle with angles $\pi/6$, $\pi/6$, $2\pi/3$, which has Newton number of 21 according to G. Wegner [Weg1992].

Packing Unit Squares into Large Squares

Let $d(a)$ be the highest density of a packing of unit squares into a square of side length a. Readily $d(a) = 1$ if a is the square of an integer. For large a, P. Erdős and R. L. Graham [ErG1975] and K. F. Roth and R. C. Vaughan [RoV1978], respectively, prove that

$$1 - \frac{c_1}{a^{15/11}} \leq d(a) \leq 1 - \frac{c_2 \|a\|^{1/2}}{a^{3/2}}$$

where c_1 and c_2 are positive absolute contants, and $\|a\|$ denotes the distance of a from the nearest integer. The dual problem is also of high interest; namely, the size of the smallest square that holds n nonoverlapping unit squares for given n (see E. Friedman [Fri2000] and M. Kearney and P. Shiu [KeS2002]).

2

Translative Arrangements

This chapter discusses finite translative arrangements of a given convex domain in $\mathbb{R}^2$. For packings, we determine the translative packing density and the asymptotic structure of the optimal packings of n translates for large n. We prove the analogous results for coverings by translates of centrally symmetric convex domain K, and, moreover, the hexagon bound for coverings of any convex shape by at least seven translates of K. The main tools are the Oler inequality for packings and the Fejes Tóth inequality for coverings. In addition, we prove the Bambah inequality for coverings when the convex hull of the centres is covered. Many of these proofs are based on constructing *Delone type simplicial complexes*. The final part of Chapter 2 discusses problems related to the Hadwiger number of K, which is the maximal number of nonoverlapping translates of K touching K.

Assuming that K is centrally symmetric, let us compare our knowledge of arrangements of translates of K (see this chapter) and of congruent copies of K (see Chapter 1). We have essentially the same results for packings. However, we know much more about coverings by translates because any two homothetic convex domains are noncrossing (see Section 1.5).

2.1. About the Minkowski Plane

Convex domains K and C are called *homothetic* if $C = x + \lambda K$ for some $\lambda > 0$. This chapter discusses problems when the geometry of the plane is considered in terms of convex domains homothetic to a given one. This view point was applied first systematically by H. Minkowski, and hence the term Minkowski plane is used. In this section we verify various well-known facts about homothetic domains.

Given an o-symmetric convex domain K, the associated Minkowski norm of any $u \in \mathbb{R}^2$ is

$$\|u\|_K = \min\{\lambda \geq 0 : u \in \lambda K\}.$$

The corresponding length of a segment pq is defined to be $\|p - q\|_K$.

Next we show that homothetic strictly convex domains behave like circular discs in many respects; say the boundaries of two homothetic strictly convex domains intersect in at most two points:

Lemma 2.1.1. *If two homothetic strictly convex domains K_1 and K_2 overlap without containing each other then ∂K_1 and ∂K_2 intersect in exactly two points p and q, and $\mathrm{int}\,K_i$ contains one of the open arcs γ_j of ∂K_j connecting p and q, where γ_1 and γ_2 are separated by the common secant pq. Assuming, in addition, that K_2 is not smaller than K_1, the tangent half lines to γ_2 at p and at q intersect.*

Remark. If K_1 and K_2 are centrally symmetric with centres z_1 and z_2, respectively, then the line $z_1 z_2$ strictly separates p and q, and the half line $z_i z_j$, $i \neq j$, intersects γ_i.

Proof. Since ∂K_2 has points both in $\mathrm{int}\,K_1$ and outside of K_1, there exist at least two common points p and q of ∂K_1 and ∂K_2. We write s to denote the common secant pq, and let $y \in \partial K_2$.

If $K_2 = K_1 + x$ then $y \in \partial K_1$ is equivalent to $y + x \in \partial K_2$. Assuming that $y \neq p, q$ lies between the lines $p(p + x)$ and $q(q + x)$, and also that $y + \alpha x \in \partial K_2$ for $\alpha > 0$, the strict convexity of K_2 yields that y and $y + \alpha x$ are strictly separated both by s and $s + x$. Therefore, γ_2 is the open arc of ∂K_2 that lies on the side of s opposite to $s + x$. If K_1 and K_2 are not translates then we may assume that $K_2 = \lambda K_1$ for $\lambda > 1$. In this case $y \in \partial K_1$ is equivalent to $\lambda y \in \partial K_2$. Assuming that $\mu y \in \partial K_2$ for $\mu > 1$, we deduce by the strict convexity of K_2 that $\mu > \lambda$ if y lies between the half lines op and oq, and $\mu < \lambda$ otherwise. Therefore, γ_2 is the open arc of ∂K_2 that lies on the side of s opposite to λs.

If K_1 and K_2 are centrally symmetric then ∂K_1 and ∂K_2 intersect on both sides of the line $z_1 z_2$, concluding the proof of Lemma 2.1.1. $\square$

For an o-symmetric convex domain K, we deduce the following statement by Lemma 2.1.1 when K is strictly convex and by approximation in the general case:

Lemma 2.1.2. *Let K be an o-symmetric convex domain, and let $p \in \partial K$. If the point q moves away from p along ∂K then $\|q - p\|_K$ increases until it equals two, and $\|q - p\|_K$ strictly increases when K is strictly convex.*

If H is an o-symmetric hexagon then the midpoints of the sides are $\pm x$, $\pm y$, and $\pm(y - x)$, where x and y are midpoints of two neighbouring sides. We deduce the following:

Lemma 2.1.3. *If H is an o-symmetric hexagon then $\Lambda + H$ is a tiling, where Λ is the lattice generated by the midpoints of the sides of $2H$.*

The rest of the statements in this section are mostly related to packings. Given a convex domain K, $x + K$ and $y + K$ intersect if and only if $y - x \in K - K$; hence, $x + K$ and $y + K$ form a packing if and only if $x + K_0$ and $y + K_0$ form a packing for $K_0 = (K - K)/2$. This observation reduces various packing problems to the case when K is o-symmetric.

Next, we show that if K is an o-symmetric convex domain then any $x \in \partial K$ is the vertex of an inscribed affine regular hexagon.

Lemma 2.1.4. *If K is an o-symmetric convex domain and $x \in \partial K$ then there exists a y such that $\pm x, \pm y, \pm(y - x) \in \partial K$. In addition, the lattice generated by $2x$ and $2y$ is a packing lattice for K.*

Proof. We choose y to be a common boundary point of K and $x + K$. For the lattice Λ generated by $2x$ and $2y$, it is equivalent to show $z \notin \mathrm{int}2K$ for $z \in \Lambda \backslash o$. There are two neighbouring vertices x' and y' of the hexagon with vertices $\pm x, \pm y, \pm(y - x)$ such that $z = 2\alpha x' + 2\beta y'$ for some nonnegative $\alpha, \beta \in \mathbb{Z}$, and we may assume that $\alpha, \beta > 0$. Since $2y' \in \partial 2K$ lies in the relative interior of the segment connecting z and $\lambda(y' - x') \in 2K$ for $\lambda = 2\alpha/(\alpha + \beta - 1)$, we conclude that $z \notin \mathrm{int}\, 2K$. $\qquad \square$

We will frequently need information about the shape of convex domains. If C is a convex domain then $r_K(C)$ denotes the maximal r such that a translate of rK is contained in C, and $R_K(C)$ denotes the minimal R such that a translate of RK contains C. In addition, the convex hull Σ of two parallel lines is called a *strip*, and we define $r_K(\Sigma)$ accordingly.

Lemma 2.1.5. *If K and C are convex domains then there exists a strip Σ that contains C and satisfies $r_K(\Sigma) \leq 2r_K(C)$.*

Proof. We may assume that $r_K(C) = 1$ and $K \subset C$. It follows that either there exist two common parallel supporting lines of C and K or there exists a triangle T whose sides are common tangents to C and K. In the first case the two supporting lines determine a strip Σ with $r_K(\Sigma) = 1$. In the second case we consider the triangle, which is determined by the midpoints of the sides of T. This triangle has a side s that intersects K, and hence the strip Σ bounded by the two tangents of K parallel to s satisfies $r_K(\Sigma) \leq 2$. $\qquad\square$

Now Lemma 2.1.5 readily yields

Lemma 2.1.6. *If K and C are convex domains then*

$$A(C) \leq 4\, r_K(C) R_K(C) \cdot A(K).$$

Fixing an o-symmetric convex domain K, let us introduce the *Busemann perimeter* $P_K(\cdot)$ with respect to $\|\cdot\|_K$: If C is a polygon then $P_K(C)$ is the sum of the lengths of its edges, and if C is a segment then $P_K(C)$ is twice the length. Then the triangle inequality yields that $P_K(C)$ is an increasing function in C; therefore, $P_K(C)$ can be defined for any convex domain C by continuity. In addition, $P_K(\cdot)$ is positive linear; namely,

$$P_K(\lambda_1 C_1 + \lambda_2 C_2) = \lambda_1 P_K(C_1) + \lambda_2 P_K(C_2) \qquad (2.1)$$

holds for positive λ_1, λ_2. Actually, $P_K(C) = S_K(C)/2$, where $S_K(\cdot)$ is the Busemann surface area, defined in Section 9.1 in any dimension. We present the optimal bounds on the "self–perimeter":

Lemma 2.1.7. *If C is a convex domain and $C_0 = (C - C)/2$ then*

$$6 \leq P_{C_0}(C) \leq 8.$$

Here equality holds in the upper bound if and only if C is a parallelogram.

Proof. We note that $P_{C_0}(C) = P_{C_0}(C_0)$ according to (2.1). There exist x, y such that $\pm x, \pm y, \pm(x - y)$ lie in ∂C_0 (see Lemma 2.1.4); hence, these six points are vertices of an inscribed affine regular hexagon H. Therefore, $P_{C_0}(C_0) \geq P_{C_0}(H) = 6$.

Turning to the upper bound, we let Q be a parallelogram of minimal area containing C_0. Then the midpoints of the sides of Q are contained in ∂C_0, which in turn yields $P_{C_0}(C_0) \leq P_{C_0}(Q) = 8$.

Readily, $P_{C_0}(C) = 8$ if C is a parallelogram. If C is not a parallelogram then neither is C_0. Therefore, there exist a diagonal d of Q and a parallel supporting line l of C_0 such that the length of $d \cap C_0$ is larger than half of the length of d, and l cuts Q into a triangle T and a pentagon M. Then the two sides of T contained in ∂Q are of equal length of, say, a with respect to $\| \cdot \|_{C_0}$, and the third side is of length less than $2a$. We conclude that $P_{C_0}(C_0) \leq P_{C_0}(M) < 8$. $\square$

2.2. Periodic Packing and Covering by Translates

Given a convex domain K, we determine its translative packing density, and if K is centrally symmetric then also its translative covering density. We discuss the two results in one section because their statements are analogous, and both are simple consequences of the existence of the related Delone triangulations in case K is strictly convex.

We recall from Section A.13 that finite arrangements of translates of K on tori correspond to periodic arrangements of translates of K in $\mathbb{R}^2$. We define the *translative packing density* $\delta_T(K)$ to be the supremum of the densities of periodic packings by translates of K, and we define the *translative covering density* $\vartheta_T(K)$ to be the infimum of the densities of periodic coverings by translates of K. In addition, let $\Delta_T = A(K)/\delta_T(K)$, and let $\Theta_T = A(K)/\vartheta_T(K)$.

Given a z-symmetric convex domain K, let $H(K) = K$ if K is a parallelogram, and let $H(K)$ be a circumscribed z-symmetric hexagon of minimal area otherwise (hence $H(K)$ is of minimal area among all circumscribed polygons of at most six sides according to Lemma 1.3.2). In addition, we write $T(K)$ to denote a triangle of minimal area such that each side of $T(K)$ is of length one with respect to $\| \cdot \|_{K-z}$.

Theorem 2.2.1. *If K is a convex domain and $K_0 = (K - K)/2$ then*

$$\Delta_T(K) = A(H(K_0)) = 8A(T(K_0)),$$

and some lattice packing of K is a densest translative periodic packing.

Since any centrally symmetric hexagon induces a lattice tiling of the plane according to Lemma 2.1.3, we deduce the following:

Corollary 2.2.2. *If a centrally symmetric hexagon D contains n nonoverlapping translates of a convex domain K then*

$$A(D) \geq n\, A(H(K_0)), \quad \text{where } K_0 = (K - K)/2.$$

Clusters based on the lattice tiling by $H(K_0)$ show that the coefficient $A(H(K_0))$ is optimal in Corollary 2.2.2. If K is centrally symmetric and n is large then D can be any convex domain (see Corollary 2.3.4).

Let us turn to coverings. Given a z-symmetric convex domain K, we write $\widetilde{H}(K)$ to denote K if K is a parallelogram, and an inscribed z-symmetric hexagon of maximal area otherwise. Actually, $\widetilde{H}(K)$ has maximal area among all inscribed polygons of at most six sides according to Corollary 1.5.5.

When K may not be centrally symmetric then we need a special type of hexagon. We call a polygon H a p-hexagon if it is either a hexagon and has two opposite equal and parallel sides or a pentagon and has two equal and parallel sides, or if H has at most four sides. We observe that centrally symmetric hexagons are p-hexagons, but unlike centrally symmetric hexagons, translates of a p-hexagon H may not tile the plane (whereas translates of H and $-H$ actually do). Let $\widetilde{H}_p(K)$ be a p-hexagon of largest area contained in K; hence, if K is centrally symmetric then we may choose $\widetilde{H}_p(K) = \widetilde{H}(K)$.

Theorem 2.2.3. $\Theta_T(K) \le A(\widetilde{H}_p(K)) \le 2A(\widetilde{T}(K))$ *holds for any convex domain K. If in addition K is centrally symmetric then*

$$\Theta_T(K) = A(\widetilde{H}(K)) = 2A(\widetilde{T}(K)),$$

and some lattice covering by K is a thinnest periodic translative covering.

Any covering of a centrally symmetric hexagon D by n translates of some convex domain K induces a periodic covering of the plane (see Lemma 2.1.3). Therefore,

$$A(D) \le n\,\Theta_T(K). \tag{2.2}$$

We will prove (see Theorem 2.8.1) that if a convex domain D is covered by $n \ge 7$ translates of a centrally symmetric domain K then $A(D) \le n\,A(\widetilde{H}(K))$.

As already mentioned, the proofs of Theorems 2.2.1 and 2.2.3 are both based on the existence of Delone triangulation. Let K be a strictly convex domain, and let $\Gamma \subset \mathbb{R}^2$ be a point set periodic with respect to a lattice Λ such that Λ defines finitely many equivalence classes of Γ. We say that $y - \lambda K$, $\lambda > 0$, is a *supporting domain* if its interior contains no point of Γ, and its boundary contains at least three points of Γ. Moreover, the convex hull of the points of Γ in $y - \lambda K$ is called a *Delone cell*. Since the common secant of any two overlapping supporting domains separates the corresponding Delone cells according to Lemma 2.1.1, the Delone cells form the *Delone cell complex* $\widetilde{\Sigma}_{\text{Del}}$, which is periodic with respect to Λ. Any edge of $\widetilde{\Sigma}_{\text{Del}}$ is contained in exactly two Delone cells, and hence $\widetilde{\Sigma}_{\text{Del}}$ tiles $\mathbb{R}^2$. Then a *Delone triangulation* Σ_{Del} of $\mathbb{R}^2$ is a triangulation of $\widetilde{\Sigma}_{\text{Del}}$ that is periodic with respect to Λ, and whose vertex set is Γ.

Let $\lambda_0 > 0$ satisfy that $\Gamma + \lambda_0 K$ cover $\mathbb{R}^2$; hence, $\lambda \leq \lambda_0$ holds for any supporting domain $y - \lambda K$. Possibly taking a sublattice of Λ, we may assume that $\lambda_0 K$ embeds isometrically into the torus $\mathbb{R}^2/\Lambda$. Now Σ_{Del} induces a finite triangulation Σ of $\mathbb{R}^2/\Lambda$, which we also call a Delone triangulation. The total sum of the angles of triangles of Σ is $\pi f_2(\Sigma)$ on the one hand, and $2\pi f_0(\Sigma)$ on the other hand; therefore,

$$f_2(\Sigma) = 2 f_0(\Sigma). \tag{2.3}$$

To prove Theorem 2.2.1, we need the following:

Lemma 2.2.4. *Given an o-symmetric strictly convex domain K, if each side of some triangle T is of length at least two with respect to $\| \cdot \|_K$, and the vertices of T lie on $y + \lambda \partial K$ for some y and $\lambda \leq 2$, then*

$$A(T) \geq 4 A(T(K)).$$

Proof. We write p, q, s to denote the vertices of T, where qs is a longest side with respect to $\| \cdot \|_K$. According to Lemma 2.1.1, we may assume that the sides pq and ps are of length two. Let s' be the point lying on the same side of the line pq as s that satisfies $\|s' - p\|_K = \|s' - q\|_K = 2$; hence, applying Lemma 2.1.1 to $y + \lambda K$ and to $s' + 2K$ yields that $s \in s' + 2K$. In addition, let $s'' = s' + p - q$, which is an endpoint of the open arc of $p + 2\partial K$ given by Lemma 2.1.1 that lie in $\mathrm{int}(s' + 2K)$. We deduce by Lemma 2.1.2 that the segment ps intersects the segment $s's''$; therefore, $A(T)$ is at least the area of the triangle pqs'. $\qquad\square$

Proof of Theorem 2.2.1. We may assume that K is strictly convex, and hence K_0 is strictly convex as well. Let $\Gamma + K$ be a periodic packing by translates of K. We may assume the packing is *saturated*; namely, any translate of K overlaps a member of $\Gamma + K$. Therefore, $\Gamma + K_0$ is also a saturated periodic packing, and we consider a Delone triangulation with respect to Γ and K_0. Since $\lambda \leq 2$ holds for any supporting domain $y - \lambda K_0$, we conclude that any Delone triangle T satisfies $A(T) \geq 4A(T(K_0))$ by Lemma 2.2.4.

Let Γ be periodic with respect to the lattice Λ such that $2K_0$ embeds isometrically into $\mathbb{R}^2/\Lambda$, and let n denote the number of equivalence classes of Γ with respect to Λ. Then Γ induces a packing of n translates of K_0 on the torus $\mathbb{R}^2/\Lambda$. Since the induced Delone triangulation on $\mathbb{R}^2/\Lambda$ has $2n$ triangles according to (2.3), we deduce that the area of torus is at least $n \, 8A(T(K_0))$. In particular, $\Delta_T(K) \geq 8A(T(K_0))$. However, the side vectors of $2T(K_0)$ generate a packing lattice Λ for K according to Lemma 2.1.4; therefore, $\Delta_T(K) = 8A(T(K_0))$.

Now, K_0 is touched by some six translates $x_1 + K_0, \ldots, x_6 + K_0$ for $x_1, \ldots, x_6 \in \Lambda$, where $x_{i+3} = -x_i$ for $i = 1, 2, 3$ and $x_{i-1} + K$ and $x_i + K_0$ touch for $i = 1, \ldots, 6$ and $x_0 = x_6$. Let l_i be a common tangent line to K_0 and $x_i + K_0$ such that $l_{i+3} = -l_i$ for $i = 1, 2, 3$. Then $l_1, \ldots, l_6$ determine an o-symmetric circumscribed hexagon H, and $\Lambda + H$ is a packing; hence, $A(H) \leq \det \Lambda = \Delta_T(K)$. Since $A(H(K_0)) \leq A(H)$, and there exists a lattice tiling by $H(K_0)$ (see Lemma 2.1.3), we conclude $\Delta_T(K) = A(H(K_0))$ and, in turn, Theorem 2.2.1. $\qquad\square$

Turning to coverings, we verify a simple property of p-hexagons:

Lemma 2.2.5. *If the hexagon H has two parallel and equal opposite sides, and T_1 and T_2 are the convex hulls of every second vertex, then*

$$A(H) = A(T_1) + A(T_2).$$

Proof. Let d_1 and d_2 be the two parallel and equal diagonals of H connecting the parallel and equal opposite sides in a way that d_i is a side of T_i, $i = 1, 2$. Thus $A(T_i) = h_i d / 2$, where d is the common Euclidean length of d_1 and d_2, and h_i is the distance of d_i from the opposite vertex of T_i. Since $A(H) = (h_1 + h_2) d / 2$, we conclude Lemma 2.2.5. $\qquad\square$

Proof of Theorem 2.2.3. We may assume that K is strictly convex. Let $\Gamma + K$ be a covering periodic with respect to some lattice Λ such that K embeds isometrically into $\mathbb{R}^2 / \Lambda$. Writing n to denote the number of equivalence classes of Γ with respect to Λ, we obtain a covering of the torus $\mathbb{R}^2 / \Lambda$ by n translates of K. The induced Delone triangulation of the torus has $2n$ triangles according to (2.3), and $\lambda \leq 1$ holds for any supporting domain $y - \lambda K$. In particular, the area of the torus is at most $2n \widetilde{T}(K)$; hence, $\Theta_T(K) \geq 2\widetilde{T}(K)$.

If K is centrally symmetric then the side vectors of $\widetilde{T}(K)$ determine a covering lattice for K; therefore, $\Theta_T(K) = 2\widetilde{T}(K)$. In addition, $\widetilde{H}(K) = 2\widetilde{T}(K)$ follows by Lemma 2.2.5.

If K is not necessarily centrally symmetric then let us consider some two Delone triangles T_1 and T_2 on the torus that intersect in a common side s. Let $s - z_1$ and $s - z_2$ be the secants of $-K$ such that $T_i - z_i$ and $s - z_j$ lie on the same side of $s - z_i$ for $j \neq i$. Then the convex hull of $T_1 - z_1$ and $T_2 - z_2$ is a p-hexagon contained in $-K$. Hence Lemma 2.2.5 yields $A(T_1) + A(T_2) \leq A(\widetilde{H}_p(K))$. It follows that the average area of the $2n$ Delone triangles is at most $A(\widetilde{H}_p(K))/2$, completing the proof of Theorem 2.2.3. $\qquad\square$

Open Problems.

   (i) Given a convex domain K, is $\Theta_T(K)$ the maximal area of centrally symmetric hexagons contained in K?

   (ii) In particular, is $\vartheta_T(K) = 3/2$ if K is a triangle? The lattice covering density of a triangle is $3/2$ (see I. Fáry [Fár1950]).

Comments. Theorem 2.2.3 in case of centrally symmetric domains and Theorem 2.2.1 are due to L. Fejes Tóth [FTL1950]. Independently C. A. Rogers [Rog1951] verified Theorem 2.2.1. Given any convex domain K, the triangle bound of Theorem 2.2.3 for periodic coverings by translates of K is proved in R. P. Bambah, C. A. Rogers, and H. J. Zassenhaus [BRZ1964], and the "p-hexagon bound" of Theorem 2.2.3 is due to G. Fejes Tóth [FTG1988].

   If K is the Euclidean unit disc then it was B. N. Delone (see [Del1934]) who initiated the study of the now so-called Delone triangulation. The analogous triangulation when K is a strictly convex domain was constructed by R. P. Bambah, C. A. Rogers, and H. J. Zassenhaus [BRZ1964].

   Let K be convex domain. I. Fáry [Fár1950] verified that $\delta_T(K) \geq 2/3$, and equality holds if and only if K is a triangle. However, if K is centrally symmetric then the minimal value of $\delta_T(K) = A(K)/A(H(K))$ is not known. Circular discs do not provide the minimum, as shown by the example of the regular octagon. By rounding off the corners of the regular octagon, K. Reinhardt [Rei1934] and K. Mahler [Mah1947] proposed a possible minimal shape whose density is $0.9024\ldots$. P. Tammela [Tam1970] proved that $\delta_T(K) > 0.8926$ for any centrally symmetric convex domain K.

   I. Fáry [Fár1950] verified that $\vartheta_T(K) \leq 3/2$ holds for any convex domain K (see the open problems). If K is centrally symmetric then $\vartheta_T(K) \leq 2\pi/\sqrt{27} = 1.2091,\ldots$ and equality holds if and only if K is an ellipse (see L. Fejes Tóth [FTL1972]).

2.3. Finite Packings of Centrally Symmetric Domains

Given an o-symmetric convex domain K, we recall that $T(K)$ denotes a triangle of minimal area such that each side of $T(K)$ is of length one with respect to $\| \cdot \|_K$, and $P_K(\cdot)$ denotes the Busemann perimeter with respect to $\| \cdot \|_K$ (see Section 2.1). The main result of this section is

Theorem 2.3.1 (Oler inequality). *Let K be an o-symmetric convex domain. If the compact convex set C contains n points such that the distance between*

any two is at least 2 with respect to $\| \cdot \|_K$ *then*

$$\frac{A(C)}{8A(T(K))} + \frac{P_K(C)}{4} + 1 \geq n.$$

Given an o-symmetric strictly convex domain K and a line l, we define an analogue of orthogonal projection onto l. We say that x' is the *projection of* $x \in \mathbb{R}^2$ *into* l *parallel to the bisecting direction* if $x' = (p + q)/2$, where p and q denote the points of l such that the sides of the triangle xpq are of equal lengths.

Proposition 2.3.2. *Given an o-symmetric strictly convex domain K and a line l, let p and q lie on the same side of l such that $\|q - p\|_K \geq 2$ and $p, q \in y + \lambda \partial K$ for some $y \in l$ and $0 < \lambda < 2$. Writing $\tilde{x}$ to denote the projection of $x \in \mathbb{R}^2$ into l parallel to the bisecting direction, $Q = \mathrm{conv}\{p, q, \tilde{p}, \tilde{q}\}$ satisfies the following:*

(i) Q does not intersect the shorter open arc pq of $y + \lambda \partial K$;

(ii) $\frac{1}{8A(T(K))} A(Q) + \frac{1}{4} \|\tilde{p} - \tilde{q}\|_K \geq \frac{1}{2}$.

Proof. We may assume that the line passing through p and parallel to l separates q and l.

To verify (ii), first we fix y and l; hence, possibly decreasing λ we may assume that $\|q - p\|_K = 2$. Now fixing p, q and the direction of l, and possibly decreasing λ, we may also assume according to Lemma 2.1.1 that l contains p. Let r be the point of the half line py such that $\|r - p\|_K = 2$, and let q' be the point on the same side of l where q lies that satisfies $\|q' - p\|_K = \|q' - r\|_K = 2$. Since $\|r - y\|_K = 2 - \lambda$, we deduce by the triangle inequality that $\|r - q\|_K \leq 2$; hence, Lemma 2.1.2 yields that the segment $q'r$ intersects the segment pq in some $s = \lambda q' + (1 - \lambda)r, 0 \leq \lambda \leq 1$. Readily,

$$\|\tilde{q} - \tilde{p}\|_K \geq \|\tilde{s} - p\|_K = 2 - \lambda.$$

However, the area of the triangle $pq'\tilde{q}'$ is at least $2A(T(K))$, and $A(Q)$ is at least the area of the triangle $ps\tilde{s}$; therefore, $A(Q)$ is at least $\lambda(2 - \lambda) 2A(T(K))$. In particular, we conclude

$$\frac{A(Q)}{8A(T(K))} + \frac{\|\tilde{p} - \tilde{q}\|_K}{4} \geq \frac{\lambda(2 - \lambda)}{4} + \frac{2 - \lambda}{4} = \frac{\frac{9}{4} - (\lambda - \frac{1}{2})^2}{4} \geq \frac{1}{2}.$$

Turning to (i), we may assume that $y = o$. Let $x_1, \ldots, x_6$ be the vertices of the affine regular hexagon inscribed into λK (see Lemma 2.1.4) such that x_1 is an endpoint of $l \cap \lambda K$ and x_2, p, q lie on the same side of l. None of the shorter arcs $x_i x_{i+1}$ of $\lambda \partial K$ contains both p and q according to Lemma 2.1.2; hence, $\tilde{q}$ lies in the segment $\tilde{x}_2 \tilde{x}_3 \subset l \cap K$. In turn, we conclude Proposition 2.3.2. $\square$

Proof of Theorem 2.3.1. We may assume that K is strictly convex. During the argument we use Lemma 2.1.1 and the corresponding remark several times without explicitly stating them. Distance and length will always be measured in terms of $\| \cdot \|_K$.

Let $x_1, \ldots, x_n \in C$ such that $\|x_i - x_j\|_K \geq 2$ for $i \neq j$, and we may also assume that there exists no $x \in C$ whose distance from each x_i is at least two. If C is a segment then the Oler inequality readily holds; therefore, let C be a convex domain that is the convex hull of $x_1, \ldots, x_n$. We will construct a *Delone type simplicial complex* Σ_{Del} and extend it to a cell decomposition Σ of C. Combining the properties of Σ_{Del} and Σ will lead to Theorem 2.3.1.

We say that $y + \lambda K$, $\lambda > 0$, is a *supporting domain* if $y \in C$ and no x_i lies in $\text{int}(y + \lambda K)$; moreover, either at least three x_is lie on $\partial(y + \lambda K)$ or $y \in \partial C$ and at least two x_is lie on $\partial(y + \lambda K)$. In particular, $\lambda < 2$. To any supporting domain, we assign the convex hull of the x_is contained in the boundary. If two supporting domains overlap then the associated convex hulls are separated by the common secant. Therefore, the family of associated convex hulls determines a cell complex $\tilde{\Sigma}_{\text{Del}}$ whose vertices are $x_1, \ldots, x_n$. Using the fact that the supporting domains cover D, we deduce that $\tilde{\Sigma}_{\text{Del}}$ is connected. Let Σ_{Del} be the simplicial complex resulting by triangulating the cells of $\tilde{\Sigma}_{\text{Del}}$ using their vertices.

Next, we construct a cell decomposition Σ of C. If $C = \text{supp } \Sigma_{\text{Del}}$ then we define $\Sigma = \Sigma_{\text{Del}}$ and $k = 0$. Otherwise, there exist some supporting domains whose centre lies in ∂C, but not in any edge of Σ_{Del}. Let $s_1, \ldots, s_k$ be the segments contained in ∂C such that, for each s_j, the endpoints of s_j are some x_is, the relative interior of s_j contains no x_i, and s_j is not an edge of Σ_{Del}. Given an s_j, let $D_{j,1}, \ldots, D_{j,m_j}$ be the supporting domains whose centres $y_{j,1}, \ldots, y_{j,m_j}$, respectively, lie in s_j in this order. The convex hull associated to any $D_{j,l}$ is either an edge $e_{j,l}$ of Σ_{Del} or a cell, and in the latter case we write $e_{j,l}$ to denote the side of the cell that cuts off $y_{j,l}$. It follows that the edges $e_{j,1}, \ldots, e_{j,m_j}$ form a broken line that connects the two endpoints of s_j, and together with s_j they bound a domain Π_j. Because Σ_{Del} is connected, we deduce that no Π_j overlaps $\text{supp} \Sigma_{\text{Del}}$; therefore, adding Π_j and s_j, $j = 1, \ldots, k$; to Σ_{Del}, we obtain a cell complex Σ. Since any edge of Σ intersecting $\text{int } C$ is contained in exactly two cells, Σ is a cell decomposition of C.

It follows by the Euler formula (A.11) that $\chi(\Sigma) = 1$. Now Σ_{Del} has k edges and k cells less than Σ; therefore, $\chi(\Sigma_{\mathrm{Del}}) = 1$ as well. We set $f_2 = f_2(\Sigma_{\mathrm{Del}})$ and write $f_{1,i}$ to denote the number of edges of Σ_{Del}, which are contained in exactly i triangles. Therefore, $3 f_2 = 2 f_{1,2} + f_{1,1}$ and $\chi(\Sigma_{\mathrm{Del}}) = 1$ yield that $(1/2) f_2 + (1/2)(2 f_{1,0} + f_{1,1}) + 1 = n$. If T is a triangle of Σ_{Del} then its vertices lie on the boundary of some supporting domain; hence, $A(T) \geq 4A(T(K))$ follows by Lemma 2.2.4. In particular,

$$\frac{A(\operatorname{supp}\Sigma_{\mathrm{Del}})}{8A(T(K))} + \frac{2f_{1,0} + f_{1,1}}{2} + 1 \geq n. \tag{2.4}$$

Next we write $e_{0,1}, \ldots, e_{0,m_0}$ to denote the edges of Σ_{Del} that are contained in ∂C, where we set $m_0 = 0$ if there exists no such edge. Writing $\mu(s)$ to denote the length of a segment s, we claim that

$$\frac{1}{4} \sum_{1 \leq l \leq m_0} \mu(e_{0,l}) \geq \frac{m_0}{2};$$
$$\frac{A(\Pi_j)}{8A(T(K))} + \frac{\mu(s_j)}{4} \geq \frac{m_j}{2} \text{ if } 1 \leq j \leq k. \tag{2.5}$$

The first inequality readily holds; hence, let $1 \leq j \leq k$. We write $Q_{j,l}$ to denote the convex hull of $e_{j,l}$ and its projection into $\operatorname{aff} s_j$ parallel to the bisecting direction. Then $Q_{j,l}, l = 1, \ldots, m_j$, tile Π_j according to Proposition 2.3.2 (i). Thus we conclude (2.5) by Proposition 2.3.2 (ii).

Since $\sum_{j=0}^{k} m_j = 2 f_{1,0} + f_{1,1}$, and ∂C is divided into the $e_{0,l}$s and the s_js, combining (2.4) and (2.5) yields Theorem 2.3.1. $\qquad\square$

The coefficients in the Oler inequality (Theorem 2.3.1) are optimal for any K independently of each other. A single point shows that the constant term is optimal, and segments show that the coefficient of the perimeter cannot be lowered. Turning to the coefficient of the area, we let the parallelogram Q be generated by two sides of $2T(K)$. Since these two sides generate a packing lattice for K, $\sqrt{n} Q$ contains the centres of n nonoverlapping translates of K for any $n \geq 1$. However, given any $\varepsilon > 0$, c_1, and c_2, if n is large then

$$\frac{1 - \varepsilon}{8A(T(K))} A\left(\sqrt{n} Q\right) + c_1 P_K\left(\sqrt{n} Q\right) + c_2 < n.$$

In the rest of the section we provide some useful consequences of the Oler inequality (Theorem 2.3.1). Given the o-symmetric convex domain K, let $\Pi(K)$ and $H(K)$ be a circumscribed o-symmetric parallelogram and a circumscribed o-symmetric polygon of at most six sides, respectively, of minimal area (compare Lemma 1.3.2). We deduce by the definition of the mixed area (see (A.15)) that if C is a polygon then

$$A(C, K) \geq \frac{A(\Pi(K))}{8} \cdot P_K(C).$$

Hence, the same inequality follows for any convex C by approximation. Therefore, the Oler inequality and $A(H(K)) = 8A(T(K))$ (see Theorem 2.2.1) yield

Corollary 2.3.3. *Let K be an o-symmetric convex domain. If the compact convex set C contains n points such that the distance between any two is at least two with respect to $\| \cdot \|_K$ then*

$$A(C) + \frac{A(H(K))}{A(\Pi(K))} \cdot 2A(C, K) + A(H(K)) \geq n \cdot A(H(K)).$$

In turn, we verify the hexagon bound for a large number of translates:

Corollary 2.3.4. *If a convex domain D contains n nonoverlapping translates of a given centrally symmetric convex domain K then*

$$A(D) \geq n \, A(H(K)),$$

provided that $n \geq n(K)$, where $n(K)$ depends on K.

Remarks. No absolute constant works as $n(K)$ (see Example 1.4.2). However, if D is a hexagon then even $n(K) = 1$ works (see Theorem 1.2.1).

Since translates of $H(K)$ tile the plane (see Lemma 2.1.3), the coefficient $A(H(K))$ is optimal.

Proof. If K is a parallelogram then $A(H(K)) = A(K)$. Thus we may assume that K is o-symmetric and is not a parallelogram. Let C_n be the convex hull of the centres of n translates. Then $C_n + K \subset D$, and the distance between any two centres is at least two with respect $\| \cdot \|_K$. We deduce by the Corollary 2.3.3 form of the Oler inequality that

$$A(C_n + K) \geq nA(H(K)) + \left(1 - \frac{A(H(K))}{A(\Pi(K))}\right) \times 2A(C_n, K)$$
$$+ A(K) - A(H(K)).$$

Since $A(H(K)) < A(\Pi(K))$, and $A(C_n, K)$ tends to infinity with n, we conclude Corollary 2.3.4. $\qquad\square$

Open Problem.

(i) Given a convex domain K, does there exist an $n(K)$ with the following property: If the convex domain D contains $n \geq n(K)$ nonoverlapping translates of K then $A(D) \geq nA(H(K_0))$ for $K_0 = (K - K)/2$.

Comments. The Oler inequality (Theorem 2.3.1) was first proved in the lengthy paper by N. Oler [Ole1961a]. A simpler proof was given by R. L. Graham, H. S. Witsenhausen, and H. J. Zassenhaus [GWZ1972], who actually verified the following statement: If Σ is a finite simplicial complex in $\mathbb{R}^2$ such that the distance between any two vertices is at least two with respect to $\| \cdot \|_K$ then

$$\frac{A(\operatorname{supp}\Sigma)}{8A(T(K))} + \frac{P_K(\operatorname{supp}\Sigma)}{4} + \chi(\Sigma) \geq f_0(\Sigma).$$

Our proof of Theorem 2.3.1 is new.

Corollary 2.3.4, and the fact that no absolute constant works as $n(K)$, are due to G. Fejes Tóth.

2.4. Asymptotic Structure of Optimal Translative Packings

Given a convex domain K, we consider the asymptotic structure of the packing minimizing the area of the convex hull of n nonoverlapping translates of K.

Theorem 2.4.1. *If K is a convex domain that is not a parallelogram, and D_n is the convex hull of n nonoverlapping translates of K such that the area of D_n is minimal, then (possibly after translation)*

$$\alpha\sqrt{n} \cdot K \subset D_n \subset \beta\sqrt{n} \cdot K,$$

where α and β are positive constants that depend only on K.

To handle the case when K is not centrally symmetric, we estimate the change of the mixed area when writing $-K$ instead of K:

Lemma 2.4.2. *If K and C are convex domains then*

$$|A(C, -K) - A(C, K)| \leq 8r_K(C) \cdot A(K).$$

Proof. We may assume that $r_K(C) = 1$. Lemma 2.1.5 yields the existence of two parallel lines l_1 and l_2 with the following properties: C lies between l_1

and l_2, and both lines touches $x + \lambda K$ for some x and $\lambda \leq 2$. We write x_i to denote a point where l_i touches, $i = 1, 2$, and we assume that $(x_1 + x_2)/2 = o$. Let s be a segment in C such that there exist two supporting lines to C at the endpoints of s, which are parallel to $x_1 x_2$; hence, $s + 4K$ contains a translate of C. Since $A(s, -K) = A(s, K)$, and, moreover, (A.19) yields $A(K, -K) \leq A(K, 2K) = 2A(K)$, we conclude

$$A(C, -K) - A(C, K) \leq A(s + 4K, -K) - A(s, K) \leq 8\,A(K).$$

Therefore, the analogous inequality for $A(C, K) - A(C, -K)$ completes the proof of Lemma 2.4.2 $\qquad \square$

Proof of Theorem 2.4.1. In the following, $c_1, c_2, \ldots$ always denote some positive constants that depend on K. Let $x_1 + K, \ldots, x_n + K$ be the nonoverlapping translates that are contained in D_n, and let C_n be the convex hull of $x_1, \ldots, x_n$. We write Λ to denote the lattice such that $\Lambda + H(K_0)$ is a tiling (see Lemma 2.1.3); hence, $\Lambda + K$ is a packing. Thus,

$$\left(\sqrt{\frac{nA(H(K_0))}{A(K)}} + c_1 \right) \cdot K$$

contains some n translates out of $\Lambda + K$, and the optimality of D_n yields

$$A(C_n) + 2A(C_n, K) \leq nA(H(K_0)) + c_2\sqrt{n}. \tag{2.6}$$

It also follows that $r_K(C_n) \leq c_3\sqrt{n}$. However, $\|x_i - x_j\|_{K_0} \geq 2$ holds for $i \neq j$; hence, applying the Corollary 2.3.3 form of the Oler inequality to C_n and K_0 leads to

$$A(C_n) + \frac{A(H(K_0))}{A(\Pi(K_0))}(A(C_n, K) + A(C_n, -K)) + A(H(K_0)) \geq nA(H(K_0)).$$

Since $|A(C_n, -K) - A(C_n, K)| \leq c_4\sqrt{n}$ by Lemma 2.4.2, we deduce

$$A(C_n) + \frac{A(H(K_0))}{A(\Pi(K_0))} 2A(C_n, K) + c_5\sqrt{n} \geq nA(H(K_0)). \tag{2.7}$$

The condition that K is not a parallelogram yields that K_0 is not a parallelogram as well; hence, $A(H(K_0))/A(\Pi(K_0)) < 1$. We conclude by (2.6) and (2.7) that

$$A(D_n, K) = A(C_n, K) + A(K, K) \leq c_6\sqrt{n}.$$

Since D_n contains a segment of length $R_{K_0}(D_n)$ with respect to $\|\cdot\|_{K_0}$, we have $R_{K_0}(D_n) \leq c_7\sqrt{n}$; hence, $r_{K_0}(D_n) \geq c_8\sqrt{n}$ follows by $A(D_n) \geq n\,A(K)$ and Lemma 2.1.6. In turn, we conclude Theorem 2.4.1. $\qquad \square$

Open Problem.
(i) Let K be a convex domain, and let D_n be a convex domain of minimal area that contains n nonoverlapping translates of K. Does any convergent subsequence of $\{D_n/\sqrt{n}\}$ tend to some hexagon?
If K is a circular disc then strong evidence is provided by the Wegner theorem (Theorem 4.3.1). If K is centrally symmetric then the statement holds for finite lattice packings (see Theorem 3.4.2 and the preceding remarks).

Comments. Theorem 2.4.1 is due to K. Böröczky Jr. and U. Schnell [BöS1998a].

2.5. Finite and Periodic Coverings

We show that the minimal density of finite coverings by n translates of a given convex domain K is essentially $\vartheta_T(K)$ if n is large.

Theorem 2.5.1. *Let K be a convex domain. If D_n is a convex domain of maximal area that can be covered by n translates of K then*

$$\lim_{n\to\infty} \frac{A(D_n)}{n} = \Theta_T(K).$$

Proof. In the following, the implied constant in $O(\cdot)$ depends only on K. For any $\varepsilon > 0$, there exists a covering $\Lambda + \{x_1 + K, \ldots, x_m + K\}$ of $\mathbb{R}^2$ of density at most $\vartheta_T(K) + \varepsilon$, where Λ is a lattice. We may assume that $\lambda_0 K$ contains $x_1 + K, \ldots, x_m + K$ for some $\lambda_0 > 0$. If λ is large then the translates in the periodic covering intersecting λK determine a finite covering with density at most $(1 + \lambda_0/\lambda)(\vartheta_T(K) + \varepsilon)$; therefore,

$$\liminf_{n\to\infty} \frac{A(D_n)}{n} \geq \Theta_T(K).$$

We define $\Theta' = \limsup_{n\to\infty} A(D_n)/n$; hence, it is sufficient to verify that $\Theta' \leq \Theta_T(K)$. Given any $\varepsilon > 0$, we claim that there exists a parallelogram R and a covering of R by some N translates of K satisfying

$$\frac{A(R)}{N} > (1 - O(\varepsilon))\,\Theta'. \tag{2.8}$$

Let $\{D_{n_k}\}$ be a subsequence satisfying $A(D_{n_k})/n_k \geq (1 - \varepsilon)\,\Theta'$ for any n_k. We may also assume that either $r_K(D_{n_k})$ tends to infinity or $r_K(D_{n_k})$ is at most some r_0 for a constant r_0. If $r_K(D_{n_k})$ tends to infinity then let R be any parallelogram satisfying $A(R)/A(R - K) > 1 - \varepsilon$. Assuming that $r_K(D_{n_k})$ is sufficiently large, we have $A(D_{n_k} \ominus (R - K)) > (1 - \varepsilon)A(D_{n_k})$, and the number of translates of K in the covering intersecting $D_{n_k} \ominus (R - K)$ is at least $(1 - \varepsilon)n_k$. Given such an n_k, let $x_1 + K, \ldots, x_{n_k} + K$ be the covering of D_{n_k}. If $y + (R - K) \subset D_{n_k}$ then we write $m(y)$ to denote the number of points out of $x_1, \ldots, x_{n_k}$ that lie in $y + (R - K)$; hence, the corresponding translates of K cover $y + R$. Since there exists a y such that $m(y)$ satisfies $A(R - K)/m(y) \geq (1 - \varepsilon)^2 A(D_{n_k})/n_k$, we conclude the claim (2.8) in this case.

Therefore, let $r_K(D_{n_k})$ be at most r_0 for a constant r_0. Then D_{n_k} is contained in some parallel strip Σ_{n_k} such that $r_K(\Sigma_{n_k}) = 2r_0$ (see Lemma 2.1.5). For a large n_k (which we specify later), we consider a covering $X_{n_k} + K$ of D_{n_k}, where X_{n_k} is of cardinality n_k. Let s_0 be a longest segment parallel to Σ_{n_k} that is contained in K, and let s_1 be a segment such that the parallelogram $P = s_0 + s_1$ is circumscribed around K; hence, two opposite sides of P touch K at the endpoints of s_0. We decompose Σ_{n_k} into the parallelograms R_i, $i \in \mathbb{Z}$, such that each is a translate of $\lceil 1/\varepsilon \rceil s_0 + 2r_0 s_1$, and R_i and R_{i+1} are consecutive tiles for any i. We fix some m such that $2r_0/m < \varepsilon$, and we may assume that the parallelograms R_i that intersect D_{n_k} are the ones with $-m \leq i \leq l$. We write $\mathcal{R}$ to denote the union of R_i such that $-m \leq i \leq 0$ or $l - m \leq i \leq l$; hence, the area of $\mathcal{R}$ is $O(m/\varepsilon)$. In particular, $A(D_{n_k}) \sim n_k \Theta'$ yields that $A(\mathcal{R}) < \varepsilon A(D_{n_k})$ for large n_k, and we fix some n_k satisfying this condition. Thus there exists some R_j, $0 \leq j \leq N$, such that

$$\frac{A(D_{n_k} \cap R_j)}{\#(X_{n_k} \cap R_j)} \geq (1 - O(\varepsilon)) \cdot \frac{A(D_{n_k})}{n_k}.$$

Let γ be maximal with the property that $D_{n_k} \cap R_j$ contains a translate of γs_1. Then $\gamma > \Theta'/(2A(P)) \geq \Theta'/(4A(K))$; hence, the choice of m yields the existence of some $\gamma_0 = (1 - O(\varepsilon))\gamma$ such that a translate of $\lceil 1/\varepsilon \rceil s_0 + \gamma_0 s_1$ is contained in $A(D_{n_k} \cap R_j)$. Therefore, we define $N = \#(X_{n_k} \cap R_j)$ and

$$R = \left(\left\lceil \frac{1}{\varepsilon} \right\rceil - 1 \right) \cdot s_0 + \gamma_0 s_1.$$

It follows that R is covered by some N translates of K, and

$$\frac{A(R)}{N} \geq (1 - O(\varepsilon)) \cdot \frac{A(D_{n_k} \cap R_j)}{N}$$

$$\geq (1 - O(\varepsilon)) \cdot \frac{A(D_{n_k})}{n_k} \geq (1 - O(\varepsilon)) \cdot \Theta'.$$

In particular, we conclude the claim (2.8).

Inequality (2.8) yields $\limsup_{n\to\infty} A(D_n)/n \le \Theta_T(K)$, completing the proof of Theorem 2.5.1. $\qquad\qquad\qquad\qquad\qquad\qquad\qquad\qquad\qquad\quad$ $\square$

2.6. The Bambah Inequality for Coverings

Given an o-symmetric convex domain K, we write $\widetilde{T}(K)$ and $\widetilde{H}(K)$ to denote an inscribed triangle and an inscribed o-symmetric polygon of at most six sides, respectively, of maximal area [compare with the Dowker theorem (Corollary 1.5.5)]. Before proving the rather intricate Fejes Tóth inequality (Theorem 2.7.1) for coverings of any convex domain by translates of K, we consider the special case when the convex hull of the centres of the translates is covered.

Theorem 2.6.1 (Bambah inequality). *Let K be an o-symmetric convex domain, and let n translates of K cover the convex hull D of the n centres. If ∂D contains m of the centres then*

$$A(D) \le \left(n - \frac{m}{2} - 1\right) \cdot A(\widetilde{H}(K)).$$

Remark. $A(\widetilde{H}(K)) = 2A(\widetilde{T}(K))$ according to Lemma 2.2.5.

Equality holds in the Bambah inequality for any $n \ge 3$ when D can be tiled by translates of $\widetilde{T}(K)$ and $-\widetilde{T}(K)$ using the n centres as vertices. Now the proof of Theorem 2.6.1 is based on the following lemma.

Lemma 2.6.2. *Let K be an o-symmetric convex domain, and let the broken line $x_1, \ldots, x_m, x_1$ bound a domain D. If $\|x_i - x_{i-1}\|_K \le 2$ for $i = 1, \ldots, m$ (with $x_0 = x_m$), and for any $x \in D$ there exists an x_i such that the segment xx_i lies in $x_i + K$, then*

$$A(D) \le (m - 2) \cdot A(\widetilde{T}(K)).$$

To prove Theorem 2.6.1 and Lemma 2.6.2, we construct a *Delone type simplicial complex* Σ_{Del}, and we extend it to a cell decomposition Σ of D. We assume that K is strictly convex; hence, we may apply Lemma 2.1.1 to homothetic copies of K. The setup that we now discuss is that a broken line $x_1, \ldots, x_m, x_1$ bound a domain D, and $x_i \in \text{int } D$ if $m < i \le n$. The segments $x_{i-1}x_i$, $i = 1, \ldots, m$ (with $x_0 = x_m$), are called the *true sides* of D. We say that a point $z \in D$ is visible from w if the segment wz is contained in D. Now

$y + \lambda K, \lambda > 0$, is called a *supporting domain* if $y \in D$ and no x_i visible from y lies in int $(y + \lambda K)$. Moreover, either at least three x_is visible from y lie on $\partial(y + \lambda K)$ or $y \in \partial D$ and at least two x_is visible from y lie on $\partial(y + \lambda K)$. A crucial observation is that if w, z_1, and z_2 are contained in a supporting domain, and z_1 and z_2 are visible from w, then

$$\mathrm{conv}\{w, z_1, z_2\} \subset D. \tag{2.9}$$

This follows from the fact that if a true side of D intersected the interior of $T = \mathrm{conv}\{w, z_1, z_2\}$ then it would have a vertex in int T.

To any supporting domain $y + \lambda K$, we assign the convex hull Π of the x_is contained in the boundary that are visible from y; hence, $\Pi \subset D$ and all points of Π are visible from y according to (2.9). We claim that the family of these convex hulls determines a cell complex $\tilde{\Sigma}_{\mathrm{Del}}$. If the claim does not hold then there exist two supporting domains $y_1 + \lambda_1 K$ and $y_2 + \lambda_2 K$ such that the associated convex hulls Π_1 and Π_2, respectively, do not intersect "properly." In particular, the common secant of $y_1 + \lambda_1 K$ and $y_2 + \lambda_2 K$ does not separate Π_1 and Π_2; hence, say Π_1 has a vertex v that lies in the interior of $y_2 + \lambda_2 K$ (see Lemma 2.1.1). In addition, there exists a $p \in \Pi_1 \cap \Pi_2 \cap (y_2 + \lambda_2 K)$. Since both v and y_2 are visible from p, we deduce that v is visible from y_2 by (2.9). This is absurd, which in turn yields the claim. The supporting domains cover D; thus, $\tilde{\Sigma}_{\mathrm{Del}}$ is connected, and its vertices are $x_1, \ldots, x_n$. Let Σ_{Del} be the simplicial complex resulting by triangulating the cells of $\tilde{\Sigma}_{\mathrm{Del}}$ using their vertices.

Next, we construct a cell decomposition Σ of D. If $D = \mathrm{supp}\, \Sigma_{\mathrm{Del}}$ then we define $\Sigma = \Sigma_{\mathrm{Del}}$ and $k = 0$. Otherwise, let $s_1, \ldots, s_k$ be the true sides of D that are not edges of Σ_{Del}. Given an s_j, let $D_{j,1}, \ldots, D_{j,m_j}$ be the supporting domains whose centres $y_{j,1}, \ldots, y_{j,m_j}$, respectively, lie in s_j in this order. The convex hull associated to any $D_{j,l}$ is either an edge $e_{j,l}$ of Σ_{Del} or a cell, and in the latter case we write $e_{j,l}$ to denote the side of the cell that cuts off $y_{j,l}$. For any $D_{j,l}$, the segment s_j intersects $\partial D_{j,l}$ at two points, and (2.9) yields

$$\mathrm{conv}\{e_{j,l}, s_j \cap D_{j,l}\} \subset D. \tag{2.10}$$

It follows that the edges $e_{j,1}, \ldots, e_{j,m_j}$ form a broken line that connects the two endpoints of s_j, and together with s_j they bound a domain Ψ_j. Since Σ_{Del} is connected, we deduce that no Ψ_j overlaps $\mathrm{supp}\, \Sigma_{\mathrm{Del}}$; therefore, adding Ψ_j and s_j, $j = 1, \ldots, k$, to Σ_{Del}, we obtain a cell complex Σ. Since any edge of Σ intersecting int D is contained in exactly two cells, Σ forms a cell decomposition of D. It follows by the Euler formula (A.11) that $\chi(\Sigma) = 1$. Now Σ_{Del} has k edges and k cells less than Σ; therefore, $\chi(\Sigma_{\mathrm{Del}}) = 1$ as well.

We conclude

$$f_2(\Sigma_{\text{Del}}) = 2n - 2 - (m - k) - \sum_{1 \leq j \leq k} m_j. \tag{2.11}$$

Proof of Lemma 2.6.2. We assume that K is strictly convex and use the presentation leading to (2.11) with $n = m$. If $y + \Lambda K$ is a supporting domain then there exists an x_i such that the segment $x_i y$ is contained in $x_i + K$; hence, $\lambda \leq 1$. We deduce

$$A(\text{supp } \Sigma_{\text{Del}}) \leq \left(m - 2 + k - \sum_{1 \leq j \leq k} m_j \right) \cdot A(\widetilde{T}(K)).$$

If $k = 0$ then we are done; therefore, let $k \geq 1$. It follows by Lemma 2.1.1 and by (2.10) that any Ψ_j is a subset of $z_j + K$, where z_j is the midpoint of s_j. Since conv Ψ_j is a polygon of at most $m_j + 1$ sides, we have $A(\Psi_j) \leq (m_j - 1)A(\widetilde{T}(K))$, which, in turn, yields Lemma 2.6.2. $\qquad\square$

Proof of Theorem 2.6.1. We assume that K is strictly convex. Writing $x_1 + K, \ldots, x_n + K$ to denote the n translates, we apply the presentation leading to (2.11) for $D = \text{conv}\{x_1, \ldots, x_n\}$. Since any supporting domain $y + \lambda K$ readily satisfies $\lambda \leq 1$, we have

$$A(\text{supp } \Sigma_{\text{Del}}) \leq \left(2n - 2 - m + k - \sum_{1 \leq j \leq k} m_j \right) \cdot A(\widetilde{T}(K)).$$

If $k = 0$ then we are done; therefore, let $k \geq 1$. We write Ψ_j' to denote the reflected image of Ψ_j through the midpoint of s_j. Now any point of Ψ_j is visible from some vertex of Ψ_j according to (2.10); thus, we may apply Lemma 2.6.2 to the union of Ψ_j and Ψ_j' (with $m = 2m_j$). We conclude $A(\Psi_j) \leq (m_j - 1)A(\widetilde{T}(K))$ and, in turn, Theorem 2.6.1. $\qquad\square$

Comments. Lemma 2.6.2 (in a more general form) was announced in R. P. Bambah and C. A. Rogers [BaR1952] and proved by R. P. Bambah, C. A. Rogers, and H. Zassenhaus [BRZ1964]. Theorem 2.6.1 is due to R. P. Bambah and A. C. Woods [BaW1971] (see also U. Betke, M. Henk, and J. M. Wills [BHW?]). Our proof of Theorem 2.6.1 is new.

2.7. The Fejes Tóth Inequality for Coverings

The Fejes Tóth inequality (Theorem 2.7.1) and its more precise and technical form (Lemma 2.7.2) are the fundamental tools for handling finite coverings by translates of a given o-symmetric convex domain K. We write $\widetilde{T}(K)$ to denote an inscribed triangle of maximal area. In addition, $\widetilde{\Pi}(K)$ and $\widetilde{H}(K)$ denote some inscribed o-symmetric polygons of at most four and six sides, respectively, of maximal area [compare with the Dowker theorem (Corollary 1.5.5)].

Theorem 2.7.1 (Fejes Tóth inequality). *If $n \geq 2$ translates of a o-symmetric convex domain K cover the convex domain D then*

$$A(D) \leq nA(\widetilde{H}(K)) + A(K) - A(\widetilde{H}(K)) - (A(\widetilde{H}(K)) - A(\widetilde{\Pi}(K))) \cdot \frac{b}{2},$$

where some b translates of K cover the boundary of D.

The proof of Theorem 2.7.1 rests on Lemma 2.7.2, which is a more precise statement about sufficiently general coverings. We say that the points $w_1, \ldots, w_n$ are in *general position* with respect to $\| \cdot \|_K$ and the convex domain D if $w_1, \ldots, w_n$ lie in int D, and there exists no y and λ such that either $\partial(y + \lambda K)$ contains four of the points or $y \in \partial D$ and $\partial(y + \lambda K)$ contains three of the points.

Next, we introduce the notion of a *free side* of a simplicial complex Σ. If e is an edge of Σ then a free side of e is a half plane bounded by the line of e such that no triangle of Σ containing e lies in that half plane. We write $b(\Sigma)$ to denote the total number of free sides of the edges; namely, $b(\Sigma)$ is the number of edges of Σ on the boundary of supp Σ, where edges that are not contained in any of the triangles are counted twice.

Lemma 2.7.2. *Let K be an o-symmetric strictly convex domain, and let $n \geq 2$ translates of K cover D such that no $n - 1$ of them form a covering and the centres are in general position with respect to $\| \cdot \|_K$ and D. Then there exists a simplicial complex Σ whose vertices are the centres of the n translates and satisfies the following:*

(i) *any triangle or edge of Σ is covered by some translate of K;*
(ii) *some $b(\Sigma)$ translates of K cover ∂D;*
(iii) $A(D) \leq A(\text{supp } \Sigma) + A(K) + A(\widetilde{\Pi}(K)) \cdot \frac{b(\Sigma)}{2}$; *and*
(iv) $A(D) \leq (n - 1) A(\widetilde{H}(K)) + A(K) - (A(\widetilde{H}(K)) - A(\widetilde{\Pi}(K)))\frac{b(\Sigma)}{2}$.

The following observation is used in the argument for Lemma 2.7.2.

Proposition 2.7.3. *Given an o-symmetric strictly convex domain K, let K_1 and K_2 be translates of K whose boundaries intersect at two points, and let*

l be a line intersecting $K_1 \cap K_2$ in a way that the centres of K_1 and K_2 lie on the same side of l. If Ξ_1 and Ξ_2 denote the part of $K_1 \backslash K_2$ and $K_2 \backslash K_1$, respectively, cut off by the other side of l then

$$A(\Xi_1) + A(\Xi_2) + A(K_1 \cap K_2) \geq A(K) - A(\widetilde{\Pi}(K)).$$

Proof. We may assume that l passes through a common point p of ∂K_1 and ∂K_2, and we write q to denote the other intersection point of ∂K_1 and ∂K_2. Now the sum $A(\Xi_1) + A(\Xi_2)$ is minimal if the secants $l \cap K_1$ and $l \cap K_2$ are of the same length, and hence the second point of intersection of l with ∂K_i is the point opposite to q, $i = 1, 2$. Therefore, we assume that this is the case, and we write P to denote the parallelogram inscribed in K_1 such that $l \cap K_1$ is a side. Since

$$A(\Xi_1) + A(\Xi_2) + A(K_1 \cap K_2) = A(K) - A(P),$$

we conclude Proposition 2.7.3. $\qquad\qquad\square$

Proof of Lemma 2.7.2. Let us outline the main idea. We will construct a *Delone type simplicial complex* Σ that satisfies (i) and (ii) by definition, and (iv) will follow from (iii) and from some combinatorial properties of Σ. Therefore, the main part of the proof is concerned with (iii). We orient the boundary of supp Σ clockwise in a way that if an edge is not contained in any triangle we count it twice, and let $z_1, \ldots, z_b$, $b = b(\Sigma)$, be the vertices of Σ on the boundary according to this orientation. In particular, $z_{i-1}z_i$ correspond to the free sides of Σ for $i = 1, \ldots, b$ and $z_b = z_0$, where z_i and z_j may denote the same vertex of Σ for $1 \leq i < j \leq b$. We will associate a region Φ_i to z_i such that Φ_i, $i = 1, \ldots, b$, and Σ together form a cell decomposition of D. The core of the argument is to verify that the average area of the Φ_is is essentially at most $A(\widetilde{\Pi}(K))/2$.

Now we start the actual proof of Lemma 2.7.2. Let $w_1 + K, \ldots, w_n + K$ be the translates that cover D. We write $x_1, \ldots, x_b$ to denote the points of ∂D in clockwise order with the following property: For any x_i, there exists some $\lambda_i > 0$ such that the interior of $x_i + \lambda_i K$ contains no point out of $w_1, \ldots, w_n$, and the boundary of $x_i + \lambda_i K$ contains two points out of $w_1, \ldots, w_n$. We call $x_i + \lambda_i K$ a *supporting domain*. In addition, $y + \lambda K$ is also called a supporting domain if $y \in \text{int } D$, three different w_i, w_j, w_k lie on the boundary of $y + \lambda K$, and no other point out of $w_1, \ldots, w_n$ lies in $y + \lambda K$. For any supporting domain, we consider the convex hull of the w_is in the supporting domain, and we define Σ to be the simplicial complex determined by these convex hulls. It is actually a simplicial complex because the common secant of intersecting supporting domains separates the corresponding w_is according to Lemma 2.1.1. Any triangle or edge of Σ is covered by a supporting domain by definition; hence, (i) readily follows. Now the free sides of Σ are in bijective correspondence with $x_1, \ldots, x_b$; therefore, $b = b(\Sigma)$ and Σ is connected.

We set $x_0 = x_b$ and $x_{b+1} = x_1$. When speaking about the arc $x_{i-1}x_i$, we mean the clockwise oriented arc from x_{i-1} to x_i on ∂D. For any x_i, $i = 0, \ldots, b - 1$, let $z_i, z_{i+1} \in \{w_1, \ldots, w_n\}$ be the points in $x_i + \lambda_i K$ such that the triangle $x_i z_{i+1} z_i$ is clockwise oriented. Since the arc $x_{i-1}x_i$ of ∂D is contained in $z_i + K$ according to the remark at Lemma 2.1.1, we conclude (ii).

The next step is to verify the claim

$$2n = f_2(\Sigma) + b(\Sigma) + 2. \tag{2.12}$$

The idea is to verify $\chi(\Sigma) = 1$ by extending Σ to a cell decomposition Σ_1 of D. First, we show that if we add the triangles $x_i z_{i+1} z_i$, $i = 0, \ldots, b - 1$, to Σ then we obtain a simplicial complex Σ_0. The only problematic case is when the common secant pq of two supporting domains K_1 and K_2 do not separate the corresponding objects. According to Lemma 2.1.1, we may assume that x_1 is the centre of K_1; moreover, x_1 and the centre y of K_2 lie on the same side of pq. It also follows that K_2 is larger then K_1; hence, the remark at Lemma 2.1.1 yields that x_1 lies in the interior of the triangle ypq. Writing l to denote a supporting line at x_1 to D, we have that l strictly separates p and q and, say, p and y lie on the same side of l. Therefore, the line $x_1 p$ separates the triangle $x_1 z_1 z_2$ from the object of Σ_0 determined by K_2; hence, Σ_0 is actually a simplicial complex. It is connected, and its vertices are $w_1, \ldots, w_n$ and $x_1, \ldots, x_b$. Now we add the cells T_i, $i = 1, \ldots, b$ to Σ_0, where T_i is bounded by the segments $x_{i-1}z_i$ and $z_i x_i$ and the arc $x_{i-1}x_i$ of ∂D. The new family forms a cell decomposition Σ_1 of D, which, in turn, yields $\chi(\Sigma_1) = 1$ according to the Euler formula (A.11). Since Σ_1 has b vertices, $3b$ edges, and $2b$ triangles more than Σ does, we deduce that $\chi(\Sigma) = 1$. In particular, the formula $3 f_2(\Sigma) = 2 f_1(\Sigma) - b(\Sigma)$ follows by counting the edges of each triangle of Σ, which, in turn, yields (2.12).

Let us turn to the proof of the crucial formula (iii). First, we consider the case $n = 2$; hence, $b = 2$. Then Σ has no triangles and only one edge $w_1 w_2$. There exists a line l through a common boundary point of $w_1 + K$ and $w_2 + K$ such that l avoids int D, and Proposition 2.7.3 yields

$$A(D) \leq 2 A(K) - (A(K) - A(\widetilde{\Pi}(K))) = A(K) + A(\widetilde{\Pi}(K)).$$

Therefore, let $n \geq 3$. We write m_i to denote the midpoint of the segment $z_i z_{i+1}$ and write Π_i to denote the domain bounded by the broken line

$x_{i-1} m_{i-1} z_i m_i x_i$ and by the arc $x_{i-1} x_i$ of ∂D. In particular,

$$A(D) = A(\operatorname{supp} \Sigma) + \sum_{i=1}^{b} A(\Pi_i). \tag{2.13}$$

Let $K_i = z_i + K$. We write $\widetilde{\Pi}_i$ to denote the part of K_i, that contains x_i and x_{i-1} and is cut off by the rays $z_i m_{i-1}$ and $z_i m_i$. Measuring the angle of a cone ω pointed at x as the area of $\omega \cap (x + K)$, we deduce that the sum of the angles of the triangles of Σ is $f_2(\Sigma) A(K)/2$. Thus, (2.12) yields

$$\sum_{i=1}^{b} A(\widetilde{\Pi}_i) = n A(K) - \frac{f_2(\Sigma)}{2} A(K) = \frac{b+2}{2} A(K).$$

Since $A(\Pi_i) = A(\widetilde{\Pi}_i) - A(\widetilde{\Pi}_i \backslash \Pi_i)$, we calculate $\sum_{i=1}^{b} A(\Pi_i)$ in the form

$$\sum_{i=1}^{b} A(\Pi_i) = \frac{b+2}{2} A(K) - \sum_{i=1}^{b} A(\widetilde{\Pi}_i \backslash \Pi_i). \tag{2.14}$$

Let σ_i denote the maximal open convex arc of ∂D that lies in int K_i and passes through x_{i-1} and x_i. Since both K_{i-1} and K_{i+1} are needed to cover D, σ_i connects a point of $K_{i-1} \cap \partial K_i$ with a point of $K_{i+1} \cap \partial K_i$, and cuts K_i into two domains. Out of these two parts, let Ω_i be the one not containing z_i. We claim that

$$2 \sum_{i=1}^{b} A(\widetilde{\Pi}_i \backslash \Pi_i) = \sum_{i=1}^{b} A(\Omega_i \backslash \Omega_{i+1}) + A(\Omega_{i+1} \backslash \Omega_i) + A(K_i \cap K_{i+1}).$$

$$\tag{2.15}$$

After we remove Π_i from $\widetilde{\Pi}_i \cap D$, the remaining set is the union of two disjoint convex sets, and we write Φ_i^+ and Φ_i^- to denote the one cut off by the segments $x_i m_i$ and $x_{i-1} m_{i-1}$, respectively. In addition, Ψ_i^+ and Ψ_i^- denote the parts of $\Omega_i \backslash \widetilde{\Pi}_i$ that are cut off by the half lines $z_i m_i$ and $z_i m_{i-1}$, respectively, where Ψ_i^+ or Ψ_i^- might be a segment. Thus,

$$A(\widetilde{\Pi}_i \backslash \Pi_i) = A(\Omega_i) + A(\Phi_i^+) - A(\Psi_i^+) + A(\Phi_i^-) - A(\Psi_i^-). \tag{2.16}$$

Since all K_{i-1}, K_i and K_{i+1} are needed to cover D, it is not hard to see that $\sigma_i \backslash \sigma_{i+1}$, $\sigma_i \cap \sigma_{i+1}$ and $\sigma_{i+1} \backslash \sigma_i$ are consecutive connected arcs in clockwise order along ∂D. In particular one half of $K_i \cap K_{i+1}$ cut off by the line $z_i z_{i+1}$ can be obtained from $\Omega_i \cap \Omega_{i+1}$ by removing Ψ_i^+ and Ψ_{i+1}^-, and adding Φ_i^+ and Φ_{i+1}^-. We conclude that

$$A(\Omega_i) + A(\Omega_{i+1}) + 2 \cdot (A(\Phi_i^+) - A(\Psi_i^+) + A(\Phi_{i+1}^-) - A(\Psi_{i+1}^-))$$
$$= A(\Omega_i \backslash \Omega_{i+1}) + A(\Omega_{i+1} \backslash \Omega_i) + A(K_i \cap K_{i+1}).$$

Therefore multiplying both sides of (2.16) by two and summing the resulting formulae for $i = 1, \ldots, b$ yield (2.15).

Let l_i be a supporting line to D at x_i; hence, the part of $K_i \backslash K_{i+1}$ (of $K_{i+1} \backslash K_i$) cut off by l_i is contained in $\Omega_i \backslash \Omega_{i+1}$ (in $\Omega_{i+1} \backslash \Omega_i$). We deduce via (2.13), (2.14), (2.15), and Proposition 2.7.3 that

$$A(D) \leq A(\text{supp } \Sigma) + \frac{b+2}{2} A(K) - \frac{b}{2} \cdot (A(K) - A(\widetilde{\Pi}(K))),$$

which, in turn, yields (iii).

Finally, the area of each triangle of Σ is at most $A(\widetilde{T}(K)) = A(\widetilde{H}(K))/2$ (see Lemma 2.2.5); therefore, (iv) follows by (iii) and (2.12). $\square$

To apply Lemma 2.7.2, we need the following statement:

Lemma 2.7.4. *Let the convex domain D in $\mathbb{R}^2$ be covered by n translates of an o-symmetric convex domain K. If $K \subset \text{int } K'$ for an o-symmetric convex domain K' then there exists a covering of D by n translates of K' such that the centres of these translates are in general position with respect to $\| \cdot \|_{K'}$ and D.*

Proof. First we claim that there exist $y_1, \ldots, y_n \in D$ such that $y_1 + K, \ldots, y_n + K$ cover D. Since the claim is continuous in K, we may assume that K is strictly convex. Let $x_1 + K, \ldots, x_n + K$ be the n translates covering D. We write y_i to denote a point closest to D with the property that $(y_i + K) \cap D$ contains $(x_i + K) \cap D$. We suppose that there exists a $y_i \notin D$, and we seek a contradiction. Let z be the closest point of D to y_i, and let $\lambda = \|y_i - z\|_K < 1$. In addition, we write l to denote a common tangent to $y_i + \lambda K$ and D, and let y_i' denote the midpoint of $l \cap (y_i + K)$. Then $D \cap (y_i + K)$ is of length less than two with respect to $\| \cdot \|_K$ and is contained in $y_i' + K$. Moreover, the distance of z from both endpoints of $l \cap (y_i + K)$ is at least $1 - \lambda$ according to the triangle inequality; hence, $\|y_i' - z\|_K < \lambda$. This contradiction verifies that $y_j \in D$ for $j = 1, \ldots, n$.

Varying $y_1, \ldots, y_n$, we obtain $\tilde{y}_1, \ldots, \tilde{y}_n \in \text{int } D$ such that even the interiors of $\tilde{y}_1 + K', \ldots, \tilde{y}_n + K'$ cover D, and we may also assume that $\tilde{y}_1, \ldots, \tilde{y}_n$ are in general position with respect to $\| \cdot \|_{K'}$ and D. In turn, we conclude Lemma 2.7.4. $\square$

Now we are ready to prove the Fejes Tóth inequality.

Proof of Theorem 2.7.1. If D can be covered by a translate of K then Theorem 2.7.1 readily follows; hence, we assume that at least two translates of K are needed to cover D. Since the Fejes Tóth inequality is continuous in K, we may assume according to Lemma 2.7.4 that K is strictly convex, and the centres of the n translates are in general position with respect to $\| \cdot \|_K$ and D. Throwing out the translates that we do not need, we may also assume that no $n - 1$ out of these n translates cover D. At this point, we may apply Lemma 2.7.2, and we let Σ be the associated simplicial complex. In particular, some $b(\Sigma)$ translates of K' cover ∂D according to (ii); hence, (iv) completes the proof of the Fejes Tóth inequality. $\square$

Comments. The Fejes Tóth inequality (Theorem 2.7.1) was proved implicitly by L. Fejes Tóth [FTL1949] for circular discs and was generalized to any centrally symmetric convex domain by his son, G. Fejes Tóth (see [FTG1987]). Their proof uses generalized Dirichlet–Voronoi cells, which can be defined with respect to any o-symmetric strictly convex domain K in the following sense: G. Fejes Tóth [FTG1987] verifies for any $p \neq q$ that the family of points whose distance from p and q are the same with respect to $\| \cdot \|_K$ is a continuous injective image of the real line, and any two such curves intersect in at most one point. (These facts follow, say, from the remark at Lemma 2.1.1.) Therefore, the generalized Dirichlet–Voronoi cells are bounded by finitely many curves out of these "bisecting curves."

2.8. Covering the Area by o-Symmetric Convex Domains

If K is an o-symmetric convex domain then Theorem 1.5.1 provides some $\widetilde{n}(K)$ such that if a convex domain D is covered by $n \geq \widetilde{n}(K)$ translates of K then

$$A(D) \leq n A(\widetilde{H}(K)),$$

where $\widetilde{H}(K)$ is an inscribed o-symmetric polygon of at most six sides of maximal area [compare with the Dowker theorem (Corollary 1.5.5)]. We now verify that one may choose $\widetilde{n}(K) = 7$.

Theorem 2.8.1. *If K is a centrally symmetric convex domain, and a convex domain D can be covered by $n \geq 7$ translates of K, then*

$$A(D) \leq n A(\widetilde{H}(K)).$$

Remark. Clusters based on the lattice tiling by $\widetilde{H}(K)$ (see Lemma 2.1.3) show that the coefficient $A(\widetilde{H}(K))$ is optimal.

Proof. According to Lemma 2.7.4, we may assume that K is an o-symmetric strictly convex domain, and the centres of the n translates are in general position with respect to $\| \cdot \|_K$ and D. In addition, we write m to denote the minimal cardinality of a subset of the n translates that covers D. Therefore, Lemma 2.7.2 can be applied, and let Σ be the associated simplicial complex. We simply write $\widetilde{T}$, $\widetilde{\Pi}$, and $\widetilde{H}$ to denote $\widetilde{T}(K)$, $\widetilde{\Pi}(K)$, and $\widetilde{H}(K)$, respectively, and measure distance always with respect to $\| \cdot \|_K$. We distinguish three cases:

Case I: $b(\Sigma) \geq 6$

In this case Lemma 2.7.2 (iv) yields

$$A(D) \leq (m-1)\,A(\widetilde{H}) + A(K) - 3 \cdot \left(A(\widetilde{H}) - A(\widetilde{\Pi}) \right).$$

Hence, Theorem 2.8.1 in this case follows by verifying

$$4A(\widetilde{H}) \geq 3A(\widetilde{\Pi}) + A(K). \tag{2.17}$$

Among the sides of $\widetilde{\Pi}$, let s be the one that cuts off the section with maximal area from K. Moreover, let $x \in \partial K$ be the point of the section where some tangent line to K is parallel to s. Since the area of the section is at most twice the area of triangle determined by x and s, the convex hull of $\widetilde{\Pi}$, x, and $-x$ is an inscribed hexagon verifying (2.17).

Case II: $b(\Sigma) \leq 4$

The heart of argument is the following claim: If a triangle T has a side of length at most λ and another side of length at most μ then

$$A(T) \leq \frac{\lambda\mu}{4} \cdot A(\widetilde{\Pi}). \tag{2.18}$$

We may assume that the common vertex of the two sides is o and that the other two vertices p and q lie on ∂K; hence, $\lambda = \mu = 1$. Then $\pm p$, $\pm q$ are vertices of an inscribed parallelogram of area $4A(T)$, which, in turn, yields (2.18). Now each edge of Σ is of length at most two, and thus $A(\text{supp}\,\Sigma) \leq 2A(\widetilde{\Pi})$ according to (2.18) and $b(\Sigma) \leq 4$. Substituting this estimate into Lemma 2.7.2 (iii) and applying (2.17) shows that

$$A(D) \leq 4A(\widetilde{\Pi}) + A(K) \leq 5A(\widetilde{H}).$$

Case III: $b(\Sigma) = 5$

According to Lemma 2.7.2 (iii) and (2.17), it is sufficient to verify that

$$A(\text{supp } \Sigma) \leq (n-4)\,A(\widetilde{H}) + \frac{A(\widetilde{\Pi})}{2}. \tag{2.19}$$

If Σ is the union of a segment and a triangle then $A(\text{supp } \Sigma) \leq A(\widetilde{\Pi})$ by (2.18); therefore, we may assume that supp Σ is a possibly nonconvex polygon. If $n \geq 8$ then we use that supp Σ is the union of three triangles T_1, T_2, and T_3, where two sides of both T_1 and T_2 are of length at most two, one side of T_3 is of length at most two and another side is of length at most four. We deduce that $A(\text{supp } \Sigma) \leq 4A(\widetilde{\Pi})$ by (2.18), which, in turn, yields (2.19). Thus, let $n = 7$. Then Σ has seven triangles, and simple considerations provide at least four triangles of Σ whose union is bounded by at most four edges of Σ. We deduce

$$A(\text{supp } \Sigma) \leq 2A(\widetilde{\Pi}) + 3A(\widetilde{T})$$

by (2.18), and, in turn, (2.19) by $A(\widetilde{T}) = A(\widetilde{H})/2$ (see Lemma 2.2.5). Therefore, the proof of Theorem 2.8.1 is now complete. □

Finally, we consider the structure of optimal coverings by a large number of translates of a centrally convex domain.

Lemma 2.8.2. *Let K be a centrally convex domain that is not a parallelogram. If n is large and D_n is a convex domain of maximal area that can be covered by n translates of K then (possibly after translation)*

$$c_1\sqrt{n} \cdot K \subset D_n \subset c_2\sqrt{n} \cdot K,$$

where c_1 and c_2 are positive constants that depend on K.

Proof. In the following, the implied constant in $O(\cdot)$ always depends only on K, and $\widetilde{\Pi}(K)$ and $\widetilde{H}(K)$ are denoted simply by $\widetilde{\Pi}$ and $\widetilde{H}$. We claim that

$$A(D_n) \leq nA(\widetilde{H}) + A(K) - A(\widetilde{H}) - \frac{A(\widetilde{H}) - A(\widetilde{\Pi})}{16} \cdot P_K(D_n), \tag{2.20}$$

where $P_K(\cdot)$ stands for the Busemann perimeter with respect to $\|\cdot\|_K$ (see Section 2.1). The Busemann perimeter of the part of ∂D_n that is covered by a translate of K is at most 8 according to Lemma 2.1.7; hence, at least

$P_K(D_n)/8$ translates of K are needed to cover ∂D_n. Therefore, the Fejes Tóth inequality (Theorem 2.7.1) yields (2.20).

Next we provide a lower bound for $A(D_n)$ using the lattice Λ such that $\Lambda + \widetilde{H}(K)$ is a tiling (see Lemma 2.1.3). Since the at most n translates of $\widetilde{H}(K)$ that are contained in $\sqrt{n}\,\widetilde{H}(K)$ cover $(\sqrt{n} - 2)\,\widetilde{H}(K)$, we deduce

$$A(D_n) \geq n A(\widetilde{H}(K)) - O\left(\sqrt{n}\right). \tag{2.21}$$

Comparing (2.21) and (2.20) shows that $P_K(D_n) = O\left(\sqrt{n}\right)$. Since D_n contains a segment of length $R_K(D_n)$ with respect to $\|\cdot\|_K$, we have $R_K(D_n) = O\left(\sqrt{n}\right)$. Therefore, $r_K(D_n) \geq c\sqrt{n}$ follows by (2.21) and Lemma 2.1.6, where c is a positive constant depending on K. $\qquad \square$

Remark 2.8.3. Let K be an o-symmetric convex domain that is not a parallelogram. Given a convex domain D and a covering lattice Λ for K, the corresponding *finite lattice covering* of D is the family of all translates of K by vectors in Λ whose interiors intersect D.

Let D_n be a convex domain of maximal area, which can be covered at most n translates of K that form a finite lattice covering. Then as n tends to infinity, $R_K(D_n)/\sqrt{n}$ stays bounded, and any convergent subsequence of $\{D_n/\sqrt{n}\}$ tends to some polygon of at most six sides. If, in addition, K is a circular disc then $\{D_n/\sqrt{n}\}$ tends to a regular hexagon.

Let us sketch a proof of this statement. The argument for Lemma 2.8.2 yields an upper bound for $R_K(D_n)/\sqrt{n}$. It follows that if Λ_n is the covering lattice associated to D_n then any convergent subsequence of $\{\Lambda_n\}$ tends to a covering lattice of maximal determinant. We make two assumptions for the rest of the sketch, which can be removed by standard arguments: K is strictly convex and $\Lambda_n = \Lambda$ for all n, where Λ is spanned by the side vectors of $\widetilde{T}(K)$. We simply write $\widetilde{H}$ and $\widetilde{T}$ to denote $\widetilde{H}(K)$ and $\widetilde{T}(K)$.

We define the simplicial complex Σ in a way that its vertices are the points $x_1, \ldots, x_k$ of Λ such that $x_i + \mathrm{int}\, K$ intersect D_n ($k \leq n$). The segment $x_i x_j$ is an edge if it is a translate of a side of $\widetilde{T}$, and $x_i x_j x_l$ is a triangle of Σ if it is a translate of $\widetilde{T}$ or $-\widetilde{T}$. In particular, Σ is connected. If a cycle of edges of Σ encircles a point of Λ then it is one of $x_1, \ldots, x_k$, and hence $\chi(\Sigma) = 1$.

We orient the boundary of supp Σ clockwise in the natural way; namely, if an edge is not contained in any triangle then it is counted twice. Let $z_1, \ldots, z_b$, $b = b(\Sigma)$, denote the vertices of Σ on the boundary according to this orientation in a way that $z_{i-1} z_i$ correspond to the free sides of Σ, $i = 1, \ldots, b$, $z_b = z_0$ (and hence a vertex may be counted twice). We write m_i to denote the midpoint of the free side $z_i z_{i+1}$ and write u_i to denote the common point of the boundaries of $z_i + K$ and $z_{i+1} + K$ that lies "left" according to the orientation. Then the segments $m_i u_i$ are pairwise disjoint, and u_i is not contained in the interior of supp Σ. In addition, we associate the part of D_n to z_i that is the part cut off the broken line $u_{i-1} m_{i-1} z_i m_i u_i$.

Next, let a_1, a_2, a_3 be the sides of $\widetilde{T}$. We write P_i to denote the convex hull of a_i and $-a_i$, and let b_i denote the number of free edges of Σ that are parallel to a_i. Combining the argument for Lemma 2.7.2 and the Fejes Tóth inequality (Theorem 2.7.1) yields

$$A(D) \le kA(\widetilde{H}) + A(K) - A(\widetilde{H}) - \sum_{i=1}^{3} (A(\widetilde{H}) - A(P_i)) \cdot \frac{b_i}{2}.$$

Therefore, we search for simplicial complexes, that have at least k vertices such that each of their edges is a translate of a side of $\widetilde{T}$ and for which the suitably weighted total length of the free edges is minimal. There always exists an extremal complex whose support is a convex polygon of at most six sides. Solving the corresponding optimization problem, we find the asymptotic shape of D_n.

Open Problems.

(i) Does the hexagon bound Theorem 2.8.1 hold for any $n \ge 2$?

It is not hard to show that the hexagon bound does hold for $n = 2$. Actually, the hexagon bound holds for $n \ge 2$ if an octagon is covered (see Theorem 1.5.1).

(ii) Given a centrally symmetric convex domain K that is not a parallelogram, let D_n be a convex domain of maximal area that can be covered by n translates of K. Does any convergent subsequence of $\{D_n/\sqrt{n}\}$ tend to some polygon of at most six sides?

The statement does hold for finite lattice coverings (see Remark 2.8.3).

Comments. Theorem 2.8.1 was proved by G. Fejes Tóth [FTG1987] for $n \ge 26$. Earlier, L. Fejes Tóth [FTL1949] settled the case of at least two circular discs.

2.9. The Hadwiger Number

The *Hadwiger number* $H(C)$ of a convex domain C is the maximal number of nonoverlapping translates of C that touch C. Since $x + C$ and $y + C$ do not overlap if and only if $x + C_0$ and $y + C_0$ do not overlap for $C_0 = (C - C)/2$, it follows that $H(C) = H(C_0)$. In addition, $H(C_0)$ is the maximal number of the points on ∂C_0 whose mutual distances are at least one with respect to $\|\cdot\|_{C_0}$.

Theorem 2.9.1. *Given a convex domain C, $H(C) = 8$ if C is a parallelogram, and $H(C) = 6$ otherwise.*

The proof of Theorem 2.9.1 is based on the notion of a *long side* of an o-symmetric convex domain C, namely, a maximal segment $s \subset \partial C$ that is longer than one with respect to $\|\cdot\|_C$. Assuming that s is parallel to some $x \in \partial C$, we have

$$\text{if } \|y - x\|_C = 1 \text{ for some } y \in \partial C \text{ then either } y \in s \text{ or } y \in -s. \quad (2.22)$$

Let us verify two auxiliary statements about long sides.

Proposition 2.9.2. *Let C be an o-symmetric convex domain, and let $x \in \partial C$. If there exist two points $z_1, z_2 \in \partial C$ on the same side of $\operatorname{lin} x$ such that $\|z_1 - x\|_C = \|z_2 - x\|_C = 1$ then z_1 and z_2 are contained in a long side of C parallel to x.*

Proof. The points $z_1, z_2, z_1 - x, z_2 - x$ all lie in ∂C, and they lie on the same side of $\operatorname{lin} x$; hence, they are collinear. Therefore, the line $z_1 z_2$ intersects ∂C in a long side. $\qquad\square$

Proposition 2.9.3. *If C is an o-symmetric convex domain then C has at most two pairs of parallel long sides.*

Proof. Let e and $-e$ be long sides of C, and let $x \in \partial C$ be parallel to e. We claim that if f is any other long side of C then

$$\text{the relative interior of } f \text{ contains either } x \text{ or } -x. \quad (2.23)$$

We suppose that the claim does not hold; hence, we may assume that the shorter arc xp of ∂C contains f, where p is the endpoint of e closer to x. Since $\|p - x\|_C = 1$, we have contradicted the monotonicity of distance along ∂C (see Lemma 2.1.2). In turn, (2.23) readily yields Proposition 2.9.3. $\qquad\square$

Proof of Theorem 2.9.1. Since some o-symmetric affine regular hexagon can be inscribed into $C_0 = (C - C)/2$ (see Lemma 2.1.4), we deduce that

$$H(C) = H(C_0) \geq 6. \quad (2.24)$$

However, if C is a parallelogram then readily $H(C) = 8$.

Therefore we assume that $x_1, \ldots, x_7$ lie on ∂C_0 according to some orientation, and $\|x_i - x_j\|_{C_0} \geq 1$ for $i \neq j$. We define $x_{i+7k} = x_i$ for any $i = 1, \ldots, 7$ and $k \in \mathbb{Z}$ and claim that if $j = i + 3$ or $j = i + 4$ then

$$x_i \text{ and } x_j \text{ lie in parallel long sides.} \tag{2.25}$$

We start with the case $x_j \neq -x_i$, when we may assume $j = i + 3$. According to Lemma 2.1.2, there is an x'_{i+1} in the shorter arc $x_i x_{i+1}$ of ∂C_0 such that $\|x'_{i+1} - x_i\|_{C_0} = 1$. Let $x'_{i+2} = x'_{i+1} - x_i$, which satisfies $\|x'_{i+2} - x'_{i+1}\|_{C_0} = \|x'_{i+2} - (-x_i)\|_{C_0} = 1$. We distinguish two possibilities. First, if x_{i+2} is contained in the open shorter arc $x'_{i+1} x'_{i+2}$ of ∂C_0 then $\|x_{i+2} - x'_{i+1}\|_{C_0} = 1$ according to Lemma 2.1.2; thus, Proposition 2.9.2 yields that x'_{i+2} and x_{i+2} are contained in a long side e of ∂C_0 parallel to x'_{i+1}. It follows by (2.22) that $x_i \in -e$, and hence x_{i+3} as a point of the shorter arc $x_{i+2}(-x_i)$ is contained in e. Second, if x_{i+2} is contained in the shorter arc $x'_{i+2}(-x_i)$ of ∂C_0 then $\|x_{i+3} - x'_{i+2}\|_{C_0} = 1$ according to Lemma 2.1.2; thus, Proposition 2.9.2 yields that x_{i+3} and $-x_i$ are contained in a long side of ∂C_0 parallel to x'_{i+2}. In turn, we deduce the claim (2.25) when $x_j \neq -x_i$. Finally, if $x_j = -x_i$ and, say, $j = i + 3$ then $x_{i+4} \neq -x_i$; hence, a long side f contains $-x_{i+4}$ and x_i. Therefore, $x_{i+3} \in -f$ yields the claim (2.25).

If any two long sides of C_0 are disjoint then $x_1, \ldots, x_7$ are contained in a pair of parallel long sides according to the claim (2.25), which is absurd as the length of any long side is at most two. If e and f are intersecting long sides of C_0 then we first consider the case when e and $-f$ are disjoint. Since any x_i is contained either in $e \cup f$ or in $(-e) \cup (-f)$ by Proposition 2.9.3 and (2.25), we may assume that $e \cup f$ contains $x_1, \ldots, x_4$. This contradicts (2.25) for $i = 1$ and $j = 4$, and therefore C_0 is the parallelogram bounded by $e, f, -e, -f$. It follows that C is a parallelogram as well, completing the proof of Theorem 2.9.1. $\qquad \square$

Comments. The problem of determining the minimal number of nonoverlapping translates that touch a given convex domain was posed in H. Hadwiger [Had1957b]. Theorem 2.9.1 is due to B. Grünbaum [Grü1961].

2.10. The Generalized Hadwiger Number

Let C be a convex domain in $\mathbb{R}^2$, and let $\alpha > 0$. We define the *generalized Hadwiger number* $H_\alpha(C)$ to be the maximal number of nonoverlapping translates of αC that touch C. In particular, $H(C) = H_1(C)$. Shrinking from the points where the translated copies of αC touch shows that $H_\alpha(C)$ is monotonically decreasing with α. When estimating $H_\alpha(C)$, we use an equivalent formulation: $H_\alpha(C)$ is the maximal cardinality of points on the boundary of

$C - \alpha C$ such that the pairwise mutual distances are at least 2α with respect to the norm defined by $C_0 = (C - C)/2$. Unlike in case of the Hadwiger number, we cannot restrict our attention to centrally symmetric domains (see Example 2.10.6).

Theorem 2.10.1. *If C is a convex domain and $C_0 = (C - C)/2$ then*

$$\left\lfloor \frac{1}{2\alpha} P_{C_0}(C) \right\rfloor \leq H_\alpha(C) \leq \frac{1+\alpha}{2\alpha} P_{C_0}(C),$$

where $\alpha \leq 2$ is assumed for the lower bound.

We deduce the following numerical bounds for the generalized Hadwiger number by Lemma 2.1.7:

Corollary 2.10.2. *If C is a convex domain and $C_0 = (C - C)/2$ then*

$$\left\lfloor \frac{3}{\alpha} \right\rfloor \leq H_\alpha(C) \leq \frac{4(1+\alpha)}{\alpha},$$

where $\alpha \leq 2$ is assumed for the lower bound. Equality holds in the upper bound if and only if C is a parallelogram and $1/\alpha$ is a positive integer.

Theorem 2.10.1 readily implies the asymptotic behavior of the generalized Hadwiger number as the parameter tends to zero:

Corollary 2.10.3. *If C is a convex domain and $C_0 = (C - C)/2$ then*

$$\lim_{\alpha \to 0} \alpha \cdot H_\alpha(C) = \frac{P_{C_0}(C)}{2}.$$

To verify Theorem 2.10.1, we extend Lemma 2.1.2 to not necessarily centrally symmetric convex domains:

Proposition 2.10.4. *Let M be a convex domain, and let $M_0 = (M - M)/2$. If the point z moves along ∂M away from some $x \in \partial M$ then $\|z - x\|_{M_0}$ increases until it reaches two.*

Proof. We may assume that M is strictly convex and that z moves clockwise. Let $z_0 \in \partial M$ be a possible position for z with $\|z_0 - x\|_{M_0} = 2\lambda < 2$, and let z be contained in the clockwise oriented arc xz_0 of ∂M. We may assume possibly after translation that $x, z_0 \in \partial(\lambda M)$. Since ∂M and $\partial(\lambda M)$ intersect in x and z_0 according to Lemma 2.1.1, we deduce that z is contained in λM, and hence $\|z - x\|_{M_0} \leq 2\lambda = \|z_0 - x\|_{M_0}$. $\square$

Proof of Theorem 2.10.1. We consider n points on the boundary of $C - \alpha C$ such that the pairwise mutual distances are at least 2α with respect to norm defined by $C_0 = (C - C)/2$. Then the convex hull Q_n of the points satisifies the inequalities

$$n \cdot 2\alpha \leq P_{C_0}(Q_n) \leq P_{C_0}(C - \alpha C) = (1 + \alpha) \cdot P_{C_0}(C), \qquad (2.26)$$

which, in turn, yield the upper bound of Theorem 2.10.1. If equality holds in the upper bound of Corollary 2.10.2 then C is a parallelogram according to Lemma 2.1.7, and we may assume that C is o-symmetric. Since C and some translates of αC tile $(1 + 2\alpha)C$, we have $\alpha = 1/k$, where k is the number of translates of αC that intersect the relative interior of a given side of C.

To prove the lower bound in Theorem 2.10.1, we assume that $\alpha \leq 2$. Let $M = C - \alpha C$. We place points $x_1, x_2, \ldots$ onto ∂M according to the clockwise orientation in a way that $\|x_{i+1} - x_i\|_{C_0} = 2\alpha$, and we let k be the minimal index such that $\|x_{k+1} - x_1\|_{C_0} < 2\alpha$. It follows by Proposition 2.10.4 that $\|x_j - x_i\|_{C_0} \geq 2\alpha$ holds if $1 \leq i < j \leq k$.

We write P to denote the convex hull of $x_1, \ldots, x_{k+1}$ and write K to denote $(1 - \alpha/2)\,C - (\alpha/2)\,C$; hence, $M = K + \alpha C_0$. In addition, $K \subset P$ because any side $x_i x_j$ of P lies in the union of $x_i + \alpha C_0$ and $x_j + \alpha C_0$, whereas neither $x_i + \alpha C_0$ nor $x_j + \alpha C_0$ overlaps K. Since $P_{C_0}(P) = (k + t) \cdot 2\alpha$ for some $0 \leq t < 1$, we conclude Theorem 2.10.1 by $P_{C_0}(K) = P_{C_0}(C)$. $\qquad\square$

Remark 2.10.5. It is natural to define $H_\infty(C)$ to be the maximal number of nonoverlapping translates of C that have a common point. We verify

$$H_\infty(C) \leq 4, \qquad (2.27)$$

where equality holds if and only if C is a parallelogram.

Let us assume that the translates $x_i + C$ touch pairwise at the common point y for $i = 1, 2, 3, 4$. Then the points $y - x_i$ lie in ∂C, and there exist parallel supporting lines to C at any two of the four points. Therefore, C is a parallelogram whose vertices are the four points, which, in turn, yields (2.27).

We note that if C is smooth then $H_\infty(C) = 2$, and if C is a triangle then $H_\infty(C) = 3$.

We have seen in Section 2.9 that $H_\alpha(C)$ and $H_\alpha((C - C)/2)$ coincide if $\alpha = 1$. However, this property may fail if $\alpha \neq 1$, as the following example of a Reuleaux triangle exhibits. We note that if T is a regular triangle of side length a then the corresponding Reuleaux triangle C is the intersection of the three circular discs of radius a that are centred at the vertices of T. In particular, C is bounded by three circular arcs.

Example 2.10.6. *If C is a Reuleaux triangle then $H_\alpha(C)$ can be both smaller or larger than $H_\alpha(C_0)$, where $C_0 = (C - C)/2$.*

We may assume that the regular triangle T is of side length two and that its circumcentre is the origin, hence, $\| \cdot \|_{C_0}$ is the Euclidean norm.

Let $\alpha_n = \sin(\pi/n)/(1 - \sin(\pi/n))$, which satisfies that n nonoverlapping circular discs of radius α_n can be placed around the unit disc in a way that each small circular disc touches the unit disc and its two neighbours. We deduce

$$H_{\alpha_n}(C_0) = n$$

for any n and that α_n is maximal with this property. Moreover, we translate $\alpha_3 C$ in the directions of the vertices of T until the three translates touch C. Then these three translates are pairwise disjoint, and hence $H_{\alpha_3+\varepsilon}(C) \geq 3$ holds for some $\varepsilon > 0$. Therefore,

$$H_\alpha(C) > H_\alpha(C_0) \quad \text{if} \quad \alpha = \alpha_3 + \varepsilon. \tag{2.28}$$

However, we claim that if n is large then

$$H_{\alpha_n}(C) < H_{\alpha_n}(C_0). \tag{2.29}$$

We verify the claim (2.29) by contradiction; namely, we suppose that $H_{\alpha_n}(C) \geq n$, and hence there exist n points $x_1, \ldots, x_n$ on $\partial(C - \alpha_n C)$ such that the distance between any two points is at least $2\alpha_n$. Let us estimate the perimeter of $\partial(C - \alpha_n C)$ in two ways: First, we have readily

$$P(C - \alpha_n C) = P((1 + \alpha_n)B^2) = n \cdot 2\alpha_n + O\left(\alpha_n^2\right). \tag{2.30}$$

Second, the boundary of $C - \alpha_n C$ consists of three circular arcs of radius 2 and of length $2\pi/3$ and three circular arcs of radius $2\alpha_n$ and of length $2\alpha_n\pi/3$. There also exist consecutive points x_i and x_j such that the shorter arc of $\partial(C - \alpha_n C)$ connecting x_i and x_j contains a circular arc of radius $2\alpha_n$ and of length $2\alpha_n\pi/6$, which, in turn, yields

$$P(C - \alpha_n C) \geq n \cdot 2\alpha_n + \left(\frac{\pi}{6} - 2\sin\frac{\pi}{12}\right) \cdot 2\alpha_n.$$

Since, for large n, this lower bound contradicts (2.30), we conclude (2.29).

Comments. The generalization of the Hadwiger number was considered first for polytopes in any dimension by L. Fejes Tóth in [FTL1970] and [FTL1975b]. Theorem 2.10.1 and Corollaries 2.10.2 and 2.10.3 are due to V. Boju and L. Funar [BoF1993].

The estimate (2.27) is due to C. M. Petty [Pet1971]. In addition, A. Bezdek, K. Kuperberg, and W. Kuperberg [BKK1995] prove that if C is homeomorphic to a circular disc then the maximal number of concurrent nonoverlapping translates of C is four.

The example of Reuleaux polygons is discussed in more detail by K. Böröczky Jr., D. G. Larman, S. Sezgin, and C. Zong [BLS2000].

2.11. Related Problems

Packings of Homothetic Domains Inside Some Convex Container

We recall that, for a convex domain K, $t_K(n)$ denotes the minimal area of a circumscribed polygon with at most n sides. Theorem 1.6.1 yields the following variant of Theorem 2.3.4:

Let K be a centrally symmetric convex domain that is neither a parallelogram nor a hexagon. If a convex domain D contains the n nonoverlapping homothetic copies $K_1, \ldots, K_n$ of K then

$$\frac{A(K_1) + \cdots + A(K_n)}{A(D)} \leq \frac{A(K)}{A(H(K))},$$

provided that

$$\frac{A(K_i)}{A(K_j)} \geq \frac{t_K(6) - t_K(7)}{t_K(5) - t_K(6)}$$

holds for $i, j = 1, \ldots, n$, and $n \geq n(K)$, where $n(K)$ depends on K.

If D is a hexagon then the inequality holds for any $n \geq 1$, a fact observed by K. Böröczky (unpublished) and first proved in print by G. Fejes Tóth [FTG1972]). We note that

$$\frac{t_K(6) - t_K(7)}{t_K(5) - t_K(6)} < 1$$

if K is strictly convex (see Lemma 1.2.3).

Covering Any Convex Shape by Homothetic Domains

Given a convex domain K, $\tilde{t}_K(n)$ denotes the maximal area of an inscribed polygon with at most n sides. Since homothetic convex domains are noncrossing, Theorem 1.6.2 yields the following generalization of Theorem 2.8.1:

Let K be a centrally symmetric convex domain that is neither a parallelogram nor a hexagon. If a convex domain D is covered by n homothetic copies $K_1, \ldots, K_n$ of K then

$$\frac{A(K_1) + \cdots + A(K_n)}{A(D)} \geq \frac{A(K)}{A(\widetilde{H}(K))},$$

provided that

$$\frac{A(K_i)}{A(K_j)} \geq \frac{\widetilde{t}_K(7) - \widetilde{t}_K(6)}{\widetilde{t}_K(6) - \widetilde{t}_K(5)}$$

holds for $i, j = 1, \ldots, n$, and $n \geq \widetilde{n}(K)$, where $\widetilde{n}(K)$ depends on K.

Again, if D is a hexagon then the inequality holds for any $n \geq 1$, a fact observed by K. Böröczky (unpublished), and the first published proof is due to G. Fejes Tóth [FTG1972]. We note that

$$\frac{\widetilde{t}_K(7) - \widetilde{t}_K(6)}{\widetilde{t}_K(6) - \widetilde{t}_K(5)} < 1$$

if K is smooth (see Lemma 1.5.4).

Packing Homothetic Copies Into a Convex Domain

Let C be a convex domain. If C contains at least two nonoverlapping translates of λC for positive λ then readily $\lambda \leq 1/2$. However, C contains four nonoverlapping translates of $C/2$ if and only if C is a parallelogram. The first nontrivial result in this direction is due to K. Doliwka and M. Lassak [DoL1995] (see K. Böröczky and Zs. Lángi [BöL?] for some missing details of the argument); namely, C contains five nonoverlapping translates of λC for $\lambda = (3 - \sqrt{5})/2$ if and only if C is either an affine regular pentagon or an affine regular decagon. In addition, K. Böröczky and Zs. Lángi [BöL03] verify that C contains six nonoverlapping translates of λC for $\lambda = (9 - \sqrt{5})/19$ if and only if C is an affine regular pentagon. It is conjectured that C does not contain seven translates of λC for $\lambda > 1/3$. However, if C is centrally symmetric then C readily contains seven nonoverlapping translates of $C/3$, and P. G. Doyle, J. C. Lagarias, and D. Randall [DLR1992] show that C does not contain six translates of λC for $\lambda > 1/3$.

A related result of Zs. Lángi [Lán2003] is that ∂C contains no six points whose pairwise distances with respect to $\| \cdot \|_{\frac{1}{2}(C-C)}$ are at least $8 - 4\sqrt{3}$, where the constant $8 - 4\sqrt{3}$ is optimal.

Next we discuss packing problems where the homothetic copies may have different sizes. Assuming that $A(C) = 1$, we write $\psi(C)$ to denote the minimal number with the following property: If the total area of a finite collection

$\{C_1, \ldots, C_k\}$ of homothetic copies of C is at most $\psi(C)$ then some translates $C_1 + x_1, \ldots, C_k + x_k$ form a packing in C. If C is a square then J. Meir and L. Moser [MeM1968] proved that $\psi(C) = 1/2$.

If C is an arbitrary convex domain then two copies of $(1/2 + \varepsilon) C$ show that $\psi(C) \leq 1/2$. However, no safe conjecture is known for the minimum of $\psi(C)$. Since there exist homothetic parallelograms $P_1 \subset C \subset P_2$ satisfying $A(P_2) \leq 4A(P_1)$, the result about squares yields that $\psi(C) \geq 1/8$.

For packing smaller homothetic squares into $[0, 1]^2$, even on-line packings are considered; namely, we learn the ith square only after the first $i - 1$ squares have been already in place. J. Januszewski and M. Lassak [JaL1997] construct on-line packings into $[0, 1]^2$ under the assumption that the total area of the small squares is at most $5/16$.

Covering a Convex Domain by Homothetic Copies

Given a convex domain C of area one, we write $\varphi(C)$ to denote the minimal number with the following property: If the total area of a finite collection $\{C_1, \ldots, C_k\}$ of homothetic copies of C is at least $\varphi(C)$ then some translates $C_1 + x_1, \ldots, C_k + x_k$ cover C. If C is a square then J. Moon and L. Moser [MoM1967] proved that $\varphi(C) = 3$, a fact later reproved independently by H. Groemer [Gro1982] and A. Bezdek and K. Bezdek [BeB1984]. It was a problem of L. Fejes Tóth to determine best lower and upper bounds for $\varphi(C)$. Readily, $\varphi(C) \geq 2$ for any convex domain C, and A. Bezdek and K. Bezdek [BeB1984] conjectured that $\varphi(C) = 2$ for triangles. This conjecture was confirmed by Z. Füredi [Für2003]. However, J Januszewski [Jan2003] proved that $\varphi(C) \leq 6.5$ for any convex domain C. It seems to be plausible that squares are extremal; namely, $\varphi(C) \leq 3$.

A related problem is to find $\varphi_0(C)$, where we allow not only homothets of C but homothets of $-C$ as well. Answering a conjecture of K. Böröczky, J. Januszewski [Jan1998] proved that $\varphi_0(C) \geq 4$ (here equality holds for triangles).

When $[0, 1]^2$ is to be covered by smaller homothetic squares, even on-line packings are considered; namely, we learn the ith square only after the first $i - 1$ squares have been already in place. J. Januszewski and M. Lassak [JaL1995a] contruct on-line coverings of $[0, 1]^2$ under the assumption that the total area of the squares is at least $(7/4)\sqrt[3]{9} + 13/8 = 5.2651 \ldots$.

Covering the Maximal Area by Given Number of Translates

Let K be an o-symmetric convex domain. We write $f(t)$ to denote the supremum of the areas of intersections of K and hexagons of area t, and we write $F(t)$ to denote the smallest concave function satisfying $F(t) \geq f(t)$.

In particular, $F(t) = f(t) = A(K)$ if and only if $t \geq A(H(K))$, and $F(t) = f(t) = t$ if and only if $t \leq A(\tilde{H}(K))$. According to G. Fejes Tóth [FTG1972] (see G. Fejes Tóth [FTG1977] for the corresponding Dowker–type theorem),

> *if D is a polygon of at most six sides then the area of the part of D that is covered by some n translates of K is at most n · $F(A(D)/n)$.*

This result yields the hexagon bound both for translative packings (see Theorem 1.2.1) and for translative coverings (see Theorem 1.5.1). We note that the factor $F(A(D)/n)$ is optimal for large n. If $f(\cdot)$ is concave then the optimality is shown by a lattice arrangement. However, if $f(\cdot)$ itself is not concave then the optimality is shown by an arrangement based on two lattice arrangements of K: One uses one of the lattice arrangements inside some cone σ, and the other one in the complement of σ. Analogous phenomena in nature with coexisting phases are discussed by G. Fejes Tóth and L. Fejes Tóth [FTF1989]).

Touching Pairs and Snakes

Let K be a convex domain in $\mathbb{R}^2$. P. Brass [Bra1996] considers the maximal number of touching pairs among n nonoverlapping translates of K and proves that this maximum is $\lfloor 4n - \sqrt{28n - 12} \rfloor$ if K is a parallelogram and $\lfloor 3n - \sqrt{12n - 3} \rfloor$ otherwise.

Next, a *translative snake* for K is a family $K_1, \ldots, K_n$ of nonoverlapping translates of K such that K_i and K_j touch if and only if $|i - j| = 1$. A translative snake is called limited if it is not the proper subset of any other translative snake. Then, K. Böröczky Jr. and V. Soltan [BöS1995] prove that the minimal cardinality of a limited translative snake is eleven if K is a parallelogram and ten otherwise.

Translative Clouds

Let us discuss a problem dual to the Hadwiger problem: Given a convex domain K, a family $x_1 + K, \ldots, x_n + K$ is called a *translative cloud* if the translates do not overlap K nor each other and any half line emanating from a point of K intersects the interior of one of the translates. Unfortunately, only a few unpublished observations are known in this direction:

- If K is a trapezoid such that the ratio of the lengths of the parallel sides is 1.5 then there exists a cloud with eight translates.
- The translates $x_1 + K, \ldots, x_n + K$ form a cloud for K if and only if $x_1 + K_0, \ldots, x_n + K_0$ form a cloud for K_0, where $K_0 = (K - K)/2$.

- If K is smooth and strictly convex then there exists a cloud of nine translates.

We conjecture that the minimal cardinality of a translational cloud is either eight or nine.

Multiple Coverings by Translates

Given $k \geq 1$ and a convex domain K, we say that n translates of K form a *k-fold covering* of a convex domain D if every point of D is contained in at least k translates, hence the density with respect to D is readily at least k. If ∂K is at least twice differentiable and is of postive curvature then we construct thin k-fold coverings for large k. Any translate of K contains at least $r^2 A(K) - r^{\frac{46}{73}+\varepsilon}$ points of the lattice $\frac{1}{r}\mathbb{Z}^2$ (see M. N. Huxley [Hux1996], chapter 18.3) for any $\varepsilon > 0$ and for large r. Therefore the periodic arrangement with respect to the lattice $\frac{1}{r}\mathbb{Z}^2$ for suitable r yields finite k-fold coverings of density at most $k + k^{\frac{23}{73}+\varepsilon}$ ($\varepsilon > 0$ arbitrary, k large).

If K is a polygon then for any $m \geq 2$ there exist k depending on K and m such that and k-fold covering by translates of K can be decomposed into m coverings. This is proved by J. Pach [Pac1986] if K is centrally symmetric, and by G. Tóth [Tót?] for any convex polygon. It is conjectured by J. Pach that this property holds for any convex domain. We note that any 33-fold covering by equal discs can be decomposed into two coverings according to P. Mani-Levitska and J. Pach [MaP?].

3

Parametric Density

This chapter, which discusses the so-called parametric density of finite packings and coverings by translates of a given convex domain K, is a natural continuation of Chapter 2. For packings, a rather complete theory is available: Let $\varrho > 0$ be the parameter. If $x_1 + K, \ldots, x_n + K$ is a packing and C_n denotes the convex hull of $x_1, \ldots, x_n$ then we define the *parametric density of the packing* to be

$$\frac{n \cdot A(K)}{A(C_n + \varrho K)}.$$

Given K, ϱ, and n, our aim is to find the maximal parametric density over all packings of n translates of K for given ϱ. It is equivalent to find the minimum of $A(C_n + \varrho K)$, where C_n contains $x_1, \ldots, x_n$ such that $x_1 + K, \ldots, x_n + K$ is a packing. We note that the case $\varrho = 1$ asks for the minimum area of a convex domain that contains n nonoverlapping translates of K.

Let $K_0 = (K - K)/2$, and let $\Pi(K_0)$ and $H(K_0)$ denote an o-symmetric parallelogram and hexagon, respectively, of minimal area containing K_0. When $\varrho < A(H(K_0))/A(\Pi(K_0))$ and n is large then Theorem 3.1.1 says that sausages are optimal; namely, if $x_1 + K, \ldots, x_n + K$ is an optimal packing for large n then $x_1, \ldots, x_n$ are aligned. In contrast, if $\varrho > A(H(K_0))/A(\Pi(K_0))$ then clusters are optimal; namely, the convex hull for the optimal packing is of large inradius for large n (see Theorem 3.2.1). Therefore, it is natural to call the quotient $A(H(K_0))/A(\Pi(K_0))$ the *critical radius* $\varrho_c(K)$ *for packings*.

The theory is far from being satisfactory for coverings: Let $\theta \geq 0$. If $x_1 + K, \ldots, x_n + K$ cover a compact convex set D then the *parametric density of the covering* is

$$\frac{n \cdot A(K)}{A(D + \theta K)}.$$

For any n, let D_n be a compact convex set that can be covered by n translates

74

of K, and $A(D_n + \theta K)$ is maximal under this condition. We will show that there exists a $\theta_c(K) \geq 0$ such that as n tends to infinity, $r_K(D_n)$ stays bounded if $\theta < \theta_c(K)$ and tends to infinity if $\theta > \theta_c(K)$. At the end of the chapter, we present various examples of both packings and coverings, which also exhibit many properties that might be typical but yet unproved in general.

The notion of parametric density was introduced by J. M. Wills [Wil1993] around 1993. For packings of centrally symmetric convex domains, the fundamental properties were established in U. Betke, M. Henk, and J. M. Wills [BHW1994]. The results were extended to packings of any convex domain in K. Böröczky Jr. and U. Schnell [BöS1998a]. For coverings, the notion we present is new. An alternative notion was introduced in U. Betke, M. Henk, and J. M. Wills [BHW1995b]. A detailed discussion about parametric density can be found in the upcoming monograph by U. Betke, M. Henk, and J. M. Wills [BHW?].

3.1. Planar Packings for Small ϱ

Let K be a convex domain, and let $K_0 = (K - K)/2$. As in Section 2.3, $\Pi(K)$ denotes a parallelogram of minimal area containing K, and we observe that $\Pi(K_0)$ can be chosen to be a translate of $\Pi(K)$. In addition, $H(K_0) = K_0$ if K_0 is a parallelogram, and $H(K_0)$ is an o-symmetric hexagon of minimal area containing K_0 otherwise. We call a packing $x_1 + K, \ldots, x_n + K$ a *sausage packing* if $x_1, \ldots, x_n$ are collinear, and we show that sausages tend to be optimal when the parameter ϱ is small.

Theorem 3.1.1. *Let K be a convex domain, and let $\varrho < A(H(K_0))/A(\Pi(K_0))$, where $K_0 = (K - K)/2$. Then any packing of n translates of K with maximal parametric density is some sausage provided that either n is large or K is centrally symmetric.*

Remark. If K is a triangle and $1/2 < \varrho < A(H(K_0))/A(\Pi(K_0)) = 3/4$ then the three translates of K by the vectors pointing to the vertices of K form a packing that is denser than any sausage packing.

We will verify that the optimal packings are clusters when $\varrho > A(H(K_0))/A(\Pi(K_0))$ and n is large (see Theorem 3.2.1). Therefore, the quotient $\varrho_c(K) = A(H(K_0))/A(\Pi(K_0))$ is called the *critical radius for packings* of K. Before proving Theorem 3.1.1, we provide the optimal bounds on the critical radius.

Lemma 3.1.2. *If K is a convex domain and $K_0 = (K - K)/2$ then*

$$\frac{3}{4} \le \frac{A(H(K_0))}{A(\Pi(K_0))} \le 1.$$

Equality holds in the upper and in the lower bound if and only if K is a parallelogram or K_0 is an affine regular hexagon, respectively.

Proof. As the upper bound is trivial, we consider the lower bound, where we may assume that K is not a parallelogram. Let T denote one of the triangles whose sides contain three nonneighbouring sides of $H(K_0)$, and we may assume that T is regular. Then $H(K_0)$ cuts off three regular triangles T_1, T_2, and T_3 of side lengths $a_1 \le a_2 \le a_3$, respectively; hence, T is of side length $a_1 + a_2 + a_3$. It follows that

$$\frac{A(T_1)}{A(H(K_0))} = \frac{a_1^2}{2a_1 a_2 + 2a_2 a_3 + 2a_3 a_1} \le \frac{1}{6}.$$

Therefore, T_1, $-T_1$ and $H(K_0)$ together form a parallelogram whose area is at most $(4/3)A(H(K_0))$. □

Proof of Theorem 3.1.1. Given $\varrho > 0$ and n, let $y_1 + K, \ldots, y_n + K$ be a sausage packing whose parametric density is maximal among sausage packings of n translates; hence, the line of $y_1, \ldots, y_n$ is parallel to a side of some parallelogram P of minimal area containing K. Writing S_n to denote the convex hull of $y_1, \ldots, y_n$, we have, by $A(\Pi(K)) = A(\Pi(K_0))$,

$$A(S_n + \varrho K) = (n - 1)A(\Pi(K_0))\varrho + A(K)\varrho^2. \tag{3.1}$$

Now let C_n be the convex hull of $x_1, \ldots, x_n$, where $x_1 + K, \ldots, x_n + K$ is a packing, and $A(C_n + \varrho K)$ is minimal under this condition. Since the mutual distances among $x_1, \ldots, x_n$ are at least two with respect to $\| \cdot \|_{K_0}$, the Corollary 2.3.3 form of the Oler inequality yields

$$A(C_n) + \frac{A(H(K_0))}{A(\Pi(K_0))} \cdot 2\,A(C_n, K_0) + A(H(K_0)) \ge n\,A(H(K_0)). \tag{3.2}$$

In particular,

$$\frac{\varrho\,A(\Pi(K_0))}{A(H(K_0))} \cdot A(C_n) + \varrho\,2A(C_n, K_0) \ge \varrho\,A(\Pi(K_0))(n - 1).$$

This inequality together with the formula (3.1), with the estimate

$$A(C_n) + \varrho\,2A(C_n, K) + \varrho^2 A(K) = A(C_n + \varrho K) \le A(S_n + \varrho K),$$

and with the formula $2\,A(C_n, K_0) = A(C_n, K) + A(C_n, -K)$ yields

$$\left(\frac{1}{\varrho} - \frac{A(\Pi(K_0))}{A(H(K_0))}\right) \cdot A(C_n) \le A(C_n, -K) - A(C_n, K). \qquad (3.3)$$

Now if K is centrally symmetric then the right-hand side is zero; hence, the optimal arrangement is a sausage for any n in this case. If K is not centrally symmetric then we estimate the right-hand side of (3.3) using the inequality (see Lemma 2.4.2)

$$|A(C_n, -K) - A(C_n, K)| \le 8\,r_K(C_n) \cdot A(K) \qquad (3.4)$$

and the left-hand side of (3.3) using $r_K(C_n)\,A(C_n, K) \le A(C_n)$ (see (A.18)). It follows that

$$\left(\frac{1}{\varrho} - \frac{A(\Pi(K_0))}{A(H(K_0))}\right) \cdot A(C_n, K)\,r_K(C_n) \le 8\,A(K) \cdot r_K(C_n). \qquad (3.5)$$

Here $A(C_n, K)$ tends to infinity with n, and hence $r_K(C_n) = 0$ holds for large n. In turn, we conclude Theorem 3.1.1. $\qquad\square$

We have seen in the remark after Theorem 3.1.1 that the condition "n is large" is necessary if $\varrho > 1/2$. Now we show that optimal sausages are optimal packings for any n if $\varrho < 3/7$.

Lemma 3.1.3. *Let K be a convex domain, and let $0 < \varrho < 3/7$. Then, for any $n \ge 2$, any packing of n translates of K with maximal parametric density is a sausage packing.*

Proof. We use the notation of the proof of Theorem 3.1.1, and we claim that

$$A(C_n, -K) - A(C_n, K) \le A(C_n). \qquad (3.6)$$

Let P be the parallelogram circumscribed around K with the following properties: Two sides are parallel to $x_1 x_2$, and these two sides contain points z_1 and z_2 of ∂K such that the other two sides of P are parallel to $z_1 z_2$. Then a translate of $-K$ is contained in P, and, in turn, a translate of P is contained in $C_n + K$. Therefore, $A(C_n, C_n + K) \ge A(C_n, -K)$, which is equivalent with (3.6).

Now $A(H(K_0))/A(\Pi(K_0)) \ge 3/4$ (see Lemma 3.1.2) yields that $1/\varrho - A(\Pi(K_0))/A(H(K_0)) > 1$ in (3.3). Thus $A(C_n) = 0$, or in other words, the packing is a sausage. $\qquad\square$

Open Problem.
(i) Is there an absolute constant N such that, for any convex domain K, if $\varrho < A(H(K_0))/A(\Pi(K_0))$ and $n \geq N$ then any optimal packing is a sausage?

Comments. The paper by U. Betke, M. Henk, and J. M. Wills [BHW1994] verified Lemma 3.1.2 and Theorem 3.1.1 for packings of centrally symmetric convex domains. The case of not necessarily symmetric convex domains is due to K. Böröczky Jr. and U. Schnell [BöS1998a].

3.2. Planar Packings for Reasonably Large ϱ

Theorem 3.1.1 says that if ϱ is small then optimal packings are eventually sausages. Now we prove the converse statement; namely, suitable clusters are optimal for larger ϱ. In this section we fix a convex domain K, and C_n always denotes the convex hull of $x_1, \ldots, x_n$, where $x_1 + K, \ldots, x_n + K$ form a packing.

Theorem 3.2.1. *Let K be a convex domain, and let $\varrho > A(H(K_0))/ A(\Pi(K_0))$, where $K_0 = (K - K)/2$. If n is large, and C_n corresponds to a packing of n translates of K with maximal parametric density, then (possibly after translation)*

$$\alpha_{\varrho,K} \sqrt{n} \cdot K \subset C_n \subset \beta_{\varrho,K} \sqrt{n} \cdot K,$$

where the positive constants $\alpha_{\varrho,K}$ and $\beta_{\varrho,K}$ depend on K and ϱ.

Remark. It follows by the proof of Theorem 3.2.1 (see (3.11) and (3.12)) and the Bonnesen inequality (A.17) that we may assume

$$\lim_{\varrho \to \infty} \alpha_{\varrho,K} = \lim_{\varrho \to \infty} \beta_{\varrho,K} = \sqrt{\frac{A(H(K_0))}{A(K)}}.$$

In particular, if ϱ and n are large then the convex hull of the optimal packing is close to be homothetic to K.

Let us summarize Theorems 3.1.1 and 3.2.1:

Corollary 3.2.2. *Let K be a convex domain, and let $\varrho > 0$ be the parameter. Then, for large n, the packing with maximal parametric density is some* *sausage packing if* $\varrho < A(H(K_0))/A(\Pi(K_0))$ *and is cluster type packing if*

$$\varrho > A(H(K_0))/A(\Pi(K_0)), \quad \text{where} \quad K_0 = (K - K)/2.$$

We recall that $\varrho_c(K) = A(H(K_0))/A(\Pi(K_0))$ is the critical radius for K. Let us fix a $\varrho > 0$, and we choose C_n in a way that $A(C_n + \varrho K)$ is minimal for given n. We claim that

$$\lim_{n \to \infty} \frac{A(C_n + \varrho K)}{n} = \begin{cases} \varrho \, A(\Pi(K_0)) & \text{if } \varrho \leq \varrho_c(K); \\ A(H(K_0)) & \text{if } \varrho \geq \varrho_c(K). \end{cases} \tag{3.7}$$

Here the case of small ϱ follows readily from the optimality of the sausage (see (3.1)), which, in turn, yields that $\liminf_{n \to \infty} A(C_n + \varrho K)/n \geq A(H(K_0))$ if $\varrho \geq \varrho_c(K)$. However, taking large parts of the lattice tiling of $H(K_0)$ shows that $\limsup_{n \to \infty} A(C_n + \varrho K)/n \leq A(H(K_0))$, completing the proof of (3.7). We note that $A(H(K_0)) = \Delta_T(K)$ according to Theorem 2.2.1.

Another way to interpret (3.7) is that there exists an asymptotic parametric density $\delta(K, \varrho) = \lim_{n \to \infty} n \, A(K)/A(C_n + \varrho K)$ for any K and ϱ; namely,

$$\delta(K, \varrho) = \begin{cases} \dfrac{A(K)}{\varrho \, A(\Pi(K_0))} & \text{if } \varrho \leq \varrho_c(K); \\ \delta_T(K) & \text{if } \varrho \geq \varrho_c(K). \end{cases} \tag{3.8}$$

Proof of Theorem 3.2.1. We write $c_1, c_2, \ldots$ to denote positive constants that depend on K and ϱ. Let $x_1 + K, \ldots, x_n + K$ be some nonoverlapping translates of K such that C_n is the convex hull of $x_1, \ldots, x_n$. To shorten the formulae, we write H and Π to denote $H(K_0)$ and $\Pi(K_0)$, respectively.

We note that $\widetilde{C}_n = (\sqrt{n \, A(H)/A(K)} + c_1) K$ contains some n points of the lattice Λ such that $\Lambda + H$ is a tiling; hence, $A(C_n + \varrho K) \leq A(\widetilde{C}_n + \varrho K)$. It follows that

$$A(C_n) + 2\varrho \, A(C_n, K) \leq n \, A(H) + 2(\varrho + c_2)\sqrt{A(H) \, A(K)} \cdot \sqrt{n}, \tag{3.9}$$

which, in turn, yields $r_K(C_n) \leq c_3 \sqrt{n}$. However, $\|x_i - x_j\|_{K_0} \geq 2$ holds for $i \neq j$; thus, the form (3.2) of the Oler inequality for C_n and K_0 leads to

$$A(C_n) + \frac{A(H)}{A(\Pi)} \cdot (A(C_n, K) + A(C_n, -K)) \geq n \, A(H) - A(H).$$

Since $|A(C_n, -K) - A(C_n, K)| \leq 8 \, r_K(C_n) \, A(K) \leq c_4 \sqrt{n} \, A(K)$ according to (3.4), we conclude

$$A(C_n) + 2 \frac{A(H)}{A(\Pi)} \cdot A(C_n, K) \geq n \, A(H) - c_5 \sqrt{A(H) \, A(K)} \cdot \sqrt{n}. \tag{3.10}$$

Now comparing (3.9) and (3.10) yields

$$A(C_n, K) \leq \frac{\varrho + c_6}{\varrho - \frac{A(H)}{A(\Pi)}} \cdot \sqrt{A(H)\,A(K)} \cdot \sqrt{n}, \qquad (3.11)$$

$$A(C_n) \sim n\,A(H). \qquad (3.12)$$

As C_n contains a segment of length $R_{K_0}(C_n)$ with respect to $\|\cdot\|_{K_0}$, we deduce that $R_{K_0}(C_n) \leq c_7\sqrt{n}$. Therefore, $r_{K_0}(C_n) \geq c_8\sqrt{n}$ follows by Lemma 2.1.6, which, in turn, yields Theorem 3.2.1. $\square$

Open Problem.
  (i) Using the notation of Theorem 3.2.1, does $C_n/\sqrt{n}$ tend to some polygon as n tends to infinity?
      If K is a parallelogram then $C_n/\sqrt{n}$ tends to K (see Theorem 3.4.1). In addition, if K is o-symmetric and the centres of the n translates are part of some packing lattice then the possible limits of $C_n/\sqrt{n}$ are polygons (see Theorem 3.4.2).

Comments. For packings of centrally symmetric convex domains, the density estimate (3.8) was proved in U. Betke, M. Henk, and J. M. Wills [BHW1994]. This was extended to packings of any convex domain (see Theorem 3.2.1) by K. Böröczky Jr. and U. Schnell [BöS1998a].

   Theorem 3.2.1 and its proof suggest for any convex domain K that if $\varrho > \varrho_c(K)$ and n is large then the structure of the optimal packing is intimately related to the densest lattice packing of K. Yet assuming that ϱ is sufficiently large and K is strictly convex, the optimal packing is not part of a densest lattice packing for any large n according to A. Schürmann [Sch2002].

3.3. Parametric Density for Coverings

In this section we present a theory for parametric density for coverings. Let K be a convex domain. Given $\theta \geq 0$, let D_n be a compact convex set covered by n translates of K such that $A(D_n + \theta K)$ is maximal and define

$$\Theta(K, \theta) = \limsup_{n \to \infty} \frac{A(D_n + \theta K)}{n}.$$

In particular, $\Theta(K, \theta)$ is increasing and continuous as a function of θ. We have seen in Theorem 2.5.1 that if $\theta = 0$ then

$$\Theta(K, 0) = \Theta_T(K) = \lim_{n \to \infty} \frac{A(D_n + \theta K)}{n}. \qquad (3.13)$$

We define the *critical radius for coverings* of K to be

$$\theta_c(K) = \max\{\theta : \Theta(K, \theta) = \Theta_T(K)\}.$$

It satisfies $0 \leq \theta_c(K) \leq 1/2$ according to Lemma 3.3.2.

Theorem 3.3.1. *Let K be a convex domain. Given $\theta \geq 0$, let D_n be a convex domain covered by n translates of K such that $A(D_n + \theta K)$ is maximal. Then, as n tends to infinity,*

(i) $r_K(D_n)$ *tends to infinity*   *if* $\theta < \theta_c(K)$;
   $r_K(D_n)$ *stays bounded*   *if* $\theta > \theta_c(K)$;
(ii) $\lim_{n \to \infty} \frac{1}{n} A(D_n + \theta K) = \Theta(K, \theta)$.

Proof. Since $\Theta(K, \theta)$ is increasing in θ, we deduce by (3.13) that

$$\liminf_{n \to \infty} \frac{A(D_n + \theta K)}{n} \geq \Theta_T(K) \tag{3.14}$$

holds for any $\theta \geq 0$. We write θ_c to abbreviate $\theta_c(K)$.

First, let $0 \leq \theta < \theta_c$, and hence the definition of the critical radius and (3.14) yield

$$\lim_{n \to \infty} \frac{A(D_n + \theta K)}{n} = \Theta_T(K). \tag{3.15}$$

We suppose that $\{D_{n_k}\}$ is a subsequence such that $r_K(D_{n_k})$ is bounded, say, by a constant r_0, and we seek a contradiction. Naturally, $A(D_{n_k}, K) > 2 A(K)$ can be assumed. Since $A(D_{n_k}) \leq 2r_0 A(D_{n_k}, K)$ holds according to (A.18), we deduce that

$$\frac{A(D_{n_k} + \theta_c K)}{A(D_{n_k} + \theta K)} = 1 + \frac{(\theta_c - \theta) \cdot 2 A(D_{n_k}, K)}{A(D_{n_k}) + A(K) + \theta \cdot 2 A(D_{n_k}, K)}$$

$$\geq 1 + \frac{\theta_c - \theta}{r_0 + 1 + \theta}.$$

Thus (3.15) yields that $\Theta(K, \theta_c) > \Theta_T(K)$. This is absurd, and hence $r(D_n)$ tends to infinity. In addition, we conclude (ii) by (3.15), not only if $\theta < \theta_c$ but if $\theta = \theta_c$ as well.

Therefore, let $\theta > \theta_c(K)$. First, we claim that if $\{D_{n_k}\}$ is a subsequence such that $r_K(D_{n_k})$ tends to infinity then

$$\lim_{n_k \to \infty} \frac{A(D_{n_k} + \theta K)}{n_k} = \Theta_T(K). \tag{3.16}$$

Let $\varepsilon > 0$, and let R be a parallelogram satisfying $A(R)/A(R - K) > 1 - \varepsilon$. We fix a large n_k and a covering $x_1 + K, \ldots, x_{n_k} + K$ of D_{n_k}. For any

$y + (R - K) \subset D_{n_k}$, we consider the $m(y)$ points out of $x_1, \ldots, x_{n_k}$ that lie in $y + (R - K)$, and hence the corresponding translates of K cover $y + R$. Now if $r_K(D_{n_k})$ is large then $A(D_{n_k} \ominus (R - K))/A(D_{n_k} + \theta K)$ is at least $1 - \varepsilon$; thus, there exists a y such that $m = m(y)$ satisfies $A(R - K)/m \geq (1 - \varepsilon) \cdot A(D_{n_k} + \theta K)/n_k$. Since R induces a periodic covering, we deduce that

$$\Theta_T(K) \geq \frac{A(R)}{m} \geq (1 - \varepsilon)^2 \cdot \frac{A(D_{n_k} + \theta K)}{n_k}.$$

Therefore, (3.14) yields the claim (3.16).

If (ii) holds for some $\theta > \theta_c(K)$ then $\Theta(K, \theta) > \Theta_T(K)$ and (3.16) yield that $r_K(D_n)$ is bounded. Therefore, it is sufficient to verify (ii). In turn, (ii) readily follows from the following claim: For any small $\varepsilon > 0$, there exist a parallelogram R that is covered by some N translates of K on the one hand and a segment $s \subset K$ that is parallel to a side of R on the other hand such that

$$\frac{A(R + \theta s)}{N} \geq (1 - O(\varepsilon)) \cdot \Theta(K, \theta), \tag{3.17}$$

where the implied constant in $O(\cdot)$ depends only on K and θ.

Let $\{D_{n_k}\}$ be a subsequence such that $A(D_{n_k} + \theta K)/n_k \geq (1 - \varepsilon) \Theta(K, \theta)$ for any n_k. Then $r_K(D_{n_k} + \theta K)$ is at most r_1 for a constant r_1 according to (3.16), and hence $D_{n_k} + \theta K$ is contained in some parallel strip Σ_{n_k} such that $r_K(\Sigma_{n_k}) = 2r_1$ (see Lemma 2.1.5). For a large n_k (which we specify later), we consider a covering $X_{n_k} + K$ of D_{n_k}, where X_{n_k} is of cardinality n_k. We define s to be a segment in K with the property that the two supporting lines of K parallel to Σ_{n_k} touch at the endpoints of s. In addition, let s_0 be a segment parallel to Σ_{n_k} such that the parallelogram $P = s_0 + s$ is circumscribed around K. We decompose Σ_{n_k} into the parallelograms $R_i, i \in \mathbb{Z}$, such that each is a translate of $\lceil 1/\varepsilon \rceil s_0 + 2r_1 s$, and R_i and R_{i+1} are consecutive tiles for any i. We fix some m such that $2r_1/m < \varepsilon$ and may assume that the parallelograms R_i that intersect D_{n_k} are the ones with $-m \leq i \leq l$. We write $\mathcal{R}$ to denote the union of R_i such that $-m \leq i \leq 0$ or $l - m \leq i \leq l$, and hence the area of $\mathcal{R}$ is $O(m/\varepsilon)$. Thus $A(D_{n_k} + \theta K) \sim n_k \Theta(K, \theta)$ yields that $A(\mathcal{R}) \leq \varepsilon A(D_{n_k} + \theta K)$ for large n_k, and we fix some n_k satisfying this condition. Therefore, there exists some $R_j, 0 \leq j \leq N$, such that

$$\frac{A((D_{n_k} + \theta K) \cap R_j)}{\#(X_{n_k} \cap R_j)} \geq (1 - O(\varepsilon)) \cdot \frac{A(D_{n_k} + \theta K)}{n_k}.$$

Let γ be maximal with the property that $(D_{n_k} + \theta K) \cap R_j$ contains a translate of γs; hence $\gamma > \Theta(K, \theta)/(2A(P)) \geq \Theta(K, \theta)/(4A(K))$. It follows by the choice of m that a translate of $\lceil 1/\varepsilon \rceil s_0 + \gamma_0 s$ is contained in $D_{n_k} \cap R_j$, where $\gamma_0 = (1 - O(\varepsilon)) \gamma - \theta$. Thus, we define $N = \#(X_{n_k} \cap R_j)$ and

$$R = \left(\left\lceil \tfrac{1}{\varepsilon} \right\rceil - 1 \right) \cdot s_0 + \gamma_0 s.$$

Then R is covered by some N translates of K, and

$$\frac{A(R + \theta s)}{N} \geq (1 - O(\varepsilon)) \cdot \frac{A((D_{n_k} + \theta K) \cap R_j)}{N}$$

$$\geq (1 - O(\varepsilon)) \cdot \frac{A(D_{n_k} + \theta K)}{n_k} \geq (1 - O(\varepsilon)) \cdot \Theta(K, \theta).$$

Therefore, we conclude the claim (3.17) and, in turn, Theorem 3.3.1. □

Given a convex domain K and $\theta > \theta_c(K)$, let us discuss the asymptotic structure of the optimal compact convex set D_n that can be covered by n translates of K. If K is a parallelogram then $\theta_c(K) = 0$, and $\lim_{n \to \infty} r_K(D_n) = 1$ when $0 < \theta < 1$, and D_n is a segment when $\theta > 70$ for large n (see Theorem 3.4.7). In addition, if K is a regular hexagon then, for any $\theta \geq \theta_c(K)$, the optimal D_n contains a translate of $K/3$ when n is large (see Theorem 3.4.8).

Lemma 3.3.2. *If K is a convex domain then $0 \leq \theta_c(K) \leq 1/2$.*

Proof. Given $1/2 < \theta < 1$, we claim that

$$\Theta(K, \theta) \geq 2\theta \cdot \Theta_T(K). \tag{3.18}$$

We may assume that K is strictly convex. Writing H to denote a p-hexagon of maximal area in K, it follows that H is a hexagon. By definition there exist a side s of H and a vector u such that $s + u$ is a side as well. Let P_0 be a parallelogram circumscribed around H such that two opposite sides are parallel to u, and the other two sides contain s and $s + u$, respectively; hence, $A(H) = (A(P) + A(P_0))/2$. We define D'_n to be the union of $P + i \cdot u$, $i = 1, \ldots, n$. Then $D'_n + \theta K$ contains two nonoverlapping parallelograms where the area of one of them is $n A(P)$, whereas the area of the other is $n A(P_0) \cdot \theta$. Therefore, the p-hexagon bound (see Theorem 2.2.3) yields

$$A(D'_n + \theta K) \geq n \cdot (A(P) + A(P_0)\theta) \geq n2\theta \cdot A(H) \geq n2\theta \cdot \Theta_T(K),$$

completing the proof of (3.18). Therefore, $\theta_c(K) \leq 1/2$. □

Open Problem. In the problems to follow, K is a convex domain, $\theta > 0$, and D_n is a compact convex set covered by n translates of K such that $A(D_n + \theta K)$ is maximal.

(i) Does $D_n/\sqrt{n}$ tend to some polygon if $0 < \theta < \theta_c(K)$?
It holds for finite lattice coverings by unit discs (see Theorem 3.4.10).

(ii) Determine the maximum of $\theta_c(K)$ over all convex domains K.

(iii) Assuming that K is centrally symmetric, are the centres of the n translates of K covering D_n aligned if $\theta > \theta_c(K)$?
It holds if K is a parallelogram and n is large (see Theorem 3.4.7).

Comments. Let K be an o-symmetric domain. U. Betke, M. Henk, and J. M. Wills [BHW1995b] presented an alternative approach to parametric density for coverings by translates of a convex domain K: They say that $x_1 + K, \ldots, x_n + K$ form a *covering configuration* if the translates cover $C_n = \mathrm{conv}\{x_1, \ldots, x_n\}$. For fixed n and for $\theta \geq 0$, the goal is to find the maximum of $A(C_n + \theta K)$ among covering configurations. If K is the unit disc B^2 and $\theta_0 = 0.2455\ldots$ then [BHW1995b] verifies that the asymptotic density is $\vartheta(B^2)$ if $\theta < \theta_0$, and larger than $\vartheta(B^2)$ if $\theta > \theta_0$ (compare with Theorem 3.4.9).

The paper by M. Meyer and U. Schnell [MeS2001] considers finite lattice coverings in the following sense: Let Λ be a planar lattice. A lattice polygon P is called a "covering polygon" if P is covered by the translates $x + K$ for all $x \in P \cap \Lambda$. Given n, we write P_n to denote the covering polygon containing n lattice points and maximizing $A(P_n + \theta K)$. Then P_n is a segment for large θ, and $P_n/\sqrt{n}$ tends to some polygon for small θ. We note that there may exist an intermediate phase when P_n is a "double sausage."

3.4. Examples

In this section we do not provide detailed proofs because the aim is to give a feeling about the essential properties of parametric density via some typical examples.

3.4.1. Packings of Parallelograms

We have an almost complete description of the optimal packing of a parallelogram P for any $\varrho > 0$.

Theorem 3.4.1. *If P is a parallelogram then $\varrho_c(P) = 1$. In addition, given $\varrho > 1$, let C_n be the convex hull of the centres n nonoverlapping translates P such that $A(C_n + \varrho P)$ is minimal.*

(i) If $n = k^2$ then C_n is a translate of $(k-1)P$;

(ii) $\lim_{n \to \infty} \frac{1}{\sqrt{n}} \cdot C_n = P$.

Proof. Since both sausages and clusters occur as optimal arrangements if $\varrho = 1$, we conclude $\varrho_c(P) = 1$. Writing $Q_n = C_n + P$, we have $A(Q_n) \geq n A(P)$ and

$$A(C_n + \varrho\, P) = A(Q_n) + 2(\varrho - 1)\, A(Q_n, P) + (\varrho - 1)^2 A(P). \quad (3.19)$$

Some elementary considerations yield integers m and l such that

$$n \leq l \cdot m < n + O\left(n^{\frac{1}{4}}\right) \quad \text{and} \quad l, m = \sqrt{n} + O\left(n^{\frac{1}{4}}\right). \quad (3.20)$$

Thus, enlarging one side of P m times and the other side l times, we obtain a parallelogram P_n that contains n nonoverlapping translates of P and satisfies

$$A(P_n) \leq A(Q_n) + O\left(n^{\frac{1}{4}}\right) \quad \text{and} \quad A(P_n, P) \leq \sqrt{A(Q_n)\, A(P)} + O\left(n^{\frac{1}{4}}\right).$$

Since $A(C_n + \varrho P) \leq A(P_n + (\varrho - 1)P)$, we deduce that

$$A(Q_n, P) \leq \left(1 + O\left(n^{-\frac{1}{4}}\right)\right) \sqrt{A(Q_n) \cdot A(P)}.$$

Therefore the Bonnesen inequality (A.17) yields (ii). If $n = k^2$ then we choose $m = l = k$ and obtain that Q_n is homothetic to kP. $\qquad\square$

Comments. Theorem 3.4.1 was observed by K. Böröczky, Jr. and U. Schnell [BöS1998a].

3.4.2. Finite Lattice Packings and the Wulff-Shape

We have seen that for finite translative packings if the parameter ϱ is larger than the "critical radius" then the ratio of the circumradius and the inradius of the optimal convex hull stays bounded as the number of translates tend to infinity (see Theorem 3.2.1). The aim of this section is to make this statement more precise for *finite lattice packings* of a given o-symmetric convex domain, namely, when the centres of the translates are points of a packing lattice.

We write S^1 to denote the Euclidean unit circle and use $|\cdot|$ to denote Euclidean length of segments. Given an o-symmetric convex domain K, the definition of $\Pi(K)$ and Theorem 2.2.1 yield

$$A(\Pi(K)) = \min_{u \in S^1} 2\, h_K(u) \cdot |u^{\perp} \cap K|; \quad (3.21)$$

$$A(H(K)) = 8\, A(T(K)). \quad (3.22)$$

According to Theorem 2.2.1 (but it is easy to verify directly), the minimal determinant of a packing lattice for K is $A(H(K))$. Let Λ be a packing lattice for K of determinant $A(H(K))$, and let $\varrho > A(H(K))/A(\Pi(K))$. We define the *Wulff-shape* associated to K, Λ, and ϱ by

$$\widetilde{W}_{K,\Lambda,\varrho} = \bigcap_{\substack{u \in S^1 \\ u^\perp \cap \Lambda \neq \{o\}}} \left\{ x : \langle x, u \rangle \leq \varrho\, h_K(u) - \frac{\det \Lambda}{2 \det(u^\perp \cap \Lambda)} \right\}. \qquad (3.23)$$

We note that $\widetilde{W}_{K,\Lambda,\varrho}$ is an o-symmetric convex polygon, which is actually a parallelogram or a hexagon if $\varrho = 1$ (see Proposition 3.4.4), and we define $W_{K,\Lambda,\varrho}$ to be the homothetic copy of $\widetilde{W}_{K,\Lambda,\varrho}$ of area $A(H(K))$.

Theorem 3.4.2. *Let K be an o-symmetric convex domain, and let $\varrho > 0$. For any n, we write C_n to denote the convex hull of n points of a packing lattice such that $A(C_n + \varrho K)$ is minimal.*

(i) If $\varrho < A(H(K))/A(\Pi(K))$ then C_n is a segment.

(ii) If $\varrho > A(H(K))/A(\Pi(K))$ then the circumradius of $C_n/\sqrt{n}$ is bounded, and any convergent subsequence $C_{n'}/\sqrt{n'}$ satisfies

$$\lim_{n' \to \infty} \frac{1}{\sqrt{n'}} C_{n'} = W_{K,\Lambda,\varrho},$$

where Λ is some packing lattice of determinant $A(H(K))$.

Remarks. Since $R_K(C_n)/\sqrt{n}$ is bounded, any subsequence of $C_n/\sqrt{n}$ contains a convergent subsequence according to the Blaschke selection theorem (Theorem A.3.1).

If there exist inequivalent packing lattices of minimal determinant (say when K is a square) then the proof of Theorem 3.4.2 provides some additional restrictions on Λ; namely, if $W_{K,\Lambda,\varrho}$ is a possible asymptotic shape and Λ' is any packing lattice of minimal determinant then $A(\widetilde{W}_{K,\Lambda,\varrho}) \leq A(\widetilde{W}_{K,\Lambda',\varrho})$.

It follows by the definition of the Wulff-shape (see (3.23)) that if ϱ and n are large then the optimal packing is close to being homothetic to K.

If Λ is a lattice then a polygon whose vertices are lattice points is called a *lattice polygon*. The proof of Theorem 3.4.2 is based on the following celebrated formula:

Lemma 3.4.3 (Pick formula). *If Λ is a planar lattice and P is a lattice polygon then*

$$\#(\Lambda \cap P) = \frac{A(P)}{\det \Lambda} + \frac{\#(\Lambda \cap \partial P)}{2} + 1.$$

Proof. We use induction on the number of lattice points in P: If this number is the smallest possible, namely, if it is three, then a translate of P is the "half" of some fundamental parallelogram, and hence $A(P) = \det \Lambda/2$. However, if P contains $k \geq 4$ lattice points then it can be dissected into lattice triangles such that each contains less than k lattice points, and we use the induction hypothesis. $\quad\square$

If Λ is a packing lattice for K, and P is a lattice polygon containing n lattice points then the Pick formula and $\det \Lambda \geq A(H(K))$ yield the Corollary 2.3.3 form of the Oler inequality:

$$A(P) + \frac{A(H(K))}{A(\Pi(K))} \cdot 2\,A(P, K) \geq (n - 1)\,A(H(K)). \tag{3.24}$$

Since the rest of the argument is rather long, we sketch only the basic ideas of the proof of Theorem 3.4.2.

If $\varrho < A(H(K))/A(\Pi(K))$ then we consider a segment S_n that contains the centre of n nonoverlapping translates of K and satisfies

$$A(S_n + \varrho K) = (n - 1)\,\varrho\,A(\Pi(K)) + \varrho^2\,A(K).$$

Since S_n corresponds to a lattice packing and $A(C_n + \varrho K) \leq A(S_n + \varrho K)$, formula (3.24) yields that C_n is a segment.

Therefore, let $\varrho > A(H(K))/A(\Pi(K))$. For any polygon P, we write $\mathcal{E}(P)$ to denote the family of edges of P and, if e is an edge, then u_e denotes the corresponding exterior unit normal. The role of the Wulff-shape is shown by the following argument: Let us only consider finite packings with respect to a fixed packing lattice Λ of minimal determinant $A(H(K))$, and let P_n denote some lattice polygon containing n lattice points. According to the Pick formula, minimizing $A(P_n + \varrho K)$ for given n and ϱ is equivalent to minimizing

$$\sum_{e \in \mathcal{E}(P_n)} \left(\varrho\, h_K(u_e) - \frac{\det \Lambda}{2 \det(u_e^{\perp} \cap \Lambda)} \right) \cdot |e|. \tag{3.25}$$

We will see that the interesting case is when the area of P_n is essentially $n\,A(H(K))$; therefore, we obtain an isoperimetric type problem. The Minkowski inequality for mixed areas will yield that the solution is asymptotically homothetic to the Wulff-shape $W_{K,\Lambda,\varrho}$.

Let us start the actual proof of Theorem 3.4.2 in the case $\varrho > A(H(K))/A(\Pi(K))$. Since suitable clusters show $A(C_n + \varrho K) \leq n\,A(H(K)) + O(\sqrt{n})$, we deduce via (3.24) the estimates

$$c_1 < \frac{r_K(C_n)}{\sqrt{n}} \leq \frac{R_K(C_n)}{\sqrt{n}} < c_2 \tag{3.26}$$

for some positive constants c_1 and c_2.

Next we state some fundamental properties of the Wulff-shape.

Proposition 3.4.4. *Let* K *be an o-symmetric convex domain, and let* Λ *be a packing lattice of minimal determinant. If* $\varrho > A(H(K))/A(\Pi(K))$ *then* $\widetilde{W}_{K,\Lambda,\varrho}$ *is an o-symmetric convex polygon that is contained in* ϱK, *and each of its sides is parallel to some lattice vector of* Λ. *In addition,* $\widetilde{W}_{K,\Lambda,\varrho}$ *has at most six sides if* $\varrho \leq 1$.

Given a packing lattice Λ of minimal determinant $A(H(K))$, the Wulff-shape may not be a lattice polygon, and possibly no dilated copy of it is a lattice polygon with respect to Λ. Thus, we construct lattice polygons that are asymptotically homothetic to the Wulff-shape. We write $e_1, \ldots, e_m$ to denote the edges of $W_{K,\Lambda,\varrho}$ and use u_i to denote the exterior unit normal to e_i, $i = 1, \ldots, m$.

Proposition 3.4.5. *Let* K *be an o-symmetric convex domain, and let* Λ *be a packing lattice of minimal determinant* $A(H(K))$. *For large* n, *there exists a lattice polytope* Q_n *for* Λ *such that* Q_n *contains* n *lattice points, and*

$$A(Q_n + \varrho K) = A(Q_n) + \sqrt{n} \cdot \sum_{i=1}^{m} \varrho \, h_K(u_i) \cdot |e_i| + o\left(\sqrt{n}\right) ;$$

$$n \det \Lambda = A(Q_n) + \sqrt{n} \cdot \sum_{i=1}^{m} \frac{\det \Lambda}{2 \det(u_i^\perp \cap \Lambda)} \cdot |e_i| + o\left(\sqrt{n}\right) .$$

Proof (Sketch). We define ω_n to be minimal with the property that $\omega_n W_{K,\Lambda,\varrho}$ contains n lattice points. We cut off a corner of $\omega_n W_{K,\Lambda,\varrho}$ of size $O(\sqrt[4]{n})$ in a way that the remaining polygon contains at least n and at most $n + O(\sqrt[4]{n})$ lattice points, and we define Q_n to be the convex hull of the lattice points remaining inside. Therefore, the formulae of Proposition 3.4.5 are the consequences of the Pick formula. $\square$

We are now ready to verify Theorem 3.4.2. Let us assume that $C_{n'}/\sqrt{n'}$ tends to the compact convex set M, and let $\Lambda_{n'}$ be the packing lattice that corresponds to $C_{n'}$. We may assume that $\Lambda_{n'}$ has a basis $v_{n'}$, $w_{n'}$ such that $\|v_{n'}\|_K = 2$ and $\|w_{n'}\|_K \leq 4$. Therefore, we may also assume that $v_{n'}$ tends to some v and $w_{n'}$ tends to some w, where v and w span a packing lattice Λ for K. Since $A(C_{n'} + \varrho K) \sim n' \det \Lambda_{n'}$, the optimality of $C_{n'}$ yields that $\det \Lambda = A(H(K))$. In particular,

$$A(M) = A(H(K)). \tag{3.27}$$

Let $\Phi_{n'}$ be the linear map satisfying $\Phi_{n'}v_{n'} = v$ and $\Phi_{n'}w_{n'} = w$; hence, $\Phi_{n'}$ tends to the identity, and $\det \Phi_{n'} \leq 1$. Therefore, $\widetilde{C}_{n'} = \Phi_{n'}C_{n'}$ is a lattice polygon for Λ containing n' lattice points, and $\widetilde{C}_{n'}/\sqrt{n'}$ tends to M. Writing $\mathcal{E}_{n'}$ to denote the family of edges of $\widetilde{C}_{n'}$ and u_e to denote the exterior unit normal to an edge e of $\widetilde{C}_{n'}$, we deduce by Proposition 3.4.5

$$A(\widetilde{C}_{n'}) + \sum_{e \in \mathcal{E}_{n'}} \varrho\, h_K(u_e) \leq A(C_{n'} + \varrho K) + o\left(\sqrt{n}\right)$$

$$\leq A(Q_n) + \sqrt{n} \sum_{i=1}^{m} \varrho\, h_K(u_i) \cdot |e_i| + o\left(\sqrt{n}\right).$$

However, the Pick formula and Proposition 3.4.5 yield

$$A(\widetilde{C}_{n'}) + \sum_{e \in \mathcal{E}_{n'}} \frac{\det \Lambda}{2 \det(u_e^\perp \cap \Lambda)} \cdot |e| = A(Q_n) + \sqrt{n} \sum_{i=1}^{m} \frac{\det \Lambda}{2 \det(u_i^\perp \cap \Lambda)} |e_i|$$

$$+ o\left(\sqrt{n}\right).$$

We conclude $A(M, \widetilde{W}_{K,\Lambda,\varrho}) \leq A(W_{K,\Lambda,\varrho}, \widetilde{W}_{K,\Lambda,\varrho})$ by taking the limit as n' tends to infinity. Since $A(M) = A(W_{K,\Lambda,\varrho})$, the Minkowski inequality (A.16) and its equality case implies that M is a translate of $W_{K,\Lambda,\varrho}$, which completes the proof of Theorem 3.4.2.

Remark 3.4.6. For finite packings of the unit Euclidean disc, $H(B^2)$ is a regular hexagon and $\Pi(B^2)$ is a square; hence, the critical radius is

$$\frac{A(H(B^2))}{A(\Pi(B^2))} = \frac{\sqrt{3}}{2} = 0.8660\ldots.$$

It is easy to see that the hexagonal lattice Λ_2 is the unique packing lattice of minimal determinant for B^2 up to congruence, and $\det \Lambda_2 = \sqrt{12}$. Some simple calculations yield that

the asymptotic shape $W_{B^2,\Lambda_2,\varrho}$ is the regular hexagon of area $\sqrt{12}$ if

$$\frac{\sqrt{3}}{2} < \varrho \leq \frac{3}{2} + \sqrt{3}.$$

This fact is in accordance with the Wegner theorem (Theorem 4.3.1) for finite packings of unit discs without any restriction; say, if the convex hull of $n = 6\binom{k}{2} + 1$ nonoverlapping unit discs is of minimal area then the convex hull of the centres is a regular hexagon of side length $2(k - 1)$.

Comments. Theorem 3.4.2 is proved in U. Betke and K. Böröczky, Jr. [BeB?], whose proof works in any dimension (see Theorem 10.11.2). The name Wulff-shape originates from crystallography: Given an o-symmetric convex domain K and a packing lattice Λ, we define the function φ for any unit vector u satisfying $u^{\perp} \cap \Lambda \neq \{o\}$ by

$$\varphi(u) = \varrho\, h_K(u) - \frac{\det \Lambda}{2 \det(u^{\perp} \cap \Lambda)}.$$

Then φ resembles the so-called Gibbs free energy for some crystals, and our task is to apply the so-called Gibbs–Curie principle; namely, given the area, determine the convex domain such that the total Gibbs free energy of its boundary (see (3.25)) is minimal. The fact that the Wulff-shape (3.23) is the optimal shape is known as the Wulff theorem in crystallography. For more detailed discussion, see the comments for Section 10.11 or U. Betke, M. Henk and J. M. Wills [BHW?].

Proposition 3.4.5, namely, the fact that the Wulff-shape is a polygon, is proved by J. M. Wills [Wil1996, Wil1997] if K is a circular disc and by U. Schnell [Sch2000a] if K is any centrally symmetric convex domain.

The Pick formula (Lemma 3.4.3) is due to G. Pick [Pic1899]. (See say P. Gritzman and J. M. Wills [GrW1993a] for a survey on relatives of the Pick formula.) A. Schürmann [Sch2000c] determined the densest finite lattice packings of the unit disc for any values of ϱ and n and verified that the centres are always part of the hexagonal lattice. A. Schürmann [Sch2000d] also showed for certain o-symmetric octagon and ϱ that the centres in the densest lattice packing of n translates are not part of the packing lattice of minimal determinant if n is large.

A related approach is introduced by J. M. Wills, who considers the so-called density deviation (see the comments to Section 10.11).

3.4.3. Coverings by Parallelograms

Coverings by translates of a parallelogram P exhibit various interesting phenomena. Since $\Theta(P, \theta) > A(P)$ for $\theta > 0$, we have $\theta_c(P) = 0$.

Theorem 3.4.7. *Given a parallelogram P and $\theta > 0$, let n translates of P cover the compact convex set D_n such that $A(D_n + \theta P)$ is maximal.*

(i) If $\theta < 1$ then $\Theta(P, \theta) = (1 + \theta) A(P)$ and $\lim_{n \to \infty} r_P(D_n) = 1$.
(ii) If $\theta > 1$ then $\Theta(P, \theta) = 2\theta A(P)$ and $\lim_{n \to \infty} r_P(D_n) = 0$.
(iii) If $\theta > 70$ then D_n is a segment for large n.

Proof. We assume that P is the unit square $[0, 1]^2$, and we use $|s|$ to denote the Euclidean length of a segment s. Given $\theta > 0$, we define $\omega(\theta) = 1 + \theta$ if $\theta \leq$

1 and $\omega(\theta) = 2\theta$ if $\theta \geq 1$. Simple considerations show that $\omega(\theta)$ satisfies the following property: Let s_1 and s_2 be two parallel secants of P that are orthogonal to some unit vector u, and let Q be the part of P between s_1 and s_2. Then

$$A(Q) + |s_1| \cdot \theta h_P(u) + |s_2| \cdot \theta h_P(u) \leq \omega(\theta), \qquad (3.28)$$

and if $\theta < 1$ then equality implies that s_1 and s_2 are opposite sides of P, and if $\theta > 1$ then equality implies that s_1 and s_2 are the same diagonal of P. Since $r_P(D_n)$ is bounded according to $\theta_c(P) = 0$ and Theorem 3.3.1, we conclude (i) and (ii) by (3.28).

Let $\theta > 70$, when we use that $2A(D_n, P)$ is the sum of the lengths of the projections of D_n into the coordinate axes. Writing p to denote the vertex $(1, 1)$ of P, and m to denote the maximal length of the secants of D_n orthogonal to the diagonal op, we may assume according to (ii) that $m \leq 1$. Let us consider the n translates of P covering D_n, where we may assume that the centre of each translate is contained in D_n. There exist $P_1, \ldots, P_{\lceil n/6 \rceil}$ among the n translates such that the distance between $P_i \neq P_j$ is larger than $\sqrt{2}$, and the orthogonal projection of any P_i into the line op lies between $(n/4)p$ and $(3n/4)p$. Then, for each P_i, any line that intersects P_i and is orthogonal to op intersects D_n in a segment of length at least $m/4$; hence, there exists a $P_i' \neq P_i$ among the n translates of P such that $P_i \cap P_i'$ has a point whose distance from the vertices of P_i is at least $m/8$. We deduce that the sum of the lengths of the projections of $P_i \cup P_i'$ into the coordinate axes is at most $4 - m/8$ for $i = 1, \ldots \lceil n/6 \rceil$; thus,

$$2A(D_n, P) \leq 2n - \frac{mn}{48}.$$

In addition, we readily have $A(D_n) \leq \sqrt{2}\, nm$. However, the segment connecting o and np is covered by some n translates of P, and hence

$$A(D_n) + 2A(D_n, P) \cdot \theta \geq 2n\,\theta.$$

Since $\theta/48 > \sqrt{2}$, we conclude $m = 0$ and, in turn, Theorem 3.4.7. $\qquad\square$

3.4.4. Coverings by Regular Hexagons

When K is a regular hexagon then we only verify that the optimal covering is never too thin.

Theorem 3.4.8. *Given a regular hexagon K and $\theta \geq 0$, let n translates of K cover the compact convex set D_n such that $A(D_n + \theta K)$ is maximal. Then $\liminf_{n \to \infty} r_K(D_n) \geq 1/2$.*

Proof. Given $0 < \omega \leq 1/2$, we define the function $\psi(\omega)$ as follows: For any strip Σ, we write l_1 and l_2 to denote the lines bounding Σ and write u to denote the exterior unit normal to the line l_1. We define $\psi(\omega)$ to be the maximum of

$$A(\Sigma \cap K) + |l_1 \cap K| \cdot \theta \, h_K(u) + |l_2 \cap K| \cdot \theta \, h_K(-u)$$

over all Σ satisfying $r_K(\Sigma) = \omega$. Then a simple argument shows that $\psi(\omega)$ is strictly increasing for $0 < \omega \leq 1/2$, and $\psi(1/2) = (2/3 + 4\theta/3)A(K)$.

Now two opposite sides of K determine a rectangle R of area $(2/3)A(K)$, and translating R parallel to its longer sides n times shows that

$$\Theta_T(K, \theta) = \lim_{n \to \infty} \frac{A(D_n + \theta K)}{n} \geq \left(\frac{2}{3} + \frac{4}{3}\theta \right) A(K) = \psi \left(\frac{1}{2} \right). \quad (3.29)$$

However, we suppose indirectly that there exists a subsequence $\{D_{n'}\}$ satisfying $\lim_{n' \to \infty} r_K(D_{n'}) < 1/2$. Thus, if n' is large then $D_{n'}$ is contained in a strip $\Sigma_{n'}$ with $r_K(\Sigma_{n'}) \leq \omega_0$, where $\omega_0 < 1/2$ is independent of n'. Therefore, $\psi(\omega_0) < \psi(1/2)$ leads to $\Theta(K, \theta) < (2/3 + 4\theta/3)A(K)$. This contradiction with (3.29) yields Theorem 3.4.8. $\qquad\square$

3.4.5. Coverings by Ellipses

To obtain results for ellipses, it is sufficient to consider coverings by unit discs. First, we define two functions $\alpha(\theta)$ and $\varphi_{B^2}(\theta)$ of $\theta > 0$, where $0 < \alpha(\theta) < \pi/2$:

$$\sin \alpha(\theta) = \frac{\sqrt{\theta^2 + 8} - \theta}{8} = \frac{2}{\theta + \sqrt{\theta^2 + 8}};$$

$$\varphi_{B^2}(\theta) = 2 \cos \alpha(\theta) \cdot \sin \alpha(\theta) + 2\theta \, \cos \alpha(\theta).$$

We note that $\varphi_{B^2}(\theta)$ is strictly monotonically increasing, and $\pi/[2\varphi_{B^2}(\theta)]$ is the asymptotic density of the "optimal sausage covering" because of the following property: If $z_1, z_2 \in B^2$ lie on the same side of the line l passing through the origin, and z_i' denotes the orthogonal projection of z_i into l, $i = 1, 2$, then

$$A(\text{conv}\,\{z_1, z_2, z_1', z_2'\}) + \theta \cdot d(z_1', z_2') \leq \varphi_{B^2}(\theta).$$

Equality holds if and only if $\text{conv}\,\{z_1, z_2, z_1', z_2'\}$ is congruent to the rectangle $R(\theta)$ whose sides are of length $d(z_1, z_2) = 2 \cos \alpha(\theta)$ and $d(z_1, z_1') = \sin \alpha(\theta)$.

We recall that $\Theta_T(B^2) = \sqrt{27}/2$ (see Theorem 2.2.3) and define $\theta^* > 0$ by

$$\varphi_{B^2}(\theta^*) = \frac{\sqrt{27}}{4}.$$

Theorem 3.4.9. *For any $\theta > 0$, let D_n be a compact convex set covered by n unit discs such that $A(D_n + \theta B^2)$ is maximal.*

(i) $\theta_c(B^2) = \theta^* = 0.2045\ldots$
(ii) If $0 \leq \theta < \theta_c(B^2)$ then there exist positive constants c_1, c_2 satisfying
   $c_1\sqrt{n} \leq r(D_n) \leq R(D_n) \leq c_2\sqrt{n}$ *for large n*
(iii) If $\theta > \theta_c(B^2)$ then $\Theta(B^2, \theta) = 2\varphi_{B^2}(\theta)$ and

$$\lim_{n \to \infty} r(D_n) = \frac{2}{\theta + \sqrt{\theta^2 + 8}}.$$

Proof (Sketch). The first step is to show

$$A(D_n + \theta B^2) > \begin{cases} 2n\, \varphi_{B^2}(\theta); \\ n\,\frac{\sqrt{27}}{2} - \sqrt{27}\,\sqrt{n}. \end{cases} \tag{3.30}$$

On the one hand, the rectangle of side lengths $2n\, \cos\alpha(\theta)$ and $\sin\alpha(\theta)$ can be covered by n unit discs. On the other hand, let us consider the tiling of $\mathbb{R}^2$ by translates of a regular hexagon H of circumradius one and hence of area $\sqrt{27}/2$. Since some translate of $\sqrt{n}\,H$ contains the centres of at most n tiles, $(\sqrt{n} - 1)H$ can be covered by n unit discs, and we conclude (3.30).

The rest of the argument resembles the proof of Lemma 2.7.2, only we use a slightly different cell decomposition. Let $w_1, \ldots, w_n$ be the centres of the n unit discs that cover D_n, and we may assume that each w_i lies in D_n. It is not so obvious but we may even assume that the centres are in general position in the following sense: The centres lie in int D_n, no four centres lie on a circle, and no three centres have the same distance from a boundary points of D_n. A circular disc B with centre p is called a supporting disc if p lies in D_n, the interior of B contains none of $w_1, \ldots, w_n$, and either ∂B contains three points out of $w_1, \ldots, w_n$ (and hence $p \in$ int D_n) or $p \in \partial D_n$ and ∂B contains two points out of $w_1, \ldots, w_n$. For any supporting disc, we consider the convex hull of the points out of $w_1, \ldots, w_n$ that lie in the supporting disc. Then there exists a simplicial complex Σ_n whose maximal cells are these convex hulls. Let $x_1, \ldots, x_b \in \partial D_n$ be the centres of the supporting discs in clockwise order whose centre lie in ∂D_n. We define $x_0 = x_b$ and $x_{b+1} = x_1$. When speaking about an arc $x_i x_j$, we mean the closed clockwise oriented arc of D_n. We note that $b = b(\Sigma_n)$ is the number of "free sides" of Σ_n, which satisfies $2n = f_2(\Sigma_n) + b(\Sigma_n) + 2$. In particular, (3.30) yields

$$A(D_n + \theta B^2) > f_2(\Sigma_n)\, \varphi_{B^2}(\theta) + b(\Sigma_n)\, \varphi_{B^2}(\theta); \qquad (3.31)$$

$$A(D_n + \theta B^2) > f_2(\Sigma_n) \frac{\sqrt{27}}{4} + b(\Sigma_n) \frac{\sqrt{27}}{4} - \sqrt{27}\sqrt{n}. \qquad (3.32)$$

For any x_i, $i = 1, \ldots, b$, let z_i and z_{i+1} be the centres (out of $w_1, \ldots, w_n$) in the corresponding supporting disc in a way that the triangle $x_i z_{i+1} z_i$ is clockwise oriented (where z_i and z_j may coincide if $|j - i| \geq 2$). Now, for any z_i, we write y_i to denote a closest point of the arc $x_{i-1} x_i$ to z_i, and we use Ξ_i to denote the domain bounded by the broken line $y_i z_i z_{i+1} y_{i+1}$ and the arc $y_i y_{i+1}$. Then the triangles of Σ_n and the domains Ξ_i, $i = 1, \ldots, b$, together form a cell decomposition of D_n.

Since we are interested in $D_n + \theta K$, we define $\Xi_i(\theta)$ to be the set of points x of $D_n + \theta B^2$ such that the closest point of D_n to x lies in Ξ_i, $i = 1, \ldots, b$. Readily, Ξ_i is contained in a circular disc of centre x_i and of radius two; hence, $A(\Xi_i(\theta)) \leq (2 + \theta)^2 \pi$. In addition, given any small $\varepsilon > 0$, there exists $\delta > 0$ such that if the total curvature of the arc $y_i y_{i+1}$ is at most δ then $A(\Xi_i(\theta)) \leq \varphi_{B^2}(\theta) + \varepsilon$. Therefore,

$$A(D_n + \theta B^2) \leq f_2(\Sigma_n) \frac{\sqrt{27}}{4} + b(\Sigma_n)(\varphi_{B^2}(\theta) + \varepsilon) + \frac{2\pi}{\delta}(2 + \theta)^2 \pi.$$

$$(3.33)$$

Now if $\theta < \theta^*$ then combining $\varphi_{B^2}(\theta) < \sqrt{27}/4$, (3.32), and (3.33) leads to $b(\Sigma_n) \leq c\sqrt{n}$, where c is a constant. In turn, (ii) readily follows. However, if $\theta > \theta^*$ then the proportion of the "boundary cells", namely, $b(\Sigma_n)/2n$, tends to one. Thus most cells Ξ_i are very close to being congruent to $R(\theta)$, which, in turn, yields (iii). $\qquad \square$

3.4.6. Finite Lattice Coverings by Unit Discs

It is more convenient to state the results of this section in terms of unit discs, but naturally they apply to finite lattice coverings by ellipses. We recall that, given a convex domain D and a covering lattice Λ for B^2, the corresponding *finite lattice covering* of D is the family of all unit discs centred at some point of Λ whose interior intersect D. We use the threshold $\theta^* = 0.2045\ldots$ from Section 3.4.5.

Theorem 3.4.10. *For $\theta > 0$, let D_n be a compact convex set such that there exists a lattice covering of D_n by at most n unit discs, and $A(D_n + \theta B^2)$ is maximal.*

(i) If $0 \leq \theta < \theta^*$ then $D_n/\sqrt{n}$ tends to a regular hexagon of area $\sqrt{27}/2$,

(ii) If $\theta > \theta^*$ then some m_n of the centres are aligned, where

$$\lim_{n \to \infty} \frac{m_n}{n} = 1 \text{ and, in addition, } \lim_{n \to \infty} r(D_n) = \frac{2}{\theta + \sqrt{\theta^2 + 8}}.$$

Proof (Sketch). The case $\theta > \theta^*$ follows similarly to the proof for Theorem 3.4.9 (iii); thus, we assume that $0 \leq \theta < \theta^*$. Let $\tilde{T}$ be a regular triangle of side length $\sqrt{3}$ and of area $\sqrt{27}/4$. Then $\tilde{T}$ is a triangle of maximal area contained in B^2, and the side vectors of $\tilde{T}$ determine the covering lattice Λ of maximal determinant $\sqrt{27}/2$.

It follows by the argument leading to Theorem 3.4.9 (ii) that $R(D_n)/\sqrt{n}$ stays bounded as n tends to infinity. We write Λ_n to denote the covering lattice associated to D_n; hence, Λ_n tends to Λ (possibly after some congruences). To make the presentation simpler, we assume that $\Lambda_n = \Lambda$ and note that the general case can be handled in a similar way. We define the simplicial complex Σ_n in a way that its vertices are the points $x_1, \ldots, x_k$ of Λ such that the interior of $x_i + B^2$ intersects D_n ($k \leq n$). The segment $x_i x_j$ is an edge if it is a translate of a side of $\tilde{T}$, and $x_i x_j x_l$ is a triangle of Σ if it is a translate of $\tilde{T}$ or $-\tilde{T}$. In particular, Σ_n is connected. If a cycle of edges of Σ_n encircles a point of Λ then it is one of $x_1, \ldots, x_k$, and hence $\chi(\Sigma_n) = 1$. We orient the boundary of supp Σ clockwise in the natural way that, say, if an edge is not contained in any triangle we count it twice. Let $z_1, \ldots, z_b$, $b = b(\Sigma)$, denote the vertices of Σ on the boundary in a way that $z_{i-1} z_i$ are the free sides of Σ, $i = 1, \ldots, b$, $z_b = z_0$ (and hence a vertex may be counted twice). We write m_i to denote the midpoint of the free side $z_i z_{i+1}$ and u_i to denote the common point of the boundaries of $z_i + B^2$ and $z_{i+1} + B^2$ that lies "left" according to the orientation. Then the segments $m_i u_i$ are pairwise disjoint, and u_i is not contained in the interior of supp Σ.

For any z_i, we write y_i to denote a closest point of ∂D_n to z_i. We associate a cell $\Xi_i \subset D_n$ to all but a constant number of free sides $z_i z_{i+1}$ in the following way: If y_i and y_{i+1} lie "left" of the line $z_i z_{i+1}$ according to the orientation then Ξ_i is cut off by the broken line $y_i z_i z_{i+1} y_{i+1}$. If y_i or y_{i+1} lies "left" of the line $z_i z_{i+1}$ according to the orientation, and ∂D_n meets the segment $z_i z_{i+1}$, then Ξ_i is cut off by the broken line $y_i z_i z_{i+1}$ or $z_i z_{i+1} y_{i+1}$, respectively. Since we are interested in $D_n + \theta K$, we define $\Xi_i(\theta)$ to be the set of points x of $D_n + \theta B^2$ such that the closest point of D_n to x lies in Ξ_i. Then the union of supp Σ_n and $\Xi_i(\theta)$ whenever it is defined covers all $D_n + \theta B^2$ but a part of bounded area (as n tends to infinity). Since each free side is of length $\sqrt{3}$, for any $\varepsilon > 0$ there exists constant N with the following property: For all but

N indices i, $\Xi_i(\theta)$ is defined and satisfies

$$A(\Xi_i(\theta)) \leq \frac{\sqrt{3}}{2} + \sqrt{3}\,\theta + \varepsilon.$$

Therefore,

$$A(D_n + \theta\,B^2) \leq A(\text{supp}\,\Sigma_n) + b(\Sigma_n) \cdot \left(\frac{\sqrt{3}}{2} + \sqrt{3}\,\theta \right) + o(b(\Sigma_n))$$

$$= n\,\frac{\sqrt{27}}{2} - b(\Sigma_n) \cdot \left(\frac{\sqrt{3}}{4} - \sqrt{3}\,\theta \right) + o(b(\Sigma_n)).$$

Since Σ_n has at most n vertices, if $b(\Sigma_n)$ is close to its possible minimum then supp Σ_n is close to be a regular hexagon. In turn, we conclude Theorem 3.4.10. $\square$

4

Packings of Circular Discs

Let us review the main results of this chapter following the history of the subject. Finite packings of equal circular discs were investigated systematically as early as 1830 by F. Bolyai and as 1890 by A. Thue. Since the middle of the twentieth century, the central problem has been to provide good density bounds with respect to convex containers in $\mathbb{S}$, where $\mathbb{S}$ is S^2, H^2, or $\mathbb{R}^2$. Given three circular discs of radius r centred at the vertices of a regular triangle of side length $2r$, we define $\sigma_{\mathbb{S}}(r)$ to be the ratio of the part of the regular triangle covered by the circular discs to the area of the triangle. In particular $\sigma_{\mathbb{R}^2}(r) = \pi/\sqrt{12}$. We say that the triangle bound holds for a finite packing of circular discs of radius r with respect to a convex container if the density is at most $\sigma_{\mathbb{S}}(r)$. Around 1940 L. Fejes Tóth proved the triangle bound when the convex container either lies in the Euclidean plane or is S^2 itself. Improving on a somewhat weaker result of J. Molnár from the 1950s, we will verify the triangle bound for packings of equal spherical circular discs inside any spherical convex container. For the hyperbolic plane the triangle bound does not hold for certain convex containers. However, if some natural conditions (which are satisfied by any circular disc) are imposed on the container then an even stronger bound (the Euclidean triangle bound $\pi/\sqrt{12}$) was proved by K. Bezdek during the 1980s.

For the optimal arrangements, L. Fejes Tóth verified that the Platonic solids with triangular faces yield densest packings on S^2, and G. Wegner determined the densest packing of n unit discs in $\mathbb{R}^2$ for approximately 96% of all n.

The approach of A. Thue was to find optimal inequalities for certain finite packings of n unit discs in the Euclidean plane. Thue's inequality was generalized to any finite packing of unit discs by H. Groemer in 1960, and we call this general form *Thue–Groemer inequality*. (Actually, N. Oler verified an even more general inequality for centrally symmetric domains, which is discussed in Chapter 2.) The Thue–Groemer inequality and its hyperbolic and spherical analogues prove to be very efficient tools for obtaining good density bounds.

4.1. Associated Cell Complexes and the Triangle Bound

The aim of this section is to construct the Dirichlet–Voronoi and Delone complexes associated to a finite packing of equal circular discs and to verify the triangle bound for certain cells. We note that the Delone complex is also discussed in Section 5.1 from the point of view of coverings.

Let $\mathbb{S}$ be $\mathbb{R}^2$, S^2, or H^2. If Π is a compact convex set in $\mathbb{S}$, and $B_1, \ldots, B_n$ are nonoverlapping circular discs, then the density of the packing with respect to Π is defined to be

$$\frac{\sum_{i=1}^{n} A(\Pi \cap B_i)}{A(\Pi)}. \tag{4.1}$$

In particular, let $T(2r)$ be the regular triangle of side length $2r$, and let $\sigma_{\mathbb{S}}(r)$ be the density of the three circular discs of radius r centred at the vertices of $T(2r)$ with respect to $T(2r)$. Given a packing of circular discs of radius r and a convex domain Π in $\mathbb{S}$, we say that *the triangle bound* holds with respect to Π if the density is at most $\sigma_{\mathbb{S}}(r)$.

The Dirichlet–Voronoi Complex

Let us consider the nonoverlapping circular discs $B_1, \ldots, B_n, n \geq 2$, of radius r inside a convex domain C in $\mathbb{S}$, where $r < \pi/3$ in the spherical case. We write x_i to denote the centre of B_i, and we define the *Dirichlet–Voronoi cell* D_i associated to x_i to be the set of points x in C whose distance from x_i is not larger than the distance from any other x_j. Thus, D_i is the intersection of C and the suitable half planes that are bounded by the perpendicular bisectors of the segments $x_i x_j$, $j \neq i$. We say that $v \in D_i$ is a *proper vertex* of D_i if v is an endpoint of a maximal segment in ∂D_i that intersects int C.

Theorem 4.1.1. *If a convex domain C in $\mathbb{S}$ contains the nonoverlapping circular discs $B_1, \ldots, B_n$, $n \geq 2$, of radius r, and some B_i is contained in the convex hull of the proper vertices of the corresponding Dirichlet–Voronoi cell D_i then*

$$\frac{A(B_i)}{A(D_i)} \leq \sigma_{\mathbb{S}}(r).$$

Equality holds if and only if D_i is a polygon circumscribed around B_i and each angle of D_i is $2\pi/3$.

Proof. We write R to denote $R(T(2r))$ and x_j to denote the centre of B_j, $j = 1, \ldots, n$. We claim that if v is a proper vertex of D_i then

$$d(v, x_i) \geq R. \tag{4.2}$$

If v is the vertex of two other Dirichlet–Voronoi cells D_k and D_m then the distances of v from x_i, x_k, and x_m are the same. Since one of the three angles $\angle x_i v x_k$, $\angle x_i v x_m$, and $\angle x_k v x_m$ is at most $2\pi/3$, and the distance between any two centres is at least $2r$, we deduce (4.2). If v is the vertex of only one other Dirichlet–Voronoi cell D_k then let l be the line touching B_i and B_k at y_i and y_k, respectively, in a way that l separates both x_i and x_k from v. As we are searching for the possible minimum of $d(v, x_i)$, we may assume that $v \in l$, and hence $\angle x_i v y_i < \pi/3$. In turn, we conclude the claim (4.2).

Next, we claim that if $T = x_i y z$ is a triangle such that $d(x_i, y) \geq R$ and $d(x_i, z) \geq R$, and the line yz does not meet int B_i, then

$$\frac{A(B_i \cap T)}{A(T)} \leq \sigma_{\mathbb{S}}(r), \tag{4.3}$$

where equality holds if and only if the sides $x_i y$ and $x_i z$ are of length R and the side yz touches B_i.

We write $B(\varrho)$ to denote the circular disc of centre x_i and of radius ϱ. Let u be the closest point of the side yz to x. If $d(x_i, u) \geq R$ then the density of B_i with respect to T is less than $A(B_i)/A(B(R)) < \sigma_{\mathbb{S}}(r)$, hence, we may assume that $d(x_i, u) < R$. In particular, it is sufficient to prove that the density of B with respect to, say, the triangle $T_z = x_i u z$ is at most $\sigma_{\mathbb{S}}(r)$. If $d(z, x_i) > R$ then we write z' to denote the point of the segment uz of distance R from x_i, and we observe that the density of B_i with respect to the triangle $x_i z' z$ is less than $A(B_i)/A(B(R)) < \sigma_{\mathbb{S}}(r)$. Thus, we may assume that $d(z, x_i) = R$. Now there exists a line through z that touches B_i at some u', and the segments $u'z$ and $x_i u$ intersect in some point w. Let $d(w, x_i) = \varrho$, and we observe that the density of B_i with respect to T_z is less than that with respect to $x_i w z$ if $u' \neq u$. In addition, the density of B_i with respect to the triangle $x_i w z$ is less than $A(B_i)/A(B(\varrho))$, and that with respect to $x_i u' w$ is larger than $A(B_i)/A(B(\varrho))$. It follows that the density of B_i with respect to T_z less than that with respect to $x_i u'z$, which is $\sigma_{\mathbb{S}}(r)$. Finally, (4.2) allows us to dissect the convex hull of the proper vertices of D_i into the type of triangles that occur in (4.3); hence, we conclude Theorem 4.1.1. □

It is easy to deduce the triangle bound in $\mathbb{R}^2$ via Theorem 4.1.1:

Corollary 4.1.2. *If a convex domain D in $\mathbb{R}^2$ contains $n \geq 2$ nonoverlapping unit discs then $A(D) > \sqrt{12}\, n$.*

Proof. We write $x_1, \ldots, x_n$ to denote the centres of the unit discs and write $D_1, \ldots, D_n$ to denote the corresponding Dirichlet–Voronoi cells. If x_i is not a vertex of the convex hull C of $x_1, \ldots, x_n$ then Theorem 4.1.1 yields that

$A(D_i) \geq \sqrt{12}$. If x_i is a vertex of C then D_i contains two nonoverlapping unit squares, each of which is built on an edge of C emanating from x_i. Therefore, $A(D_i) \geq 2 + \pi/2 > \sqrt{12}$, completing the proof of Corollary 4.1.2.   □

We will show that the triangle bound holds also on S^2 (see Theorem 4.4.2). However, the triangle bound may not hold in H^2 in general, but it does hold with respect a suitable class of convex domains (see Theorem 4.6.1). Next we consider compact surfaces of constant curvature (see Section A.5.4).

Corollary 4.1.3. *The corresponding triangle bound holds for any packing of embedded circular discs of radius r on a compact surface of constant curvature.*

Let $T(2r_k)$ be the regular triangle of angle $2\pi/k$ in $\mathbb{S}$ for an integer $k \geq 3$, where $\mathbb{S}$ is $\mathbb{R}^2$, S^2, or H^2. Then there exists a compact quotient X_k of $\mathbb{S}$ such that the density of some packing of embedded circular discs of radius r_k on X_k is $\sigma_{\mathbb{S}}(r_k)$.

Proof. Given a packing of n embedded discs of radius r on a compact surface X of constant curvature 0, 1, or -1, it is the image of a periodic packing of circular discs of radius r in $\mathbb{S}$ that is $\mathbb{R}^2$, S^2, or H^2, respectively. The corresponding Dirichlet–Voronoi cells in $\mathbb{S}$ are convex polygons. We choose one Dirichlet–Voronoi cell from each of the n equivalence classes, and hence their total area is $A(X)$. Since each Dirichlet–Voronoi cell is the convex hull of its vertices, Theorem 4.1.1 yields that the density of the packing on X is at most $\sigma_{\mathbb{S}}(r)$.

However, the suitable X_k is a compact surface of constant curvature that can be triangulated into isometric copies of $T(2r_k)$ (see Example A.5.10).   □

We recall from Section 1.1 that $\delta(B^2)$ is the supremum of the densities of periodic packings of unit discs in $\mathbb{R}^2$, or equivalently, the supremum of the densities of finite packings of unit discs on tori.

Corollary 4.1.4. $\delta(B^2) = \pi/\sqrt{12}$ *and the hexagonal lattice of determinant $\sqrt{12}$ induces the densest periodic packing by unit discs.*

The Delone Complex

The notion of the Delone complex is rather simple for a packing of n circular discs of radius $r < \pi/4$ on S^2; therefore, we discuss this case independently. We assume that the packing is *saturated*; namely, no more spherical circular discs of radius r can be inserted without overlapping some disc already in place. The convex hull of the centres of the spherical discs in $\mathbb{R}^3$ is a three-polytope P of n vertices whose interior contains the origin of $\mathbb{R}^3$. Projecting the faces of P radially onto S^2 yields the spherical *Delone cell complex* associated to the packing. Any triangulation of this cell complex using the n vertices is called a *Delone triangulation* of S^2. We observe that the vertices of any Delone cell are contained in the boundary of a spherical circular disc that does not contain any other vertex and whose radius is less than $2r$.

In general, let $\mathbb{S}$ be $\mathbb{R}^2$, H^2, or S^2, and let $x_1, \ldots, x_n, n \geq 2$, be contained in a convex domain C in $\mathbb{S}$ such that $d(x_i, x_j) \geq 2r$ holds for any $i \neq j$ for some given positive r. In addition, we assume that $x_1, \ldots, x_n$ is saturated with respect to C and r; namely, for any $x \in C$ there exists an x_i satisfying $d(x, x_i) < 2r$. We call a circular disc B a *supporting disc* if its centre y is contained in C, the interior of B contains no point of $\{x_1, \ldots, x_n\}$, and either ∂B contains at least three points of $\{x_1, \ldots, x_n\}$ or $y \in \partial C$ and ∂B contains at least two points of $\{x_1, \ldots, x_n\}$. We note that the radius of any supporting disc is less than $2r$. For each supporting disc, we consider the convex hull of the points out of $x_1, \ldots, x_n$ that are contained in the disc, and we let Σ' be the family of all the segments and convex polygons resulting this way. Since the common secant of two overlapping supporting discs separates the corresponding points of $\{x_1, \ldots, x_n\}$ (see Lemma A.5.4), we deduce that Σ' determines a cell complex, which we call the *Delone complex* associated to $\{x_1, \ldots, x_n\}$ and C. The definition of Σ' yields that the set of vertices of Σ' is $\{x_1, \ldots, x_n\}$. A triangulation Σ of Σ' using the same set of vertices is called a *Delone simplicial complex*, whose triangles are the *Delone triangles*. If $C = S^2$ then we obtain the same complex as before. We note that the Delone cell complex is the dual of the Dirichlet–Voronoi cell complex in a sense that, say, the vertices of the Dirichlet–Voronoi cell complex are the centres of the supporting discs.

Lemma 4.1.5. *If T is a Delone triangle for a saturated family of points with respect to r and a convex domain C in $\mathbb{S}$ then*

$$A(T) \geq A(T(2r)),$$

and equality holds if and only if T is congruent to $T(2r)$.

Proof. Given a triangle M in $\mathbb{S}$, let $R_0(M)$ be the radius of the circle that passes through the vertices of M. It is sufficient to show that if each side of a triangle M is at least $2r$, and $R_0(M) < 2r$, then the minimum of $A(M)$ is obtained exactly when M is congruent to $T(2r)$. We write p_1, p_2, and p_3 to denote the vertices of M in a way that $d(p_1, p_2) \leq d(p_1, p_3) \leq d(p_2, p_3)$. We may decrease the length of $p_1 p_3$ until $d(p_1, p_3) = d(p_1, p_2)$, and then may decrease $d(p_1, p_3) = d(p_1, p_2)$ until it becomes $2r$ because these operations decrease $R_0(T)$ and $A(T)$. Now the angle at p_1 is at least the angle γ of $T(2r)$ and less than 2γ. Since $A(T)$ attains its minimum when the angle at p_1 is either minimal or maximal according to Lemma A.5.2, we conclude Lemma 4.1.5. $\qquad\qquad\square$

Let us consider the circular discs of radius r that are centred at the points in Lemma 4.1.5. Based on Lemma 4.1.5, it is easy to show that the density of the packing with respect to any Delone triangle is at most $\sigma_{\mathbb{S}}(r)$. We do not discuss this result in detail because we do not need it.

Formulae and Estimates for $\sigma_{\mathbb{S}}(r)$

We note that

$$\sigma_{\mathbb{S}}(r) = \frac{(\kappa \, A(T(2r)) + \pi) \, A(B(r))}{2\pi \, A(T(2r))},$$

where $B(r)$ denotes a circular disc of radius r in $\mathbb{S}$, and $\kappa = 0, 1, -1$ if $\mathbb{S}$ is $\mathbb{R}^2$, S^2, or H^2, respectively. Thus the laws of sine and cosine yield

$$\sigma_{\mathbb{R}^2}(r) = \frac{\pi}{\sqrt{12}}; \tag{4.4}$$

$$\sigma_{S^2}(r) = \frac{6(1 - \cos r) \arcsin \frac{1}{2\cos r}}{6 \arcsin \frac{1}{2\cos r} - \pi}; \tag{4.5}$$

$$\sigma_{H^2}(r) = \frac{6(\operatorname{ch} r - 1) \arcsin \frac{1}{2\operatorname{ch} r}}{\pi - 6 \arcsin \frac{1}{2\operatorname{ch} r}}. \tag{4.6}$$

With respect to $\sigma_{H^2}(r)$, we verify

$$\frac{\pi}{\sqrt{12}} < \sigma_{H^2}(r) < \frac{3}{\pi}. \tag{4.7}$$

Simple applications of L'Hospital's rule yield that the lower bound is the limit of $\sigma_{H^2}(r)$ at $r = 0$, and the upper bound is the limit at $r = \infty$. In particular, it is sufficient to show that $\sigma_{H^2}(r)$ is monotonically increasing. Let us introduce the variable $t = \arcsin(1/(2\operatorname{ch} r))$, $0 < t < \pi/6$, which is a decreasing function of r. We write $\sigma_{H^2}(r)$ in the form

$$\sigma_{H^2}(r) = \varphi(t) = \frac{f(t)}{g(t)}, \quad \text{where} \quad f(t) = \frac{1}{2\sin t} - 1 \quad \text{and} \quad g(t) = \frac{\pi}{6t} - 1.$$

We show that $\varphi(t)$ is monotonically decreasing by introducing another auxiliary function; namely, let

$$\psi(t) = \frac{f'(t)}{g'(t)} = \frac{3}{\pi} \cdot \frac{t^2 \cos t}{\sin^2 t}.$$

Then $\varphi(t)$ and $\psi(t)$ can be extended to be continuous functions on $(0, \pi/6]$ by setting $\varphi(\pi/6) = \psi(\pi/6) = \pi/\sqrt{12}$, and ψ is strictly monotonically decreasing because $\sin t < t$ and $\cos t < (1 + \cos^2 t)/2$ yield

$$\psi'(t) = \frac{2t}{\sin^3 t} \cdot \left(\sin t \cdot \cos t - t \cdot \frac{1 + \cos^2 t}{2} \right) < 0.$$

If $\varphi(t)$ is not monotonically decreasing then it has a local maximum at some $t_0 \in (0, \pi/6]$; hence, $\varphi(t_0) = \psi(t_0)$ follows by

$$\varphi'(t) = \frac{g'(t)}{g(t)} (\psi(t) - \varphi(t)).$$

In particular, there exists $t < t_0$ such that $\varphi(t) \leq \varphi(t_0)$ and $\varphi'(t) \geq 0$. Thus, the formula for $\varphi'(t)$ leads to $\psi(t) \leq \varphi(t) \leq \varphi(t_0) = \psi(t_0)$, contradicting $\psi(t) > \psi(t_0)$. In turn, we conclude (4.7).

An analogous argument to this one shows that $\sigma_{S^2}(r)$ is monotonically decreasing and tends to $\pi/\sqrt{12}$ as r tends to zero. Therefore,

$$\sigma_{S^2}(r) < \frac{\pi}{\sqrt{12}} \tag{4.8}$$

holds for any $0 < r < \pi/3$.

Comments. The presentation of this section is based on L. Fejes Tóth [FTL1964]. We note that Dirichlet–Voronoi cells were first considered by G. L. Dirichlet (see [Dir1850]) and by G. F. Voronoi (see [Vor1908]). Moreover, Delone cells were first considered by B. N. Delone (see [Del1934]).

Corollary 4.1.2 has a long story. Some version of it was proved by A. Thue in [Thu1892] and [Thu1910] (see also [Thu1977]), and the case when D is a polygon whose angles are at most $2\pi/3$ each is due to B. Segre and K. Mahler [SeM1944]. Finally Corollary 4.1.2 is proved by L. Fejes Tóth [FTL1949]. Actually, the triangle bound holds for a wider class of containers than the convex ones. We say that a domain D is a $c(\varrho)$-*domain* if for any boundary point p there exists a circular disc of radius ϱ that is tangent to D at p and does not overlap D in a neighbourhood of p. Then J. Molnár [Mol1960] proved that if $\varrho \geq \varrho_0 = 4.76$ then the triangle bound holds for any packing of at least two unit discs inside a $c(\varrho)$-domain in $\mathbb{R}^2$.

A. Heppes [Hep2001] investigated the uncovered part of $c(\varrho)$-domains in $\mathbb{R}^2$, S^2, or H^2. We write $H_3(\varrho)$ to denote the area enclosed by three mutually tangent circular discs of radius ϱ. Then A. Heppes [Hep2001] proved for any positive ϱ that if the $c(\varrho)$-domain D contains n nonoverlapping circular discs of radius ϱ then the area of the uncovered part of D is at least $2(n-1)$ $H_3(\varrho)$. We note that A. Heppes [Hep1999] found the optimal packings of $n \leq 4$ equal embedded circular discs in any torus. In addition, R. L. Graham, B. D. Lubachevsky, and F. H. Stillinger [GLS1997] produced various efficient packings of equal embedded circular discs on some tori. Moreover, C. Bavard [Bav1996] determined the hyperbolic compact surfaces where there exists one embedded circular disc of radius r whose density is $\sigma_{H^2}(r)$.

Corollary 4.1.4 or equivalent statements have been proved by various authors, namely, by A. Thue (see [Thu1892] and [Thu1910], or [Thu1977]), L. Fejes Tóth [FTL1940], and B. Segre and K. Mahler [SeM1944].

4.2. The Absolute Version of the Thue–Groemer Inequality

According to Lemma 4.1.5, the regular triangle $T(\varrho)$ of side length ϱ is the optimal Delone triangle for a saturated packing of circular discs of radius $\varrho/2$. In this section we develop this observation into a global formula for a compact convex set that contains the centres of n nonoverlapping circular discs of radius $\varrho/2$. To shorten many formulae, we write $|\cdot|$ to denote the area.

Theorem 4.2.1. *Let C be a compact convex set in $\mathbb{R}^2$, H^2, or S^2, and let $\varrho > 0$ (where $\varrho \leq \pi/2$ in the spherical case). If C contains n points such that the distance between any two is at least ϱ then*

$$\frac{1}{2|T(\varrho)|} \cdot |C| + \frac{1}{2\varrho} \cdot P(C) + 1 \geq n.$$

Equality holds if and only if either C is a segment of length $(n-1)\varrho$ or C is a polygon that can be triangulated using n vertices into regular triangles of edge length ϱ.

Remarks. If equality holds then the corresponding packing of n circular discs of radius $\varrho/2$ is called a *Groemer packing*.

Segments of length $(n-1)\varrho$ show that the coefficient of the perimeter is optimal, and, say, $T(\varrho)$ shows that the coefficient of the area cannot be lowered. In $\mathbb{R}^2$ and H^2 there exist convex domains for which equality holds in Theorem 4.2.1 for any $n \geq 3$ and ϱ.

We call the Euclidean version of Theorem 4.2.1 the *Thue–Groemer inequality*: If the compact convex set C contains the centres of n nonoverlapping unit discs then

$$\frac{|C|}{\sqrt{12}} + \frac{P(C)}{4} + 1 \geq n. \tag{4.9}$$

We note that the Oler inequality (Theorem 2.3.1) generalizes the Thue–Groemer inequality to Minkowski planes.

Theorem 4.2.1 provides an efficient tool for investigating finite packings of equal circular discs; for example, it leads directly to the Euclidean triangle bound (Corollary 4.1.2): If D is a convex domain containing $n \geq 2$ nonoverlapping unit discs, and C is the convex hull of the vertices, then (4.9) and $P(C) \geq 4$ yield

$$|D| \geq |C| + P(C) + \pi \geq \sqrt{12}\,n + \left(1 - \frac{\sqrt{3}}{2}\right) P(C) + \pi - \sqrt{12} > \sqrt{12}\,n.$$

We note that the proof of the density bound $\pi/\sqrt{12}$ for packings of hyperbolic circular discs of radius r into th r-convex domains (see Theorem 4.6.1) will be also based on Theorem 4.2.1.

To prove Theorem 4.2.1, we need the following estimate:

Proposition 4.2.2. *Let p and q be points of a circle of radius less than ϱ in $\mathbb{R}^2$, H^2, or S^2. In addition, we assume that p and q lie on the same side of a line l passing through the centre, and we write $\tilde{p}$ and $\tilde{q}$ to denote their projection into l, respectively, and Q to denote the convex hull of $p, q, \tilde{q}, \tilde{p}$. If $d(p, q) \geq \varrho$ then*

$$\frac{|Q|}{2|T(\varrho)|} + \frac{d(\tilde{p}, \tilde{q})}{2\varrho} \geq \frac{1}{2}, \tag{4.10}$$

and equality holds if and only if $p, q \in l$ and $d(p, q) = \varrho$.

Proof of Proposition 4.2.2 in $\mathbb{R}^2$. We may assume that $p \in l$; hence, the left-hand side of (4.10) is $(1/\sqrt{3})\sin 2\alpha + (1/2)\cos \alpha$ for $\alpha = \angle \tilde{q}\,pq$. Since this function of α is strictly concave, we conclude Proposition 4.2.2. $\quad\square$

In the hyperbolic and spherical cases, the proof of Proposition 4.2.2 is more tedious and is postponed until after the proof of the Thue–Groemer inequality.

Proof of Theorem 4.2.1 Assuming Proposition 4.2.2. We may assume that the family of n points is saturated with respect to C and $\varrho/2$ and that C is the convex hull of the points; and is two dimensional. Let Σ be the associated Delone cell complex. We write $e_1 \ldots, e_m$ to denote the edges on Σ, and $\nu(e_i)$

to denote the number of two-cells in Σ containing e_i. In addition, $P_1, \ldots, P_k$ denote the cells of Σ, and $s(P_j)$ denotes the number of sides of P_j. Now $|P_j| \geq (s(P_j) - 2)\,|T(\varrho)|$ according to the triangle bound Lemma 4.1.5, and readily $\sum_{j=1}^{k} s(P_j) = \sum_{i=1}^{m} \nu(e_i)$. We deduce that the Euler characteristic $\chi(\Sigma) = k - m + n$ of Σ satisfies

$$\frac{|\mathrm{supp}\Sigma|}{2|T(\varrho)|} + \frac{1}{2}\sum_{i=1}^{m}(2 - \nu(e_i)) + \chi(\Sigma) \geq n, \tag{4.11}$$

and equality holds if and only if Σ can be triangulated using the n vertices in a way that each triangle is congruent to $T(\varrho)$.

We call a segment s on the boundary of C a *boundary segment* if it contains exactly two out of the n points that are the endpoints of s, and s is not an edge of Σ. There exists a chain $e_{1,s}, \ldots, e_{l_s,s}$ of edges of Σ such that each is the secant of some supporting disc that is centred at a point of s, and $e_{1,s}, \ldots, e_{l_s,s}$ together with s bound a possibly nonconvex polygon $\Pi(s)$ whose relative interior is disjoint from the relative interior of any two-cell in Σ. For each $e_{i,s}$, we consider the convex hull $Q_{i,s}$ of $e_{i,s}$ and its orthogonal projection into s, and we let $\Sigma(s)$ be the cell complex determined by $Q_{i,s}, i = 1, \ldots, l_s$. We define the cell complex Σ_0 to be the union of Σ and of the complexes $\Sigma(s)$ for any boundary segment s. The definitions of Σ and Σ_0 yield that Σ_0 is a cell decomposition of C; hence, $\chi(\Sigma_0) = 1$ according to the Euler formula (A.11). It follows by adding the complexes $\Sigma(s)$ one by one to Σ that $\chi(\Sigma) = 1$ as well. However, applying Proposition 4.2.2 to each $Q_{i,s}$ leads to

$$\frac{|C|}{2|T(\varrho)|} + \frac{P(C)}{2\varrho} \geq \frac{|\mathrm{supp}\Sigma|}{2|T(\varrho)|} + \frac{1}{2}\sum_{i=1}^{m}(2 - \nu(e_i)).$$

Hence, (4.11) yields Theorem 4.2.1. $\square$

Proof of Proposition 4.2.2 in H^2 and S^2. We discuss only the hyperbolic case in detail and comment on the spherical case at the end. Let Ω be the left-hand side of (4.10); hence, Proposition 4.2.2 is equivalent with $\Omega \geq 1/2$. Our plan is to check the minimums of Ω along suitable one-parameter families.

We start with an auxiliary statement about triangles. For given $c > 0$ and $0 < \alpha < \pi/2$, let T be the triangle such that one side is of length c, the opposite angle is a right angle, and another angle is α. We write β to denote the third angle of T and use b to denote the side opposite to β. It follows that $|T| = \pi/2 - \alpha - \beta$, and the laws of sines and cosines yield

$$\tan \alpha \tan \beta \,\mathrm{ch}\, c = 1 \quad \text{and} \quad \mathrm{th}\, b = \mathrm{th}\, c \cos \alpha.$$

We define

$$\varphi_c(\alpha) = \frac{|T|}{2|T(\varrho)|} + \frac{b}{2\varrho}.$$

Using the formulae $(\tan t)' = 1 + \tan^2 t$ and $(\operatorname{th} t)' = 1 - \operatorname{th}^2 t$, we deduce that

$$\frac{\partial}{\partial \alpha} \varphi_c(\alpha) = \frac{1}{1 + \sin^2 \alpha \operatorname{sh}^2 \varrho} \left(\frac{\operatorname{ch} \varrho - 1}{2|T(\varrho)|} - \frac{\operatorname{sh} \varrho \operatorname{ch} \varrho}{2\varrho} \sin \alpha - \frac{\operatorname{sh}^2 \varrho}{2|T(\varrho)|} \sin^2 \alpha \right).$$
(4.12)

To prove Proposition 4.2.2, we may assume that $d(p, q) = \varrho$, and p is not farther from l than q. We write r to denote the radius of the semicirle bounded by l and use ψ to denote the angle of l and the perpendicular bisector of the segment pq. In addition, we write T_0 to denote the triangle whose sides are of lengths r, r, and ϱ, and ω denotes the half of the angle opposite to the side of length ϱ; hence, $\omega \leq \psi \leq \pi/2$. Let us fix r and vary ψ. Since

$$\Omega = \begin{cases} \varphi_r(\psi - \omega) + \dfrac{|T_0|}{2|T(\varrho)|} - \varphi_r(\psi + \omega) & \text{if } \omega \leq \psi \leq \pi/2 - \omega; \\[2ex] \varphi_r(\psi - \omega) + \dfrac{|T_0|}{2|T(\varrho)|} + \varphi_r(\pi/2 - \psi - \omega) & \text{if } \pi/2 - \omega \leq \psi \leq \pi/2, \end{cases}$$

we deduce by (4.12) that

$$\frac{\partial}{\partial \psi} \Omega = \frac{\operatorname{sh} r \operatorname{ch} r \cdot \sin \omega \cos \psi}{(1 + \sin(\psi + \omega) \operatorname{sh}^2 r)(1 + \sin(\psi - \omega) \operatorname{sh}^2 r)}$$

$$\times \left(-\frac{\operatorname{sh}^2 r}{\varrho} \cdot \sin^2 \psi + \frac{2 \operatorname{sh} r \cos \omega}{|T(\varrho)|} \cdot \sin \psi + \frac{1 + \operatorname{sh}^2 r \sin^2 \omega}{\varrho} \right).$$

Now the expression in the parentheses is quadratic with negative main coefficient in $\sin \psi$ and is positive for $\sin \psi = 0$. Therefore, it is sufficient to check whether $\Omega \geq 1/2$ if $\psi = \omega$ or $\psi = \pi/2$.

If $\psi = \omega$ then $p \in l$, and we write α to denote angle of the segment pq and l. The conditions in Proposition 4.2.2 yield $0 \leq \alpha < \gamma(\varrho)$, where $\gamma(\varrho)$ is the angle of $T(\varrho)$. Since $\Omega = \varphi_\varrho(\alpha)$, and $\Omega = 1/2$ holds if $\alpha = 0$ or $\alpha = \gamma(\varrho)$, we conclude Proposition 4.2.2 by (4.12).

Thus let $\psi = \pi/2$, and hence the perpendicular bisector of pq is perpendicular to l and divides Q into two equal halves Q' and Q''. We write x to denote the distance of the midpoint of pq from l, and we write θ to denote the acute angle of Q' and η to denote the side of Q' contained in l. We draw the diagonal of Q' cutting θ, apply the laws of sines and cosines in the two resulting triangles, and deduce

$$\text{th}\,\frac{\varrho}{2} = \text{ch}\,x\,\text{th}\,\eta \quad \text{and} \quad \text{sh}\,\frac{\varrho}{2}\,\text{th}\,x\tan\theta = 1.$$

In particular, $|Q| = \pi - 2\theta$ and $d(\tilde{p}, \tilde{q}) = 2\eta$ yield

$$\frac{\partial}{\partial x}\,\Omega = \frac{1}{1 + \text{ch}^2\frac{\varrho}{2}\,\text{sh}^2 x} \cdot \left(\frac{\text{sh}\,\frac{\varrho}{2}}{|T(\varrho)|} - \frac{\text{sh}\,\frac{\varrho}{2}\,\text{sh}\,\frac{\varrho}{2}\,\text{sh}\,x}{\varrho} \right).$$

The expression in the parentheses is a monotonically decreasing function of x, and hence it is sufficient to check Ω if either $x = 0$ or x is the height h of $T(\varrho)$. If $x = h$ then Q contains a copy of $T(\varrho)$; thus, $\Omega > 1/2$. However, if $x = 0$ then $\Omega = 1/2$, completing the proof of Proposition 4.2.2 in the hyperbolic case.

Finally, in the spherical case, we may actually assume that $p \in l$. Thus, the formula analogous to (4.12) yields Proposition 4.2.2 on S^2. □

Comments. It was A. Thue who recognized that a statement like (4.9) should hold in the Euclidean plane, and he proved some analogous results in [Thu1892] and [Thu1910] (see also [Thu1977]). His work was long forgotten, and later H. Groemer [Gro1960] and N. Oler [Ole1961a] proved (4.9). (N. Oler actually proved the more general Theorem 2.3.1.) The equality condition for the Thue–Groemer inequality was described in H. Groemer [Gro1960].

The essential case of the Thue–Groemer inequality is when the compact convex set is the convex hull of the points. This case can be generalized to any finite simplicial complex Σ in H^2, $\mathbb{R}^2$, or S^2 satisfying that the distance between any two vertices is at least ϱ. Writing $\nu(e)$ to denote the number of triangles of Σ containing a given edge e of Σ, we have

$$\frac{|\text{supp}\,\Sigma|}{2|T(\varrho)|} + \frac{1}{2\varrho} \cdot \sum_{e\,\text{edge}} (2 - \nu(e))|e| + \chi(\Sigma) \geq f_0(\Sigma). \tag{4.13}$$

The Euclidean version of (4.13) is proved by J. H. Folkman and R. L. Graham [FoG1969], and the hyperbolic and the spherical version is due to K. Böröczky Jr. [Bör2002].

4.3. Minimal Area for Packings of *n* Unit Discs

Given $n \geq 2$, we discuss packings of n unit discs such that the area of the convex hull is minimal. The Thue–Groemer inequality and its equality case suggest that the optimal packing might be a Groemer–packing; namely, the convex hull C_n of the centres can be triangulated using the n centres as vertices into regular triangles of edge length two. However, the Groemer packings of highest densities are the ones where $P(C_n)$ is minimal. Since C_n has at most six sides, we deduce $A(C_n) \leq (\sqrt{3}/24)\, P(C_n)^2$ according to the isoperimetric inequality for polygons (Theorem A.5.8). Therefore, the Thue–Groemer inequality (4.9) yields $P(C_n) \geq 2\lceil\sqrt{12n-3}-3\rceil$. If, in addition, $P(C_n) = 2\lceil\sqrt{12n-3}-3\rceil$ then the Groemer packing is called a *Wegner packing*. A typical example is when C_n is a regular hexagon. On the one hand there may not exist any Wegner packing for a given n (see Theorem 4.3.3); on the other hand there exist two noncongruent Wegner packings, say, of 18 unit discs.

Theorem 4.3.1 (Wegner inequality). *If D_n is the convex hull of n nonoverlapping unit discs then*

$$A(D_n) \geq \sqrt{12}\cdot(n-1) + (2-\sqrt{3})\cdot\left\lceil\sqrt{12n-3}-3\right\rceil + \pi.$$

Equality holds if and only if the packing is a Wegner packing.

We note that Theorem 4.3.1 readily yields the triangle bound Corollary 4.1.2. The lower bound of Theorem 4.3.1 is a very good estimate even if strict inequality holds:

Lemma 4.3.2. *For any $n \geq 2$, there exists a Groemer packing of n unit discs whose convex hull D_n satisfies*

$$A(D_n) \leq \sqrt{12}\cdot(n-1) + (2-\sqrt{3})\cdot\left\lceil\sqrt{12n-3}-2\right\rceil + \pi.$$

Lemma 4.3.2 will be verified after the proof of Theorem 4.3.3. We do not prove the Wegner theorem because the proof is rather involved. The main idea is to consider the so-called P-hull of the circular discs instead of the convex hull. The P-hull is contained in the convex hull, and it coincides with the convex hull for Groemer packings. However, the P-hull is rather sensitive to the way the circular discs are arranged, and it allows an inductional argument for n to prove the estimate of Theorem 4.3.1.

Next we characterize all n such that a Wegner packing of n unit discs exists, and we show that these numbers constitute $95.83\ldots\%$ of $\mathbb{N}$:

Theorem 4.3.3.

(i) *Given $n \geq 2$, some Wegner packing of n unit discs exists if and only if $\lceil\sqrt{12n-3}\rceil^2 + 3 - 12n \neq (3k-1)\cdot 9^m$ for positive $k, m \in \mathbb{Z}$.*

(ii) Given $N \geq 2$, let $f(N)$ be the number of $2 \leq n \leq N$ such that there exists some Wegner packing of n unit discs. Then

$$\lim_{N \to \infty} \frac{f(N)}{N} = \frac{23}{24}.$$

Proof. Let C_n be the convex hull of the centres in a Groemer packing of n unit discs. If the lengths of three nonneighbouring sides are $2a, 2b, 2c$ then the other three nonneighbouring sides are of lengths $2(a + d), 2(b + d),$ $2(c + d)$, and the latter three sides determine a regular triangle of side length $2(a + b + c + d)$ for some integer d. It follows that the perimeter of C_n is $2p$ for $p = 2a + 2b + 2c + 3d$. Moreover,

$$n = \frac{A(C_n)}{\sqrt{12}} + \frac{p}{2} + 1,$$

$$A(C_n) = \sqrt{3} \cdot ((a + b + c + d)^2 - a^2 - b^2 - c^2).$$

Thus the isoperimetric deficit $p^2 - \sqrt{12}\, A(C_n)$ can be expressed as

$$\begin{aligned}
(p + 3)^2 + 3 - 12n &= p^2 - \sqrt{12}\, A(C_n) \\
&= (2a - b - c)^2 + 3(b - c)^2 + 3d^2.
\end{aligned}$$

Here the first expression is never of the form $3k - 1$, whereas $x^2 + 3y^2 + 3z^2$, $x, y, z \in \mathbb{Z}$, is never of the form $(3k - 1) \cdot 9^m$. Therefore, we conclude the necessity condition in (i).

If $n \leq 30$ then the condition in (i) is also sufficient, as can be checked by hand. So let $n \geq 31$, and let $p = \lceil \sqrt{12n - 3} - 3 \rceil$ satisfy that $A = (p + 3)^2 + 3 - 12n$ is not of the form $(3k - 1) \cdot 9^m$. Then $3A$ is not of the form $(9k - 3) \cdot 9^m$, and thus $3A$ can be written in the form

$$3A = 3x^2 + \tilde{y}^2 + \tilde{z}^2 \tag{4.14}$$

for some $x, \tilde{y}, \tilde{z} \in \mathbb{Z}$ (see the comments). Checking remainders modulo 3, we deduce that $\tilde{y} = 3y$ and $\tilde{z} = 3z$ for some $y, z \in \mathbb{Z}$, and hence

$$(p + 3)^2 + 3 - 12n = x^2 + 3y^2 + 3z^2.$$

Changing x to $-x$ if necessary, we may assume that $p \equiv x \pmod 3$. Since a square is 0 or 1 $\pmod 4$, and the role of y and z is symmetric, we may assume that $p \equiv z \pmod 2$, and hence $x \equiv y \pmod 2$. We define

$$a = \frac{p + 2x - 3z}{6}, \quad b = \frac{p + 3y - x - 3z}{6}, \quad c = \frac{p - 3y - x - 3z}{6}, \quad d = z.$$

The divisibility properties imply that a, b, c, d are integers. We need the existence of a hexagon (namely, $C_n/2$) whose sides are parallel to the respective sides of some regular hexagon, and whose side lengths are $a, b, c, a + d$, $b + d$, and $c + d$ in this order. It is equivalent that all these numbers are positive, which, in turn, holds if

$$2|x| + 3|z| < p \quad \text{and} \quad |x| + 3|y| + 3|z| < p. \tag{4.15}$$

Since $p < \sqrt{12n - 3} - 2$, we have

$$x^2 + 3y^2 + 3z^2 < (p + 3)^2 - (p + 2)^2 = 2p + 5. \tag{4.16}$$

Here $p \geq 17$ follows by $n \geq 31$, and hence the conditions in (4.15) follow by (4.16) and the Cauchy–Schwarz inequality (A.1). Therefore, there exists a Groemer packing of n unit discs with p boundary discs, completing the proof of (i).

Turning to (ii), let $g(N) = N - f(N)$. It is sufficient to show that, for any $\varepsilon > 0$ and for N large,

$$g(N) = \frac{1 + O(\varepsilon)}{24} \cdot N, \tag{4.17}$$

where the implied constant in $O(\cdot)$ is some absolute constant. Given $n \geq 2$, we set $s = \lceil \sqrt{12n - 3} \rceil$ and define t by the formula

$$12n - 3 = s^2 - t.$$

The condition $s = \lceil \sqrt{12n - 3} \rceil$ is equivalent to $s^2 - t > (s - 1)^2$ and hence to $s > t + 1/2$. However, $n \leq N$ is equivalent to $s \leq \sqrt{12N - 3 + t}$. Therefore, $g(N)$ is the number of "good" pairs $s, t \in \mathbb{N}$ such that $t = l \cdot 9^m$ for some positive $l, m \in \mathbb{Z}$ with $l \equiv -1 \pmod 3$, $n = (s^2 - t + 3)/12$ is an integer, and

$$\frac{t + 1}{2} < s \leq \sqrt{12N - 3 + t}. \tag{4.18}$$

If N is large then $t < (1 + \varepsilon)2\sqrt{12N}$ and $m < (1 + \varepsilon)\log_9 2\sqrt{12N}$ follows by (4.18). Now n is an integer if and only if either $l \equiv -4 \pmod{12}$ and $s \equiv \pm 3 \pmod{12}$ or $l \equiv -1 \pmod{12}$ and $s \equiv 0, 6 \pmod{12}$. We observe that if t is fixed and $t < (1 - \varepsilon)2\sqrt{12N}$ then a "good" pair s of t occurs uniformly and with density $1/6$. Therefore, given large N and $1 \leq m \leq (1 - \varepsilon)\log_9 2\sqrt{12N}$, the number of "good" pairs $s, t \in \mathbb{N}$ is

$$(1 + O(\varepsilon)) \sum_{\substack{2 \leq l \leq \frac{(1-\varepsilon)2\sqrt{12N}}{9^m} \\ l \equiv -1, -4 \pmod{12}}} \frac{1}{6} \cdot \left(\sqrt{12N} - \frac{l \cdot 9^m}{2} \right) = (1 + O(\varepsilon)) \cdot \frac{N}{3 \cdot 9^m}.$$

We conclude that

$$g(N) = (1 + O(\varepsilon)) \cdot \sum_{m \geq 1} \frac{N}{3 \cdot 9^m} = (1 + O(\varepsilon)) \cdot \frac{N}{24},$$

completing the proof of Theorem 4.3.3. □

Proof of Lemma 4.3.2. If there exists no Wegner packing for some n then $p = \lceil \sqrt{12n - 3} - 3 \rceil$ is divisible by 3 according to Theorem 4.3.3 (i). Thus $(p + 1 + 3)^2 + 3 - 12n$ is not equal $(3k - 1) \cdot 9^m$ for any positive $k, m \in \mathbb{Z}$, and the proof of the sufficiency of the condition in Theorem 4.3.3 (i) yields the existence of a Groemer packing of n unit discs with $p + 1$ boundary discs. In turn, we conclude Lemma 4.3.2. □

Open Problem.

(i) Prove or disprove that, for any n, the packing of n unit discs minimizing the area of the convex hull of the discs is the Groemer packing of minimal perimeter. In other words, if D_n is the optimal convex hull of n nonoverlapping unit discs then equality holds either in Theorem 4.3.1 or in Lemma 4.3.2.

Comments. See L. E. Dickson [Dic1939] p. 97 for condition (4.14).

L. Fejes Tóth conjectured in [FTL1975a] that if $n = 6\binom{k}{2} + 1$ for some $k \geq 2$ then the optimal packing of n unit discs is given by the regular hexagon of side length $2(k - 1)$. Theorem 4.3.1 is due to G. Wegner [Weg1986], who also verifies that $n = 121$ is the smallest n such that no Wegner packing of n discs exists, and exhibits the two noncongruent Wegner packings of 18 discs. Lemma 4.3.2 and Theorem 4.3.3 are proved by K. Böröczky Jr. and I. Z. Ruzsa [BöR?].

4.4. Packings on the Sphere

There are two types of natural containers for finite packings of equal spherical circular discs: One is the sphere S^2 itself, and the other type is a spherical convex domain that is a proper subset of S^2. We start our discussion with packings with respect to S^2.

Readily, the optimal arrangement for two circular discs is given by antipodal centres, and for three circular discs, the centres are evenly spaced on a great cirle. Let us now consider $n \geq 4$ nonoverlapping spherical circular discs of radius r. The case of three circular discs shows that $r < \pi/3$, and hence we can form the Dirichlet–Voronoi cell decomposition. Since any Dirichlet–Voronoi cell is a spherical polygon, the density inside any of them is at most $\tau_{S^2}(r)$ according to Theorem 4.1.1. We conclude the following:

Theorem 4.4.1. *The density of any packing of $n \geq 4$ spherical circular discs of radius r on S^2 is at most $\sigma_{S^2}(r)$, and equality holds if and only if $n = 4, 6, 12$, and the packing is determined by the regular tetrahedron, octahedron, and icosahedron, respectively.*

The triangle bound of Theorem 4.4.1 can be interpreted in the following way: If r_n is the maximal radius of $n \geq 4$ equal nonoverlapping circular discs on S^2 then

$$r_n \leq \arccos \frac{1}{2 \sin \frac{n\pi}{6n-12}}, \qquad (4.19)$$

and equality holds for $n = 4, 6, 12$.

When $n = 5$, readily, we have $r_5 \geq r_6 = \pi/4$. We suppose that five nonoverlapping circular discs of obtuse radius can be placed on S^2, and we seek a contradiction. The opposite hemispheres to any two of the centres intersect in a two-gon of acute angle. In turn, this two-gon, which contains the other three centres, can be divided into two spherical triangles of diameter $\pi/2$, and one of these triangles contain at least two centres. This is absurd, and hence we conclude

$$r_5 = \pi/4 = r_6. \qquad (4.20)$$

A similar argument yields that any five nonoverlapping spherical circular discs of radius $\pi/4$ satisfy that two of the centres are antipodal, and the corresponding great circle contains the other three centres.

Actually, r_n is also known for $n = 7, \ldots, 11$ (hence for all $n \leq 12$) and for $n = 24$. We do not provide (the rather involved) proofs (see the comments for references). The centres in the optimal configuration are vertices of some semiregular polytope if $n = 8$ or 24, and one square and three (or four) regular triangles meet at each vertex if $n = 8$ (if $n = 24$). In addition, $r_{11} = r_{12}$, and any optimal packing of eleven spherical circular discs is obtained by removing one from the icosahedral packing of twelve discs. We note that the optimal packings of other numbers of equal spherical circular discs seem to have little symmetry.

Next, we show that the triangle bound holds for packings of equal spherical circular discs not only with respect to S^2 but with respect to any spherical convex domain that contains the discs.

Theorem 4.4.2. *If a spherical convex domain $C \neq S^2$ contains $n \geq 2$ circular discs of radius r, $0 < r < \pi/3$, then*

$$A(C) > n \cdot \frac{A(B(r))}{\sigma_{S^2}(r)}.$$

Since the spherical triangle bound satisfies $\sigma_{S^2}(r) < \pi/\sqrt{12}$ according to (4.8), we deduce the following:

Corollary 4.4.3. *If the spherical convex domain C contains $n \geq 2$ circular discs of radius r, $0 < r < \pi/3$, then*

$$\frac{n\,A(B(r))}{A(C)} < \frac{\pi}{\sqrt{12}}.$$

Remark. The constant $\pi/\sqrt{12}$ is optimal in Corollary 4.4.3 because the metric on small patches of S^2 can be arbitrarily close to being Euclidean.

To prove Theorem 4.4.2, one essential step is to consider the case $n = 2$.

Proposition 4.4.4. *If the spherical convex domain C is the convex hull of two touching circular discs of radius r, $0 < r < \pi/3$, then*

$$\frac{2\,A(B(r))}{A(C)} < \sigma_{S^2}(r).$$

Proof. We define

$$\alpha(r) = \arcsin \frac{1}{2\cos r} \quad \text{and} \quad \beta(r) = \arcsin \frac{1}{\sqrt{2}\cos r}.$$

Now the common tangent of the two circular discs and the line passing through the centres divide C into four congruent parts. Each of these parts can be dissected into two convex domains, namely, a sector of some circular disc of radius r with angle $\pi - 2\beta(r)$ and a quadrilateral such that three angles are right angles and the fourth angle, which is enclosed by two sides of length r, is $2\beta(r)$. Therefore, Proposition 4.4.4 is equivalent to the inequality (cf. (4.5))

$$\frac{\frac{1}{2} A(B(r))}{\frac{\pi - 2\beta(r)}{2\pi} A(B(r)) + 2\beta(r) - \frac{\pi}{2}} < \frac{\frac{\alpha(r)}{2\pi} A(B(r))}{\alpha(r) - \frac{\pi}{6}}.$$

Here $A(B(r)) = 2\pi(1 - \cos r)$, and hence the inequality can be rewritten in the form

$$\frac{2}{\pi}\beta(r) - \frac{1}{2\cos r} > 1 - \frac{\pi}{6\alpha(r)} \cdot \frac{1}{\cos r}.$$

Let us introduce the variable $s = 1/(2\cos r)$. Since $\alpha(r) > \pi/6$, Proposition 4.4.4 follows if

$$f(s) = \frac{2}{\pi} \arcsin \sqrt{2}s - \frac{6}{\pi} \arcsin s + s > 0 \tag{4.21}$$

holds for $1/2 < s \leq 1/\sqrt{2}$. We observe that $f(0) = 0$, and the derivative

$$f'(s) = 1 + \frac{2}{\pi} \cdot \frac{7 - 16s^2}{\sqrt{1 - s^2} \cdot \sqrt{1 - 2s^2} \cdot (3\sqrt{1 - s^2} + \sqrt{2} \cdot \sqrt{1 - 2s^2})}$$

is positive for $s \leq \sqrt{7}/4$. If $s \in [\sqrt{7}/4, 1/\sqrt{2}]$ then we consider the intervals $[\sqrt{7}/4, 0.68]$ and $[0.68, 1/\sqrt{2}]$, and we deduce $f(s) > 0$ by the inequalities

$$\frac{2}{\pi} \arcsin\left(\sqrt{2} \cdot \frac{\sqrt{7}}{4}\right) - \frac{6}{\pi} \arcsin 0.68 + \frac{\sqrt{7}}{4} = 0.0032 > 0,$$

$$\frac{2}{\pi} \arcsin\left(\sqrt{2} \cdot 0.68\right) - \frac{6}{\pi} \arcsin \frac{1}{\sqrt{2}} + 0.68 = 0.0031 > 0.$$

In turn, we conclude Proposition 4.4.4. □

Proof of Theorem 4.4.2. Let $C \neq S^2$ be a spherical convex domain, and let the nonoverlapping circular $B_1, \ldots, B_n$, $n \geq 2$, discs of radius r and of centre $x_1, \ldots, x_n$ be contained in C. We have $r \leq \pi/4$ because $C \neq S^2$. Theorem 4.4.2 will follow by showing that the triangle bound holds for each Dirichlet–Voronoi cell.

We may assume that C is the convex hull of the discs and write D_i to denote the Dirichlet–Voronoi cell of x_i. If $B_i \cap \partial C$ does not contain any circular arc then the triangle bound holds according to Theorem 4.1.1. Therefore, we may assume that $B_1 \cap \partial C$ contains a circular arc; hence, x_1 is a vertex of the convex hull K of $x_1, \ldots, x_n$. Let us observe that if a circular disc B of radius r is moved away from B_1 along a great circle passing through x_1 then the convex hull of B_1 and B increases as a set. In particular, Proposition 4.4.4 yields directly Theorem 4.4.2 if K is a segment. Thus, we may assume that int $K \neq \emptyset$, and x_1x_2 and x_1x_3 are the two sides of K meeting at x_1. For $i = 2, 3$, let Q_i be the quadrilateral such that Q_i does not overlap K, two sides of length r meet at x_1, one of these sides lies in the segment x_1x_i, and every angle but the one at x_1 is a right angle. Then Q_2 and Q_3 are contained in D_1, and the total area of Q_2, Q_3, and the part of B_1 not covered by Q_2 and Q_3 is just half the area of the convex hull of two touching circular discs of radius r. Therefore, the triangle bound holds for D_1 according to Proposition 4.4.4, which, in turn, yields Theorem 4.4.2. □

Comments. The problem of finding the maximal radius of n nonoverlapping equal spherical circular discs is due to the Dutch biologist P. M. L. Tammes [Tam1930]. He investigated the distribution of orifices on pollen grains and found that, in some cases, the orifices are spread onto an approximately spherical surface in a way that the minimal distance between any two orifices is as large as possible (see L. Fejes Tóth [FTL1964] or T. Tarnai [Tar1984] and [Tar?]). P. M. L. Tammes himself solved the case of five circular discs.

Theorem 4.4.1 and (4.19) are due to L. Fejes Tóth (see say [FTL1964]). The estimate (4.19) was slightly improved by R. M. Robinson [Rob1961] when n is large.

Packings of $n = 7, 8, 9$ circular discs on S^2 were treated by K. Schütte and B. L. van der Waerden [ScW1951]. The cases $n = 10, 11$ were solved by L. Danzer (see [Dan1986]). Independent proofs were provided by L. Hárs [Hár1986] when $n = 10$, and by K. Böröczky [Bör1983] when $n = 11$. Finally, if $n = 24$ then the maximal radius is due to R. M. Robinson [Rob1961]. (See L. Fejes Tóth [FTL1964] and [FTL1972] for thorough discussions of these results.) Contrary to some expectations, T. Tarnai and Zs. Gáspár [TaG1991] constructed an arrangement of 23 points on S^2 showing that $r_{23} > r_{24}$. We note that if $n = 7, 9, 10$ then the optimal arrangement is not so symmetric (see L. Fejes Tóth [FTL1964] pp. 231–232 for a description).

For other numbers of circular discs only estimates are known. The quest for efficient packings has been encouraged by the presence of powerful computers (see the home page [Slo] of N. J. A. Sloane). Moreover, efficient packings with high symmetry on S^2 (say packings on the projective plane) are discussed by T. Tarnai [Tar?]. When $n = 13$ and $n = 14, 15, 16$ then K. Böröczky and L. Szabó [BöS2003a] and [BöS2003b], respectively, give quite good upper bounds. We will only discuss the case of thirteen circular discs in some detail (see Theorem 4.5.1) because it is related to the Newton–Gregory problem (see the comments for Section 4.5).

Recalling that r_n is the maximal radius of n equal nonoverlapping circular discs on S^2, we have the known asymptotic estimates of

$$\sqrt{\frac{2\pi}{\sqrt{3}}} \cdot \frac{1}{\sqrt{n}} - O\left(\frac{1}{n^{5/6}}\right) < r_n < \sqrt{\frac{2\pi}{\sqrt{3}}} \cdot \frac{1}{\sqrt{n}} - \frac{f(n)}{n^{3/2}}, \quad (4.22)$$

where $f(n)$ tends to infinity. Here the lower bound is due to R. van der Waerden in [Wae1952], and the upper bound is due to K. Böröczky Jr [Bör2001]. Corollary 4.4.3 was proved by J. Molnár [Mol1952]. Let K be a spherical convex domain, $K \neq S^2$. G. Fejes Tóth (personal communication) pointed out the following estimate:

If $n \geq 3$ congruent copies of K can be packed on S^2 then there exists a polygon of $\lceil 6(n - 2)/n \rceil$ sides that contains K, and whose area is at most $4\pi/n$.

This result can be proved similarly to Theorem 1.2.1. The corresponding Dowker–type theorem is due to J. Molnár [Mol1955]. If K is a certain spherical regular pentagon then efficient packings are known (see the survey by T. Tarnai [Tar?]).

The polyhedral version of the Tammes problem is investigated by K. Bezdek, G. Blekherman, R. Connelly and B. Csikós [BBC2001].

4.5. Packings of Spherical Discs of Radius π/6

For thirteen nonoverlapping equal circular discs on S^2, determining the optimal radius seems out of reach. However, the question of whether the radius of the circular discs can be chosen to be $\pi/6$ has attracted considerable attention because of the equivalent *Newton–Gregory problem* (see the comments), namely, whether thirteen unit balls in $\mathbb{R}^3$ can touch a given unit ball without overlapping.

The twelve circular discs of radius $\pi/6$ on S^2 that are centred at the vertices of suitable icosahedron are pairwise disjoint. We prove the following:

Theorem 4.5.1. *At most twelve circular discs of radius $\pi/6$ can be packed on S^2.*

Remark. It follows that the Newton number of three-balls is twelve (cf. Section 8.1).

In this section we write $|\cdot|$ to denote the area on the sphere. In addition, distance, triangle, etc. always mean spherical distance, spherical triangle, etc. We use the laws of sines and cosines to calculate areas of polygons of given angles and side lengths. Let us recall that r_n is the maximal radius of n equal nonoverlapping spherical circular discs. To prove Theorem 4.5.1, one of the basic tools is the triangle bound of Lemma 4.1.5; namely, if T is a Delone triangle for a saturated packing of circular discs of radius r_n then

$$|T| \geq |T(2r_n)|. \tag{4.23}$$

It follows that $2r_{12}$ is the spherical distance between neighbouring vertices of the inscribed regular icosahedron and, in addition, that

$$r_{13} < r_{12}. \tag{4.24}$$

We prove Theorem 4.5.1 by contradiction; therefore, we suppose that

$$r_{13} \geq \frac{\pi}{6}. \tag{4.25}$$

Let us consider a Delone triangulation for some saturated family of $n \geq 13$ points on S^2 such that the mutual distances between the points are at least

$2r_{13}$. Then the area of each Delone triangle T satisfies

$$|T| \geq |T(2r_{13})| \geq \left| T\left(\frac{\pi}{3}\right) \right| = 0.5512\ldots$$

according to the triangle bound (4.23). Now the number of Delone triangles is $2n - 4$ by the Euler formula, and hence $24 \cdot |T(\pi/3)| > 4\pi$ yields $n \leq 13$. However, $22 \cdot |T(\pi/3)| < 4\pi$, and thus the triangle bound itself does not contradict (4.25). Since

$$16 \cdot \left| T\left(\frac{\pi}{3}\right) \right| + 3.75 > 4\pi, \tag{4.26}$$

Theorem 4.5.1 is a consequence of the following improvement of the triangle bound for six of the triangles:

Lemma 4.5.2. *Let us consider a Delone triangulation for thirteen points $p_1, \ldots, p_{13}$ on S^2 such that the mutual spherical distances are at least $2r_{13}$. If $r_{13} \geq \pi/6$ then the total area of some six of the Delone triangles is larger than 3.75.*

The rest of the section is devoted to the proof of Lemma 4.5.2. A useful tool is to decompose S^2 into convex cells in a way such that each cell is the union of some Delone triangles, and each side is of length $2r_{13}$. We colour the segment between two of the points $p_1, \ldots, p_{13}$ red if the distance of the points is $2r_{13}$, and we call the geometric graph whose edges are the red segments and whose vertices are $p_1, \ldots, p_{13}$ the red graph. We verify that

(a) *the red graph has no isolated vertex;*
(b) *each red edge is a side of some Delone triangle; and*
(c) *the red edges divide S^2 into triangles, rhombi, and pentagons.*

For any red edge e, the circular disc whose diameter is e contains no other p_i than the endpoints; therefore, e is an edge of the Delone triangulation. The next step is to show a statement weaker than (c); namely,

$$\text{some red edges decompose } S^2 \text{ into convex spherical polygons.} \tag{4.27}$$

If this is not the case then $p_1, \ldots, p_{13}$ can be perturbed one by one into a family $p'_1, \ldots, p'_{13}$ such that the mutual distances among $p'_1, \ldots, p'_{13}$ are larger than $2r_{13}$ using the following operations: At each stage, we move either a p_i that is the endpoint of exactly one remaining red edge or a p_i such that the remaining red edges emanating from p_i are contained in a closed hemisphere, which contains p_i on its boundary. Since the existence of $p'_1, \ldots, p'_{13}$ contradicts the maximality of r_{13}, we conclude (4.27).

Any convex red polygon P has at most five sides because $2r_{13} \geq \pi/3$. We suppose that P contains a p_i different from its vertices, and we seek a contradiction. The number k of the sides of P satisfies $k \leq 5$, and the triangle bound (4.23) yields $|P| \geq k \cdot |T(2r_{13})|$. However, the area of P is at most the area of the regular k-gon Q_k of side length $2r_{13}$ according to the isoperimetric inequality (Theorem A.5.8) for polygons. Since $r_{13} < r_{12}$ yields that the circumradius of Q_k is less than $2r_{13}$, we deduce $|Q_k| < k \cdot |T(2r_{13})|$. This is absurd, which completes the proof of (a), (b), and (c).

Given $\varphi < \pi/2$, we write $S(\varphi)$ to denote the square of side length φ and write $\alpha_3(\varphi)$ and $\alpha_4(\varphi)$ to denote the angles of $T(\varphi)$ and $S(\varphi)$, respectively. We claim that

$$\alpha_3(\varphi) < \alpha_4(\varphi) < 2\alpha_3(\varphi), \ 0 < \varphi < \frac{\pi}{3};$$
$$\alpha_3\left(\frac{\pi}{3}\right) + \alpha_4\left(\frac{\pi}{3}\right) = \pi. \tag{4.28}$$

Here the inequalities readily follow by Lemma A.5.2. To verify the formula for $\varphi = \pi/3$, we consider the great circles determined by two opposite sides of $S(\pi/3)$. The two great circles divide S^2 into four two-gons, and one of them is decomposed into $S(\pi/3)$ and two copies of $T(\pi/3)$. In turn, we conclude (4.28).

Returning to the arrangement of thirteen points, we note that each angle of a Delone triangle is at most $2\alpha_3(2r_{13})$ because the circumradius is at most $2r_{13}$. Given a red rhombus or a red pentagon, one or two diagonals, respectively, of it occur as edges of the Delone triangulation. We call these diagonals Delone diagonals. Lemma A.5.2 yields for any red rhombus R that the Delone diagonal connects two vertices, where the angle α is at least $\alpha_4(2r_{13})$ and the area of R is a decreasing function of α.

It can be deduced via the Euler formula (A.10) that the Delone triangulation on the thirteen vertices has a vertex, say p_1, of degree at least six. We will find the six triangles for Lemma 4.5.2 near p_1.

Case I: No Delone Diagonal of Any Red Pentagon Ends at p_1

We pair two Delone triangles meeting at p_1 if their union is a red rhombus, and in this way we obtain an arrangement of rhombi and triangles at p_1 as described in Proposition 4.5.3 (with $\varphi = 2r_{13}$).

Proposition 4.5.3. *Let $\varphi \geq \pi/3$, and let the rhombi $R_1, \ldots, R_k$ and the triangles $T_1, \ldots, T_l$ form a neighbourhood of a common vertex p without overlapping. In addition, we assume that $2k + l \geq 6$, the sides meeting at p are of lengths φ, and the angle of a rhombus (triangle) at p is at least $\alpha_4(\varphi)$ $(\alpha_3(\varphi))$.*

Then $2k + l = 6$, *and*

$$|R_1| + \cdots + |R_k| + |T_1| + \cdots |T_l| \geq 2 \cdot \left| T\left(\frac{\pi}{3}\right) \right| + 2 \cdot \left| S\left(\frac{\pi}{3}\right) \right| = 3.82\dots.$$

Proof. First we assume that $k \leq 2$. Then the sum of the angles at p is at least $2\alpha_3(\pi/3) + 2\alpha_4(\pi/3) = 2\pi$; therefore, $\varphi = \pi/3$, and we have two squares congruent to $S(\pi/3)$ and two regular triangles congruent to $T(\pi/3)$.

Thus let $k \geq 3$. Considering the sum of the angles at p shows that $k = 3$ and $l = 0$. For $i = 1, 2, 3$, we write ψ_i to denote the angle of R_i at p, and T_i' to denote the triangle containing p that is cut off from R_i by the diagonal of R_i not containing p. We may assume that $\psi_1 \leq \psi_2 \leq \psi_3$. Since decreasing φ while keeping ψ_1, ψ_2, and ψ_3 constant decreases the areas of the rhombi R_1, R_2, and R_3, we may also assume that $\varphi = \pi/3$. Now Lemma A.5.3 yields that decreasing ψ_1 while keeping $\psi_1 + \psi_2$ constant decreases $A(T_1') + A(T_2')$. Applying the same argument to T_2' and T_3', we may finally assume that R_1 and R_2 are congruent to $S(\pi/3)$, and hence R_3 is the union of two copies of $T(\pi/3)$. $\qquad\qquad\qquad\qquad\qquad\qquad\qquad\qquad\qquad\qquad \square$

It follows by Proposition 4.5.3 that exactly six Delone triangles meet at p_1, and they are the ones for Lemma 4.5.2.

Case II: A Delone Diagonal of Some Red Pentagon Ends at p_1

Certain useful lower bounds on the area of a red pentagon are provided by Proposition 4.5.5, the proof of which, in turn, is based on the following proposition.

Proposition 4.5.4. *For a quadrilateral* $Q = pqrs$, *we assume that the vertices* p *and* q *are fixed and the lengths of the sides* qr, rs, *and* sp *are equal and fixed. If the angle* α *of* Q *at* p *increases from its infimum to its supremum then the area of* Q *increases until* Q *is symmetric through the perpendicular bisector of* pq, *and the area decreases after that position. In addition, the angle at* r *increases, and the angles at* q *and* s *decrease.*

Proof. Lemma A.5.2 yields directly the statements about the angles. Since the four vertices of Q lie on a circle if and only if Q is symmetric, it is sufficient to verify the following claim: If r lies inside the circle σ passing through p, q, and s, and Q_0 is obtained from Q by increasing α slightly, then $|Q_0| < |Q|$.

We write p_0, q_0, r_0, and s_0 to denote the new positions of p, q, r, and s; namely, they are the vertices of Q_0. Let t be a point on the circle σ such that

r lies in the interior of the quadrilateral $P = spqt$, and together with Q, we deform P into the quadrilateral $P_0 = s_0 p_0 q_0 t_0$ keeping the side lengths. Let us consider the points r_q and r_s in P_0, where the triangle $q_0 r_q t_0$ (the triangle $s_0 r_s t_0$) is congruent to the triangle qrt (the triangle srt). Now Lemma A.5.2 yields that $d(q_0, r_s) > d(q_0, r_0)$ and $d(s_0, r_q) > d(s_0, r_0)$; hence, we deduce again by Lemma A.5.2 that the interiors of Q_0 and the triangles $q_0 r_q t_0$ and $s_0 r_s t_0$ are disjoint. In particular, it is sufficient to prove that $|P_0| < |P|$.

Finally, we place the circular cap of radius ϱ on top of each side of P_0, where ϱ is the radius of the circle passing through p, q, and s. According to the isoperimetric inequality (Theorem A.5.7), the union of these four arcs encloses an area, which is less than the area of the circular disc of radius ϱ; hence $|P_0| < |P|$. $\qquad\qquad\square$

Proposition 4.5.5. *Let P be a red pentagon. Then*

(i) $|P| \geq |T(\pi/3)| + |S(\pi/3)| = 1.91\ldots$;
(ii) $|P| > 2.16$ if the angle of P at the common endpoint of the Delone diagonals is at most 3.

Proof. We write $q_1, \ldots, q_5$ to denote the vertices of P, where q_1 is the common endpoint of the Delone diagonals, and α_i to denote the angle of P at q_i. We will only use the following two properties of P:

- the sides of P are of equal length $\varphi \geq \pi/3$,
- $\alpha_2 \leq \alpha_3$ and $\alpha_5 \leq \alpha_4$.

Here the second property is the consequence of the fact that $q_1 q_3$ and $q_1 q_4$ are Delone diagonals. This property and Proposition 4.5.4 yield that $\alpha_3, \alpha_4 \leq \alpha_1$.

We assume that P is of minimal area under these two properties, where $\alpha_i = \pi$ is allowed for (i), and $\alpha_1 \leq 3$ is assumed for (ii). It follows that P is not the regular pentagon according to the isoperimetric inequality for pentagons (see Theorem A.5.8). In the following, when two neighbouring vertices of P are moved then it is done in a way such that $|P|$ decreases according to Proposition 4.5.4.

To prove (i), moving q_1 and q_5 shows that $\alpha_1 = \pi$, and moving q_3 and q_4 yields that, say, $\alpha_3 = \pi$ as well. Hence, P is a triangle of side lengths 2φ, 2φ, and φ. Decreasing the side of length φ to $\pi/3$ decreases the area of the triangle by Lemma A.5.2; therefore, $\varphi = \pi/3$, and P is the union of a regular triangle and a square.

For (ii), moving q_3 and q_4 shows that P is symmetric in the perpendicular bisector of the side $q_4 q_5$, say. If we keep the diagonal $q_1 q_3$ and the symmetry of P but decrease φ then we obtain a decreasing family of sets, and hence $\varphi = \pi/3$. Now moving q_1 and q_5 may destroy the symmetry but decrease the area unless $\alpha_1 = 3$. Therefore, $|P| = 2.169\dots$. $\qquad\square$

Let us continue with the proof of Lemma 4.5.2 in Case II. If at least two red pentagons meet at p_1 then Proposition 4.5.5 (i) provides six Delone triangles with total area at least 3.82; hence, we assume that p_1 is the vertex of exactly one red pentagon P.

First, we consider the case when only one Delone diagonal of P ends at p_1. We write $p_1, \dots, p_5$ to denote the vertices of P in cyclic order in a way that the Delone diagonals are $p_1 p_3$ and $p_5 p_3$, and we let $R = p_1 p_2 p_3' p_5$ be the rhombus such that p_3' is the vertex opposite to p_1. Since $p_1 p_2 p_3$ is a Delone triangle, the point p_3 lies in the circular disc determined by p_1, p_2, p_5, and hence the common angle of R and P at p_1 is at least $\alpha_4(2r_{13})$. Next, we write B to denote the circular disc such that p_2, p_3, p_5 lies on the boundary and c to denote the centre of B. We claim that c and p_2 lie on the same side of $p_3 p_5$. If not then p_4 is contained in the interior of the disc B, and hence p_2 lies in the interior of the circular disc determined by p_4, p_3, p_5. This is absurd because $p_4 p_3 p_5$ is a Delone triangle, which, in turn, yields the claim. Therefore, the triangle $p_2 p_3 p_5$ has larger area than the triangle $p_2 p_3' p_5$ according to Lemma A.5.2; thus, the total area of the Delone triangles $p_1 p_2 p_3$ and $p_1 p_3 p_5$ is at least $|R|$. In particular, Proposition 4.5.3 yields Lemma 4.5.2.

Finally, we assume that p_1 is the common endpoint of the Delone diagonals of P. If the angle of P at p_1 is at most 3 then the total area of the three Delone triangles inside P together with any other three Delone triangles is greater than $2.16 + 3 |T(\pi/3)| > 3.81$, confirming Lemma 4.5.2. If the angle of P at p_1 is larger than 3 then the number of Delone triangles that are not contained in P and meet at p_1 is three, and two of these triangles form a red rhombus R_0 whose larger angle is at p_1. Now the angle of R_0 at p_1 is at least $2\pi - \alpha(\pi/3) - 3$; hence, $|R_0| \geq |\tilde{R}_0| = 1.34$, where $\tilde{R}_0$ is a rhombus with an angle $2\pi - \alpha(\pi/3) - 3$ and of side length $\pi/3$. However, we have already seen that $|P| > 1.91$, and the area of the sixth triangle at p_1 is larger than 0.55; therefore, the total area of the six Delone triangles at p_1 is larger than 3.8. In particular, the proof for Lemma 4.5.2 and, in turn, the proof for Theorem 4.5.1 are now complete. $\qquad\square$

Comments. In 1694 Isaac Newton and David Gregory discussed the problem whether thirteen solid unit three-balls can touch a given unit ball, or equivalently, whether the maximal number of nonoverlapping spherical discs of radius $\pi/6$ on S^2 is thirteen. Newton had the right idea; namely, this maximal number is twelve but the first and still incomplete solutions were provided only in 1874–1875 by the physicists C. Bender [Ben1874], S. Günter [Gün1875], and R. Hoppe [Hop1874]. The first "rigorous proof" was given by K. Schütte and B. L. van der Waerden [ScW1953], which was simplified by J. Leech [Lee1956] into an argument based on an elegant idea. The proof in [Lee1956] considers the network of segments between any two centres whose distance is at most arccos($1/7$) and estimates the area of the resulting possibly nonconvex spherical polygons. There is one *a priori* possible arrangement where the estimates do not work, but this case is eliminated by graph theoretic arguments. The sketch of J. Leech [Lee1956] is reproduced in M. Aigner and G. Ziegler [AiZ1999] and Ch. Zong [Zon1999a]. Actually, the proof is left out from the second edition [AiZ2001] of the book by M. Aigner and G. Ziegler because the authors realized that a proper presentation of the argument of J. Leech is rather involved. The proof of Theorem 4.5.1 presented here is due to K. Böröczky [Bör2003a]. An independent proof applying similar ideas is provided by W. Y. Hsiang [Hsi2001]. Further arguments are provided by K. M. Anstreicher [Ant?], who mixed geometric tools and linear programming, and by O. R. Musin [Mus?], who improved substantially the linear programming bound (6.1).

No improvement of the estimate $r_{13} \leq \pi/6 = 0.5235$ was known until recently, when K. Böröczky and L. Szabó [BöS2003a] proved $r_{13} \leq 0.5137$. The conjectured optimal arrangement, which is described in K. Schütte and B. L. van der Waerden [ScW1951], shows that $r_{13} \geq 0.4986$.

4.6. Packings of Hyperbolic Circular Discs

For arrangements in the hyperbolic plane, the choice of practical container for finitely many nonoverlapping equal circular discs is unclear. In the ensuing we discuss "sausage packings": Let n circular discs of radius r be arranged in a way such that the centres are collinear and consecutive discs touch each other. Then the convex hull is the union of the n discs and $2(n - 1)$ triangles, and hence the density is larger than $(\mathrm{ch}\, r - 1)/\mathrm{ch}\, r$. Since $\sigma_{H^2}(r) < 3/\pi$ according to (4.7), the triangle bound does not hold with respect to the convex hull if r is large. Therefore, we consider a restricted class of convex containers.

Let $r > 0$. We call a compact set D a th$r$-*convex domain* if for any two points $x, y \in D$, D contains both hypercycle arcs of associated distance r connecting x and y. In particular, a th$r$-convex domain is a convex domain such that the curvature is at least th$r$ at every boundary point whenever it exists (see Lemma A.5.5), but we do not use this fact in this chapter. Typical examples are circular discs of any radii, or given a compact convex set C, the family of points whose distance from C is at most r (the *parallel domain* of radius r).

According to Corollary 4.1.2, the density of a packing of $n \geq 2$ Euclidean circular discs of radius r with respect to the convex hull is at most $\pi/\sqrt{12}$, and this is the optimal bound. In the hyperbolic case we verify the same estimate with respect to the th r-convex hull.

Theorem 4.6.1. *If a* th r*-convex domain* D *in* H^2 *contains* $n \geq 2$ *nonoverlapping circular discs of radius* r *then*

$$A(D) > 4\sqrt{3}(\operatorname{ch} r - 1) \cdot n.$$

Remark. In particular, the density of the packing with respect to D is less than $\pi/\sqrt{12}$. Since $\sigma_{H^2}(r) > \pi/\sqrt{12}$ according to (4.7), the triangle bound holds inside th r-convex domains. Actually, the density of a circular disc with respect to the corresponding Dirichlet–Voronoi cell might be equal to $\sigma_{H^2}(r)$ for certain values of r (see Theorem 4.1.1); therefore, it is the effect of the boundary that lowers the density below $\pi/\sqrt{12}$.

As any circular disc is th r-convex (see Corollary A.5.6), combining Corollary 4.1.2, Corollary 4.4.3, and Theorem 4.6.1 leads to the following:

Corollary 4.6.2. *If at least two nonoverlapping equal circular discs are placed in a larger circular disc in* $\mathbb{R}^2$, S^2, *or* H^2 *then the density of the packing is less than* $\pi/\sqrt{12}$.

Let us discuss the optimality of Theorem 4.6.1. First, the density bound $\pi/\sqrt{12}$ equivalent to Theorem 4.6.1 is optimal because if r is small then a circular disc of radius $\sqrt{r}$ contains nonoverlapping circular discs of radius r such that the density is arbitrary close to $\pi/\sqrt{12}$. For large r, Theorem 4.2.1 provides a very weak estimate on the one hand, and still th r-convexity is close to be the optimal condition in Theorem 4.6.1 on the other hand. More precisely, if $r \geq 1.75$ and a th r-convex domain contains at least two nonoverlapping circular discs of radius r then the density of the packing is at most $2\pi/r$ (which follows from $A(T(2r)) > 2\pi/3$ and Theorem 4.2.1). By contrast, given $0 < \varepsilon < 1/3$ and large r, let n circular discs of radius r be arranged in a way that the centres are collinear and consecutive discs touch each other. Then some elementary calculations yield that the density with respect to the th ϱ-convex hull of the circular discs is at least $1 - \varepsilon$ for $\varrho = r - \ln(r/(\pi\varepsilon))$.

Proof of Theorem 4.6.1. The argument is based on the absolute version Theorem 4.2.1 of the Thue–Groemer inequality. We write C to denote the convex hull of the centres of $n \geq 2$ nonoverlapping circular discs of radius r. We may assume that D is the parallel domain of C of radius r, and hence its area is (cf. (A.4))

$$\operatorname{ch} r \cdot A(C) + \operatorname{sh} r \cdot P(C) + 2\pi(\operatorname{ch} r - 1).$$

According to Theorem 4.2.1, Theorem 4.6.1 follows from the inequality

$$\left(\frac{1}{2A(T(2r))} \cdot A(C) + \frac{1}{4r} \cdot P(C) + 1 \right) \cdot 2\pi(\operatorname{ch} r - 1)$$

$$< \frac{\pi}{\sqrt{12}} \cdot (\operatorname{ch} r \cdot A(C) + \operatorname{sh} r \cdot P(C) + 2\pi(\operatorname{ch} r - 1)),$$

which is, in turn, equivalent to

$$\left(\frac{\sqrt{12}}{A(T(2r))} - \frac{\operatorname{ch} r}{\operatorname{ch} r - 1} \right) \cdot A(C) < \left(\frac{r \operatorname{sh} r}{\operatorname{ch} r - 1} - \sqrt{3} \right) \cdot \left(\frac{P(C)}{r} - \omega_r \right),$$

$$(4.29)$$

where

$$\omega_r = (4\sqrt{3} - 2\pi) \cdot \left(\frac{\operatorname{sh} r \cdot r}{\operatorname{ch} r - 1} - \sqrt{3} \right)^{-1}.$$

The coefficient of $P(C)$ in (4.29) is positive because

$$f(r) = \frac{r \operatorname{sh} r}{\operatorname{ch} r - 1} \text{ is monotonically increasing and hence } \frac{r \operatorname{sh} r}{\operatorname{ch} r - 1} > 2.$$

$$(4.30)$$

We also deduce using (4.30) that

$$\omega_r < 6 - 2\sqrt{3} < 4, \qquad (4.31)$$

which, in turn, yields (4.29) if $n = 2$.

Therefore, let $n \geq 3$, and hence $P(C) \geq 6r$. It follows by the inequality $\arcsin t < \pi t / 3$ for $0 < t < 1/2$ that

$$\sigma_{H^2}(r) = \frac{\pi - A(T(2r))}{A(T(2r))} \cdot (\operatorname{ch} r - 1)$$

$$= \frac{6 \arcsin \frac{1}{2 \operatorname{ch} r}}{A(T(2r))} \cdot (\operatorname{ch} r - 1) < \frac{\pi(\operatorname{ch} r - 1)}{A(T(2r)) \operatorname{ch} r},$$

which, together with $\sigma_{H^2}(r) > \pi/\sqrt{12}$ (see (4.7)), leads to the estimate

$A(T(2r)) < \sqrt{12}(\operatorname{ch} r - 1)/\operatorname{ch} r$. In particular, the coefficient of $A(C)$ in (4.29) is positive. We note that $\arcsin t > t$ and $\sigma_{H^2}(r) < 3/\pi$ (see (4.7)) yield the lower bound

$$A(T(2r)) > \frac{\pi(\operatorname{ch} r - 1)}{\operatorname{ch} r}. \tag{4.32}$$

We define R by the formula $P(C) = 2\pi \operatorname{sh} R$, and hence $A(C) \le 2\pi(\operatorname{ch} R - 1)$ holds by the isoperimetric inequality (Theorem A.5.7). Therefore, (4.29) (and, in turn, Theorem 4.6.1) follows if the inequality

$$2\pi(\operatorname{ch} R - 1)\left(\frac{\operatorname{ch} r - 1}{A(T(2r))} - \frac{\operatorname{ch} r}{\sqrt{12}}\right) < (2\pi \operatorname{sh} R - \omega_r r)\left(\frac{\operatorname{sh} r}{\sqrt{12}} - \frac{\operatorname{ch} r - 1}{2r}\right) \tag{4.33}$$

holds for any R satisfying $2\pi \operatorname{sh} R \ge 6r$.

For fixed $\theta > 0$, checking the derivative with respect to R shows that $(\operatorname{sh} R - \theta)/(\operatorname{ch} R - 1)$ is monotonically increasing if $(\operatorname{ch} R - 1)/\operatorname{sh} R < \theta$ and monotonically decreasing if $(\operatorname{ch} R - 1)/\operatorname{sh} R > \theta$. Since $(\operatorname{ch} R - 1)/\operatorname{sh} R$ is monotonically increasing, it is sufficient to consider the cases $2\pi \operatorname{sh} R = 6r$ and $R = \infty$ in (4.33).

First, let $2\pi \operatorname{sh} R = 6r$. Then $\operatorname{ch} R = \sqrt{1 + 9r^2/\pi^2}$, and (4.31) yields

$$2\pi \operatorname{sh} R - \omega_r r \ge 6r - \omega_r r > \sqrt{12} \cdot r.$$

According to (4.32), it is sufficient to prove that

$$\left(2 - \frac{\pi}{\sqrt{3}}\right) \operatorname{ch} r \left(\sqrt{1 + \frac{9r^2}{\pi^2}} - 1\right) < \left(\operatorname{sh} r - \frac{\sqrt{3}(\operatorname{ch} r - 1)}{r}\right) \cdot r. \tag{4.34}$$

We split the proof of (4.34) into two cases. If $r \le 1.2$ then the bounds $(\operatorname{ch} r - 1)/r < (1/2)\operatorname{sh} r$ (see (4.30)) and $\sqrt{1 + 9r^2/\pi^2} < 1 + 9r^2/(2\pi^2)$ show that it is sufficient to consider the inequality

$$\left(2 - \frac{\pi}{\sqrt{3}}\right) \cdot \frac{9}{2\pi^2} < \left(1 - \frac{\sqrt{3}}{2}\right) \cdot \frac{\operatorname{th} r}{r}.$$

Now this inequality holds for $r = 1.2$, and $\operatorname{th} r/r$ is monotonically decreasing; therefore, (4.34) holds if $r \le 1.2$.

However, if $r \ge 1.2$ then $(\operatorname{ch} r - 1)/r \le f(1.2) \cdot \operatorname{sh} r$ holds by (4.30). Using the bound $\sqrt{1 + 9r^2/\pi^2} < 1 + 3r/\pi$, (4.34) follows from

$$\left(2 - \frac{\pi}{\sqrt{3}}\right) \cdot \frac{3}{\pi} < (1 - f(1.2)\sqrt{3}) \cdot \operatorname{th} r.$$

Since this inequality holds for $r = 1.2$, and $\operatorname{th} r$ is monotonically increasing, we conclude (4.34) for all r, and, in turn, (4.33) when $2\pi \operatorname{sh} R = 6r$.

Thus, let R tend to infinity. Since $(\operatorname{ch} r - 1)/(2r) < \operatorname{sh} r/4$ by (4.30), we are done if

$$\frac{\operatorname{ch} r - 1}{A(T(2r))} < \frac{1}{\sqrt{12}} \cdot \operatorname{ch} r + \left(\frac{1}{\sqrt{12}} - \frac{1}{4}\right) \cdot \operatorname{sh} r. \qquad (4.35)$$

If we estimate the left-hand side by (4.32), it is sufficient to verify

$$\frac{1}{\pi} < \frac{1}{\sqrt{12}} + \left(\frac{1}{\sqrt{12}} - \frac{1}{4}\right) \cdot \operatorname{th} r.$$

This inequality holds for $r = 1.1$, and hence also for $r \geq 1.1$. Therefore, let $r \leq 1.1$. In this case we rewrite (4.35) in the form

$$\frac{\operatorname{ch} r - 1}{\frac{1}{\sqrt{12}} \cdot \operatorname{ch} r + \left(\frac{1}{\sqrt{12}} - \frac{1}{4}\right) \cdot \operatorname{sh} r} < A(T(2r)).$$

We have equality for $r = 0$; thus, it is sufficient to verify the same inequality for the derivatives, namely, that

$$\frac{(\sqrt{12} - 3) \cdot (\operatorname{ch} r - 1) + \sqrt{12} \cdot \operatorname{sh} r}{\operatorname{ch}^2 r + (2 - \sqrt{3}) \cdot \operatorname{ch} r \cdot \operatorname{sh} r + \left(\frac{7}{4} - \sqrt{3}\right) \cdot \operatorname{sh}^2 r} < \frac{6 \cdot \operatorname{sh} r}{\operatorname{ch} r \sqrt{4 \operatorname{ch}^2 r - 1}}.$$

We may forget about $(7/4 - \sqrt{3}) \cdot \operatorname{sh}^2 r$ in the dominator; hence, we prove

$$(\sqrt{12} - 3)(\operatorname{ch} r - 1)\sqrt{4 \operatorname{ch}^2 r - 1} + \sqrt{12} \operatorname{sh} r \sqrt{4 \operatorname{ch}^2 r - 1}$$
$$< 6 \operatorname{ch} r \cdot \operatorname{sh} r + (12 - 6\sqrt{3}) \operatorname{sh}^2 r.$$

Here $r \leq 1.1$ yields

$$(\sqrt{12} - 3)(\operatorname{ch} r - 1)\sqrt{4 \operatorname{ch}^2 r - 1} < (11 - 6\sqrt{3}) \operatorname{sh}^2 r$$

because this inequality holds for $r = 1.1$, and the quotient of the left-hand side and the right-hand side is monotonically increasing. Thus, the only thing left to verify is

$$\sqrt{12}\sqrt{4 \operatorname{ch}^2 r - 1} < 6 \operatorname{ch} r + \operatorname{sh} r.$$

Squaring and rearranging show that this formula actually identically holds for any r, and, in turn, we deduce (4.35). Therefore, the proof of (4.33) (hence also of Theorem 4.6.1) is finally complete. $\qquad \square$

Comments. Corollary 4.6.2 was conjectured by L. Fejes Tóth [FTL1972] and proved by K. Bezdek [Bez1982] in the hyperbolic case. Theorem 4.6.1 is due to K. Bezdek [Bez1984] (see also K. Böröczky Jr. [Bör2002]).

We note that the density of infinite packings in the hyperbolic plane is an intricate notion. One cannot define it simply using elementary cells like the Delone cells or the Dirichlet–Voronoi cells (see K. Böröczky [Bör1974]), but it is possible to define using ergodic theory for "homogeneous packings" (see L. Bowen and Ch. Radin [BoR2003] and [BoR?]). Since the triangle bound holds for every Dirichlet–Voronoi cell, the results of L. Bowen and Ch. Radin [BoR2003] show that the density of any such homogeneous infinite packing by circular discs of radius r is at most $\sigma_{H^2}(r)$ and equal to $\sigma_{H^2}(r)$ if the packing is coming from a side-to-side tiling by regular triangles of side length $2r$. Therefore, we have a stronger bound for finite packings inside th r-convex domains (see Theorem 4.6.1) than for infinite packings.

4.7. Related Problems

Minimal Perimeter for Packings of Unit Discs

Let us consider $n \geq 2$ nonoverlapping unit discs such that the perimeter of the convex hull D_n is minimal. Since $P(D_n) = P(C_n) + 2\pi$ holds for the convex hull C_n of the centres, it is equivalent to find the minimum of $P(C_n)$ under the condition that the compact convex set C_n contains n points whose mutual distances are at least two. If $n = 3$ and $n = 7$ then the Thue–Groemer inequality (4.9) and the isoperimetric inequality (Theorem A.5.8) for polygons yield that C_n is a regular triangle or a regular hexagon, respectively, of side length two. Conjecturally optimal configurations for small n and the connection of the problem to the pattern formation of lotus receptacles is discussed by T. Tarnai [Tar2001]. Moreover, as n tends to infinity, $D_n/\sqrt{n}$ tends to a circular disc of area $\sqrt{12}$ (see K. Böröczky Jr. [Bör1994] and Ch. Zong [Zon1995a], or Theorem 1.8.1). We recall that if the area of a convex domain containing n nonoverlapping unit discs is minimal then the convex domain is of hexagonal shape according to the Wegner theorem (Theorem 4.3.1) for approximately 96% of all n. Therefore, optimizing with respect to the area and with respect to the perimeter lead to different asymptotic shapes.

We note that the structure of an optimal packing of n unit discs with respect to the perimeter resembles hexagonal packings for large n. Yet no optimal packing is part of the hexagonal lattice packing for $n \geq 371$ according to A. Schürmann [Sch2002].

Minimal Diameter for Packings of Unit Discs

Finding the minimal diameter of $n \geq 3$ nonoverlapping unit discs is equivalent to determining the convex polygon C_n of minimal diameter that contains n

points whose mutual distances are at least two. More precisely, $\operatorname{diam} C_n + 2$ is the minimal diameter for packings of n unit discs.

It is not hard to show that C_n is the regular n-gon if $n = 3, 4, 5$ and a regular pentagon of circumradius two if $n = 6$ (see P. Bateman and P. Erdős [BaE1951]). The case $n = 7$ was solved by P. Bateman and P. Erdős [BaE1951], and the case $n = 8$ by A. Bezdek and F. Fodor [BeF1999]; namely, C_n is the regular $(n - 1)$-gon of side length two. The proof for $n = 7, 8$ is based on the fact that if each side of a p-gon is at least 2 and p is a prime then the diameter of the p-gon is minimized by the regular p-gon. This fact was proved in K. Reinhardt [Rei1922] and rediscovered later independently by S. Vincze [Vin1950] and D. G. Larman and N. K. Tamvakis [LaT1984]. We note that, for $n = 9$, the optimal arrangement is not given by the regular octagon (see A. Bezdek and F. Fodor [BeF1999]).

If n tends to infinity then $C_n / \sqrt{n}$ tends to a circular disc of diameter $2\sqrt{\sqrt{12}/\pi}$ (see the remark after Theorem 1.8.1). Although the structure of an optimal arrangement corresponding to D_n resembles hexagonal packings for large n, it is not part of the hexagonal lattice packing according to A. Schürmann [Sch2002].

Packings of Unit Discs inside a Larger Disc

The problem of minimal radius of a circular disc, that holds n nonoverlapping unit discs is solved for various small n. The cases $n \leq 10$ are discussed by U. Pirl [Pir1969], the case $n = 11$ is due to J. B. M. Melissen [Mel1994a], and the cases $n = 12, 13, 14$ and $n = 19$ are solved by F. Fodor in [Fod2000], [Fod2003a], [Fod2003b], and [Fod1999], respectively.

The most recent list of good constructions is contained in R. L. Graham, B. D. Lubachevsky, K. J. Nurmela, and P. R. J. Östegård [GLN1998]. In addition, if R_n is the minimal radius of a circular disc that contains n nonoverlapping unit discs then asymptotically $R_n \sim \sqrt{\sqrt{12}/\pi}\,\sqrt{n}$ (see Lemma 1.1.2).

Packings of Unit Discs inside a Square

The problem of optimal packings of n congruent circular discs in a square has received special attention since L. Moser [Mos1960] proposed it (see P. G. Szabó and E. Specht [SzS?]). If $n \leq 5$ then the solution is simple, and if $n = 6$ then an elegant argument due to J. B. M. Melissen [Mel1994b] is available. Actually, B. L. Schwartz [Sch1970] solved the problem for $n = 6$ earlier but his proof is rather complicated. If $7 \leq n \leq 27$ then K. J. Nurmela and P. R. J. Östegård [NuÖ1999] gave computer-aided solutions along a unified approach. Naturally, some specific cases were known before: If $n = 8$ then the solution is due to A. Meir and J. Schaer [MeS1965] (and this was actually the original problem proposed by L. Moser), and if $n = 9$ then the solution is due to J. Schaer [Sch1965]. For the cases $10 \leq n \leq 20$, R. Peikert, D. Würtz, M. Monagan, and C. de Groot [PWM1992] gave computer-aided proofs. The known algorithms have been further developed by M. Cs. Markót, who has solved the case $n = 28$ in [Mar?].

It is an interesting question when n is a square whether the solution is given by the square grid. The answer is clearly no for large n, and actually already for $n = 49$ according to a configuration due to K. J. Nurmela and P. R. J. Östegård [NuÖ1997]. However, the square grid is optimal for $n = 4, 9, 16, 25$ (see K. J. Nurmela and P. R. J. Östegård [NuÖ1999]). If $n = 36$ then K. Kirchner and G. Wengerodt [KiW1987] provided a rather troublesome (positive) solution. It would be nice to have a more clear (possible computer-aided) argument available.

Since the optimal packing of unit discs in a large region is based on the hexagonal lattice, it might be very difficult to find an infinite sequence of n where the optimal solution is known. Denoting by s_n the minimal side length of a square that holds n nonoverlapping unit discs, K. J. Nurmela, P. R. J. Östegård, and R. aus dem Spring [NÖS1999] construct packings showing that $\sqrt{12}\,n \geq s_n^2 - (3 + \sqrt{3})s_n + 2 + 2\sqrt{3}$, and they conjecture that equality holds for infinitely many n. We note that the Thue–Groemer inequality (4.9) yields $\sqrt{12}\,n < s_n^2 - (4 - 2\sqrt{3})s_n + (4 - 2\sqrt{3})$.

The hunt for conjecturally optimal configurations never ceases. The current records for $n \leq 200$ are contained in P. G. Szabó and E. Specht [SzS?].

A related problem is to find optimal packings of n equal circular discs into a rectangle. Given any rectangle, the problem is solved by M. Ruda [Rud1969] for $n \leq 5$. If $n = 6$ and the ratio of the longer side to the shorter side is at most $2/\sqrt{3}$ then the solution is due to J. B. M. Melissen [Mel1997]. If the shape of the rectangle is not prescribed then B. D. Lubachevsky and R. Graham [LuG2003] found dense packings of $n \leq 5000$ circular discs. Based on these examples, [LuG2003] conjectures that the ratio of the shorter side to the longer side of the optimal rectangle tends to $2 - \sqrt{3}$ as n tends to infinity.

Packings of Unit Discs inside an Equilateral Triangle

The problem of finding the smallest equilateral triangle containing n nonoverlapping unit discs is readily equivalent to determining the minimal t_n such that the equilateral triangle of side length t_n holds n points whose mutual distances are at least two. This problem was first investigated by F. Bolyai before 1830, who presented a reasonably dense (but actually not optimal) packing for $n = 19$ (see P. G. Szabó and E. Specht [SzS?] or the second edition of [Bol1832], Volume 2, pp. 119–122).

It follows by the Thue–Groemer inequality (4.9) that $t_{n(k)} = 2k$ for the triangular number $n(k) = (k + 1)(k + 2)/2$. It is a celebrated problem of P. Erdős and N. Oler [Ole1961b] whether $t_{n(k)-1} = 2k$; namely, if an equilateral triangle holds $n(k) - 1$ nonoverlapping unit discs then it can hold $n(k)$ as well. The conjecture is verified by R. Milano [Mil1987] if $n = 5$, by J. B. M. Melissen [Mel1993] if $n = 9$, and by Ch. Payan [Pay1997] if $n = 14$.

In addition to these results, the optimal arrangement is known for $n \leq 4$ (see R. Milano [Mil1987]), for $n = 7, 8, 12$ (see J. B. M. Melissen [Mel1993]), and for $n = 11$ (see J. B. M. Melissen [Mel1994c]). The best constructions to date are presented by R. L. Graham and B. D. Lubachevsky [GrL1995]. Since, for large n, the density of the optimal packing is close to $\pi/\sqrt{12}$, we have asymptotically $t_n \sim \sqrt{8n}$.

Packings of Unit Discs with Minimal Second Moment

Among packings of n unit discs with centres $x_1, \ldots, x_n$, we now search for the minimum of the *second moment* $\sum_i (x_i - c)^2 = 1/n \sum_{i<j} (x_i - x_j)^2$, where $c = (x_1 + \ldots + x_n)/n$ is the centroid of the centres. R. L. Graham and N. J. A. Sloane [GrS1990] presented conjecturally optimal arrangements of $n \leq 500$ unit discs. In addition T. Y. Chow [Cho1995] determined the optimal packings of $n \leq 500$ unit discs assuming that the centres are points of the hexagonal lattice. It is not hard to see that if n is large then the convex hull of the centres in an optimal arrangement is close to some circular disc of area $n\sqrt{12}$.

Parametric Density

Let $\varrho > 0$, and let $\mathbb{S}$ be either $\mathbb{R}^2$ or H^2. For any convex domain C in $\mathbb{S}$, we write $C(\varrho)$ to denote the parallel domain of radius ϱ, namely, the set of points whose distance from C is at most ϱ. In particular, $C(\varrho)$ is a convex domain.

We define the *parametric density* of n nonoverlapping circular discs of radius r in $\mathbb{S}$ to be $nA(B(r))/A(C_n(\varrho))$, where C_n is the convex hull of the centres. If $\varrho = r$ then $C_n(\varrho)$ is the convex hull of the discs in the Euclidean case and the th r-convex hull of the discs in the hyperbolic case. Let us assume that $A(C_n(\varrho))$ is minimal (and hence the parametric density is maximal) among all packings of n circular discs of radius r in $\mathbb{S}$. Then the absolute version Theorem 4.2.1 of the Thue–Groemer inequality and some elementary considerations yield a certain critical radius ϱ_c with the following property: If $\varrho < \varrho_c$ then C_n is a segment (hence $C_n(\varrho)$ is a "sausage"), and if $\varrho > \varrho_c$ then C_n is two dimensional where

$$\varrho_c = \frac{\sqrt{3}}{2} r \text{ in } \mathbb{R}^2; \tag{4.36}$$

$$\text{th}\varrho_c = \frac{A(T(2r))}{2r} \text{ in } H^2. \tag{4.37}$$

In the Euclidean case the notion of parametric density was introduced by J. M. Wills [Wil1993] (see also U. Betke, M. Henk and J. M. Wills [BHW1994]), and a whole theory is centred around it (see Chapter 3 for results and history). In the hyperbolic case the notion is new and arose in discussions with J. M. Wills and A. Schürmann.

Touching Pairs and Snakes

H. Harborth [Hei1974] verified that the maximal number of touching pairs among n nonoverlapping unit discs in $\mathbb{R}^2$ is $\lfloor 3n - \sqrt{12n - 3} \rfloor$, and the optimal arrangements are characterized by R. C. Heitmann and Ch. Radin [HeR1980]. The paper by Ch. Radin [Rad1981] discusses a related problem of physics that leads to the same solution.

In the hyperbolic plane L. Bowen [Bow2000] solves the analogous problem if the common diameter of the n discs is the side length of an equilateral triangle whose copies tile H^2.

A snake of unit discs in $\mathbb{R}^2$ is a family $B_1, \ldots, B_n$ of nonoverlapping unit discs such that B_i and B_j touch if and only if $|i - j| = 1$. A snake is called limited if it is not the proper subset of any other snake. T. Bisztriczky, K. Böröczky Jr., H. Harborth, and L. Piepmeyer [BBH1991] verify that the minimal cardinality of a limited snake is ten.

Multiple Packings of Unit Discs

Given $k \geq 1$, an arrangement of n Euclidean unit discs is called a *k-fold packing* if every point of the plane is in the interior of at most k discs; hence, the density of a k-fold packing with respect to any container is readily at most k. G. Fejes Tóth [FTG1976] obtained the following improvement: Any k-fold packing of n unit discs into a hexagon H satisfies

$$A(H) \geq n \cdot 6 \tan \frac{\pi}{6k}; \qquad (4.38)$$

namely, the density is at most k times the density of the unit disc with respect to the circumscribed regular $6k$-gon. If $k = 1$ then (4.38) is just the optimal (hexagon) bound. If $k = 2$ then (4.38) yields that the density is at most $1.95420\ldots$, whereas A. Heppes [Hep1955] constructed two-fold packings of density $8\pi/(7\sqrt{15}) = 1.8540\ldots$.

Finally, we construct dense k-fold packings for large k. Any unit disc contains at most $r^2\pi + r^{46/73+\varepsilon}$ points of the lattice $1/r\ \mathbb{Z}^2$ (see M. N. Huxley [Hux1996], chapter 18.3) for any $\varepsilon > 0$ and for large r. Therefore, the periodic arrangement with respect to the lattice $1/r\ \mathbb{Z}^2$ for suitable r yields finite k-fold packings of density at least $k - k^{23/73+\varepsilon}$ (for $\varepsilon > 0$ arbitrary and k large).

Packing Circular Discs of Different Sizes in $\mathbb{R}^2$

It was L. Fejes Tóth who observed around 1950 that packings of circular discs of slightly different sizes cannot be denser than the packing density $\pi/\sqrt{12}$ of equal discs. About ten years later, K. Böröczky (unpublished; see also G. Blind [Bli1969]) proved the following qualitative estimate: If a hexagon D contains $n \geq 2$ nonoverlapping circular discs of radius at most 1 and at least

$$\sqrt{\frac{6\tan\frac{\pi}{6} - 7\tan\frac{\pi}{7}}{5\tan\frac{\pi}{5} - 6\tan\frac{\pi}{6}}} = 0.7429\ldots$$

then the ratio of the total area of the circular discs to the area of D is at most $\pi/\sqrt{12}$. This result has been recently extended to any convex domain D by G. Blind and R. Blind [BlB2002].

However, we have seen in Remark 1.6.3 that there exist packings of circular discs of radius either 1 or 0.6457 inside large hexagons such that the density is larger than $\pi/\sqrt{12}$, an example due to J. Molnár [Mol1959]. Actually, L. Danzer (unpublished) constructed packings of circular discs of four different sizes in large hexagons such that the density is larger than $\pi\sqrt{12}$, and the radii of the circular discs lie in the interval [0.654, 1].

Finally, we consider packings of circular discs of very different sizes. If a convex domain D with $r(D) \geq 2$ contains nonoverlapping circular discs whose radii lie in the interval $[r, 1]$, then the density of the packing is at most $1 - cr^{0.97}$ for some positive absolute constant c according to D. G. Larman [Lar1966a]. This bound was improved to $1 - cr^{0.967}$ by A. Megeney [Meg1996]. Morever, if D is a hexagon and the number of different radii is at most k, then L. Fejes Tóth [FTL1972] proved that the density is at most $1 - (1 - \pi/\sqrt{12})^k$. This latest bound also holds when D is any convex domain and the number of circular discs with maximal radius is at least two.

Minkowski Arrangements

We say that a family of (not necessarily equal) circular discs forms a *Minkowski arrangement* if none of the centres lie in the interior of any other element of the family. Then replacing each disc by the concentric circular disc of half radius leads to a packing.

Here we discuss two results where the discs are essentially equal in a densest Minkowski arrangement of given number of discs. In the Euclidean plane L. Fejes Tóth proved in [FTL1965] and in [FTL1967a] that the density of the discs with respect to their union is less than $2\pi/\sqrt{3}$. We note that this is equivalent to the triangle bound for the packing formed by the discs of half sizes. In the spherical case L. Fejes Tóth [FTL1999] provides upper bounds for the density of a Minkowski arrangement of n discs. This bound is actually sharp if $n = 3, 4, 6$, or 12. For these values of n, the discs in the densest Minkowski arrangement are equal, and the arrangement itself is determined by the regular triangle, tetrahedron, octahedron, and icosahedron, respectively.

Packings of Possibly Different Circular Discs on S^2

We say that a packing of finitely many circular discs (of possibly different radii) on S^2 is *solid* if whenever the discs are rearranged in a way that they still form a packing then the new packing is congruent to the original one. In addition, we call a spherical tiling Archimedean if each tile is a regular polygon and for any pair of vertices there exists an isometry of the tiling that maps one vertex to the other. L. Fejes Tóth [FTL1968] conjectured that the inscribed circular discs of an Archimedean tiling form a solid packing if and only if three tiles meet at each vertex, namely, when the tiling is *trihedral*. The conjecture has been verified in a series of papers by various authors. In particular, the task is completed by A. Florian [Flo2001] concerning the solidity for trihedral tilings and by A. Florian and A. Heppes [FlH2003] concerning the nonsolidity for the nontrihedral tilings (see [Flo2001] and [FlH2003] for references).

Let us consider the spherical Archimedean tiling that consists of n congruent squares and two antipodal regular n-gons for some $n \geq 3$. If $n \geq 4$ then the Archimedean tiling determines the densest packing of spherical circular discs under the assumption that the radius of every disc is either the inradius of the regular n-gon or the inradius of the square in the Archimedean tiling. This statement is due to A. Heppes and G. Kertész [HeK1997] when $n \neq 5$ and to A. Florian and A. Heppes [F1H1999] when $n = 5$. If $n = 3$ then the Archimedean tiling does not determine the densest packing (see [HeK1997]).

5

Coverings by Circular Discs

This chapter is mostly concerned with the triangle bound for coverings of convex domains by equal circular discs. In $\mathbb{S}$, where $\mathbb{S}$ is S^2, H^2, or $\mathbb{R}^2$, let us consider three circular discs of radius r centred at the vertices of a regular triangle $\widetilde{T}(r)$ of circumradius r. We write $\tau_{\mathbb{S}}(r)$ to denote the ratio of the area of the part of $\widetilde{T}(r)$ covered by the circular discs to the area of $\widetilde{T}(r)$, hence, say $\tau_{\mathbb{R}^2}(r) = 2\pi/\sqrt{27}$ independently of the value of r. Given $n \geq 2$ circular discs of radius r that cover a convex domain in $\mathbb{S}$, we say that the *triangle bound* holds if the density of the covering is at least $\tau_{\mathbb{S}}(r)$. Following L. Fejes Tóth, we prove the triangle bound in the Euclidean case, and in the spherical case if all of S^2 is covered. We also verify the triangle bound for coverings of any spherical convex domain, which strengthens a result of J. Molnár, namely, that the density is at least $2\pi/\sqrt{27}$. The hyperbolic case is more subtle because the triangle bound does not hold with respect to any convex domain if r is large. Therefore, we impose certain condition on the convex domain that is covered (a condition satisfied, say, by any circular disc), and we prove an even stronger bound than the triangle bound, namely, that the density is at least $2\pi/\sqrt{27}$.

A common generalization of packing and covering asks for the maximal area that circular discs of given number and of given radius can cover. We prove the corresponding triangle bound due to L. Fejes Tóth, a bound equivalent to his celebrated moment inequality. Finally, we discuss the maximal perimeter of a convex domain in $\mathbb{R}^2$ that can be covered by n unit discs. Following G. Fejes Tóth, we show that the optimal convex domain is a "sausage."

5.1. The Delone Complex and the Triangle Bound

Let $\mathbb{S}$ be $\mathbb{R}^2$, S^2, or H^2, and we write κ to denote its curvature, which is 0, 1, and -1, respectively. The main goal of this section is to introduce the so-called triangle bound and to construct the Delone simplicial cell complex associated to a finite covering by equal circular discs. The latter is a very powerful tool; essentially, the mere existence yields the triangle bound for coverings of compact surfaces of constant curvature (see Theorem 5.1.1). For coverings of a convex domain C in $\mathbb{S}$, we only establish the main combinatorial properties of the Delone simplicial cell complex (see Lemma 5.1.4), and the density estimates will require additional work in the coming sections. We remark that the Delone simplicial cell complex is discussed from the point of view of packings in Section 4.1.

We write $B(r)$ to denote a circular disc of radius r and write $\widetilde{T}(r)$ to denote the inscribed regular triangle in $\mathbb{S}$. Now $\widetilde{T}(r)$ is covered by the three circular sectors that are cut off by $\widetilde{T}(r)$ from the three circular discs of radius r centred at the vertices of $\widetilde{T}(r)$. Let $\tau_{\mathbb{S}}(r)$ be the ratio of the total area of the three circular sectors to the area of $\widetilde{T}(r)$; therefore, the laws of sines and cosines yield

$$\tau_{S^2}(r) = \frac{6(1 - \cos r)\arctan\frac{1}{\sqrt{3}\cos r}}{6\arctan\frac{1}{\sqrt{3}\cos r} - \pi};$$

$$\tau_{\mathbb{R}^2}(r) = \frac{2\pi}{\sqrt{27}};$$

$$\tau_{H^2}(r) = \frac{6(\operatorname{ch} r - 1)\arctan\frac{1}{\sqrt{3}\operatorname{ch} r}}{\pi - 6\arctan\frac{1}{\sqrt{3}\operatorname{ch} r}}.$$

We note that

$$\frac{\sqrt{12}}{\pi} < \tau_{H^2}(r) < \frac{2\pi}{\sqrt{27}} < \tau_{S^2}(r), \tag{5.1}$$

which can be verified analogously to the corresponding bounds for $\sigma_{\mathbb{S}}(r)$ for packings (see Section 4.1). Here $2\pi/\sqrt{27}$ is the limit of $\tau_{S^2}(r)$ and of $\tau_{H^2}(r)$ if r tends to zero, and $\sqrt{12}/\pi$ is the limit of $\tau_{H^2}(r)$ if r tends to infinity.

Given a finite covering of either a convex domain in $\mathbb{S}$ by circular discs of radius r or a compact surface of constant curvature κ by metric discs of radius r, we say that the *triangle bound* holds if the density is at least $\tau_{\mathbb{S}}(r)$. We note that finite coverings of compact surfaces of constant curvature κ correspond canonically to periodic coverings of $\mathbb{S}$ (see Section A.5.4).

Theorem 5.1.1. *If a compact quotient X of $\mathbb{S}$, where $\mathbb{S}$ is $\mathbb{R}^2$, S^2, or H^2, is covered by n metric discs of radius r then $A(X) \leq nA(B(r))/\tau_{\mathbb{S}}(r)$. However, if $\widetilde{T}(r_k)$ is a regular triangle of angle $2\pi/k$ in $\mathbb{S}$ for some integer $k \geq 3$ then there exist a compact quotient X_k of $\mathbb{S}$ and a covering of X_k by embedded metric discs of radius r_k with density $\tau_{\mathbb{S}}(r)$.*

We recall from Section 1.1 that $\vartheta(B^2)$ is the infimum of the densities of periodic coverings of $\mathbb{R}^2$ by unit discs, or equivalently, the infimum of the densities of finite coverings of tori by equal metric discs.

Corollary 5.1.2. *$\vartheta(B^2) = 2\pi/\sqrt{27}$, and the hexagonal lattice of determinant $\sqrt{27}/2$ induces the densest periodic covering by unit discs.*

Theorem 5.1.1 is proved by investigating periodic coverings of $\mathbb{S}$. Given a periodic covering of $\mathbb{S}$ by circular discs of radius r (where $r < \pi/2$ in S^2), we write $\mathcal{C}$ to denote the family of the centres. We call a circular disc B a supporting disc if its interior does not contain any of the centres in $\mathcal{C}$, and ∂B contains at least three of the centres in $\mathcal{C}$. To any supporting disc B, we assign the convex hull of the centres in $\mathcal{C}$ that are contained in B and call the convex hull a Delone cell. Since the common secant of two intersecting supporting discs separates the associated convex hulls (see Lemma A.5.4), the Delone cells form a *Delone cell complex*. It is easy to show that any side of a Delone cell is the side of another Delone cell as well; therefore, the Delone cells tile $\mathbb{S}$. We observe that any symmetry of the covering is also a symmetry of the Delone cell complex, and we call any periodic triangulation of the Delone cell complex a *Delone triangulation* of $\mathbb{S}$. A triangle of any such triangulation is called a *Delone triangle*. Since any supporting disc is of radius at most r, each Delone triangle is of circumradius at most r. We note that if $\mathbb{S} = S^2$ then the spherical Delone cell complex can be obtained by taking the convex hull of the centres in $\mathbb{R}^3$ and projecting radially the faces of the resulting three-polytope.

Lemma 5.1.3. *If T is a triangle of circumradius at most r in $\mathbb{S}$ then*

$$A(T) \leq A(\widetilde{T}(r)),$$

with equality if and only if T is congruent to $\widetilde{T}(r)$.

Proof. We may assume that $T \subset B(r)$ has maximal area, and hence the vertices of T lie on $\partial B(r)$. If p and q are vertices of T then the third vertex is the midpoint of one of the arcs of $\partial B(r)$ connecting p and q according to Lemma A.5.3. Therefore, T is regular. $\square$

Proof of Theorem 5.1.1. Let κ be the curvature of $\mathbb{S}$, where $\mathbb{S}$ is $\mathbb{R}^2$, S^2, or H^2, and let X be a compact surface of constant curvature κ that is covered by n metric discs of radius r. The covering of X induces a periodic covering of $\mathbb{S}$ by circular discs of radius r, and we construct a periodic Delone simplicial complex. Let $T_1, \ldots, T_k$ represent the equivalence classes of the Delone triangles; hence, their total area is $A(X)$. To any pair (T_i, v), where v is a vertex of T_i, we assign the sector determined by the two sides of T_i meeting at v of the circular disc of radius r centred at v. In particular, the total area of the $3k$ sectors is $nA(B(r))$. Moreover, we define the density associated to a T_i to be the ratio of the total area of the three circular sectors corresponding to the vertices of T_i to the area of T_i. Because the sum of the angles of T_i is $\pi + \kappa A(T_i)$, the associated density is

$$\left(\frac{\pi}{A(T_i)} + \kappa \right) \frac{A(B(r))}{2\pi}.$$

However, $R(T_i) \leq r$, and hence Lemma 5.1.3 yields that the density associated to T_i is at least $\tau_{\mathbb{S}}(r)$. In turn, we conclude $A(X) \leq nA(B(r))/\tau_{\mathbb{S}}(r)$.

For $r = r_k$, the suitable X_k is the compact surface of Example A.5.10, completing the proof of Theorem 5.1.1.   □

Next we construct the Delone simplicial complex for finite coverings of a convex domain C in $\mathbb{S}$, where $C \neq S^2$. Let C be covered by $n \geq 2$ different circular discs of radius r in a way that intC contains the centres $x_1, \ldots, x_n$ of the discs. We call a circular disc B a *supporting disc* if its centre lies in C, its interior contains no x_i, and either ∂B contains at least three x_is or the centre of B lies in ∂C and ∂B contains at least two x_is. To any supporting disc, we assign the convex hull of the x_is in the disc. Since the common secant of two intersecting circular discs separates the associated convex hulls (see Lemma A.5.4), we obtain the so-called Delone cell complex with vertices $x_1, \ldots, x_n$. Any triangulation of this cell complex using the same vertices is called a *Delone simplicial cell complex*, and a triangle of the triangulation is called a *Delone triangle*.

For any edge e of a simplicial complex Σ, a *free side* of e (if it exists at all) is a half plane bounded by the line of e such that no triangle of Σ containing e lies in that half plane. We write $b(\Sigma)$ to denote the total number of free sides of the edges.

Lemma 5.1.4. *Let a convex domain $C \neq S^2$ in $\mathbb{S}$ be covered by n different circular discs of radius r whose centres lie in intC. If Σ is an associated Delone simplicial complex Σ then*

    (i) Σ is connected;
    (ii) $2n = b(\Sigma) + f_2(\Sigma) + 2$;
    (iii) the at most $b(\Sigma)$ discs centred in the free sides cover ∂C;
    (iv) each edge and triangle of Σ is of circumradius at most r.

Proof. Let Σ_{Del} denote the associated Delone cell complex, whose structure is easier to understand in terms of the Dirichlet–Voronoi complex Σ_{DV}. Thus we write D_i to denote the set of points in C whose distance from x_i is not larger than the distance from any other x_j, and we define Σ_{DV} to be the cell decomposition of C determined by $D_1, \ldots, D_n$. In particular, the Euler formula (A.11) yields $\chi(\Sigma_{\text{DV}}) = 1$.

The vertices of Σ_{DV} of degree at least three are the centres of the supporting discs. In addition, $x_i x_j$, $i \neq j$, is an edge of Σ_{Del} if and only if either D_i and D_j share a common edge or an edge of Σ_{DV} in $D_i \cap \partial C$ and an edge of Σ_{DV} in $D_j \cap \partial C$ intersect. We deduce that Σ_{Del} is connected, and $\chi(\Sigma_{\text{Del}}) = \chi(\Sigma_{\text{DV}}) = 1$.

Turning to Σ, we see that (i) and (iii) readily follow. Since $2f_1(\Sigma) = 3f_2(\Sigma) + b(\Sigma)$, (ii) is a consequence of $\chi(\Sigma) = 1$. Finally, we conclude (iv) as each supporting disc is of radius at most r. □

When covering a convex domain by circular discs of equal radius, the triangle bound does hold in the Euclidean and the spherical cases (see Theorems 5.3.1 and 5.4.2), but it may not hold in H^2 (see Section 5.5).

Comments. This section presents ideas of L. Fejes Tóth (see [FTL1964] and [FTL1972]), except for Corollary 5.1.2, which is due to R. Kershner [Ker1939]. Instead of associated density with respect to a Delone triangle T (see the proof of Theorem 5.1.1), it is more natural to consider the sum of the areas of the parts of the discs that are contained in T to the area of T. K. Böröczky [Bör?a] proves that the triangle bound also holds for this natural notion of density in H^2 and in $\mathbb{R}^2$.

5.2. The Bambah Inequality and the Fejes Tóth Inequality

In this section we consider finite coverings of convex domains by equal discs in $\mathbb{S}$, where $\mathbb{S}$ is $\mathbb{R}^2$, S^2, or H^2. First, we state the Bambah inequality for special types of coverings, and afterwards we prove the Fejes Tóth inequality which is an efficient tool for estimating density. We recall that $\widetilde{T}(r)$ is an regular triangle in $\mathbb{S}$ of circumradius r. Replacing Lemma 2.1.1 by Lemma A.5.4, we have that the argument leading to Theorem 2.6.1 yields

Theorem 5.2.1 (Bambah inequality). *If $n \geq 3$ circular discs of radius r in $\mathbb{S}$ cover the convex hull D of the discs, and m centres lie in ∂D, then*

$$A(D) \leq (2n - m - 2) \cdot A(\widetilde{T}(r)).$$

Equality holds if and only if D can be tiled by congruent copies of $\widetilde{T}(r)$ using the n centres as vertices.

Given two secants of $B(r)$ that have a common endpoint and the same length and are orthogonal to each other, we write $\Psi(r)$ to denote the cap cut off by one of these secants from $B(r)$. Notably in the Euclidean case, the four caps that are cut off by the sides of an inscribed square are congruent to $\Psi(r)$.

Theorem 5.2.2 (Fejes Tóth inequality). *If a convex domain D in $\mathbb{S}$ is covered by $n \geq 2$ circular discs of radius r, and $\phi(r)$ denotes 1, $\cos r$, $\operatorname{ch} r$ if $\mathbb{S}$ is $\mathbb{R}^2$, S^2, or H^2, respectively, then*

$$A(D) \leq \left(n - \frac{b}{2} - 1 \right) \cdot 2\phi(r) \, A(\widetilde{T}(r)) + A(B(r)) + \frac{b}{2} \cdot (A(B(r)) - 4A(\Psi(r))),$$

where some b discs of radius r cover ∂D.

Remark. The Euclidean version says that if a convex domain D in $\mathbb{R}^2$ is covered by $n \geq 2$ unit discs then

$$A(D) \leq \frac{\sqrt{27}}{2} \cdot n - \left(\frac{\sqrt{27}}{4} - 1 \right) \cdot b + \pi - \frac{\sqrt{27}}{2}. \tag{5.2}$$

The Fejes Tóth inequality (Theorem 5.2.2) provides all the information we need in the Euclidean and spherical cases. However, the refined inequality (Lemma 5.2.3) will be essential for coverings in H^2. Actually, the proof of Theorem 5.2.2 is based on Lemma 5.2.3 all in $\mathbb{R}^2$, S^2, and H^2.

In H^2 our main interest lies in th r-convex sets (see Section A.5.1 for a detailed discussion). We recall that a convex domain D is th r-convex if, for any boundary point x of D, D lies on the convex side of some hypercycle of associated distance r containing x. Equivalently, the curvature is at least th r at every boundary point whenever it exists. To handle coverings of th r-convex sets, we introduce the following relatives of rectangles and squares: A *hyper rectangle* is a convex domain, which is bounded by two segments of length $2a$ that share a common perpendicular bisector l and by two hypercycle arcs whose points are of distance a from l. In addition, we define $\tilde{Q}(r)$ to be a hyper rectangle of maximal area contained in $B(r)$.

Lemma 5.2.3. *For $r > 0$, let the convex domain $D \neq S^2$ in $\mathbb{S}$ be covered by $n \geq 2$ circular discs of radius r in a way that each circular disc is needed to* cover D, each centre is contained in int D, and no three of the n centres lie on any circle centred at some point of ∂D. In addition, let Σ be an associated Delone simplicial complex.

(i) If $\phi(r)$ is 1, $\cos r$, $\operatorname{ch} r$ in case $\mathbb{S}$ is $\mathbb{R}^2$, S^2, H^2, respectively, then

$$A(D) \leq A(\operatorname{supp} \Sigma) \cdot \phi(r) + A(B(r)) + \frac{b(\Sigma)}{2} \cdot (A(B(r)) - 4A(\Psi(r)));$$

(ii) moreover, if D is a $\operatorname{th} r$-convex domain in H^2 then

$$A(D) \leq A(\operatorname{supp} \Sigma) \cdot \operatorname{ch} r + A(B(r)) + \frac{b(\Sigma)}{2} \cdot A(\tilde{Q}(r)).$$

The proof of Lemma 5.2.3 requires the following statement:

Proposition 5.2.4. *Let B_1 and B_2 be circular discs of radius r with intersecting interiors in $\mathbb{S}$, and let l be a line intersecting $B_1 \cap B_2$ in a way that the centres of B_1 and B_2 lie on the same side of l. If Ξ_1 and Ξ_2 denotes the part of $B_1 \backslash B_2$ and $B_2 \backslash B_1$, respectively, cut off by the other side of l then*

$$A(\Xi_1) + A(\Xi_2) + A(B_1 \cap B_2) \geq 4 A(\Psi(r)).$$

Analogously, if $\mathbb{S} = H^2$ and l is a hypercycle of associated distance r, and the centres of B_1 and B_2 lie on the convex side of l, then

$$A(\Xi_1) + A(\Xi_2) + A(B_1 \cap B_2) \geq A(B(r)) - A(\tilde{Q}(r)).$$

Proof. Let $L = A(\Xi_1) + A(\Xi_2) + A(B_1 \cap B_2)$, and let l_0 be the line of the common secant of B_1 and B_2. We may assume that l passes through a common point p of ∂B_1 and ∂B_2. Now the sum $A(\Xi_1) + A(\Xi_2)$ is minimal if $l \cap B_1$ and $l \cap B_2$ have the same length; hence, we may also assume that l is orthogonal to the line l_0.

If l is a line then L is twice the sum of the areas of the two caps of B_1 cut off by l and by l_0. Assuming that l and l_0 are orthogonal, the sum of the areas of the caps is minimal when the secants of l and l_0 are of the same length, which, in turn, yields Proposition 5.2.4 in this case.

If l is a hypercycle of associated distance r in H^2 then we may replace l with the hypercycle l' that passes through p and whose points are of distance $d(p, q)/2$ from the perpendicular bisector of the common secant of B_1 and B_2. Then L is the difference of the area of B_1 and the inscribed hyper rectangle determined by l', p, and q, and hence Proposition 5.2.4 follows by the maximality of $A(\tilde{Q}(r))$. $\square$

Proof of Lemma 5.2.3. We first prove (i) in the Euclidean case and then sketch the necessary changes for the spherical and hyperbolic cases at the end of the proof. The family of points on ∂D whose minimal distance from the vertices of Σ is attained for two vertices of Σ is in a bijective correspondence with the free sides of Σ. Thus their number is $b = b(\Sigma)$, and we write $x_1, \ldots, x_b$ to denote these points on ∂D in clockwise order. We set $x_0 = x_b$ and $x_{b+1} = x_1$. When speaking about the arc $x_{i-1} x_i$, we mean the clockwise oriented arc from x_{i-1} to x_i on ∂D. For $i = 1, \ldots, b$, we write z_i and z_{i+1} to denote two vertices of Σ closest to x_i in a way such that the triangle $x_i z_{i+1} z_i$ is clockwise oriented and does not overlap $\operatorname{supp} \Sigma$. We note that z_i and z_j may coincide for $i \neq j$. In addition, let B_i be the circular disc in the covering whose centre is z_i.

First, we consider the case $n = 2$ and hence $b = 2$. Then Σ has no triangles and only one edge $z_1 z_2$. There exists a supporting line l to D that passes through $B_1 \cap B_2$; thus, (i) follows by Proposition 5.2.4.

Therefore, let $n \geq 3$. We write m_i to denote the midpoint of the segment $z_i z_{i+1}$, and assign to each z_i the domain Π_i bounded by the broken line $x_{i-1} m_{i-1} z_i m_i x_i$ and the arc $x_{i-1} x_i$ on ∂D. In particular,

$$A(D) = A(\operatorname{supp} \Sigma) + \sum_{i=1}^{b} A(\Pi_i).$$

Next, let $\widetilde{\Pi}_i$ be the sector of B_i that contains x_i and x_{i-1} and is cut off by the two rays starting from z_i and passing through m_{i-1} and m_i, respectively. Since the total sum of the angles of Σ is $f_2(\Sigma) \pi$, we deduce by Lemma 5.1.4 (ii) that

$$\sum_{i=1}^{b} A(\widetilde{\Pi}_i) = n \cdot A(B(r)) - \frac{f_2(\Sigma) \pi}{2\pi} \cdot A(B(r)) = \frac{b+2}{2} \cdot A(B(r)),$$

which, in turn, yields

$$A(D) = A(\operatorname{supp} \Sigma) + \frac{b+2}{2} \cdot A(B(r)) - \sum_{i=1}^{b} A(\widetilde{\Pi}_i \setminus \Pi_i). \qquad (5.3)$$

Let σ_i denote the maximal open convex arc of ∂C that lies in $\operatorname{int} B_i$ and passes through x_{i-1} and x_i. Now both B_{i-1} and B_{i+1} are needed to cover D; hence, σ_i connects a point of $B_{i-1} \cap \partial B_i$ with a point of $B_{i+1} \cap \partial B_i$ and cuts B_i into two domains. We write Ω_i to denote the part not containing z_i, and we claim that

$$2\sum_{i=1}^{b} A(\widetilde{\Pi}_i \setminus \Pi_i) = \sum_{i=1}^{b} A(\Omega_i \setminus \Omega_{i+1}) + A(\Omega_{i+1} \setminus \Omega_i) + A(B_i \cap B_{i+1}).$$

$$(5.4)$$

Removing Π_i from $\widetilde{\Pi}_i \cap D$, we are left with two disjoint convex sets, and we write Φ_i^+ and Φ_i^- to denote the ones cut off by the segments $x_i m_i$ and

$x_{i-1}m_{i-1}$, respectively. In addition, we write Λ_i^+ and Λ_i^- to denote the parts of $\Omega_i \setminus \tilde{\Pi}_i$ cut off by the half lines $z_i m_i$ and $z_i m_{i-1}$, respectively (where Λ_i^+ or Λ_i^- might be the empty set). It follows that

$$A(\tilde{\Pi}_i \setminus \Pi_i) = A(\Omega_i) + A(\Phi_i^+) - A(\Lambda_i^+) + A(\Phi_i^-) - A(\Lambda_i^-). \quad (5.5)$$

Since all B_{i-1}, B_i and B_{i+1} are needed to cover D, it is not hard to see that $\sigma_i \setminus \sigma_{i+1}$, $\sigma_i \cap \sigma_{i+1}$ and $\sigma_{i+1} \setminus \sigma_i$ are consecutive connected arcs in clockwise order along ∂D. In particular one half of $B_i \cap B_{i+1}$ cut off by the line $z_i z_{i+1}$ can be obtained from $\Omega_i \cap \Omega_{i+1}$ by removing Λ_i^+ and Λ_{i+1}^-, and adding Φ_i^+ and Φ_{i+1}^-. We deduce that

$$A(\Omega_i) + A(\Omega_{i+1}) + 2\left(A(\Phi_i^+) - A(\Lambda_i^+) + A(\Phi_{i+1}^-) - A(\Lambda_{i+1}^-)\right)$$
$$= A(\Omega_i \setminus \Omega_{i+1}) + A(\Omega_{i+1} \setminus \Omega_i) + A(B_i \cap B_{i+1}).$$

Therefore multiplying both sides of (5.5) by two and summing the resulting formulae for $i = 1, \ldots, b$ yields (5.4).

Finally, if l_i is a supporting line to D at x_i then the part of $B_i \setminus B_{i+1}$ (of $B_{i+1} \setminus B_i$) cut off by l_i is contained in $\Omega_i \setminus \Omega_{i+1}$ (in $\Omega_{i+1} \setminus \Omega_i$). We conclude via Proposition 5.2.4 that

$$A(D) \le A(\text{supp } \Sigma) + \frac{b+2}{2} \cdot A(B(r)) - \frac{b}{2} \cdot 4A(\Psi(r)),$$

which, in turn, yields Lemma 5.2.3 in the Euclidean case.

For spherical coverings, the only differences in the argument are that $A(B(r)) = 2\pi \cdot (1 - \cos r)$ and the total sum of the angles of Σ is $f_2(\Sigma)\pi + A(\text{supp } \Sigma)$, and hence (5.3) is replaced by

$$A(D) = A(\text{supp } \Sigma) \cdot \cos r + \frac{b+2}{2} \cdot A(B(r)) - \sum_{i=1}^{b} A(\tilde{\Pi}_i \setminus \Pi_i).$$

Moreover, the analogous alterations yield (i) in the hyperbolic case. Finally, if C is th r-convex then we simply take l_i to be a supporting hypercycle of associated distance r to C at x_i, completing the proof of Lemma 5.2.3.   □

Now we can easily verify the Fejes Tóth inequality.

Proof of Theorem 5.2.2. If one circular disc of radius r can cover D then we are done; thus we may assume that this is not the case. If the centre x of one of the circular discs lies outside of D then we may replace this disc by the circular disc of radius r whose centre is the closest point of D to x. Possibly increasing the common radius r and varying the centres, we may assume that each centre is contained in int D, no proper subset of the discs covers D, and no three of the n centres lie on any circle centred at some point of ∂D. Now we may apply Lemma 5.2.3. Any triangle of Σ is of area at most $A(\widetilde{T}(r))$ according to Lemma 5.1.3; hence, $A(\text{supp } \Sigma) \leq (2n - b - 2) \cdot A(\widetilde{T}(r))$ follows by Lemma 5.1.4. Therefore, Lemma 5.2.3 yields the inequality of Theorem 5.2.2 with $b = b(\Sigma)$, where some b unit discs cover ∂D according to Lemma 5.1.4. $\square$

Comments. The Euclidean case of Theorem 5.2.1 is due to R. P. Bambah and A. C. Woods [BaW1971]. Moreover, the Euclidean case of Theorem 5.2.2 is contained implicitly in the paper by L. Fejes Tóth [FTL1949], and some gaps in the proof were filled by G. Fejes Tóth [FTG1987].

5.3. Covering the Area by Euclidean Discs

Let us observe that if a convex domain D in $\mathbb{R}^2$ cannot be covered by one unit disc then neither can ∂D be covered by one unit disc. Therefore, the Euclidean version (5.2) of the Fejes Tóth inequality yields the following:

Theorem 5.3.1. *If a convex domain D in $\mathbb{R}^2$ is covered by $n \geq 2$ unit discs then*

$$A(D) < \frac{\sqrt{27}}{2} \cdot n.$$

Remark. The constant $\sqrt{27}/2$ is optimal, which is shown by clusters based on the tiling of the plane by regular hexagons of area $\sqrt{27}/2$.

Finally, we discuss optimal finite coverings by unit discs:

Lemma 5.3.2. *If the convex domain D_n has maximal area among the convex domains covered by $n \geq 2$ unit discs then*

$$c_1\sqrt{n} < r(D_n) \leq R(D_n) < c_2\sqrt{n},$$

where c_1 and c_2 are positive constants.

Remark. For finite lattice coverings by unit discs, $D_n/\sqrt{n}$ tends to a regular hexagon of area $\sqrt{27}/2$ according to Remark 2.8.3.

Proof. Let H be a regular hexagon of area $\sqrt{27}/2$, and hence of circumradius one. Since $(\sqrt{n} - 2) H$ is met by at most n tiles of the tiling of $\mathbb{R}^2$ by translates of H, we deduce

$$A(D_n) \geq A((\sqrt{n} - 2)H) > \frac{\sqrt{27}}{2} \cdot n - 2\sqrt{27}\sqrt{n}. \qquad (5.6)$$

We apply the Fejes Tóth inequality (5.2) to D_n. Readily, $b \geq (1/2\pi)P(D_n)$; therefore, comparing (5.2) and (5.6) shows that $P(D_n) \leq c_2\sqrt{n}$ for some constant c_2, and hence $R(D_n) \leq c_2\sqrt{n}$. In addition, $r(D_n) \geq c_1\sqrt{n}$ is a consequence of (5.6) and the inequality $r(D_n)P(D_n) \geq A(D_n)$ (see (A.8)). $\qquad \square$

Open Problem.

   (i) For $n = 6 \cdot \binom{k}{2} + 1$, $k \geq 2$, let us consider the convex domain of maximal area that can be covered by n unit discs. Is the convex hull of the centres the regular hexagon of side length $\sqrt{3}(k-1)$?

Comments. Theorem 5.3.1 was proved by L. Fejes Tóth [FTL1949], and some gaps of the proof were filled by G. Fejes Tóth [FTG1987]. G. Fejes Tóth [FTG1995b] provides a substantial generalization, namely, if D is a connected bounded set that is the complement of a family of circular discs of radius at least $16.7473\ldots$ and D is covered by $n \geq 2$ unit discs, then $A(D) \leq n \cdot \sqrt{27}/2$.

5.4. Coverings by Spherical Discs

When discussing coverings of a spherical convex domain D by equal circular discs, we distinguish the cases when D is S^2 itself and when it is a proper subset of S^2. First, we consider coverings of S^2 by $n \geq 2$ circular discs of radius r, where $r \leq \pi/2$. If $n \leq 3$ then the centres are contained in a closed hemisphere; hence, $r = \pi/2$. If $n \geq 4$ then we may assume that $r < \pi/2$, and Theorem 5.1.1 gives us:

Theorem 5.4.1. *If $n \geq 4$ circular discs of radius $r < \pi/2$ cover S^2 then the density of the covering is at least $\tau_{S^2}(r)$. Equality holds if and only if $n = 4, 6, 12$ and the covering is determined by the regular tetrahedron, octahedron and icosahedron, respectively.*

Remark. If we write R_n to denote the minimal radius of $n \geq 4$ equal circular discs that cover S^2, it is equivalent to say that

$$R_n \geq \arccos \frac{1}{\sqrt{3}\tan\frac{n\pi}{6n-12}}. \qquad (5.7)$$

In addition, to the previous cases, the optimal solution is known if $n = 5, 7, 10, 14$. The proofs for $n = 7, 10, 14$ discs are rather involved, and we do not discuss them in this book (see the comments for references). For five circular discs, we will provide a proof based on Steiner symmetrization in a more general context (see Theorem 6.6.1). We note that if $n = 5, 6, 7$ then the optimal arrangement is determined by the Euclidean bipyramid over the regular triangle, the square, and the regular pentagon, respectively. If $n = 10, 12, 14$ then the centres in the optimal arrangement can be partitioned into antipodal pairs, and there exists a pair of antipodal centres such that some two circles around them contain the remaining centres.

So far we have verified the triangle bound for coverings of S^2. Now we prove it for coverings of any spherical convex domain.

Theorem 5.4.2. *Any covering of a spherical convex domain $D \neq S^2$ by $n \geq 2$ circular discs of radius $r < \pi/2$ is of density larger than $\tau_{S^2}(r)$.*

Since $\tau_{S^2}(r) > 2\pi/\sqrt{27}$ according to (5.1), Theorems 5.4.1 and 5.4.2 yield

Corollary 5.4.3. *If $n \geq 2$ circular discs of radius $r < \pi/2$ cover a convex domain D on S^2 then the density of the covering is larger than $2\pi/\sqrt{27}$.*

As $\vartheta(B^2) = 2\pi/\sqrt{27}$ (see Corollary 5.1.2), spherical coverings by circular discs of small radius show that the constant $2\pi/\sqrt{27}$ is optimal in Corollary 5.4.3.

Proof of Theorem 5.4.2. We may assume that D cannot be covered by one circular disc of radius r, and hence $b \geq 2$ holds in the Fejes Tóth inequality (Theorem 5.2.2). Therefore, Theorem 5.4.2 follows by verifying

$$\frac{A(B(r))}{\tau_{S^2}(r)} \cdot n > 2A(\widetilde{T}(r)) \cos r \cdot \left(n - \frac{b}{2} - 1 \right)$$
$$+ A(B(r)) + (A(B(r)) - 4A(\Psi(r))) \cdot \frac{b}{2} \qquad (5.8)$$

for any $b \geq 2$. In turn, (5.8) is a consequence of the estimates

$$2 A(\widetilde{T}(r)) \cos r < \frac{A(B(r))}{\tau_{S^2}(r)}, \qquad (5.9)$$

$$A(B(r)) + (A(B(r)) - 4 A(\Psi(r))) < 2 \cdot \frac{A(B(r))}{\tau_{S^2}(r)}. \qquad (5.10)$$

We define $\pi/6 < \alpha(r) < \pi/2$ by the formula $\cos r = 1/(\sqrt{3} \tan \alpha(r))$; hence, $A(\widetilde{T}(r)) = 6\alpha(r) - \pi$ and $A(B(r))/\tau_{S^2}(r) = 2\pi(6\alpha(r) - \pi)/(6\alpha(r))$. Thus

(5.9) can be written in the form

$$\tan\alpha(r) - \frac{6}{\sqrt{3}\pi}\,\alpha(r) > 0. \tag{5.11}$$

Since equality holds in (5.11) for $\alpha = \pi/6$, and the left-hand side is an increasing function of $\alpha \geq \pi/6$, we conclude (5.9).

To verify (5.10), we define $\pi/4 < \beta(r) < \pi/2$ by $\cos r = 1/\tan\beta(r)$. Then $A(B(r)) = 2\pi \cdot (1 - \cos r)$ yields

$$\Psi(r) = 2\beta(r) \cdot (1 - \cos r) - \left(2\beta(r) - \frac{\pi}{2}\right) = \frac{\pi}{2} - 2\beta(r) \cdot \cos r.$$

In particular, (5.10) is equivalent to the inequality

$$\frac{\pi}{3\alpha(r)} \cdot \frac{1}{\cos r} + \frac{4}{\pi} \cdot \beta(r) < \frac{1}{\cos r} + 2. \tag{5.12}$$

If $\alpha(r) \geq \pi/3$ then (5.12) readily holds by $\beta(r) < \pi/2$. Thus, we assume that $\alpha(r) \leq \pi/3$, and hence $x = 1/\cos r$ satisfies that $1 \leq x \leq 3$. We rewrite (5.12) in the form

$$\frac{\pi}{3\arctan\frac{x}{\sqrt{3}}}\left(x - \frac{6}{\pi}\arctan\frac{x}{\sqrt{3}}\right) < x - \frac{4}{\pi}\arctan x. \tag{5.13}$$

Here $x - (6/\pi)\arctan x/\sqrt{3} \geq 0$ holds for $x \geq 1$, as can be checked by differentiation. For $\lambda > 0$, let us define the auxiliary function

$$f_\lambda(x) = \lambda \cdot \left(x - \frac{6}{\pi}\arctan\frac{x}{\sqrt{3}}\right) - \left(x - \frac{4}{\pi}\arctan x\right),$$

whose second derivative with respect to x is

$$f_\lambda''(x) = \frac{2x}{\pi \cdot (3+x^2)^2} \cdot \left(\frac{3\sqrt{3} \cdot \lambda}{2} - \left(1 + \frac{2}{1+x^2}\right)^2\right).$$

We verify (5.13) by dividing the interval $[1, 3]$ into four subintervals; thus, let the interval $[x_1, x_2]$ be either $[1, 1.1]$, $[1.1, 1.4]$, $[1.4, 2.3]$, or $[2.3, 3]$. We define

$$\lambda = \frac{\pi}{3\arctan\frac{x_1}{\sqrt{3}}}.$$

Now numerical evaluation shows that $f_\lambda''(x_1) > 0$, and hence $f_\lambda''(x) > 0$ holds for $x \geq x_1$. Since $f_\lambda(x_1)$ and $f_\lambda(x_2)$ are negative by numerical evaluation, the convexity of f_λ yields (5.13) for $[x_1, x_2]$. In turn, we conclude Theorem 5.4.2. $\qquad\square$

Comments. Theorem 5.4.1 and (5.7) are due to L. Fejes Tóth (see, e.g., [FTL1964]). The cases of five and seven spherical circular discs were settled

by K. Schütte [Sch1955], and G. Fejes Tóth [FTG1969] solved the problem
of ten and fourteen discs. The home page [Slo] of N. J. A. Sloane contains a
list of efficient coverings by reasonably large number of equal circular discs.

A. Heppes [Hep1998] determined the optimal covering of the projective
plane P^2 by four equal circular discs, or equivalently, the optimal covering
of S^2 by eight equal circular discs that form four antipodal pairs. Moreover,
efficient coverings of S^2 with high symmetry are discussed by T. Tarnai [Tar?].

Let us recall that R_n is the minimal radius of n equal circular discs that
cover S^2. If n tends to infinity then

$$\sqrt{\frac{8\pi}{\sqrt{27}} \cdot \frac{1}{\sqrt{n}}} + \frac{g(n)}{n^{\frac{3}{2}}} < R_n < \sqrt{\frac{8\pi}{\sqrt{27}} \cdot \frac{1}{\sqrt{n}}} + O\left(\frac{1}{n^{\frac{5}{6}}}\right), \qquad (5.14)$$

where $g(n)$ tends to infinity. Here the upper bound is due to G. Fejes Tóth in
[FTG1970], and the lower bound is due to K. Böröczky Jr [Bör2001].

Corollary 5.4.3 is proved by J. Molnár [Mol1953].

5.5. Coverings by Hyperbolic Discs

After the results for covering convex sets in the Euclidean plane and on the
sphere, one may hope that the triangle bound holds for covering a convex
domain in the hyperbolic plane by circular discs of any radius r. Let us show
that this is not the case. We define $a(r)$ to be the common length of two equal
secants of $B(r)$ that are orthogonal to each other at a common endpoint, and
let $\Psi(r)$ to be a cap of $B(r)$ that is cut off by some secant of length $a(r)$.
Since the secant of length $a(r)$ and the tangents at the endpoints enclose an
isosceles triangle whose equal sides are just r, we deduce that $A(\Psi(r)) < \pi$.
Now we place $n \geq 2$ circular discs of radius r with the centres aligned, and
the common secant of any two consecutive discs is of length $a(r)$. For any
such common secant, we consider the two lines perpendicular to the secant at
the endpoints and throw away the four caps congruent to $\Psi(r)$, which are cut
off by these two lines. We then obtain a convex domain D_n and a covering
of D_n of density at most $A(B(r))/[A(B(r)) - 6\pi]$, which tends to one as r
tends to infinity. However, $\tau_{H^2}(r) > \sqrt{12}/\pi = 1.1026\ldots$ (see (5.1)); hence,
the triangle bound does not hold for large r.

Therefore, we consider a restricted class of convex domains: We recall that
a convex domain D is th r-convex if any boundary point x of D is contained
in some hypercycle of associated distance r such that D lies on the convex
side of the curve (see Section A.5.1).

Theorem 5.5.1. *If a th r-convex domain D is covered by $n \geq 2$ circular discs of radius r then*

$$A(D) \leq \sqrt{27}\,(\operatorname{ch} r - 1) \cdot n.$$

Remark. This is equivalent saying that the density of the covering is larger than $2\pi/\sqrt{27}$. Since $\tau_{H^2}(r) < 2\pi/\sqrt{27}$ according to (5.1), we conclude the triangle bound for any covering of a th r-convex domain by $n \geq 2$ circular discs of radius r.

As any circular disc is th r-convex (see Corollary A.5.6), Theorem 5.5.1 together with Theorem 5.3.1 and Corollary 5.4.3 yields the following absolute bound for coverings by equal circular discs:

Corollary 5.5.2. *If at least two nonoverlapping equal circular discs cover a larger circular disc in $\mathbb{R}^2$, S^2, or H^2 then the density of the covering is at least $2\pi/\sqrt{27}$.*

Concerning the optimality of Theorem 5.5.1, the following statements will be verified at the end of the proof of Theorem 5.5.1.

Remark 5.5.3.

(i) The density bound $2\pi/\sqrt{27}$ equivalent to Theorem 5.5.1 is optimal as a bound for arbitrary r.

(ii) th r-convexity of D is essentially the optimal condition in Theorem 5.5.1 for large r. More precisely, let $\varrho = r - \ln 6/\varepsilon$ for some $0 < \varepsilon < 1$. If r is large and $n \geq 2$ then there exists a covering of some the ϱ-convex domain by n circular discs of radius r such that the density is at most $1 + \varepsilon$.

(iii) Given r and n, there exist a th r-convex domain D and a covering of D by n circular discs of radius r such that the density is less than $\pi/\eta = 3.5644\dots$ where $\operatorname{sh}\eta = 1$. Therefore, the density bound $2\pi/\sqrt{27}$ of Theorem 5.5.1 is optimal up to an absolute constant factor for any r and n.

(iv) If $r \leq 0.6425$ then the density of a covering of any convex domain by $n \geq 2$ circular discs of radius r is at least $2\pi/\sqrt{27}$ (i.e., there is no need for th r-convexity).

In the course of the proof Theorem 5.5.1, we frequently need inequalities of the form $f(x) > g(x)$ for $x > 0$ and for differentiable functions f and g.

If $f(0) \geq g(0)$, and $f'(x) > g'(x)$ can be verified by rearrangement and possibly using simple estimates like $\operatorname{sh} x > x$, then we say that $f(x) > g(x)$ follows by differentiation.

The proof of Theorem 5.5.1 is based on Lemma 5.2.3 (ii); therefore, we need some estimates on the quantities occurring in that inequality. Let us start with $A(\tilde{Q}(r))$:

Proposition 5.5.4. *If $r > 0$ then*

   (i) $A(\tilde{Q}(r)) < 4 \cdot (\operatorname{ch} r - 1)$;

   (ii) $A(\tilde{Q}(r)) < \frac{8}{3} \cdot \operatorname{ch} r$;

   (iii) $A(\tilde{Q}(r)) < \frac{7}{3} \cdot \operatorname{ch} r$ *if* $\operatorname{ch} r \leq 3.5$.

Proof. For $0 < a < r$, let R_a be a hyper rectangle inscribed into $B(r)$ such that the common perpendicular bisector l of the two straight line sides intersects R_a in a segment of length $2a$. Then $A(R_a) = 4a \cdot \operatorname{sh} \varrho$ according to (A.4), where ϱ is the distance of the hypercycle arcs from l. Now the formula $\operatorname{ch} r = \operatorname{ch} \varrho \cdot \operatorname{ch} a$ yields

$$\frac{d\varrho}{da} = -\frac{\operatorname{sh} a \cdot \operatorname{ch} \varrho}{\operatorname{ch} a \cdot \operatorname{sh} \varrho}$$

as a varies. Therefore, differentiating $a \cdot \operatorname{sh} \varrho$ with respect to a shows that $A(R_a)$ is maximal if and only if

$$\operatorname{ch}^2 r = \frac{\operatorname{ch}^3 a}{\operatorname{ch} a - a \cdot \operatorname{sh} a}. \tag{5.15}$$

From now on, let a satisfy (5.15); hence,

$$A(\tilde{Q}(r)) = \frac{4 a^{\frac{3}{2}} \sqrt{\operatorname{sh} a}}{\sqrt{\operatorname{ch} a - a \cdot \operatorname{sh} a}}. \tag{5.16}$$

It follows by differentiation that $\operatorname{ch} a - a \cdot \operatorname{sh} a$ is monotonically decreasing, and hence there exists a unique positive a_0 ($a_0 = 1.1996\ldots$) satisfying

$$\operatorname{ch} a_0 - a_0 \cdot \operatorname{sh} a_0 = 0.$$

In particular, $a < a_0$.

First we prove (i), which is equivalent to

$$a^{\frac{3}{2}} \sqrt{\operatorname{sh} a} < \operatorname{ch}^{\frac{3}{2}} a - \sqrt{\operatorname{ch} a - a \cdot \operatorname{sh} a}.$$

Differentiation yields that $\operatorname{ch} a - a \operatorname{sh} a$ is at most $1 - (a/2) \operatorname{sh} a$, and hence its square root is at most $1 - (a/4) \operatorname{sh} a$. Since $a^{3/2} \sqrt{\operatorname{sh} a}$ is at most $a \cdot \operatorname{sh} a$, it is sufficient to prove

$$1 + \frac{3}{4}a \cdot \operatorname{sh} a < \operatorname{ch}^{\frac{3}{2}} a.$$

Squaring the derivatives of both sides leads to the inequality

$$\operatorname{sh}^2 a + a^2 \operatorname{ch}^2 a + 2a \operatorname{ch} a \cdot \operatorname{sh} a < 4 \operatorname{ch} a \cdot \operatorname{sh}^2 a.$$

As this inequality follows by using $\operatorname{ch}^2 a = 1 + \operatorname{sh}^2 a$ on the left-hand side, and $4 \operatorname{ch} a > 2 \operatorname{ch} a + 2 + a^2$ on the right-hand side, we conclude (i).

Let us turn to the other two estimates for $A(\tilde{Q}(r))$. We note that

$$A(\tilde{Q}(r)) = 4 q(a) \cdot \operatorname{ch} r \quad \text{for } q(a) = \left(\frac{a}{\operatorname{ch} a}\right)^{\frac{3}{2}} \cdot \sqrt{\operatorname{sh} a}.$$

Here differentiation yields that $a/\operatorname{ch} a$ is monotonically increasing for $a \le a_0$; thus, $q(a)$ is monotonically increasing as well. We define a_1 by the property that $\operatorname{ch} r = 3.5$ corresponds to $a = a_1$. We deduce by numerical calculations that $q(a_0) < 8/3$ and $q(a_1) < 7/3$, which, in turn, yield (ii) and (iii). $\quad\square$

To estimate the expression $A(\operatorname{supp} \Sigma)$ that occurs in Lemma 5.2.3, we use two bounds. The first is concerned with $\tilde{T}(r)$.

Proposition 5.5.5. *If $r > 0$ then*

(i) $A(\tilde{T}(r)) < 6\sqrt{3} \cdot \frac{\operatorname{ch} r - 1}{3 \operatorname{ch} r + 1}$;

(ii) $A(\tilde{T}(r)) \cdot \operatorname{ch} r < \pi \cdot (\operatorname{ch} r - 1)$;

(iii) $A(\tilde{T}(r)) \cdot \operatorname{ch} r < 3.025 \cdot (\operatorname{ch} r - 1)$ *if* $\operatorname{ch} r \le 3.5$.

Proof. We note that

$$A(\tilde{T}(r)) = \pi - 6 \cdot \arctan \frac{1}{\sqrt{3} \operatorname{ch} r} = 6 \cdot \arctan \frac{\sqrt{3}(\operatorname{ch} r - 1)}{3 \operatorname{ch} r + 1}. \quad (5.17)$$

Therefore, the inequality $\arctan t < t$ for $t > 0$ yields (i).

To verify (ii) and (iii), let us introduce the variable $t = 1/(\sqrt{3} \operatorname{ch} r)$. Since (ii) is equivalent to $\arctan t > (\pi\sqrt{3}/6)t$ for $0 < t < 1/\sqrt{3}$, the strict concavity of $\arctan t$ yields (ii). Similarly, (iii) is equivalent to $\arctan t > (3.025\pi/6)t + \pi - 3.025/6$ for $1/(3.5\sqrt{3}) \le t < 1/\sqrt{3}$, which holds again by the strict concavity of $\arctan t$ and by numerical calculations. $\quad\square$

To estimate $A(\operatorname{supp} \Sigma)$, we will also need another bound. Since each of the $b = b(\Sigma)$ free sides of Σ is of length at most $2r$, we deduce by the isoperimetric inequality Theorem A.5.8 for polygons that $A(\operatorname{supp} \Sigma)$ is at most the area of the regular b-gon with side length $2r$; namely,

$$A(\operatorname{supp} \Sigma) \le 2b \cdot \left(\arccos \frac{\cos \frac{\pi}{b}}{\operatorname{ch} r} - \frac{\pi}{b} \right). \tag{5.18}$$

Proposition 5.5.6. *Let* $0 < c_1 < 1$ *and* $c_2 > 0$. *Then, for* $b \ge 3$,

$$\arccos \left(c_1 \cdot \cos \frac{\pi}{b} \right) - \frac{c_2}{b} \le \max \left\{ \arccos c_1, \arccos \frac{c_1}{2} - \frac{c_2}{3} \right\}.$$

Proof. The function $f(b) = \arccos c_1 \cdot \cos \pi/b - c_2/b$ of b satisfies

$$f'(b) = \frac{1}{b^2} \cdot \left(c_2 - c_1 \pi \cdot \left(\frac{1 - c_1^2}{\sin^2 \frac{\pi}{b}} + 1 \right)^{-1/2} \right).$$

Now the expression in the parentheses is monotonically increasing; therefore, $f(b)$ is either decreasing, or increasing, or decreasing up to some b_0 and increasing after b_0. Thus $f(b)$ is at most the maximum of $f(3)$ and $\lim_{b \to \infty} f(b)$. □

Proof of Theorem 5.5.1. If D can be covered by one circular disc of radius r then we are done; thus we assume that this is not the case. If the centre x of one of the circular discs lies outside of D then we may replace this disc by the circular disc of radius r whose centre is the closest point of D to x. Possibly increasing the common radius r and varying the centres, we may assume that each centre is contained in int D, no proper subset of the discs covers D, and no three of the n centres lie on any circle centred at some point of ∂D. Now we may apply Lemma 5.2.3. Let Σ be an associated Delone simplicial complex, and we simply write f_2 and b instead of $f_2(\Sigma)$ and $b(\Sigma)$, respectively. Then Σ has n vertices, and Lemma 5.1.4 yields $n = f_2/2 + b/2 + 1$. According to Lemma 5.2.3 (ii), it is sufficient to prove

$$A(\operatorname{supp} \Sigma) \cdot \operatorname{ch} r + 2\pi \, (\operatorname{ch} r - 1) + \frac{b}{2} \cdot A(\tilde{Q}(r))$$

$$\le \sqrt{27} \left(\frac{f_2}{2} + \frac{b}{2} + 1 \right) \cdot (\operatorname{ch} r - 1). \tag{5.19}$$

If $n = 2$ then $b = 2$, $f_2 = 0$, and $A(\operatorname{supp} \Sigma) = 0$. In particular, we deduce by Proposition 5.5.4 (i) that (5.19) is a consequence of

$$2\pi (\operatorname{ch} r - 1) + 4(\operatorname{ch} r - 1) \le 2 \cdot \sqrt{27}(\operatorname{ch} r - 1),$$

which readily holds. Therefore, let $n \ge 3$, and hence $b \ge 3$ follows from the connectivity of supp Σ (see Lemma 5.1.4).

$$Case\ I: \operatorname{ch} r \geq 3.5$$

According to Proposition 5.5.4 (i) and Proposition 5.5.5 (ii), (5.19) follows from the inequality

$$\pi f_2(\operatorname{ch} r - 1) + 2\pi(\operatorname{ch} r - 1) + 2b(\operatorname{ch} r - 1)$$
$$\leq \frac{\sqrt{27}}{2}(b + f_2 + 2)(\operatorname{ch} r - 1).$$

Thus (5.19) holds if

$$f_2 \leq \frac{\sqrt{27} - 4}{2\pi - \sqrt{27}} \cdot b - 2.$$

Therefore, let

$$f_2 \geq \frac{\sqrt{27} - 4}{2\pi - \sqrt{27}} \cdot b - 2.$$

Now we apply Proposition 5.5.4 (ii) and (5.18), and hence (5.19) is the consequence of the inequality

$$2b \left(\arccos \frac{\cos \frac{\pi}{b}}{\operatorname{ch} r} - \frac{\pi}{b} \right) \operatorname{ch} r \tag{5.20}$$
$$+ 2\pi (\operatorname{ch} r - 1) + \frac{4}{3} b \operatorname{ch} r < \frac{\sqrt{27}}{2}(b + f_2 + 2)(\operatorname{ch} r - 1).$$

We deduce by

$$f_2 \geq \frac{\sqrt{27} - 4}{2\pi - \sqrt{27}} \cdot b - 2,$$

by dividing through (5.20) by $2b \cdot \operatorname{ch} r$, and by regrouping the terms that it is sufficient to verify

$$\arccos \frac{\cos \frac{\pi}{b}}{\operatorname{ch} r} - \frac{\pi}{\operatorname{ch} r} \cdot \frac{1}{b} < \frac{\sqrt{27} \cdot (\pi - 2)}{2 \cdot (2\pi - \sqrt{27})} \cdot \left(1 - \frac{1}{\operatorname{ch} r} \right) - \frac{2}{3}. \tag{5.21}$$

According to Proposition 5.5.6, this estimate follows from the inequalities

$$\arccos \frac{1}{\operatorname{ch} r} + \frac{\sqrt{27} (\pi - 2)}{2 (2\pi - \sqrt{27})} \cdot \frac{1}{\operatorname{ch} r} < \frac{\sqrt{27} (\pi - 2)}{2 (2\pi - \sqrt{27})} - \frac{2}{3}; \tag{5.22}$$

$$\arccos \frac{1}{2 \operatorname{ch} r} - \frac{\pi}{3 \operatorname{ch} r} + \frac{\sqrt{27}(\pi - 2)}{2(2\pi - \sqrt{27})} \cdot \frac{1}{\operatorname{ch} r} < \frac{\sqrt{27}(\pi - 2)}{2(2\pi - \sqrt{27})} - \frac{2}{3}.$$

$$(5.23)$$

Let us introduce the variable $s = 1/\operatorname{ch} r$. Since the derivatives of both

$$\arccos s + \frac{\sqrt{27} \cdot (\pi - 2)}{2 \cdot (2\pi - \sqrt{27})} \cdot s \quad \text{and}$$

$$\arccos \frac{s}{2} + \left(\frac{\sqrt{27} \cdot (\pi - 2)}{2 \cdot (2\pi - \sqrt{27})} - \frac{\pi}{3} \right) \cdot s$$

are positive for $0 < s \leq 1/3.5$, and both inequalities (5.22) and (5.23) hold for $\operatorname{ch} r = 3.5$, we conclude (5.19).

Case II: $2.5 \leq \operatorname{ch} r \leq 3.5$

We handle this case very similarly to Case I. We deduce by Proposition 5.5.4 (i) and Proposition 5.5.5 (iii) that (5.19) is the consequence of the inequality

$$3.025 \, f_2 \cdot (\operatorname{ch} r - 1) + 2\pi(\operatorname{ch} r - 1) + 2b \cdot (\operatorname{ch} r - 1)$$

$$\leq \frac{\sqrt{27}}{2} (b + f_2 + 2) \cdot (\operatorname{ch} r - 1).$$

Thus (5.19) holds if

$$f_2 \leq \frac{\sqrt{27} - 4}{6.05 - \sqrt{27}} \cdot b - \frac{4\pi - 2\sqrt{27}}{6.05 - \sqrt{27}}.$$

Therefore, let

$$f_2 \geq \frac{\sqrt{27} - 4}{6.05 - \sqrt{27}} \cdot b - \frac{4\pi - 2\sqrt{27}}{6.05 - \sqrt{27}}.$$

Now we apply Proposition 5.5.4 (iii) and (5.18), and hence (5.19) is the consequence of the inequality

$$2b \left(\arccos \frac{\cos \frac{\pi}{b}}{\operatorname{ch} r} - \frac{\pi}{b} \right) \operatorname{ch} r + 2\pi(\operatorname{ch} r - 1) + \frac{7}{6} b \cdot \operatorname{ch} r$$

$$< \frac{\sqrt{27}}{2} (b + f_2 + 2)(\operatorname{ch} r - 1). \qquad (5.24)$$

We deduce by the condition on f_2, by dividing through (5.24) by $2b \cdot \operatorname{ch} r$, and by regrouping the terms that it is sufficient to verify

$$\arccos \frac{\cos \frac{\pi}{b}}{\operatorname{ch} r} - \left(\frac{\pi}{\operatorname{ch} r} - \frac{\sqrt{27}}{4} \left(2 - \frac{4\pi - 2\sqrt{27}}{6.05 - \sqrt{27}} \right) \right.$$

$$\left. \times \left(1 - \frac{1}{\operatorname{ch} r} \right) \right) \cdot \frac{1}{b} < \frac{\sqrt{27}}{4} \left(1 + \frac{\sqrt{27} - 4}{6.05 - \sqrt{27}} \right) \left(1 - \frac{1}{\operatorname{ch} r} \right) - \frac{7}{12}.$$

This inequality can be proved similarly to (5.21) in Case I.: First, one eliminates b using Proposition 5.5.6, then verifies the two resulting inequalities by introducing the new variable $s = 1/\operatorname{ch} r$, where $0 < s \le 1/2.5$.

Case III: $\operatorname{ch} r \le 2.5$

We deduce by $f_2 \ge A(\operatorname{supp}\Sigma)/A(\widetilde{T}(r))$ and by Proposition 5.5.4 (i) that (5.19) is the consequence of the inequality

$$A(\operatorname{supp}\Sigma) \operatorname{ch} r + 2\pi(\operatorname{ch} r - 1) + 2b(\operatorname{ch} r - 1)$$

$$< \frac{\sqrt{27}}{2} \left(b + \frac{A(\operatorname{supp}\Sigma)}{A(\widetilde{T}(r))} + 2 \right) (\operatorname{ch} r - 1), \qquad (5.25)$$

which can be rewritten in the form

$$A(\operatorname{supp}\Sigma) \cdot \left(\operatorname{ch} r - \frac{\sqrt{27}(\operatorname{ch} r - 1)}{2 A(\widetilde{T}(r))} \right) + (2\pi - \sqrt{27}) \cdot (\operatorname{ch} r - 1)$$

$$< b \left(\frac{\sqrt{27}}{2} - 2 \right) \cdot (\operatorname{ch} r - 1). \qquad (5.26)$$

Proposition 5.5.5 (i) yields that the coefficient of $A(\operatorname{supp}\Sigma)$ is at most $(\operatorname{ch} r - 1)/4$, and we apply (5.18) to $A(\operatorname{supp}\Sigma)$. Therefore, after dividing by $b(\operatorname{ch} r - 1)/2$, (5.26) will follow if the inequality

$$\arccos \frac{\cos \frac{\pi}{b}}{\operatorname{ch} r} - \frac{2\sqrt{27} - 3\pi}{b} < \sqrt{27} - 4 \qquad (5.27)$$

is verified. Now b can be eliminated using Proposition 5.5.6 as before, and the two resulting inequalities are the consequences of $1/\operatorname{ch} r \ge 1/2.5$. We conclude (5.19) and, in turn, Theorem 5.5.1. □

Proof of the statements of Remark 5.5.3. (i) simply follows from the observation that if r is small then a circular disc of radius $\sqrt{r}$ can be covered by circular discs of radius r in a way that the density is arbitrarily close to $2\pi/\sqrt{27}$.

The proofs of (ii) and (iii) use the quadrilateral $L(\varrho)$, which has three right angles, and the acute angle is enclosed by two sides of length ϱ.

To verify (ii) for given $0 < \varepsilon < 1$, we define $\varrho = r - \ln(6/\varepsilon)$ for large r. Let the quadrilateral $p_1 p_2 p_3 p_4$ satisfy that the opposite angles at p_2 and p_4 are right angles, and the angle at p_1 is the acute angle of $L(\varrho)$. In addition, the diagonal $p_1 p_3$ is of length r, and the side $p_1 p_4$ is of length ϱ. We write a^* to denote the length of the side $p_1 p_2$, and hence $\varrho < a^* < r$.

Now we place $n \geq 2$ circular discs of radius r so that the centres are aligned and the common secants of consecutive discs are of length $2a^*$. For each secant, we draw the two hypercycles of associated distance ϱ, which are orthogonal to the secant at the endpoints, and whose convex sides contain the secant. We define D to be the family of points of the union of the n discs that lie on the convex side of each of the $2(n-1)$ hypercycles, and we note that D is a th ϱ-convex domain. For any of the given discs and for a corresponding secant, the laws of sines and cosines yield a circular sector of angle $\varepsilon\pi/2$ that contains the parts of the disc cut off by the secant and the two orthogonal hypercycles (assuming that r is large enough). Therefore, the density of the covering is less than $1 + \varepsilon$.

Turning to (iii), we may assume that $n \geq 4$. Let $k = \lfloor (n-2)/2 \rfloor$, and let $\tilde{a}(r)$ be the length of the shorter sides of $L(r)$, which satisfies $\operatorname{sh}\tilde{a}(r) = \operatorname{th} r$. Given a segment of length $2k\,\tilde{a}(r)$, we define the th r-convex domain D to be the family of points whose distance from the segment is at most r. Then D can be dissected into two semicircular discs of radius r and $2k$ congruent copies of the following convex domain $\Pi(r)$: $\Pi(r)$ is bounded by three segments and an arc of some hypercycle of associated distance r, where the middle segment is of length $2\tilde{a}(r)$ and the other two segments are orthogonal to it and are of length r. If $\operatorname{ch} r \geq (\sqrt{5}+1)/2$ then the shorter diagonal of $L(r)$ is of length at most r; hence, D can be covered by n circular discs of radius r, and the density of the covering is at most the maximum of 3 and $A(B(r))/A(\Pi(r))$. Now differentiation shows that $A(B(r))/A(\Pi(r)) = \pi\,(\operatorname{ch} r - 1)/[\tilde{a}(r)\operatorname{sh} r]$ is increasing if $\operatorname{ch} r \geq (\sqrt{5}+1)/2$; hence, it is less than its limit π/η at infinity where $\operatorname{sh}\eta = 1$. Finally, it is not hard to construct coverings of suitably small density if $\operatorname{ch} r \leq (\sqrt{5}+1)/2$.

Let $r \leq 0.6425$. We prove (iv) similarly to the proof of Theorem 5.5.1 in the case $\operatorname{ch} r \leq 2.5$, only using Lemma 5.2.3 (i) instead of (ii) for the covering of some convex domain by n circular discs of radius r. We define $\psi(r)$ by the relation

$$A(B(r)) - 4\,A(\Psi(r)) = \psi(r) \cdot (\operatorname{ch} r - 1).$$

Hence

$$\psi(r) = \frac{8\operatorname{ch} r}{\operatorname{ch} r - 1} \cdot \arctan\frac{\operatorname{ch} r - 1}{\operatorname{ch} r + 1}.$$

Introducing the variable $t = (\operatorname{ch} r - 1)/(\operatorname{ch} r + 1)$, we find that $\psi(r)$ is increasing.

We replace the term $2b \, (\operatorname{ch} r - 1)$ by $(\psi(r) \, b/2) \, (\operatorname{ch} r - 1)$ on the left-hand side of (5.25), and it is sufficient to prove the resulting inequality. Therefore, we replace the right-hand side of (5.27) by $\sqrt{27} - \psi(r)$. According to Proposition 5.5.6, the density of the covering is at least $2\pi/\sqrt{27}$ if

$$\arccos \frac{1}{2\operatorname{ch} r} - (\sqrt{12} - \pi) \le \sqrt{27} - \psi(r),$$

which holds for $r \le 0.6425\ldots$ by numerical calculations. $\square$

Comments. Theorem 5.5.1 is proved by K. Böröczky Jr. [Bör?b].

We note that the density of infinite coverings of the hyperbolic plane is an intricate notion. One cannot define it simply using elementary cells like the Delone cells or the Dirichlet–Voronoi cells (see K. Böröczky [Bör1974]), but it is possible to define this density using ergodic theory for "homogeneous coverings" (see L. Bowen and Ch. Radin [BoR2003] and [BoR?]). Since the triangle bound holds for every Delone cell (see K. Böröczky [Bör?a]), the results of L. Bowen and Ch. Radin [BoR2003] show that the density of any infinite covering by circular discs of radius r is at least $\tau_{H^2}(r)$ (whenever density exists) and equal to $\tau_{H^2}(r)$ if the covering is coming from a side-to-side tiling by regular triangles of circumradius r. Therefore, we have a stronger bound for finite coverings of th r-convex domains (see Theorem 5.5.1) than for infinite coverings.

5.6. The Moment Inequality and Covering the Maximal Area

We prove the triangle bound for the following problem, which is a common generalization of packing and covering: Find the maximal area that circular discs of given number and given radius can cover. In, addition we discuss the equivalent *moment inequality*, which is the basis of the arguments for many deep results (see the comments for references).

Theorem 5.6.1. *Given a polygon C of at most six sides in $\mathbb{R}^2$, the area of the part of C covered by some n circular discs of radius r is at most $n \cdot A(H \cap B)$, where H is a regular hexagon of area $A(C)/n$, and B is the concentric circular disc of radius r.*

Next we present a version for any compact surface of constant curvature. Let $\mathbb{S}$ be $\mathbb{R}^2$, S^2, or H^2, and let X be a compact surface of constant curvature that is a factor of $\mathbb{S}$ (see Section A.5.4). Moreover, let n be a positive integer that satisfies

$$n \geq 4 \text{ when } X = S^2, \text{ and } n \geq 2 \text{ when } X = P^2.$$

Recalling that any cell decomposition of X into n cells has at most $3(n - \chi(X))$ edges (see (A.13)), we define the *characteristic triangle* of the pair (X, n) to be as isosceles triangle T in $\mathbb{S}$ such that the two equal sides meet at a vertex v and enclose an angle α where

$$A(T) = \frac{A(X)}{6(n - \chi(X))} \text{ and } \alpha = \frac{n\pi}{3(n - \chi(X))}.$$

Here $\chi(S^2) = 2$, $\chi(P^2) = 1$, and $\chi(X) \leq 0$ otherwise; hence, the definition makes sense. We write $B(r, y)$ to denote the circular discs of radius r and of centre y in $\mathbb{S}$ and write $d(\cdot, \cdot)$ to denote distance on X.

Theorem 5.6.2. *Given a compact surface X of constant curvature 0, 1, or -1, the area of the union of n metric discs of radius r on X is at most $6(n - \chi(X))A(T \cap B(r, v))$, where T is the characteristic triangle of the pair (X, n).*

For Theorem 5.6.1, we write $\bar{R}$ and $\bar{r}$ to denote the circumradius and the inradius of H, respectively. For Theorem 5.6.2, we write $\bar{R}$ to denote the common length of the sides of T meeting at v and write $\bar{r}$ to denote the distance of v from the opposite side of T. According to Theorem 5.6.2, if n metric discs of radius r cover X then $r \geq \bar{R}$, and if n isometrically embedded circular discs of $\mathbb{S}$ form a packing on X then $r \leq \bar{r}$. These results have been proved earlier as Theorem 5.1.1 and Corollary 4.1.3, and the analogous results with respect to a hexagon in $\mathbb{R}^2$ are subjects of Theorem 5.3.1 and Corollary 4.1.2.

Before proving Theorems 5.6.1 and 5.6.2, we present two consequences, both of which have fundamental applications in various areas of mathematics (see the comments). These results are commonly known as *moment inequalities*.

Corollary 5.6.3. *Let C be a polygon of at most six sides in $\mathbb{R}^2$, and let f be a decreasing nonnegative function. If $x_1, \ldots, x_n \in C$, and the Jordan measurable $D_1, \ldots, D_n \subset C$ do not overlap, then*

$$\sum_{i=1}^{n} \int_{D_i} f(\|x - x_i\|) \, dx \leq n \cdot \int_{H} f(\|x\|) \, dx,$$

where H is an o-symmetric regular hexagon of area $A(D)/n$.

Corollary 5.6.4. *Let X be a compact surface of constant curvature 0, 1, or -1, and let f be a decreasing nonnegative function. If the Jordan measurable $D_1, \ldots, D_n \subset X$ do not overlap, and $x_1, \ldots, x_n \in X$, then*

$$\sum_{i=1}^{n} \int_{D_i} f(d(x, x_i)) \, dx \leq 6(n - \chi(X)) \int_T f(d(x, v)) \, dx,$$

where T is the characteristic triangle of the pair (X, n).

Remark 5.6.5.

(i) The analogues of Corollaries 5.6.3 and 5.6.4 hold if f is a bounded increasing function, only with $\geq$ instead of $\leq$.

(ii) The tiling of $\mathbb{R}^2$ by regular hexagons shows that the coefficients of n on the right-hand sides of Theorem 5.6.1 and Corollary 5.6.3 are optimal. Moreover, equality holds in Theorem 5.6.2 and Corollary 5.6.4 if $x_1, \ldots, x_n$ are vertices of a decomposition of X into regular triangles (see Example A.5.10), and $D_1, \ldots, D_n$ are the images of the corresponding Dirichlet–Voronoi cells in $\mathbb{S}$.

(iii) Using ideas leading to Theorem 1.4.1 yields Theorem 5.6.1 and Corollary 5.6.3 even if C is any convex domain and $n \geq n_0$, where n_0 is an absolute constant (compare, Corollary 4.1.2 and Theorem 5.3.1).

Proofs of Theorems 5.6.1 and 5.6.2. During the argument, the area of the empty set is assumed to be zero. We start with Theorem 5.6.2 when $X = S^2$. Since $6(n - 2)A(T \cap B(r, v))$ is $A(S^2)$ if $r \geq \bar{R}$, we may assume that $r < \bar{R}$. We write $x_1, \ldots, x_n$ to denote the centres of the n circular discs of radius r and write D_i to denote the Dirichlet–Voronoi cell of x_i with respect to $x_1, \ldots, x_n$, namely, the family of points of X that are not farther from x_i than from any other x_j. For any x_j, we consider the hemisphere containing x_i that is determined by the perpendicular bisector of the segment $x_i x_j$. Then D_i is the intersection of these hemispheres. In particular, $D_1, \ldots, D_n$ are polygons, that form a cell decomposition of S^2. We further decompose each D_i into triangles, such that every triangle is the convex hull of x_i and a side of D_i, and we let $\{T_1, \ldots, T_k\}$ be the family of these triangles, where $k \leq 6(n - 2)$. We write y_j to denote the vertex of T_j that is some x_i and write σ_j to denote the two-gon determined by the two sides of T_j meeting at y_j. In addition, we define

$$M_j = (\sigma_j \cap B(\bar{R}, y_j)) \backslash T_j \quad \text{and} \quad t_j = A(M_j)$$

for $j = 1, \ldots, k$, and $t_j = 0$ if $k < j \leq 6(n - 2)$.

The main idea of the proof is to decompose the union of the n circular discs of radius r into $\{T_j \cap B(r, y_j)\}$ for $j = 1, \ldots, k$. Moreover, we calculate $A(T_j \cap B(r, y_j))$ in the form

$$A(T_j \cap B(r, y_j)) = A(\sigma_j \cap B(r, y_j)) - A(M_j \cap B(r, y_j))$$

and estimate $A(M_j \cap B(r, y_j))$ in terms of $t_j = A(M_j)$. Therefore, let $a(t)$ be the area of the intersection of $B(r, v)$ and some cap of $B(\bar{R}, v)$ with area t if $0 < t < A(B(\bar{R}, v))$, and let $a(0) = 0$. Writing t^* to denote the area of the cap cut off from $B(\bar{R}, v)$ by T and writing σ^* to denote the two-gon determined by the two sides of T meeting at v, we have

$$A(T \cap B(r, v)) = A(\sigma^* \cap B(r, v)) - a(t^*).$$

Let us verify the basic properties of $a(t)$, namely, that

$$a(t) \text{ is increasing and convex} \tag{5.28}$$

if $0 \le t < A(B(\bar{R}, v))/2$. Readily, $a(t)$ is increasing. Next if a cap of $B(\bar{R}, v)$ with area $t < [A(B(\bar{R}, v))]/2$ is bounded by a secant of length $2q$, and in turn this secant intersects $B(r, v)$ in a secant of length $2p$, then $a'(t) = \sin p/\sin q$ (see (A.4)). Since

$$\frac{\cos q}{\cos p} = \frac{\cos \bar{R}}{\cos r} \tag{5.29}$$

we have $\partial q/\partial p = (\sin p \cos q)/(\cos p \sin q)$, and hence

$$\frac{\partial}{\partial p} \frac{\sin p}{\sin q} = \frac{\cos^2 p \sin^2 q - \sin^2 p \cos^2 q}{\sin^3 q \cos p} = \frac{\cos^2 p - \cos^2 q}{\sin^3 q \cos p} > 0.$$

Therefore, $\sin p/\sin q$ is an increasing function of p, which in turn yields (5.28).

Next, for any $j = 1, \ldots, k$, we claim the estimate

$$A(M_j \cap B(r, y_j)) \ge a(t_j). \tag{5.30}$$

If M_j does not overlap $B(r, y_j)$ then some cap of $B(\bar{R}, y_j))$ not overlapping $B(r, y_j)$ contains M_j, and hence $a(t_j) = 0$. Therefore, we assume that ∂M_j intersects $\partial B(r, y_j)$ at two points w_1 and w_2. We define t'_j by $a(t'_j) = A(M_j \cap B(r, y_j))$ and define Ω_j to be the cap of $B(\bar{R}, y_j)$ with area t'_j such that $w_1, w_2 \in \Omega_j$, and the bounding secant of Ω is symmetric through the perpendicular bisector of the segment $w_1 w_2$. Then Ω_j contains $M_j \backslash B(r, y_j)$; hence, we deduce $t'_j \ge t_j$ and, in turn, (5.30).

Now the total areas of the M_j's satisfy

$$\sum_{j=1}^{6(n-2)} t_j \geq \sum_{j=1}^{k} A(\sigma_j \cap B(\bar{R}, y_j)) - A(T_j)$$

$$= nA(B(\bar{R}, v)) - A(S^2) = 6(n-2)t^*.$$

We deduce that the area covered by the n circular discs of radius r is

$$\sum_{i=1}^{k} A(T_j \cap B(r, y_j)) \leq \sum_{i=1}^{k} A(\sigma_j \cap B(r, y_j)) - a(t_j)$$

$$= nA(B(r, v)) - \sum_{j=1}^{6(n-2)} a(t_j)$$

$$\leq nA(B(r, v)) - 6(n-2)a(t^*),$$

which is $6(n-2)A(T \cap B(r, v))$. Hence, Theorem 5.6.2 follows if $X = S^2$. If $X = P^2$ then we simply represent the objects on P^2 as quotients by the antipodal map on S^2. Otherwise, we write X as the quotient of $\mathbb{S}$ by a lattice Λ, where $\mathbb{S}$ is either $\mathbb{R}^2$ or H^2. Using the notation analogous to the case $X = S^2$, let $\Gamma \subset \mathbb{S}$ be the inverse image of $\{x_1, \ldots, x_n\}$. We choose an inverse image $x_i' \in \Gamma$ of x_i, and we write D_i to denote the Dirichlet–Voronoi cell of x_i' with respect to Γ. Again, $T_1, \ldots, T_k$ form a triangulation of the union of $D_1, \ldots, D_n$, and now $k \leq 6(n - \chi(X))$. The only essential differences in the proof are that σ_j is the convex cone bounded by the two rays that start from y_j and are determined by the two sides of T_j meeting at y_j. Moreover if $\mathbb{S} = \mathbb{R}^2$ and $\mathbb{S} = H^2$, then $a'(t) = p/q$ and $a'(t) = \operatorname{sh} p/\operatorname{sh} q$, respectively, and (5.29) is replaced by $q^2 - p^2 = \bar{R}^2 - r^2$ and $\operatorname{ch} q/\operatorname{ch} p = \operatorname{ch} \bar{R}/\operatorname{ch} r$, respectively.

Finally, for Theorem 5.6.1, we use that $k \leq 6n$ according to Lemma A.5.9, completing the proofs of Theorems 5.6.1 and 5.6.2. $\qquad\square$

Proofs of Corollaries 5.6.3 and 5.6.4. We may assume that each D_i is the family of points of X that are not farther from x_i than from any other x_j. For any $r > 0$, we define the step function γ_r by $\gamma_r(t) = 1$ if $0 \leq t \leq r$, and $\gamma_r(t) = 0$ if $t > r$. On any given interval, f can be arbitrarily well approximated by functions of the form $\sum_{i=1}^{k} \alpha_i \gamma_{r_i}$ where each α_i is positive. Therefore, we may assume that f coincides with some step function γ_r, and hence the expressions on the left-hand sides of Corollaries 5.6.3 and 5.6.4 are the areas covered by the n metric discs of radius r centred at $x_1, \ldots, x_n$. In turn, Theorems 5.6.1 and 5.6.2 yield Corollaries 5.6.3 and 5.6.4, respectively. $\qquad\square$

Comments. Most of the results and proofs in this section are due to L. Fejes Tóth (see [FTL1964] or [FTL1972]) except for the hyperbolic versions of Theorem 5.6.2 and Corollary 5.6.4, whose details are worked out by B. Orvos-Nagyné Farkas [ONF?]. A thorough reference, which discusses also applications to polyhedral approximation and isoperimetric type problems in the three-space, is by A. Florian [Flo1993]. The importance of the results is underlined by the large number of proofs that have been published (see [Flo1993] and P. M. Gruber [Gru?a]). Actually, some versions of Theorem 5.6.1 exist also on S^2 and in H^2, namely, when D is replaced by the union of n polygons (see M. Imre [Imr1964] and B. Bollobás and N. Stern [BoS1972]).

Careful examination of the preceding arguments shows that if $\bar{r} \leq r \leq \bar{R}$ in Theorem 5.6.2, or f is strictly monotonically decreasing on $(\bar{r}, \bar{R})$ in Theorem 5.6.4, then equality holds only for the regular tilings described in Remark 5.6.5 (ii) (see L. Fejes Tóth [FTL1964] or [FTL1972]). Stability versions and asymptotic properties of the optimal arrangements for Theorem 5.6.1 and Corollary 5.6.3 are provided by P. M. Gruber [Gru2001] and by G. Fejes Tóth [FTG2001], and for the spherical and hyperbolic cases of Theorem 5.6.2 and Corollary 5.6.4 by K. Böröczky Jr. and G. Fejes Tóth [BöF2002] and by B. Orvos-Nagyné Farkas [ONF?], respectively.

According to the spherical case of Remark 5.6.5 (ii), if $n = 4, 6, 12$ then the arrangements determined by the regular tetrahedron, octahedron, and icosahedron, respectively, are optimal in Theorem 5.6.2 for any r. L. Fejes Tóth [FTL1971] proved that, given $n \neq 4, 6, 12$ in Theorem 5.6.2, there exist r_1 and r_2 such that no optimal arrangement of the n centres with respect to r_1 is congruent to any optimal arrangement with respect to r_2. Conjecturally optimal configurations are presented for any r by P. W. Fowler and T. Tarnai in [FoT1996] if $n \leq 11$, and in [FoT1999] if $n = 13$ (see T. Tarnai [Tar?] for related references).

The asymptotic formula (as n tends to infinity) resulting from the moment inequality has been generalized to higher dimensions in various forms. These generalizations have applications in many areas in mathematics, such as information science, volume approximation, and numerical integration. For comprehensive surveys about all these related problems, see P. M. Gruber [Gru2002, Gru?a, Gru?b].

5.7. Covering the Perimeter by Unit Discs

We have seen (see Lemma 5.3.2) that clusters maximize the area of a convex domain that can be covered by n unit discs for large n. In this section we consider the maximal perimeter of a convex domain that can be covered by $n \geq 2$ unit circular discs, and we prove that a "sausage" is optimal:

Theorem 5.7.1. *If the boundary of a convex domain C can be covered by $n \geq 1$ unit circular discs then*

$$P(C) \leq 4 \cdot \left(\sqrt{n^2 - 1} + \arcsin \frac{1}{n} \right).$$

Equality holds if and only if the centres are aligned and the distance between any two consecutive centres is $2 \sqrt{1 - 1/n^2}$.

Remark. It follows that if a convex domain C can be covered by $n \geq 1$ unit circular discs then the same conclusions hold.

Let C_n be a compact convex set whose boundary can be covered by $n \geq 1$ unit circular discs and whose perimeter is maximal under this condition. We determine C_n by induction on $n \geq 1$. If $n = 1$ then C_1 is readily a unit disc. Let $n \geq 2$, and we assume that Theorem 5.7.1 holds for $n - 1$. We observe that the arrangement described in Theorem 5.7.1 shows that

$$P(C_n) \geq 4 \cdot \left(\sqrt{n^2 - 1} + \arcsin \frac{1}{n} \right) > 4n; \tag{5.31}$$

hence C_n is not a segment.

We write $B_1, \ldots, B_n$ to denote the n unit discs covering C_n and write x_i to denote the centre of B_i. Let D_i be the *Dirichlet–Voronoi cell* corresponding to x_i, namely, the set of points in C_n that are not farther from x_i than from any other x_j. Then $D_i \subset B_i$, and int $D_i \neq \emptyset$ because $P(C_n) > P(C_{n-1})$ holds according to (5.31) and the induction hypothesis. We write Σ to denote the cell decomposition of C_n determined by $D_1, \ldots, D_n$. The greatest part of the proof of Theorem 5.7.1 deals with the combinatorics of the arrangement: First, we show that each vertex of Σ is of degree three (see Proposition 5.7.2), and next that the combinatorial structure of $D_1, \ldots, D_n$ (say in this order) is the same as for the sausage arrangement described in Theorem 5.7.1 (see Proposition 5.7.3). After that, Proposition 5.7.4 shows that $D_i \cap D_{i+1}$ is the common secant of B_i and B_{i+1} and that any two tangent lines at a convex curve in some $D_i \cap C_n$ intersect in the way one expects. Then Proposition 5.7.5 allows us to reduce the problem into finding the maximum of a concave function in one variable.

In the course of the proof of Theorem 5.7.1, we use the following well-known facts: If p and q divide a unit circle into two arcs where one of the arcs is of length φ, and a point s moves along the other arc, then

$\varphi = 2 \cdot \angle psq$ and

$d(p, s) + d(q, s)$ is maximal exactly if $d(p, s) = d(q, s)$. (5.32)

In addition, we write $|\sigma|$ to denote the arc length of a convex curve σ.

Proposition 5.7.2. *Each vertex of Σ has degree three and lies in ∂C_n.*

Proof. We write m to denote the number of vertices of Σ in ∂C_n. We deduce by the Euler formula (A.11) and by $2 f_1(\Sigma) \geq 3 f_0(\Sigma)$ that $m \leq 2n - 2$, and $m = 2n - 2$ is equivalent to Proposition 5.7.2.

We suppose that $m \leq 2n - 3$, and we seek a contradiction with $P(C_n) > 4n$ (see (5.31)). We note that $n \geq 3$. The m vertices of Σ in the boundary divide ∂C_n into m convex arcs $\sigma_1, \ldots, \sigma_m$. We write $\beta(\sigma_i)$ to denote the total curvature of σ_i (see Section A.5.1); hence, $\sum_{i=1}^{m} \beta(\sigma_i) \leq 2\pi$. We claim that

$$|\sigma_i| \leq \begin{cases} \dfrac{2}{\cos \frac{1}{2} \beta(\sigma_i)} & \text{if } \beta(\sigma_i) \leq \dfrac{\pi}{2}; \\[2ex] \beta(\sigma_i) + 2\sqrt{3} - \dfrac{2\pi}{3} & \text{if } \beta(\sigma_i) \geq \dfrac{\pi}{2}. \end{cases} \tag{5.33}$$

Let σ_i be one of the arcs, and we write p_1 and p_2 to denote the first and the second endpoints of σ_i, respectively, according to the clockwise orientation. In addition, the tangent line to σ_i at p_j is denoted by l_j, $j = 1, 2$. If $\beta(\sigma_i) \leq \pi/2$ then $\angle p_1 q p_2 = \pi - \beta(\sigma_i)$ holds for the intersection point q of l_1 and l_2, where q is the midpoint of σ_i if σ_i is a segment. Therefore, $d(p_1, p_2) \leq 2$, and (5.32) yields (5.33).

If $\beta(\sigma_i) \geq \pi/2$ then let D_h be the Dirichlet–Voronoi cell containing σ_i, and let r_j be the point of ∂B_h on the exterior side of l_j, and where the tangent to ∂B_h is parallel to l_j, $j = 1, 2$. Then the length of the clockwise oriented arc $r_1 r_2$ of ∂B_h is $\beta(\sigma_i)$. We define γ_j, $j = 1, 2$, by the property that l_j cuts off an arc of length $2\gamma_j$ from ∂B_h; hence,

$$|\sigma_i| \leq \beta(\sigma_i) + \sum_{j=1,2} (2 \sin \gamma_j - \gamma_j).$$

Since the maximum of $2 \sin \gamma - \gamma$ on $[0, \pi/2]$ is attained at $\gamma = \pi/3$, we conclude (5.33).

Next we claim that if $m \leq 2n - 3$ for some $n \geq 3$, and the sum of the nonnegative real numbers $\beta_1, \ldots, \beta_m$ is at most 2π, then

$$\Omega = \sum_{\beta_i > \frac{\pi}{2}} \left(\beta_i + 2\sqrt{3} - \frac{2\pi}{3} \right) + \sum_{\beta_i \leq \frac{\pi}{2}} \frac{2}{\cos \frac{1}{2} \beta_i} < 4n. \tag{5.34}$$

We may assume $\sum_{i=1}^{m} \beta_i = 2\pi$. If $\beta_i > \pi/2$ for any i then $\Omega \leq 6\sqrt{3} < 12$, and we are done. Therefore, we may assume that $\beta_1 \leq \pi/2$. Since $1/\cos(t/2)$ is convex on $[0, \pi/2]$, we may also assume that whenever $\beta_i \leq \pi/2$ and $i \geq 2$ then either $\beta_i = \pi/2$ or $\beta_i = 0$. We write k to denote the number of

β_i satisfying $\beta_i \geq \pi/2$ and $i \geq 2$; hence, $k \geq 1$. Since $2/\cos(\beta/2) < \beta + 2\sqrt{3} - 2\pi/3$ holds for $\beta = \pi/2$, we deduce that

$$\Omega \leq 2\pi - \beta_1 + k \cdot \left(2\sqrt{3} - \frac{2\pi}{3}\right) + \frac{2}{\cos\frac{1}{2}\beta_1} + (m - k - 1) \cdot 2$$

$$\leq 4n - 8 + 2\pi + \frac{2}{\cos\frac{1}{2}\beta_1} - \beta_1 + 2\sqrt{3} - \frac{2\pi}{3} - 2.$$

It follows by the convexity of $2/\cos(\beta_1/2) - \beta_1$ that it is maximal either at $\beta_1 = 0$ or at $\beta_1 = \pi/2$, which, in turn, yields the claim (5.34).

Since $P(C_n) = \sum_{i=1}^{m} |\sigma_i|$, we conclude $P(C_n) < 4n$ by (5.34) and (5.33). This contradicts (5.31), completing the proof of Proposition 5.7.2. □

Next we determine the combinatorial structure of $D_1, \ldots, D_n$.

Proposition 5.7.3. *The Dirichlet–Voronoi cells $D_1, \ldots, D_n$ can be numbered in a way that $D_i \cap D_j = \emptyset$ if and only if $|j - i| > 1$.*

Proof. We deduce by Proposition 5.7.2 that it is sufficient to verify for each D_i that $D_i \cap \partial C_n$ consists of at most two convex arcs. We prove this statement by contradiction. Thus, we suppose that, for some D_i, $D_i \cap \partial C_n$ consists of the pairwise disjoint convex arcs $\sigma_1, \ldots, \sigma_m$ in this order, where $m \geq 3$. Let $y_i \in \sigma_i$, and let l_i be a line tangent to C_n at y_i. Since $m \geq 3$, there exist two consecutive lines, say l_1 and l_2, such that l_1 and l_2 intersect at a point p, and $\angle y_1 p y_2 \geq \pi/3$. Now the triangle $y_1 p y_2$ contains some Dirichlet–Voronoi cell D_j with $j \neq i$, and $P(C_n) > P(C_{n-1})$ yields that $D_j \not\subset B_i$; hence, $p \notin B_i$. Let z_1 and z_2 be the closest points of $B_i \cap py_1$ and $B_i \cap py_2$, respectively, to p; hence, we deduce $d(z_1, z_2) \leq \sqrt{3}$ by $\angle z_1 p z_2 \geq \pi/3$.

It follows by $D_j \not\subset B_i$ that the line $z_1 z_2$ intersects C_n in a segment s, and we write C' to denote the half of C_n determined by s that contains $D_i \cap \partial C_n$. Now C' is covered by $\{B_h\}_{h \neq j}$, and hence $P(C') \leq P(C_{n-1})$. However, let us consider the arc σ' of ∂C_n cut off by D'. Since the angle of the tangent half lines at the endpoints of σ' is at least $\pi/3$, (5.32) yields that $|\sigma'| \leq 2|s|$. Therefore,

$$P(C_n) = P(C') + |\sigma'| - |s| \leq P(C_{n-1}) + \sqrt{3}.$$

This is absurd because $P(C_n) > P(C_{n-1}) + \sqrt{3}$ according to (5.31) and the induction hypothesis. □

Proposition 5.7.3 provides points $p_1, \ldots, p_{n-1}, \tilde{p}_{n-1}, \ldots, \tilde{p}_1$ in this order on ∂C_n such that $D_i \cap D_{i+1}$ is the segment $p_i \tilde{p}_i$ for $i = 1, \ldots, n - 1$. Let

l_i and $\widetilde{l}_i$ be tangent lines to C_n at p_i and $\widetilde{p}_i$, respectively. The points p_i and $\widetilde{p}_i$ divide l_i and $\widetilde{l}_i$ into two half lines, respectively, namely, $l_i = l_i^+ \cup l_i^-$ and $\widetilde{l}_i = \widetilde{l}_i^+ \cup \widetilde{l}_i^-$. Here l_i^+ and $\widetilde{l}_i^+$ lie on the same side of the line $p_i \widetilde{p}_i$ where D_{i+1} lies.

Proposition 5.7.4. *Let $i = 1, \ldots, n - 1$.*

(i) $\{p_i, \widetilde{p}_i\} = \partial B_i \cap \partial B_{i+1}$;

(ii) l_1 *and* $\widetilde{l}_1$ *do not intersect in* int B_1, *whereas* l_{n-1} *and* $\widetilde{l}_{n-1}$ *do not intersect in* int B_n;

(iii) *if* $n \geq 3$ *and* $i \leq n - 2$ *then* $l_i^+ \cap l_{i+1}^- \neq \emptyset$ *and* $\widetilde{l}_i^+ \cap \widetilde{l}_{i+1}^- \neq \emptyset$;

(iv) *if* $n \geq 3$ *and* $i \leq n - 2$ *then the sum of the angle of* l_i^+ *and* l_{i+1}^- *and the angle of* $\widetilde{l}_i^+$ *and* $\widetilde{l}_{i+1}^-$ *is larger than* π.

Proof. The basis of the argument is the claim that

$$\text{no open unit disc contains any } D_i. \tag{5.35}$$

We verify (5.35) by contradiction; thus, we suppose that a D_i is contained in the interior of the unit disc B_i'.

Possibly reversing the order of $D_1, \ldots, D_n$, we may assume that either $i = 1$ and l_1^- and $\widetilde{l}_1^-$ intersect or are parallel or that $i \geq 2$ and l_{i-1}^- and $\widetilde{l}_{i-1}^-$ intersect or are parallel. We claim that there exist unit discs $B_1', \ldots, B_i'$ such that

$$D_1 \cup \ldots \cup D_i \subset \text{int } B_1' \cup \ldots \cup \text{int } B_i'.$$

We note that l_{j-1}^- and $\widetilde{l}_{j-1}^-$ intersect or are parallel for $2 \leq j \leq i$. Thus, if B_j' is known then B_{j-1}' can be constructed by moving B_{j-1} a little in the direction of l_{j-1}^-.

Now let h be the maximal index, $i \leq h \leq n$, such that

$$D_1 \cup \ldots \cup D_h \subset \text{int } B_1' \cup \ldots \cup \text{int } B_h'.$$

The maximality of $P(C_n)$ yields that there exists no $y \in B_1' \backslash C_n$ such that conv$\{y, C_n\}$ is covered by $B_1', \ldots, B_h'$ and $B_{h+1}, \ldots, B_n$. We deduce that $n \geq h + 2$, and l_{h+1}^- and $\widetilde{l}_{h+1}^-$ intersect in B_1'. Then by moving B_{h+1} a little in the direction of l_{h+1}^+, the resulting B_{h+1}' satisfies

$$D_1 \cup \ldots \cup D_{h+1} \subset \text{int } B_1' \cup \ldots \cup \text{int } B_{h+1}'.$$

This is absurd, and we conclude (5.35).

Let us verify the statements of Proposition 5.7.4. First, we suppose that, say, some p_i is not a common point of ∂B_i and ∂B_{i+1}, and we seek a contradiction.

Since p_i lies on the common secant of B_i and B_{i+1}, we deduce that $p_i \in \mathrm{int} B_i \cap B_{i+1}$. Now either l_i^+ or l_i^- encloses an angle of at least $\pi/2$ with the half line $p_i \widetilde{p}_i$; say, l_i^- does. Then the length of the arc of ∂B_i between the half lines l_i^- and $p_i \widetilde{p}_i$ is less than π, and hence D_i can be covered by an open unit disc, which contradicts (5.35).

Since (ii) readily follows from (5.35), we turn to (iii). We suppose, say, that l_i^+ and l_{i+1}^- do not intersect, and we seek a contradiction. We observe that l_i^- and l_{i+1}^- are parallel or intersect; thus, either the angle of l_i^- and $\widetilde{l}_i^-$ or the angle of l_{i+1}^- and $\widetilde{l}_{i+1}^+$ is obtuse or a right angle; say, the angle of l_i^- and $\widetilde{l}_i^-$ is such. Since $d(p_i, \widetilde{p}_i) < 2$, we deduce that D_i can be covered by an open unit disc, contradicting (5.35).

Finally, let us verify (iv) by contradiction, and hence we suppose that the angle of l_i^+ and l_{i+1}^- plus the angle of $\widetilde{l}_i^+$ and $\widetilde{l}_{i+1}^-$ is at most π. Then either the angle of l_i^- and $\widetilde{l}_i^-$ or the angle of l_{i+1}^+ and $\widetilde{l}_{i+1}^+$ is obtuse or a right angle; therefore, the argument can be finished as before. $\qquad\square$

Next we introduce some additional notions. We write α_i^+ to denote the angle of the half lines l_i^+ and $p_i p_{i+1}$ if $1 \le i \le n-2$, and we write α_i^- to denote the angle of the half lines l_i^- and $p_i p_{i-1}$ if $2 \le i \le n-1$. The angles $\widetilde{\alpha}_i^+$ and $\widetilde{\alpha}_i^-$ are defined analogously at $\widetilde{p}_i$. If $2 \le i \le n-1$ then let

$$\gamma_i = \alpha_i^+ + \widetilde{\alpha}_i^+ + \alpha_{i+1}^- + \widetilde{\alpha}_{i+1}^-.$$

It is convenient to define $\alpha_1^-, \widetilde{\alpha}_1^-$, and γ_1 and $\alpha_n^+, \widetilde{\alpha}_n^+$ and γ_n to be zero.

According to Proposition 5.7.4 (ii), the part of $\partial B_0 \backslash B_1$ between l_1^- and $\widetilde{l}_1^-$ is nonempty, and we write φ_0 to denote its length. Similarly, φ_n denotes the length of the part of $\partial B_n \backslash B_{n-1}$ between l_{n-1}^+ and $\widetilde{l}_{n-1}^+$. Finally, for $i = 1, \ldots, n-1$, φ_i stands for the length of $\partial B_i \cap B_{i+1}$, which is the same as the length of $\partial B_{i+1} \cap B_i$.

Proposition 5.7.5. *The sequence $\varphi_0, \ldots, \varphi_n$ is concave, and if $1 \le i \le n$ then*

$$2 \cdot (\varphi_0 + \varphi_n) \le (\varphi_{i-1} + \varphi_i - 2\gamma_i) + (\varphi_{n-i} + \varphi_{n-i+1} - 2\gamma_{n-i+1}).$$

Proof. We claim that, for $i = 1, \ldots, n-1$, we have

$$\varphi_{i-1} + \varphi_{i+1} \le 2\varphi_i - 2 \cdot (\alpha_i^+ + \widetilde{\alpha}_i^+ + \alpha_i^- + \widetilde{\alpha}_i^-). \tag{5.36}$$

We write φ_i^- to denote length of the part of $\partial B_i \backslash B_{i+1}$ between l_i^- and $\widetilde{l}_i^-$ and, φ_i^+ to denote the length of the part of $\partial B_{i+1} \backslash B_i$ between l_i^+ and $\widetilde{l}_i^+$. If neither l_i nor $\widetilde{l}_i$ intersects the interior of $B_i \cap B_{i+1}$ then some elementary considerations based on (5.32) yield that $2\varphi_i = \varphi_i^- + \varphi_i^+$. We deduce by (5.32) that

$$\varphi_{i-1} \le \varphi_i^- - 2 \cdot (\alpha_i^- + \widetilde{\alpha}_i^-) \text{ and } \varphi_{i+1} \le \varphi_i^+ - 2 \cdot (\alpha_i^+ + \widetilde{\alpha}_i^+),$$

and hence we conclude (5.36). Now the case when, say, l_i intersects the interior of $B_i \cap B_{i+1}$ can be reduced to the previous case by rotating l_i around p_i, which completes the proof of (5.36).

We deduce by (5.36) that the sequence $\varphi_0, \ldots, \varphi_n$ is concave, which, in turn, yields Proposition 5.7.5 if $i = 1$ or $i = n$. Therefore, we assume that $2 \le i \le n - 1$. Since $\alpha_{i-1}^- + \widetilde{\alpha}_{i-1}^- + \alpha_i^+ + \widetilde{\alpha}_i^+$ is nonnegative, adding (5.36) and its version for $i - 1$ results in

$$\varphi_{i-2} + \varphi_{i+1} \le \varphi_{i-1} + \varphi_i - 2 \cdot (\alpha_{i-1}^+ + \widetilde{\alpha}_{i-1}^+ + \alpha_i^- + \widetilde{\alpha}_i^-).$$

In particular, the concavity of $\varphi_0, \ldots, \varphi_n$ yields Proposition 5.7.5. $\quad\square$

By now we have a very good understanding of the structure of the covering of C_n by $B_1, \ldots, B_n$; therefore, we are able to estimate $P(C_n)$ in terms of $\varphi_0, \ldots, \varphi_n$.

Proof of Theorem 5.7.1. Let $2 \le i \le n - 1$. According to Proposition 5.7.4 (iii), we may define q_i ($\widetilde{q}_i$) to be the point of intersection of l_{i-1}^+ and l_i^- (of $\widetilde{l}_{i-1}^+$ and $\widetilde{l}_i^-$). If the two intersecting half lines are opposite and collinear then we take the midpoint of the common segment as the point of intersection. We note that $\gamma_i < \pi$ according to Proposition 5.7.4 (iv) and that

$$|\partial C_n \cap D_i| \le d(p_{i-1}, q_i) + d(q_i, p_i) + d(\widetilde{p}_{i-1}, \widetilde{q}_i) + d(\widetilde{q}_i, \widetilde{p}_i).$$

Since both $\alpha_{i-1}^+ + \alpha_i^-$ and $\widetilde{\alpha}_{i-1}^+ + \widetilde{\alpha}_i^-$ are at most γ_i, (5.32) yields

$$|\partial C_n \cap D_i| \le \frac{d(p_{i-1}, p_i) + d(\widetilde{p}_{i-1}, \widetilde{p}_i)}{\cos \frac{1}{2} \gamma_i}. \tag{5.37}$$

We may assume that p_{i-1}, p_i, $\widetilde{p}_i$, and $\widetilde{p}_{i-1}$ are in clockwise order on ∂B_i, and each arc on ∂B_i is oriented clockwise. Let s be the point of the arc $p_{i-1}\widetilde{p}_i$ of ∂B_i such that the arc $p_{i-1}s$ has the same length as $p_i\widetilde{p}_i$ has. Then $d(p_{i-1}, p_i) = d(s, \widetilde{p}_i)$ holds, and the arc $\widetilde{p}_{i-1}s$ is of length $\varphi_{i-1} + \varphi_i$. We deduce by (5.32) that

$$|\partial C_n \cap D_i| \le \frac{4 \cdot \cos \frac{1}{4}(\varphi_{i-1} + \varphi_i)}{\cos \frac{1}{2} \gamma_i},$$

which, in turn, yields by the addition formula for cosines that

$$|\partial C_n \cap D_i| \le 4 \cdot \cos \frac{\varphi_{i-1} + \varphi_i - 2\gamma_i}{4}. \tag{5.38}$$

If $i = 1, n$ then $\gamma_i = 0$, and a similar argument as before leads to

$$|\partial C_n \cap D_1| \le \varphi_0 + 4 \cdot \cos \frac{\varphi_0 + \varphi_1}{4};$$
$$|\partial C_n \cap D_n| \le \varphi_n + 4 \cdot \cos \frac{\varphi_{n-1} + \varphi_n}{4}. \tag{5.39}$$

Now the cosine function is strictly decreasing and strictly concave on $[0, \pi/2]$; hence, we conclude by Proposition 5.7.5 the estimate

$$P(C_n) \le \varphi_0 + \varphi_n + 4n \cdot \cos \frac{\varphi_0 + \varphi_n}{4}. \tag{5.40}$$

It follows by differentiation with respect to $(\varphi_0 + \varphi_n)/4$ that the only maximum point of the right-hand side on $[0, \pi/2]$ satisfies $\sin[(\varphi_0 + \varphi_n)/4] = 1/n$; thus, the lower bound of (5.31) is the value of $P(C_n)$. In particular, equality holds in all the preceding inequalities.

We deduce by equality in (5.38) that $\gamma_i = 0$, and the segments $p_{i-1}\widetilde{p}_{i-1}$ and $p_i\widetilde{p}_i$ share a common perpendicular bisector. In addition, equality in both inequalities of (5.39) yields that D_0 is symmetric through the perpendicular bisector of the segment $p_1\widetilde{p}_1$ and contains the arc of length φ_0 on ∂B_1, and the analogous condition holds for D_n. Therefore, Theorem 5.7.1 follows by $\sin[(\varphi_0 + \varphi_n)/4] = 1/n$. □

Comments. Theorem 5.7.1 is due to G. Fejes Tóth, P. Gritzmann, and J. M. Wills [FGW1984].

5.8. Related Problems

Covering Some Specific Shapes in $\mathbb{R}^2$

We discuss optimal coverings of circular discs, equilateral triangles, and rectangles by n equal circular discs. Naturally, for large n, the asymptotic density of the optimal coverings is $2\pi/\sqrt{27}$.

When covering a circular disc, the optimal solution is found by K. Bezdek for $n \le 7$, where the nontrivial cases are $n = 5$ (see [Bez1983]) and $n = 6$ (see [Bez1979]). The cases $n = 8, 9, 10$ are solved by G. Fejes Tóth [FTGa]. Thin coverings for larger n can be found in K. J. Nurmela [Nur1999].

For equilateral triangles, the optimal coverings are known if $n \le 6, n = 9$, and $n = 10$ (see J. B. M. Melissen [Mel1997]). In addition, K. J. Nurmela [Nur2000] presents efficient coverings by at most 36 circular discs.

The problem of thin coverings of a squares by n equal circular discs was initiated by T. Tarnai and Zs. Gáspár [TaG1995]. If $n \le 5$ or $n = 7$ then A. Heppes and J. B. M. Melissen [HeM1997] solved the problem. For many other values of n, K. J. Nurmela [Nur1999] provides thin coverings. Actually A. Heppes and J. B. M. Melissen [HeM1997] solved the problem of covering any rectangle by at most five equal circular discs. In addition, they determined the optimal covering of a rectangle by seven circular discs if the ratio of the longer side to the shorter one is at most $1.3445\ldots$ or at least $3.4301\ldots$. If this ratio is at least $3.1180\ldots$ and $n = 6$ then the problem is solved by J. B. M. Melissen and P. C. Schuur [MeS2000].

Multiple Coverings by Unit Discs

Given $k \ge 1$, an arrangement of n Euclidean unit discs is called a *k-fold covering* of a convex domain D if every point of D is contained in at least k discs; hence, the density with respect to D is readily at least k. G. Fejes Tóth [FTG1976] obtained the following improvement: Any k-fold covering of a hexagon H by n unit discs satisfies

$$A(H) \le n \cdot 3 \sin \frac{\pi}{3k}; \tag{5.41}$$

namely, the density is at least k times the density of the unit disc with respect to the inscribed regular $6k$-gon. If $k = 1$ then (5.41) is just the optimal (hexagon) bound. If $k = 2$ then (5.41) yields that the density is at least $2.0954\ldots$, whereas L. Danzer [Dan1960] constructed periodic twofold coverings of density $2.347\ldots$

Next, we construct thin k-fold coverings for large k. Any unit disc contains at least $r^2\pi - r^{46/73+\varepsilon}$ points of the lattice $(1/r)\,\mathbb{Z}^2$ (see M. N. Huxley [Hux1996], chapter 18.3) for any $\varepsilon > 0$ and for large r. Therefore, the periodic arrangement with respect to the lattice $(1/r)\,\mathbb{Z}^2$ for suitable r yields finite k-fold coverings of density at most $k + k^{23/73+\varepsilon}$ (with $\varepsilon > 0$ arbitrary and k large).

We note that any 33-fold covering by unit discs can be decomposed into two coverings according to P. Mani-Levitska and J. Pach [MaP?].

Coverings by Circular Discs of Different Sizes in $\mathbb{R}^2$

It was L. Fejes Tóth who observed around 1950 that coverings by circular discs of slightly different sizes cannot be less dense than the covering density $2\pi/\sqrt{27}$ of equal discs. Later K. Böröczky (unpublished; see also G. Blind [Bli1969]) proved the following qualitative estimate: If a hexagon D is covered by $n \geq 2$ circular discs of radius at most 1 and at least

$$\sqrt{\frac{7 \sin \frac{2\pi}{7} - 6 \sin \frac{2\pi}{6}}{6 \sin \frac{2\pi}{6} - 5 \sin \frac{2\pi}{5}}} = 0.7921 \ldots$$

then the ratio of the total area of the circular discs to the area of D is at least $2\pi/\sqrt{27}$. According to Theorem 1.6.2, this result holds for any convex domain D provided that $n \geq n_0$ for an absolute constant n_0. It would be interesting to know whether one can take $n_0 = 2$.

However, we have seen in Remark 1.6.4 that large hexagons can be covered by circular discs of radius either 1 or 0.7149 in a way that the density is less than $2\pi/\sqrt{27}$, an example due to J. Molnár [Mol1959].

Concerning coverings of a polygon P of at most six sides by circular discs of very different sizes, L. Fejes Tóth [FTL1972] proves that if the radii lie in the interval $[r, 1]$ for $r > 0$, then the density of the covering is at least $(r^2\pi/3)/\sin(r^2\pi/3)$. Moreover, according to G. Fejes Tóth [FTG1995b], if circular disc of two distinct radii cover a rectangle than the density of the covering is at least $1.0189\ldots$, which lower bound is optimal.

Solid Coverings of S^2

We say that a covering of S^2 by finitely many circular discs (of possibly different radii) is *solid* if whenever the discs are rearranged in a way that they still cover S^2 then the new covering is congruent to the original one. In addition, we call a spherical tiling Archimedean if each tile is a regular polygon, and for any pair of vertices there exists an isometry of the tiling that maps one vertex to the other. L. Fejes Tóth [FTL1968] conjectured that the circumscribed circular discs of an Archimedean tiling form a solid covering if and only if three tiles meet at each vertex, namely, when the tiling is *trihedral*. A. Florian and A. Heppes [FlH2003] show that nontrihedral tilings lead to nonsolid coverings. Concerning trihedral tilings, L. Fejes Tóth [FTL1968] himself settled the case when the Archimedean tiling is coming from a regular three-polytope, and G. Fejes Tóth [FTG1974] verified the conjecture in many other cases.

Part 2
Arrangements in Higher Dimensions

6

Packings and Coverings by Spherical Balls

Packings of equal balls on the sphere have been investigated intensively since the middle of the 20th century. We write $M(d, \varphi)$ to denote the maximal number of nonoverlapping spherical balls of radius φ on S^d for $d \geq 3$ and $\varphi < \pi/2$, and we list the known optimal packings. The case $\varphi \geq \pi/4$ has been completely understood (see Theorems 6.1.1 and 6.2.1, and the Comments to Sections 6.1 and 6.2 for references). More precisely an optimal packing of n equal spherical balls is provided by the Euclidean regular $(n - 1)$-simplex if $3 \leq n \leq d + 2$, and by the Euclidean $(d + 1)$-crosspolytope if $d + 3 \leq n \leq 2d + 2$. In particular $M(d, \pi/4) = 2d + 2$. The optimal packings are unique up to congruency if $n \leq d + 2$ or $n = 2d + 2$. If $\varphi < \pi/4$ then the following values are known:

$$M(3, \tfrac{\pi}{6}) = 24 \qquad\qquad M(3, \tfrac{\pi}{10}) = 120$$

$$M(4, \tfrac{1}{2}\text{arc cos } \tfrac{1}{5}) = 16 \qquad M(5, \tfrac{1}{2}\text{arc cos } \tfrac{1}{4}) = 27$$

$$M(6, \tfrac{1}{2}\text{arc cos } \tfrac{1}{3}) = 56 \qquad M(7, \tfrac{\pi}{6}) = 240$$

$$M(20, \tfrac{1}{2}\text{arc cos } \tfrac{1}{9}) = 112 \qquad M(20, \tfrac{1}{2}\text{arc cos } \tfrac{1}{7}) = 162$$

$$M(21, \tfrac{1}{2}\text{arc cos } \tfrac{1}{11}) = 100 \qquad M(21, \tfrac{1}{2}\text{arc cos } \tfrac{1}{6}) = 275$$

$$M(21, \tfrac{1}{2}\text{arc cos } \tfrac{1}{4}) = 891 \qquad M(22, \tfrac{1}{2}\text{arc cos } \tfrac{1}{5}) = 552$$

$$M(22, \tfrac{1}{2}\text{arc cos } \tfrac{1}{3}) = 4,600 \qquad M(23, \tfrac{\pi}{6}) = 196,560$$

Of these values, we only prove $M(3, \pi/10) = 120$ (Corollary 6.4.2) due to K. Böröczky [Bör1978]. The rest of the results require the linear programming bound that we summarise at the end of this introduction. $M(3, \pi/6) = 24$ has been proved recently by O. R. Musin [Mus?], and the other entries are due to V. I. Levenšteĭn [Lev1979]. Moreover $M(7, \pi/6)$ and $M(23, \pi/6)$ were also determined by A. M. Odlyzko and N. J. A. Sloane [OdS1979], and the optimal arrangements related to $M(6, (1/2)\text{arc cos}(1/3))$, $M(7, \pi/6)$, $M(22, (1/2)\text{arc cos}(1/3))$ and $M(23, \pi/6)$ are unique up to congruency according to E. Bannai and N. J. A.

Sloane [BaS1981]. We note that $M(d, \pi/6)$ is the so-called Newton or kissing number; namely, the maximal number of nonoverlapping unit balls in $\mathbb{R}^{d+1}$ that touch a given unit ball (see Section 8.1).

For the density of packings of spherical balls of radius $\varphi < \pi/4$, we present the obvious lower bound resulting from the greedy algorithm, which is still the essentially best lower bound available. However, we show that the density of a packing is at most $2^{-0.5d+o(d)}$ (see R. A. Rankin [Ran1955] or Theorem 6.3.1), an upper bound that is relatively simple to verify and essentially optimal if φ is close to $\pi/4$. If $\varphi \leq 31°$ then the upper bound $2^{-0.599d+o(d)}$ was proved by G. A. Kabatjanskii and V. I. Levenšteǐn [KaL1978] based on the linear programming bound.

Fewer optimal arrangements are known when S^d is covered by n equal spherical balls for $d \geq 3$. Besides the simple cases $n \leq d + 2$ (see Theorem 6.5.1), only the following cases have been solved: When $n = d + 3$ (see Theorem 6.6 or K. Böröczky Jr. and G. Wintsche [BöW2003]), and when $d = 3$ and $n = 8$ (see L. Dalla, D. G. Larman, P. Mani–Levitska, and Ch. Zong [DLM2000] or Theorem 6.7.1). The last result says that the eight vertices of the four-dimensional regular crosspolytope determine an optimal covering on S^3. However, sound density estimates are available for coverings. We verify that there exists a covering of S^d by n equal spherical balls whose density is at most $d \ln d$, and even at most $d \ln \ln d$ if the number of balls is polynomial in d (see Theorem 6.8.1). We note that covering S^{d-1} by equal balls is intimately connected to polyhedral approximation of the solid unit ball B^d (see J. Matoušek [Mat2002] or K. Böröczky Jr. and G. Wintsche [BöW2003]).

The last section discusses the simplex bound of K. Böröczky [Bör1978] for ball packings in compact hyperbolic manifolds of any dimension. We have included this bound in this chapter because it is the only notable result about finite arrangements in higher dimensional hyperbolic spaces.

Some Related Notions

We call the intersection of finitely many hemispheres a *spherical polytope* if it is contained in an open hemisphere and it has nonempty interior on S^d. A spherical polytope P is the radial projection of some Euclidean d-polytope Q such that aff Q does not contain the origin of $\mathbb{R}^{d+1}$, and the faces of P are the radial projections of the faces of Q. As usual zero- and one-dimensional faces are called vertices and edges, respectively. If Q is a d-simplex then P is called a spherical d-simplex. In addition, a spherical d-simplex is regular if all edges are of the same length. We say that a finite tiling Σ of S^d by spherical simplices is a *triangulation* of S^d if the intersection of any subfamily of the tiles is a common face, provided it is nonempty. An i-cell of Σ is an i-dimensional face of some tile if $i < d$ and it is a tile if $i = d$. We write $f_i(\Sigma)$ to denote the number of i-cells of Σ.

Finally, we write $B(\varphi, x)$ to denote the spherical ball of radius φ and of centre x, and we write $|\cdot|$ to denote the volume (Hausdorff d-measure) on S^d. Given a $\sigma \subset S^d$, its *boundary* and *interior* on S^d are denoted by $\partial\sigma$ and int σ, respectively; moreover, σ is called *Jordan measurable on S^d* if $|\partial\sigma| = 0$.

The Linear Programming Bound

First we introducce some notation. The *Gegenbauer or ultraspherical* polynomials $G_k(t)$ associated S^d are defined by the initial conditions $G_0(t) = 1$ and $G_1(t) = t$, and the recurrence

$$(k + d - 2) \cdot G_k(t) = (2k + d - 1) \cdot tG_{k-1}(t) - (k - 1) \cdot G_{k-2}(t).$$

The normalization is chosen to ensure $G_k(1) = 1$. To indicate the role of the Gegenbauer polynomials, we note that for any fixed $u \in S^d$, the function $F_k(x) = G_k(\langle u, x \rangle)$ on S^d is a spherical harmonics that satisfies

$$\int_{S^d} F_k(x)F_m(x)\,dx = \int_{-1}^{1} G_k(t)G_m(t)(1 - t^2)^{\frac{d}{2}-1}dt = 0 \text{ for } m \neq k.$$

Let $x_1, \ldots, x_n \in S^d$ be the centers of nonoverlapping spherical balls of radius $\varphi < \pi/4$, and let $t_{ij} = \langle x_i, x_j \rangle$. In particular $t_{ij} \leq \cos 2\varphi$ for $i \neq j$. According to a result of I. J. Schoenberg [Sch1942], the symmetric $n \times n$ matrix $[G_k(t_{ij})]$ is positive semidefinite for any k, and hence $\Sigma_{i,j}G_k(t_{ij}) \geq 0$.

Let us now consider a polynomial $f(t) = \Sigma_{k=0}^N c_k G_k(t)$ satisfying $c_0 > 0, c_k \geq 0$ for $k > 0$ and $f(t) \leq 0$ for $t \in [-1, \cos 2\varphi]$. The preceding discussion yields

$$nf(1) \geq \sum_{i,j} f(t_{ij}) = \sum_{k=0}^{N} c_k \sum_{i,j} G_k(t_{ij}) \geq \sum_{i,j} c_0 G_0(t_{ij}) = n^2 c_0.$$

We conclude the *linear programming bound*

$$M(d, \varphi) \leq \frac{f(1)}{c_0} = \frac{c_0 + \cdots + c_N}{c_0} \tag{6.1}$$

due to P. Delsarte [Del1972]. To find good explicit bound for given d and φ is now a matter of careful computer search of suitable f. G. A. Kabatjanskii and V. I. Levenšteĭn [KaL1978] investigated further properties of the Gegenbauer polynomials, and used the linear programming bound to prove the upper bound $2^{-0.599d+o(d)}$ for the density of the packing if $\varphi \leq 31°$. We refer to J. H. Conway and N. J. A. Sloane [CoS1999], T. Ericson and V. Zinoview

[ErZ2001], and Ch. Zong [Zon1999] for detailed discussion with applications. O. R. Musin [Mus?] proves $M(3, \pi/6) = 24$ by improving the Delsarte method.

6.1. Packing $d + 2$ Balls on S^d

As one may expect, the vertices of the regular Euclidean $(d + 1)$-simplex determine the optimal packing of $d + 2$ equal spherical balls on S^d.

Theorem 6.1.1. *Given $2 \le n \le d + 2$, if n spherical balls of radius φ are packed on S^d then $\varphi \le (1/2) \arccos[-1/(n-1)]$, and equality holds if and only if the centres are vertices of a regular $(n-1)$-simplex of circumradius one in $\mathbb{R}^{d+1}$.*

Proof. Let the unit vectors $u_1, \ldots, u_n$ in $\mathbb{R}^{d+1}$ represent the centres of the n spherical balls. Since

$$n^2 = n \sum_{i=1}^{n} u_i^2 = \left(\sum_{i=1}^{n} u_i \right)^2 + \sum_{1 \le i < j \le n} (u_i - u_j)^2 \ge \sum_{1 \le i < j \le n} (u_i - u_j)^2,$$

the minimum of $(u_i - u_j)^2$ over pairs $u_i \ne u_j$ is at most $2n/(n-1)$, and equality holds if and only if $(u_i - u_j)^2 = 2n/(n-1)$ for all $i \ne j$. In turn, we conclude Theorem 6.1.1. $\square$

Comments. Theorem 6.1.1 follows from the paper by H. W. E. Jung [Jun1901] (see Theorem A.8.1 for the Jung inequality).

6.2. Packing at Most $2d + 2$ Balls on S^d

In this section we show that if $d + 3$ balls of equal radius are packed on S^d then an even $2d + 2$ balls of the same radius can be packed on S^d.

Theorem 6.2.1. *If $n \ge d + 3$ balls of radius φ are packed on S^d then $\varphi \le \pi/4$. In addition, if $\varphi = \pi/4$ then $n \le 2d + 2$, and $n = 2d + 2$ holds if and only if the centres are vertices of a regular crosspolytope in $\mathbb{R}^{d+1}$.*

Remark. If $\varphi = \pi/4$ then we prove that the balls can be partitioned into $n - d - 1$ subfamilies (each containing at least two balls) such that any two balls coming from different families touch.

Proof. Theorem 6.2.1 (and the remark after it) follows from the following claim: If $u_1, \ldots, u_n$ are nonzero vectors that span $\mathbb{R}^m$, $m \geq 2$, such that $n \geq m + 2$ and $\langle u_i, u_j \rangle \leq 0$ for $i \neq j$, then there exist pairwise orthogonal linear subspaces $L_1, \ldots, L_{n-m}$ that span $\mathbb{R}^m$, and each L_i contains $\dim L_i + 1$ vectors out of $u_1, \ldots, u_n$ that span L_i. We prove the claim by induction on m, where the case $m = 2$ follows from the fact that the angle between any two u_is is at least $\pi/2$. If u_i and u_n point to opposite directions for some $i < n$ then $u_j \in u_n^\perp$ holds for $j \neq i, n$, and hence the claim follows by the induction hypothesis. If no u_i points to the direction opposite to u_n then we write u_i, $i = 1, \ldots, n - 1$, as $u_i = v_i + \alpha_i u_n$, where v_i is orthogonal to u_n, and $\alpha_i \leq 0$. We deduce $\langle v_i, v_j \rangle \leq 0$ for $i \neq j$, and even $\langle v_i, v_j \rangle < 0$ if both $\alpha_i, \alpha_j < 0$. We apply the induction hypothesis to the nonzero vectors $v_1, \ldots, v_{n-1}$ in $u_n^\perp$ and obtain the suitable pairwise orthogonal linear subspaces $L'_1, \ldots, L'_{n-m}$ of $u_n^\perp$. We may assume that L'_1 contains all v_i (if any) such that $\alpha_i < 0$; therefore, we define $L_i = L'_i$ for $i \geq 2$ and L_1 to be the span of L'_1 and u_n. □

Comments. Theorem 6.2.1 was proposed as a problem by H. Davenport and Gy. Hajós [DaH1951]. Numerous solutions arrived in a relatively short time, namely, the ones by J. Aczél [Acz1952] and by T. Szele [Sze1952] and the unpublished ones due to M. Bognár, Á. Császár, T. Kővári, and I. Vincze. Independently, R. A. Rankin [Ran1955] proved Theorem 6.2.1.

6.3. Density Bounds for Packings on S^d

Theorem 6.2.1 says that at most $2d$ spheres of radius $\pi/4$ can be packed into S^d according to Theorem 6.2.1. However, if $\varphi < \pi/4$ is fixed and d tends to infinity then $M(d, \varphi)$ is exponential in d:

Theorem 6.3.1. *If $d \geq 3$ and $\varphi < \pi/4$ then*

$$\sqrt{2\pi} \cdot \sin^{-d} 2\varphi < M(d, \varphi) < 23 \, d^{\frac{3}{2}} \sin^{-d} \varphi \cdot 2^{-0.5d}.$$

Proof. The argument combines notions of the Euclidean space $\mathbb{R}^{d+1}$ and of S^d. We observe that if $\psi < \pi/2$ and $x \in S^d$ then the orthogonal projection of the spherical ball $B(\psi, x)$ into the tangent hyperplane of S^d in $\mathbb{R}^{d+1}$ at x is a Euclidean d-ball of radius $\sin \psi$, and the exterior normals at the points of $B(\psi, x)$ enclose an angle of at most ψ with the exterior normal at x. Therefore,

$$\sin^d \psi \cdot \kappa_d < |B(\psi, x)| < \sin^d \psi \cdot \frac{\kappa_d}{\cos \psi}. \tag{6.2}$$

In addition, the d-measure of S^d is $(d+1)\kappa_{d+1}$, where the quotient of κ_{d+1} and κ_d satisfies (see (A.29))

$$\sqrt{\frac{2\pi}{d+2}} < \frac{\kappa_{d+1}}{\kappa_d} < \sqrt{\frac{2\pi}{d+1}}. \tag{6.3}$$

To prove the lower bound for $M(d, \varphi)$, we simply apply the greedy algorithm. Let us consider a system of nonoverlapping spherical balls of radius φ such that no extra spherical ball of radius φ can be inserted without overlapping at least one of the given balls. Then S^d is covered by the spherical balls of radius 2φ whose centres coincide with the given centres. Since $(d+1)\kappa_{d+1}/\kappa_d > \sqrt{2\pi d}$ holds according to (6.3), we conclude by (6.2) the estimate

$$M(d, \varphi) > \sqrt{2\pi d} \cdot \cos 2\varphi \cdot \sin^{-d} 2\varphi.$$

This formula yields the lower bound in Theorem 6.3.1 if $\cos 2\varphi > 1/\sqrt{d}$, and the case $\cos 2\varphi \leq 1/\sqrt{d}$ is consequence of the bound $M(d, \varphi) \geq 2d$.

Turning to the upper bound for $M(d, \varphi)$, we let $B(\varphi, x_1), \ldots, B(\varphi, x_n)$ form a packing. We define the function σ_i for $i = 1, \ldots, n$ and $x \in \mathbb{R}^{d+1}$ by

$$\sigma_i(x) = \begin{cases} 1 - \dfrac{(x - x_i)^2}{2\sin^2 \varphi} & \text{if } \|x - x_i\| \leq \sqrt{2}\sin\varphi; \\ 0 & \text{if } \|x - x_i\| \geq \sqrt{2}\sin\varphi. \end{cases}$$

If, say, $x_1, \ldots, x_m$ are the centres of Euclidean distance at most $\sqrt{2}\sin\varphi$ from some $x \in \mathbb{R}^{d+1}$ then

$$m\sum_{i=1}^{m}(x - x_i)^2 = \sum_{1 \leq i < j \leq m}(x_i - x_j)^2 + \left(mx - \sum_{i=1}^{m} x_i\right)^2$$
$$\geq 2m(m-1) \cdot \sin^2\varphi; \tag{6.4}$$

therefore, $\sum_{i=1}^{n}\sigma_i(x) \leq 1$.

First, we assume that $\sin\varphi \leq d/[(d+1)\sqrt{2}]$, and let α be the acute angle satisfying $\sin\alpha = \sqrt{2}\sin\varphi$. A quick look at (6.4) shows that $\sum_{i=1}^{n}\sigma_i(x)$ is very small if the Euclidean distance of x and the mean of $x_1, \ldots, x_m$ is large. Therefore, we do not consider σ_i on S^d but on the sphere $\cos\alpha\, S^d$. In particular,

$$|\cos\alpha\, S^d| \geq \int_{\cos\alpha\, S^d} \sum_{i=1}^{n} \sigma_i(x)\, dx = n \int_{\cos\alpha\, S^d} \sigma_1(x)\, dx. \tag{6.5}$$

To provide a lower bound for the integral of σ_1 on $\cos\alpha\, S^d$, we define the acute angle β by the relation $\sin\beta = [d/(d+1)]\sin\alpha$. We claim that if the angle of $x \in \cos\alpha\, S^d$ and x_1 is at most β then

$$\sigma_1(x) > 1/d. \tag{6.6}$$

According to the law of cosines,

$$(x - x_1)^2 \le 1 + \cos^2\alpha - 2\cos\alpha\cos\beta = \sin^2\beta + (\cos\beta - \cos\alpha)^2$$

holds, where $\sin\alpha \le d/(d+1)$ and $\cos\alpha \ge \sqrt{1 - d^2/(d+1)^2}$ yield

$$\frac{\cos\beta - \cos\alpha}{\sin\alpha} = \frac{\sin^2\alpha - \sin^2\beta}{\sin\alpha(\cos\beta + \cos\alpha)} < \frac{1 - \frac{d^2}{(d+1)^2}}{2\sqrt{1 - \frac{d^2}{(d+1)^2}}} < \frac{1}{\sqrt{2(d+1)}}.$$

In turn, we conclude the claim (6.6). Therefore, (6.2) implies

$$\int_{\cos\alpha\, S^d} \sigma_1(x)\,dx > \frac{1}{d}\cos^d\alpha \cdot \sin^d\beta \cdot \kappa_d > \kappa_d \cos^d\alpha \cdot \frac{\sqrt{2}^d}{e \cdot d}\sin^d\varphi.$$

Substituting this estimate into (6.5) and applying (6.3) result in

$$M(d,\varphi) < ed\sqrt{2\pi(d+1)} \cdot \sin^{-d}\varphi \cdot 2^{-0.5d}.$$

Finally, if $\sin\varphi > d/[(d+1)\sqrt{2}]$ then we define the acute angle ψ by the relation $\sin\psi = [d/(d+1)]\sin\phi$, and we observe that

$$M(d,\varphi) \le M(d,\psi) < e^2 d\sqrt{2\pi(d+1)} \cdot \sin^{-d}\varphi \cdot 2^{-0.5d}.$$

Therefore, Theorem 6.3.1 follows by $e^2\sqrt{2\pi(d+1)} < 23\sqrt{d}$. $\qquad\square$

The lower bound in Theorem 6.3.1 is the best one known unless φ is very close to $\pi/4$, yet it is unclear how good the bound is when φ is small. The upper bound in Theorem 6.3.1 is a very good one if φ is close to $\pi/4$; say, it gives $23\,d^{3/2}$ if $\varphi = \pi/4$, whereas $M(d,\pi/4) = 2d + 2$ according to Theorem 6.2.1. However, an improved estimate is known if $\varphi \le 31.4987°$; namely, the *Kabatjanskii–Levenšteĭn bound* says

$$M(d,\varphi) < \sin^{-d}\varphi \cdot 2^{-0.599 \cdot d + o(d)} \tag{6.7}$$

where the pace of how $o(d)/d$ tends to zero depends only on φ.

Open Problem.

(i) Given some acute angle φ, improve on the bounds of Theorem 6.3.1 and (6.7). The case $\varphi = \pi/6$ is of special interest because $M(d, \pi/6)$ is the Newton number in $\mathbb{R}^{d+1}$.

Comments. The lower bound in Theorem 6.3.1 is due to C. E. Shannon [Sha1959] and J. M. Wyner [Wyn1965]. The upper bound in Theorem 6.3.1 was proved by C. A. Rogers (unpublished), and a more precise (and more involved) estimate was later verified by R. A. Rankin [Ran1955]; this bound coincides with Theorem 6.3.1 up to a constant factor.

The proof of (6.7) due to G. A. Kabatjanskii and V. I. Levenšteĭn [KaL1978] is based on the linear programming bound (6.1).

We call an arrangement of spherical balls a *k-fold packing* if no point is contained in the interior of k of the balls. Various estimates for k-fold packings are presented in N. M. Blachman and L. Few [BlF1963], Sh. I. Galiev [Gal1996], and V. Blinovsky [Bli1999].

6.4. The Simplex Bound for Packings on S^3

Given $\pi/4 \leq \varphi < \pi/2$, Theorems 6.1.1 and 6.2.1 determine the maximal number of nonoverlapping spherical balls of radius φ on S^d. Therefore, let $\varphi < \pi/4$. We write $T^3(2\varphi)$ to denote a regular spherical three-simplex of edge length 2φ, and define $\sigma_{S^3}(\varphi)$ to be the ratio of the volume of the part of $T^3(2\varphi)$ occupied by the four balls of radius φ centred at the vertices of $T^3(2\varphi)$ to the volume of $T^3(2\varphi)$. The aim of this section to prove the *simplex bound* for ball packings on S^3:

Theorem 6.4.1 (Coxeter–Böröczky bound). *The density of any packing of balls of radius $\varphi < \pi/4$ on S^3 is at most $\sigma_{S^3}(\varphi)$. Equality holds if and only if $\varphi = \pi/10$, and the centres are the* 120 *vertices of the* 600-*cells inscribed into B^4.*

Theorems 6.1.1 and 6.2.1 yield that the regular simplex and the regular crosspolytope in $\mathbb{R}^d$ determine the optimal packing of $d + 1$ and $2d$, respectively, spherical balls on S^{d-1}. In addition, the twenty vertices of the icosahedron determine an optimal packing on S^2 according to Theorem 4.4.1. Since adding the 600-cells to the list, we have enumerated all Euclidean regular simplicial polytopes (see Section A.7), Theorem 6.4.1 yields

Corollary 6.4.2. *Given a simplicial regular polytope of n vertices in $\mathbb{R}^d$, $d \geq 3$, its vertices determine the optimal packing of n equal spherical balls on S^{d-1}.*

In this section, segment, ball, etc. always mean spherical segment, spherical ball, etc., respectively. In addition, a *two-sphere* is a copy of S^2 in S^3, which is the analogue of two-planes in the Euclidean setting. In particular, the perpendicular bisector of a segment is a two-sphere that passes through the midpoint and is orthogonal to the segment. Theorem 6.4.1 is proved by considering the Dirichlet–Voronoi cells, which we now construct. Let $x_1, \ldots, x_n$ denote the centres of the n nonoverlapping balls of radius $\varphi < \pi/4$ on S^3, where we may assume that every open hemisphere contains some x_i. The Dirichlet–Voronoi cell D_i of x_i is the set of points of S^3 that are not farther from x_i than from any x_j. We observe that D_i is the intersection of the hemispheres that contain x_i and is bounded by the perpendicular bisectors of certain segments $x_i x_j$, $x_j \neq x_i$. In particular, D_i is a spherical polytope and $D_1, \ldots, D_n$ title S^3.

We define the density of a ball B inside an object X to be $|B \cap X|/|X|$. Naturally, Theorem 6.4.1 follows if we prove the simplex bound for the density of each ball inside its Dirichlet–Voronoi cell. The argument uses simplices of special type; namely, we call a spherical three-simplex $v_0 v_1 v_2 v_3$ an *orthoscheme* if the edge $v_0 v_1$ is orthogonal to the two-sphere $v_1 v_2 v_3$, and the two-sphere $v_0 v_1 v_2$ is orthogonal to the edge $v_2 v_3$. We note that the order of vertices is important. A typical example is related to the regular simplex: Let $\tilde{v}_i$ be the circumcentre of an i-face of $T^3(2\varphi)$, $i = 0, 1, 2, 3$, in a way that $\tilde{v}_j$ is contained in the i-face corresponding to $\tilde{v}_i$ for $0 \leq i < j \leq 3$, hence, $\tilde{v}_3$ is a vertex of $T^3(2\varphi)$. Now $T^3(2\varphi)$ can be decomposed into twenty-four orthoschemes congruent to the orthoscheme $S(\varphi) = \tilde{v}_0 \tilde{v}_1 \tilde{v}_2 \tilde{v}_3$, and the density of $B(\varphi, \tilde{v}_0)$ inside $S(\varphi)$ is $\sigma_{S^3}(\varphi)$.

Let us describe the idea of the proof of the simplex bound inside a Dirichlet–Voronoi cell D_i. We start with the case when the *foot condition* holds, namely, the foot of the perpendicular from x_i to the two-sphere of any face lies in the relative interior of the face, and the foot of the perpendicular from x_i to the great circle of any edge lies in the relative interior of the edge. Then to any triple (v, e, F), where F is a face, $e \subset F$ is an edge, and $v \in e$ is a vertex of D_i, we associate an orthoscheme $x_i v_1 v_2 v_3$ such that $v = v_3$, and v_2 and v_1 are the foots of the perpendiculars from x_i to e and F, respectively. These orthoschemes form a tiling of D_i, and $d_{S^3}(x_i, v_j) \geq d_{S^3}(\tilde{v}_0, \tilde{v}_j)$, $j = 1, 2, 3$. Therefore, Lemma 6.4.4 yields that the density of $B(\varphi, x_i)$ inside the orthoscheme $x_i v_1 v_2 v_3$ is at most $\sigma_{S^3}(\varphi)$. In general, some faces of D_i violate the foot condition. We will verify that the distance of x_i from the two-sphere of each of these faces is larger then $d_{S^3}(\tilde{v}_0, \tilde{v}_3)$; hence, we intersect D_i by the ball of centre x_i and of radius $d_{S^3}(\tilde{v}_0, \tilde{v}_3)$, and we prove the simplex bound inside the resulting convex body.

We may assume that all objects occurring in this argument are contained in a fixed open hemisphere. Therefore, given two points x and y, we can speak about the great circle xy, the segment xy, and the ray xy (meaning the semi-circle starting from x and passing through y). We note that if a great circle l is orthogonal to a two-sphere Π then l is orthogonal to any great circle in Π that it meets. We frequently use the following consequence of the law of cosines: If each side of a spherical triangle is less than $\pi/2$ then at least two angles are acute. Given two triangles sharing a common side, let us define the angle of them. We choose an interior point p of the common side, and we start one–one rays from p in both triangles such that both rays are orthogonal to the common side. Then the angle of the two triangles is the angle of the two rays, which is independent of the choice of p.

As a preparation for the proof of Lemma 6.4.4, we verify a simple auxiliary statement:

Proposition 6.4.3. *Let x, y_0, z be points of S^3 satisfying that $\angle xy_0z = \pi/2$ and $d_{S^3}(x, z) < \pi/2$, and let l be a great circle that does not contain x and is orthogonal to the segment y_0z at y_0. If a variable point y moves away from y_0 along l in a way that $d_{S^3}(x, y) < \pi/2$ then the angle $\angle zxy$ strictly increases.*

Proof. For any position of y, there exists a spherical triangle $y'y_0'z'$ in S^2 such that the spherical lengths of the sides $z'y'$, $y'y_0'$, and $y_0'z'$ are $\angle zxy$, $\angle yxy_0$, and $\angle y_0xz$, respectively, and the angle at y_0' is the right angle. Since $\angle yxy_0$ is increasing, the law of cosines yields Proposition 6.4.3. $\square$

The proof of Theorem 6.4.1 relies on comparing densities in orthoschemes:

Lemma 6.4.4. *If the orthoschemes $S = v_0v_1v_2v_3$ and $S' = v_0'v_1'v_2'v_3'$ on S^3 satisfy $0 < \varphi \le d_{S^3}(v_0, v_i) \le d_{S^3}(v_0', v_i') < \pi/2$ for $i = 1, 2, 3$, then*

$$\frac{|S \cap B(\varphi, v_0)|}{|S|} \ge \frac{|S' \cap B(\varphi, v_0')|}{|S'|},$$

and equality holds if and only if $d_{S^3}(v_0, v_i) = d_{S^3}(v_0', v_i')$ for $i = 1, 2, 3$.

Proof. First, we present a formula for the volume of spherical tetrahedra. Let T be the spherical tetrahedron with vertices $y_0y_1y_2y_3$ of diameter less than $\pi/2$. We write α_0 to denote the angle of the triangles $y_0y_1y_2$ and $y_0y_1y_3$, and let p_α denote the point of the edge y_2y_3 such that the angle of the triangles $y_0y_1y_2$ and $y_0y_1p_\alpha$ is α. In addition, let $\mathcal{R}_{y_0y_1}^\alpha(T)$ be the Jordan measurable set that results from rotating the triangle $y_0y_1p_\alpha$ around y_0y_1 on S^d.

Since the spherical volume of the tetrahedron $y_0 y_1 p_\alpha p_{\alpha + \Delta \alpha}$ is asymptotically $\Delta \alpha / (2\pi) \cdot |\mathcal{R}^\alpha_{y_0 y_1}(T)|$ as $\Delta \alpha$ tends to zero, we deduce

$$|T| = \frac{1}{2\pi} \int_0^{\alpha_0} \left| \mathcal{R}^\alpha_{y_0 y_1}(T) \right| \, d\alpha. \tag{6.8}$$

Moreover, similar considerations yield for any ball $B(r, y_0)$ that

$$|T \cap B(r, y_0)| = \frac{1}{2\pi} \int_0^{\alpha_0} \left| \mathcal{R}^\alpha_{y_0 y_1}(T) \cap B(r, y_0) \right| \, d\alpha. \tag{6.9}$$

To verify Lemma 6.4.4, we may assume $d_{S^3}(v_0, v_i) < d_{S^3}(v_0, v_i')$ for one $i \in \{1, 2, 3\}$, and $v_j' = v_j$ if $j \neq i$. We abbreviate $\mathcal{R}^\alpha_{v_0' v_j'}(S')$ as $\mathcal{R}^\alpha_{0j}$.

Case I: $d_{S^3}(v_0, v_3) < d_{S^3}(v_0, v_3')$ and $v_j' = v_j$ if $j = 0, 1, 2$

We assume that v_3 is contained in the segment $v_2 v_3'$, and we write p_α to denote the point of the edge $v_2 v_3'$ such that the angle of the triangles $v_0 v_1 v_2$ and $v_0 v_1 p_\alpha$ is α. Let $\varrho(\alpha) = d_{S^3}(v_0, p_\alpha)$.

We claim that if $\alpha_1 < \alpha_2$ then the density of $B(\varphi, v_0)$ inside $\mathcal{R}^{\alpha_1}_{01}$ is larger than that inside $\mathcal{R}^{\alpha_2}_{01}$. Since $B(\varrho(\alpha_1), v_0)$ contains $\mathcal{R}^{\alpha_1}_{01}$, the density of $B(\varphi, v_0)$ inside $\mathcal{R}^{\alpha_1}_{01}$ is larger than that inside $B(\varrho(\alpha_1), v_0)$. However, the intersection $\mathcal{R}^{\alpha_2}_{01} \backslash \mathcal{R}^{\alpha_1}_{01}$ and the face $v_1 v_2 v_3$ lie outside of $B(\varrho(\alpha_1), v_0)$, and hence the density of $B(\varphi, v_0)$ inside $\mathcal{R}^{\alpha_2}_{01} \backslash \mathcal{R}^{\alpha_1}_{01}$ is smaller than that inside $B(\varrho(\alpha_1), v_0)$. Therefore, we conclude the claim, which, in turn, yields Case I of Lemma 6.4.4 by (6.8) and (6.9).

Case II: $d_{S^3}(v_0, v_2) < d_{S^3}(v_0, v_2')$ and $v_j' = v_j$ if $j = 0, 1, 3$

Since $d_{S^3}(v_0, v_2) < d_{S^3}(v_0, v_2')$, and the triangles $v_1 v_2 v_3$ and $v_1 v_2' v_3$ have right angle at v_2 and at v_2', respectively, we may assume that the segments $v_1 v_2'$ and $v_3 v_2$ intersect in a point q. We verify Lemma 6.4.4 in two steps: First, we show that the density of $B(\varphi, v_0)$ inside the tetrahedron $v_0 v_1 q v_3$ is larger than that inside the tetrahedron $v_0 v_1 v_2' v_3$, and then we show that the density of $B(\varphi, v_0)$ inside the tetrahedron $v_0 v_1 v_2 v_3$ is larger than that inside the tetrahedron $v_0 v_1 q v_3$.

For the first step, we apply (6.8) to $T = S'$, $y_0 = v_0$, and $y_1 = v_3$, and let p_α denote the point of the edge $v_1 v_2'$ such that the angle of the triangles $v_0 v_3 v_1$ and $v_0 v_3 p_\alpha$ is α. We claim that if $\alpha_1 < \alpha_2$ then

$$\frac{\left| B(\varphi, v_0) \cap \mathcal{R}^{\alpha_1}_{03} \right|}{\left| \mathcal{R}^{\alpha_1}_{03} \right|} > \frac{\left| B(\varphi, v_0) \cap \mathcal{R}^{\alpha_2}_{03} \right|}{\left| \mathcal{R}^{\alpha_2}_{03} \right|}. \tag{6.10}$$

Since the segment $v_2' v_3$ is orthogonal to the two-sphere $v_0 v_1 v_2'$, and the ray $v_0 p_\alpha$ rotates away from the ray $v_0 v_2'$ as α decreases, we deduce by Proposition 6.4.3 that

$$\angle p_{\alpha_1} v_0 v_3 > \angle p_{\alpha_2} v_0 v_3.$$

Similarly, the segment $v_0 v_1$ is orthogonal to the two-sphere $v_1 v_2' v_3$, and the ray $v_0 p_\alpha$ rotates away from the ray $v_0 v_1$ as α increases; hence,

$$\angle p_{\alpha_1} v_3 v_0 < \angle p_{\alpha_2} v_3 v_0.$$

Therefore, $\partial \mathcal{R}_{03}^{\alpha_2}$ intersects the open segment $v_3 p_{\alpha_1}$ in a point p', and hence the density of $B(\varphi, v_0)$ inside $\mathcal{R}_{03}^{\alpha_2}$ is less than that inside $\mathcal{R}_{03}^{\alpha_2} \cap \mathcal{R}_{03}^{\alpha_1}$. Now the open segment $p_{\alpha_1} v_2'$ avoids the circular disc that lies in the two-sphere $v_0 v_1 v_2'$, whose centre is v_0 and whose radius is $d_{S^3}(v_0, p_{\alpha_1})$. It follows that the open segment $p_{\alpha_1} v_3$ avoids spherical ball with centre v_0 and of radius $d_{S^3}(v_0, p_{\alpha_1})$, and thus $\angle v_0 p_{\alpha_1} v_3$ is at least $\pi/2$. We deduce that $\angle v_0 p' v_3$ is obtuse, and hence any ray that starts from v_0 and intersects the interior of $\mathcal{R}_{03}^{\alpha_2}$ will intersect $\mathcal{R}_{03}^{\alpha_1}$ in a segment of length larger than $\varphi' = d_{S^3}(v_0, p')$. Therefore, the density of $B(\varphi, v_0)$ inside $\mathcal{R}_{03}^{\alpha_2} \cap \mathcal{R}_{03}^{\alpha_1}$ is less than that inside $B(\varphi', v_0)$. However, $\mathcal{R}_{03}^{\alpha_2} \backslash \mathcal{R}_{03}^{\alpha_1}$ is contained in $B(\varphi', v_0)$, completing the proof of (6.10).

Combining the formulae (6.8) and (6.9) and the inequality (6.10) yields that the density of $B(\varphi, v_0)$ inside S' is less than that inside the tetrahedron $v_0 v_1 q v_3$. However, q is contained in the edge $v_2 v_3$ of S, and hence we deduce by the argument for Case I that the density of $B(\varphi, v_0)$ inside the tetrahedron $v_0 v_1 q v_3$ is less than that inside S. In turn, we conclude Lemma 6.4.4 in Case II.

Case III: $d_{S^3}(v_0, v_1) < d_{S^3}(v_0, v_1')$ and $v_j' = v_j$ if $j = 0, 2, 3$

We may assume that the open segments $v_0 v_1'$ and $v_1 v_2$ intersect at a point s. Now the density of $B(\varphi, v_0)$ inside the tetrahedron $v_0 s v_2 v_3$ is less than that inside S according to the argument in the first step of Case II. Since the density of $B(\varphi, v_0)$ inside S' is readily less than that in the tetrahedron $v_0 s v_2 v_3$, the proof of Lemma 6.4.4 is now complete. $\qquad \square$

We are ready to prove the simplex bound.

Proof of Theorem 6.4.1. We define $r_i(\varphi) = d_{S^3}(\tilde{v}_0, \tilde{v}_i)$ for $i = 0, 1, 2, 3$, which is the circumradius of i-faces of $T^3(2\varphi)$. In particular, $r_1(\varphi) = \varphi$. Let us consider a Dirichlet–Voronoi cell D_j. For a face F of D_j, we write $\omega(F)$ to denote the foot of the perpendicular from x_j to the two-sphere of F. Since D_j contains $B(\varphi, x_j)$, we deduce

$$d_{S^3}(\omega(F), x_j) \geq \varphi.$$

If e is an edge of D_j then we write $\omega(e)$ to denote the foot of the perpendicular from x_j to the great circle of e. Now the two-sphere that is orthogonal to the great circle of e at $\omega(e)$ contains at least three centres out of $x_1, \ldots, x_n$ such that these centres are at the same distance $r_2(e) = d_{S^3}(\omega(e), x_j)$ from $\omega(e)$. There exist two out of these centres, say x_k and x_m, such that $\angle x_k \omega(e) x_m$ is at most $2\pi/3$, and hence

$$d_{S^3}(\omega(e), x_j) \geq r_2(\varphi).$$

Finally, if w is a vertex of D_j then it is of the same distance $d_{S^3}(w, x_j)$ from at least four centres out of $x_1, \ldots, x_n$. We apply Theorem 6.1.1 to the unit vectors in the tangent (Euclidean) hyperplane to S^3 at w. It follows that there exist two out of the at least four centres, say $x_{k'}$ and $x_{m'}$, such that $\angle x_{k'} w x_{m'}$ is at most $\arccos(-1/4)$; hence,

$$d_{S^3}(w, x_j) \geq r_3(\varphi). \tag{6.11}$$

Next we claim that if $\omega(F)$ is not contained in the relative interior of the face F of D_j then

$$d_{S^3}(\omega(F), x_j) > r_3(\varphi). \tag{6.12}$$

Now $\omega(F)$ is the midpoint of the segment $x_j x_m$ for some x_m, and the assumption on $\omega(F)$ yields that the distance of some x_k from $\omega(F)$ is at most $d_{S^3}(\omega(F), x_j)$. Thus, there exists a triangle (either $x_k \omega(F) x_j$ or $x_k \omega(F) x_m$) such that two sides enclose an angle of at most $\pi/2$, the lengths of these two sides are at most $d_{S^3}(\omega(F), x_j)$, and the third side is at least 2φ. However, there exists an isosceles triangle whose equal sides are of length $r_3(\varphi)$ and enclose an angle of $\arccos(-1/4)$, and the third side is of length 2φ. Therefore, we conclude (6.12) by the spherical law of cosines.

To verify Theorem 6.4.1, it is sufficient to consider the density of $B(\varphi, x_j)$ inside $D_j \cap B(r_3(\varphi), x_j)$. In turn, we decompose $D_j \cap B(r_3(\varphi), x_j)$ into finitely many Jordan measurable sets such that each set is the union of certain segments $x_j x$ with $x \notin \operatorname{int} B(r_3(\varphi), x_j)$, and we show that the density of $B(r_3(\varphi), x_j)$ inside each of the sets is at most $\sigma_{S^3}(\varphi)$. According to (6.11), no vertex of D_j lies in the interior of $B(r_3(\varphi), x_j)$. Let Σ be the union of the segments connecting x_j to the points of $D_j \cap \partial B(r_3(\varphi), x_j)$. Then either Σ is of volume zero or the density of $B(\varphi, x_j)$ inside σ is $|B(\varphi, x_j)|/|B(r_3(\varphi), x_j)|$, which is readily less than $\sigma_{S^3}(\varphi)$.

Next, let F be a face of D_j that intersects int $B(r_3(\varphi), x_j)$; hence, $\omega(F)$ is contained in the relative interior of F according to (6.12). If F is not contained in $B(r_3(\varphi), x_j)$ then let us consider a maximal (circular) arc σ of $F \cap \partial B(r_3(\varphi), x_j)$. We write $\Sigma(\sigma)$ to denote the union of the triangles $x_j\omega(F)x$ for $x \in \sigma$. Then the density of $B(\varphi, x_j)$ inside $\Sigma(\sigma)$ is the limit of the density of $B(\varphi, x_j)$ inside some orthoschemes $x_j\omega(F)v_2^{(n)}v_3^{(n)}$, where $d_{S^3}(x_j, v_3^{(n)}) = r_3(\varphi)$ and $d_{S^3}(x_j, v_2^{(n)})$ tends to $r_3(\varphi)$ as n tends to infinity. Here $d_{S^3}(x_j, v_2^{(n)})$ can be assumed to be increasing, and hence Lemma 6.4.4 yields that the density of $B(\varphi, x_j)$ inside $\Sigma(\sigma)$ is less than $\sigma_{S^3}(\varphi)$.

Therefore, it is sufficient to consider the density of $B(\varphi, x_j)$ inside the tetrahedron determined by x_j, $\omega(F)$, and $e \cap B(r_3(\varphi), x_j)$, where e is an edge of F that intersects int $B(r_3(\varphi), x_j)$. The triangle $x_j\omega(F)\omega(e)$ divides this tetrahedron into two equal orthoschemes, and we deduce by Lemma 6.4.4 that the density of $B(\varphi, x_j)$ inside both orthoschemes is at most $\sigma_{S^3}(\varphi)$. In turn, we conclude the simplex bound for the density of $B(\varphi, x_j)$ inside D_j, and hence also for the density of the ball packing.

Finally, we assume that the density of the ball packing is $\sigma_{S^3}(\varphi)$, and thus the density of each ball inside its Dirichlet–Voronoi cell is $\sigma_{S^3}(\varphi)$, as well. We deduce by the previous proof that each vertex of D_j is of distance $r_3(\varphi)$ from x_j, and D_j can be dissected into orthoschemes congruent to $S(\varphi)$. Here each orthoscheme is of the form $x_j\omega(F)\omega(e)w$, where F is a face of D_j, e is a side of F, and w is an endpoint of e. Therefore, any vertex w of some Dirichlet–Voronoi cell is the "terminal vertex" of twenty-four orthoschemes that together form a regular tetrahedron of edge length 2φ. The vertices of each such regular tetrahedron are among the centres $x_1, \ldots, x_n$, and the tetrahedra form a triangulation of S^3. For any x_j, the opposite faces of the regular tetrahedron meeting at x_j form a triangulation of $\partial B(2\varphi, x_j)$ into equal regular spherical triangles of edge length 2φ, and $\partial B(2\varphi, x_j)$ bounds a Euclidean three-ball of radius $\sin 2\varphi$. Since $\varphi < \pi/4$, we deduce that the centres in $\partial B(2\varphi, x_j)$ are vertices of an icosahedron. This determines the value of φ and hence the whole triangulation of S^3. Therefore, $x_1, \ldots, x_n$ are the vertices of the 600-cells, completing the proof of Theorem 6.4.1. $\qquad \square$

Comments. Theorem 6.4.1 and Corollary 6.4.2 were conjectured by H. S. M. Coxeter [Cox1963] and proved by K. Böröczky [Bör1978], who actually proved the simplex bound in S^d for any d. Recently N. N. Andreev [And1999] gave another proof of the optimality of the 600-cell using the linear programming bound (6.1). We note that Dirichlet–Voronoi cells were first considered by G. L. Dirichlet (see say [Dir1850]) and by G. F. Voronoi (see [Vor1908]).

6.5. Covering S^d by at Most $d + 2$ Balls

Two hemispheres readily cover S^d where hemispheres can be considered spherical balls of radius $\pi/2$. In contrast, to cover S^d with spherical balls of acute radii, we need many more balls than two:

Theorem 6.5.1. *No $d + 1$ spherical balls of acute radii cover S^d. Moreover, if $d + 2$ spherical balls of radius φ cover S^d then $\varphi \geq \arccos 1/(d + 1)$ with equality if and only if the centres are vertices of a regular simplex.*

The proof of Theorem 6.5.1 depends on the following observation:

Lemma 6.5.2. *Spherical balls of radius $\varphi < \pi/2$ cover S^d if and only if the convex hull of the centres in $\mathbb{R}^{d+1}$ contains a Euclidean ball that is centred at the origin and is of radius $\cos \varphi$.*

Proof. First, we assume that spherical balls of radius $\varphi < \pi/2$ cover S^d, and we write P to denote the convex hull of the centres. The condition $\varphi < \pi/2$ yields that P contains the origin in its interior. If F is a facet of P then aff F cuts off a spherical ball $B(\psi, y)$ with $\psi < \pi/2$. We have $\psi \leq \varphi$ because some spherical ball of radius φ around a vertex of P contains y; hence, the Euclidean distance of aff F from the origin is at least $\cos \varphi$. However, let P be a polytope whose vertices lie on S^d and contains the Euclidean ball of radius $\cos \varphi$ centred at the origin. For any $y \in S^d$, P intersects the convex hull (in $\mathbb{R}^{d+1}$) of $B(\varphi, y)$; hence, $B(\varphi, y)$ contains a vertex v of P. Therefore, $y \in B(\varphi, v)$, completing the proof of Lemma 6.5.2. $\qquad\square$

Proof of Theorem 6.5.1. Let us consider $d + 1$ spherical balls of acute radii on S^d. Then the centres are contained in a Euclidean hyperplane H of $\mathbb{R}^{d+1}$, and the point of S^d farthest from H is not contained in any of the $d + 1$ spherical balls.

Turning to coverings by $d + 2$ unit balls, we let T^{d+1} be a regular simplex inscribed into B^{d+1}. We observe that the inscribed ball of T^{d+1} is centred at the origin and is of radius $1/(d + 1)$. According to Lemma 6.5.2, it is sufficient to prove that if S is a Euclidean simplex contained in B^{d+1} then

$$r(S) \le \frac{1}{d+1}, \tag{6.13}$$

and equality implies that S is congruent to T^{d+1}. We may assume that $r(S)$ is maximal among simplices contained in B^{d+1}. Since the inscribed ball B of S touches each facet, we deduce that each vertex of S lies on S^d. We suppose that S is not regular, and we seek a contradiction. It follows that S has three vertices u, v, w such that $d(u, v) \ne d(u, w)$, and let H be the hyperplane that bisects the edge vw perpendicularly. We write X' to denote the image of any object X by the Steiner symmetrization with respect to H. Then $B' \subset S'$ is a ball of radius $r(S)$, and hence S' is also an optimal simplex contained in B^{d+1}. However, the vertex u' of S' lies in the interior of B^{d+1}, which is a contradiction. In turn, we conclude (6.13); hence, Theorem 6.5.1 as well.

$\square$

Open Problem.

(i) Given any acute φ, prove that the spherical volume covered by $d + 2$ spherical balls of radius φ on S^d is maximal if the centres are vertices of a regular simplex in $\mathbb{R}^{d+1}$.

The case $d = 2$ is contained in Theorem 5.6.2. The result would yield the analogue of the moment inequality (Corollary 5.6.4). In turn, it would imply that the inscribed regular simplex has maximal mean width among all simplices contained in the unit ball of $\mathbb{R}^d$, which is an open problem for $d \ge 4$.

Comments. Theorem 6.5.1 has been folklore (see, e.g., Sh. I. Galiev [Gal1996]).

6.6. Covering S^d by $d + 3$ Balls

When S^d is covered by $d + 3$ equal spherical balls, the optimal arrangement tends to be as symmetric as possible. It is quite apparent in case d is even because then the centres are vertices of the most symmetric Euclidean $(d + 1)$-polytope of $d + 3$ vertices.

Theorem 6.6.1. *If $d + 3$ equal spherical balls with minimal possible radius cover S^d then the centres are the vertices of a $\lfloor (d + 1)/2 \rfloor$- and a $\lceil (d + 1)/2 \rceil$-dimensional regular simplex of circumradius one in $\mathbb{R}^{d+1}$.*

Proof. We plan to apply Lemma 6.5.2. More precisely, we seek the polytope inside the unit ball with the largest inscribed ball regardless of the place of the centre of the inscribed ball, and we show that the centre of the inscribed ball is actually the origin for the optimal polytope. Therefore, let $P \subset B^{d+1}$ be a Euclidean polytope with at most $d + 3$ vertices such that $r(P)$ is maximal, and let B be an inscribed ball of P.

If P has only $d + 2$ vertices then it is a $(d + 1)$-simplex. We choose a vertex v and a point $u \in S^d$ such that the segment vu intersects the interior of P. Then the convex hull P' of P and u has $d + 3$ vertices, and each facet of P' that touches B contains v. Therefore, B can be enlarged from v in a way that the resulting ball is still contained in P', which is a contradiction.

Therefore, P has actually $d + 3$ vertices. According to the Radon theorem (see (A.2)), P is the convex hull of a k-dimensional simplex S and an m-dimensional simplex T such that $k + m = d + 1$ holds for $k, m \geq 1$, and $S \cap T \neq \emptyset$. It follows that S and T intersect in one point. We write q to denote this point that is not a vertex of P.

Next, we suppose that a vertex v of P is contained in the interior of B^{d+1}; say, v is a vertex of T. We choose a v' outside of P such that the ray $v'v$ intersects the interior of P, and for any facet F of P that does not contain v, the point v' lies in the open half space determined by aff F that intersects P. Now we define P' to be the convex hull of P and v', and we observe that P' has $d + 3$ vertices as well. We choose a vertex w of T different from v. Since any facet of P contains either v or w, we deduce that any facet of P' contains either v' or w. However, no facet of P' containing v' touches B, and thus B can be enlarged from w in a way that the resulting ball is still contained in P'. This contradiction yields that each vertex of P lies on S^d.

Finally, we suppose that there exist an edge vw of say S and a vertex u of T such that $d(u, v) \neq d(u, w)$. Let H be the hyperplane that bisects the edge vw perpendicularly, and we write X' to denote the image of any object X by the Steiner symmetrization with respect to H. We observe that $P' \subset B^{d+1}$, and P' is a polytope with $d + 3$ vertices. Since $B' \subset P'$, we deduce that P' is an optimal polytope contained in B^{d+1} as well. However, the vertex u' of P' lies in the interior of B^{d+1}, which is absurd. Therefore, S and T are regular k, and m-simplices, respectively, whose affine hulls are orthogonal, and meet at the common circumcentre of S and T.

We deduce that P has a unique inscribed ball that is centred at the origin and touches each facet. Any facet C of P is the convex hull of a $(k - 1)$-face F of S and an $(m - 1)$-face G of T. We observe that the distances of F and G from the origin are $1/k$ and $1/m$, respectively, and $F \subset (\text{lin } G)^\perp$ and $G \subset (\text{lin } F)^\perp$. Denoting the cicumcentres of F and G by p and q, respectively, we have that the distance of the origin from C is the distance of the origin from the segment pq. Because the segments op and oq are orthogonal and of lengths $1/k$ and $1/m$, respectively, we deduce that $r(P) = 1/(k^2 + m^2)$. The condition $k + m = d + 1$ yields that the sum $k^2 + m^2$ is minimal if k and m are as close to $(d + 1)/2$ as possible. Therefore, the dimensions of S and T are $\lfloor (d + 1)/2 \rfloor$ and $\lceil (d + 1)/2 \rceil$ in this or in the reverse order. In turn, we conclude Theorem 6.6.1 by Lemma 6.5.2. $\qquad\square$

Comments. Theorem 6.6.1 is proved by K. Böröczky Jr. and G. Wintsche [BöW2003].

6.7. Covering S^3 by Eight Balls

We show that the eight vertices of the regular four-crosspolytope determine an optimal covering of S^3 by eight equal spherical balls.

Theorem 6.7.1. *If eight spherical balls of radius φ cover S^3 then $\varphi \geq \pi/3$, and equality holds if and only if the centres are vertices of a regular crosspolytope in $\mathbb{R}^4$.*

The proof of Theorem 6.7.1 uses tools from Euclidean geometry. First, we verify an extremal property of spherical regular simplices.

Lemma 6.7.2. *Given a spherical ball B on S^3, the regular simplex inscribed into B has maximal volume among spherical simplices contained in B.*

Proof. We write o to denote the origin of $\mathbb{R}^4$. For any subset σ of S^3, we write $\Omega(\sigma)$ to denote the union of the Euclidean segments of the form ox, $x \in \sigma$; hence, $|\sigma| = 4V(\Omega(\sigma))$.

Let $T = x_1 x_2 x_3 x_4$ be a spherical simplex of maximal volume contained in B, and thus $x_i \in \partial B$, $i = 1, 2, 3, 4$. We suppose that T is not regular and seek a contradiction. We may assume that the sides $x_1 x_3$ and $x_1 x_4$ of T are of different lengths. Let H be the hyperplane in $\mathbb{R}^4$ that is the perpendicular bisector of the Euclidean segment $x_3 x_4$. We write x_1' and x_2' to denote the projections of x_1 and x_2 into H, respectively, and σ' to denote the spherical simplex on S^3 cut out by the cone pos $\{x_1, x_2, x_3', x_4'\}$. In particular, $\sigma' \subset B$. Now applying the Steiner symmetrization with respect to H to the cone pos $\{x_1, x_2, x_3, x_4\}$, we obtain the cone pos $\{x_1, x_2, x_3', x_4'\}$, and hence the Steiner symmetrial of $\Omega(\sigma)$ is contained in $\Omega(\sigma')$. Since $x_1' \in \text{int } B^3$, we deduce that this containment is strict, which, in turn, yields $|\sigma| < |\sigma'|$. This is absurd, completing the proof of Lemma 6.7.2.

□

Proof of Theorem 6.7.1. Let eight spherical balls of radius φ cover S^3 with centres $p_1, \ldots, p_8$, where the vertices of the regular crosspolytope show that $\varphi \leq \pi/3$ can be assumed. We write P to denote the convex hull of the centres, and we triangulate the facets of P using the vertices of P in a way that the radial projection of the three-simplices to S^3 results in a triangulation Σ of S^d with vertices $p_1, \ldots, p_8$.

First, we assume that $f_3(\Sigma) > 16$. Since each two-cell of Σ is contained in exactly two three-cells, and each three-cell has four faces, we deduce that $f_2(\Sigma) = 2 f_3(\Sigma)$. However,

$$f_0(\Sigma) - f_1(\Sigma) + f_2(\Sigma) - f_3(\Sigma) = 0$$

holds according to the Euler–Poincaré formula (see the Comments). Therefore, $f_0(\Sigma) = 8$ and $f_3(\Sigma) > 16$ yield $f_1(\Sigma) > 24$. It follows that there exists a vertex of Σ, say p_1, that is connected to any other vertex by an edge of Σ. We write B to denote the Euclidean ball centred at the origin and of radius $1/2$, which is contained in P according to $\varphi \leq \pi/3$ and Lemma 6.5.2. Let H be the hyperplane that touches B at $-p_1/2$, and let H^+ be the half space that contain p_1. Since $p_i \in B^d$, and each segment $p_1 p_i$, $i = 2, \ldots, 8$, avoids the interior of B, we deduce that each p_i lies in H^+, and if $p_i \in H$ then $p_1 p_i$ touches B. In particular, the at least four vertices of P that are connected to p_1 by an edge are not contained in H; hence, H contains at most three vertices of P. It follows that $-p_1/2$ is contained in at least two facets of P, which is absurd.

Therefore, $f_3(\Sigma) \leq 16$. If P is the regular crosspolytope then Σ has sixteen three-cells that are regular simplices, and the circumscribed spherical ball of each three-cell is of radius $\pi/3$. In general each three-cell of Σ is contained in some spherical ball of radius φ, and, hence the condition $\varphi \leq \pi/3$ and Lemma 6.7.2 yield that P is actually the regular crosspolytope. In turn, we conclude Theorem 6.7.1.

□

The proof of Theorem 6.7.1 was based on some clever tricks. A possible general approach to solve various related problems would be the simplex bound for spherical coverings: Given an acute angle φ, we consider some regular spherical simplex $\widetilde{T}^d(\varphi)$ of circumradius φ on S^d and the $d + 1$ spherical balls of radius φ centered at the vertices. We define $\tau_{S^d}(\varphi)$ to be the ratio of the sum of the volumes of the sectors of the balls inside $\widetilde{T}^d(\varphi)$ over the volume of $\widetilde{T}^d(\varphi)$.

Conjecture 6.7.3 (Simplex bound for spherical coverings). *Given an acute angle φ and $d \geq 3$, the density of any covering of S^d by spherical balls of radius φ is at most $\tau_{S^d}(\varphi)$.*

The case $d = 2$ is known to hold (see Theorem 5.4.2).

Open Problems.

(i) Prove that the optimal covering of S^3 by 120 equal spherical balls is determined by the 600 cells.

(ii) Prove for some $d \geq 4$ that the optimal covering of S^d by $2d$ equal spherical balls is determined by the regular $(d+1)$-dimensional crosspolytope.

Problems (i) and (ii) would directly follow from Conjecture 6.7.3.

(iii) For $d \geq 3$ and $d + 2 \leq m \leq 2d + 2$, let m equal spherical balls of minimal radius cover S^d. Prove that the centres are the vertices of a suitable number mutually orthogonal $\lfloor (d+1)/(m-d-1) \rfloor$- and $\lceil (d+1)/(m-d-1) \rceil$-dimensional regular simplices of circumradius one in $\mathbb{R}^{d+1}$.

Comments. Lemma 6.7.2 is due to K. Böröczky [Bör1987] (in S^d for any d), and Theorem 6.7.1 was proved by L. Dalla, D. G. Larman, P. Mani–Levitska, and Ch. Zong [DLM2000]. For the Euler–Poincaré formula, consult e.g. G. Ziegler [Zie1995].

6.8. Economic Coverings of S^d by Equal Balls

The aim of this section to show that there exists an economic covering of the sphere by equal spherical balls of any given acute radius.

Theorem 6.8.1. *Given $0 < \varphi < \pi/2$, certain at most*

$$c \cos \varphi \cdot \sin^{-d} \varphi \cdot d^{\frac{3}{2}} \ln(2 + d \cos^2 \varphi)$$

spherical balls of radius φ cover S^d, where c is an absolute constant.

Remark. The density of the covering is at most $cd \ln(2 + d \cos^2 \varphi)$.

Let us discuss the optimality of the estimates of Theorem 6.8.1 for large d and some consequences of Theorem 6.8.1.

Example 6.8.2. *Given $\gamma \geq 1$ and an acute angle $\varphi \geq \arccos(\gamma/\sqrt{d+1})$, some $c(\gamma)d$ spherical balls of radius φ cover S^d, where $c(\gamma)$ depends only on γ. Readily, at least $d + 2$ spherical balls of radius φ are needed.*

Example 6.8.3. *If $n \leq d^{\ln d}$ (say, n is polynomial in d) then suitable n spherical balls of equal radius cover S^d in a way that the density is at most $cd \ln \ln d$ for some absolute constant c.*

Lemma 6.8.4. *If spherical balls of radius $\varphi \leq \arcsin(1/\sqrt{d+1})$ cover S^d then the density is at least cd for some positive absolute constant c.*

Proof. We estimate the density of the covering of S^d by projecting it locally to Euclidean hyperplanes in $\mathbb{R}^{d+1}$. Let P be the convex hull of the centres of the spherical balls in $\mathbb{R}^{d+1}$, and let Σ' be a triangulation of ∂P using the vertices of P. Projecting radially the simplices of Σ to S^d, we obtain a triangulation, called a Delone triangulation, of S^d using the centres of the balls as vertices. Let T be one of the spherical d-simplices of the Delone triangulation, and hence its spherical circumradius is at most φ. We consider the $d + 1$ spherical balls centred at the vertices of T, and we define the *associated density* to be the ratio of the sum of the volumes of the sectors of the balls, lying in the simplex, to the volume of the simplex. Now the Euclidean d-simplex T' whose vertices coincide with the vertices of T is of circumradius at most $\sin \varphi$. According to the Coxeter–Few–Rogers bound (see (8.3)), the associated density of the $d + 1$ Euclidean balls of radius $\sin \varphi$ in aff T' centred at the vertices of T' is at least $c'd$ for some positive absolute constant c'. Since $\cos \varphi \geq \sqrt{1 - 1/(d+1)}$, we conclude that the associated density for the spherical simplex T is at least cd. The same estimate for each Delone d-simplex yields the density bound cd for the covering of S^d. $\quad\square$

The proof of Theorem 6.8.1 consists of two steps: First, given any acute φ, we construct a covering of S^d by spherical balls of radius φ whose density is of order $d \ln d$ (see Lemma 6.8.5). After that we improve the efficiency when φ is close to $\pi/2$.

Lemma 6.8.5. *If $\varphi < \pi/2$ then there exists a covering of S^d by spherical balls of radius φ whose density is at most*

$$\left(1 + \frac{c \cdot \ln \ln d}{\ln d}\right) \cdot d \ln d,$$

where c is an absolute constant.

Since the proof of Lemma 6.8.5 is probabilistic, we provide some basic estimates on the probability measure $\Omega(\varphi) = |B(\varphi, x)|/|S^d|$ of $B(\varphi, x)$.

Lemma 6.8.6. *Let* $0 < \varphi < \pi/2$.

(i) $\Omega(\varphi) > \frac{1}{\sqrt{2\pi(d+1)}} \cdot \sin^d \varphi$;

(ii) $\Omega(t\varphi) < t^d \cdot \Omega(\varphi)$ *if* $1 < t < \frac{\pi}{2\varphi}$;

(iii) *if* $\varphi \leq \arccos \frac{1}{\sqrt{d+1}}$ *then*

$$\frac{1}{3\sqrt{2\pi(d+1)}} \cdot \frac{1}{\cos\varphi} \cdot \sin^d \varphi < \Omega(\varphi) < \frac{1}{\sqrt{2\pi d}} \cdot \frac{1}{\cos\varphi} \cdot \sin^d \varphi;$$

(iv) $\frac{1}{3e\sqrt{2\pi}} < \Omega(\varphi) < \frac{1}{2}$ *if* $\varphi \geq \arccos \frac{1}{\sqrt{d+1}}$.

Proof. Knowing that $|S^d| = (d+1)\kappa_{d+1}$ and $\sqrt{(d+2)/2\pi} > \kappa_d/\kappa_{d+1} > \sqrt{(d+1)/2\pi}$ (see (A.29)), it is sufficient to verify corresponding bounds for $|B(\varphi, x)|$. The boundary of any $B(\psi, x)$ in S^d is the boundary of a d-ball of radius $\sin \psi$ in $\mathbb{R}^{d+1}$; hence, using polar coordinates on S^d results in the formula

$$|B(\varphi, x)| = d\kappa_d \int_0^\varphi \sin^{d-1} \psi \, d\psi. \tag{6.14}$$

To prove (i), we observe that

$$\frac{\partial}{\partial \varphi} |B(\varphi, x)| = d\kappa_d \sin^{d-1} \varphi$$
$$> \kappa_d \cdot (\sin^{d-1} \varphi + (d-1)\varphi \cos\varphi \sin^{d-2} \varphi).$$

Since the last expression is $\kappa_d \cdot (\partial/\partial\varphi)(\varphi \cdot \sin^{d-1}\varphi)$, it follows that

$$|B(\varphi, x)| > \kappa_d \varphi \sin^{d-1}\varphi > \kappa_d \sin^d \varphi,$$

which, in turn, yields (i). In addition, (ii) follows by

$$\frac{\partial}{\partial t} |B(t\varphi, x)| = d\kappa_d \, \varphi \sin^{d-1}(t\varphi) < d\kappa_d \, \varphi \, t^{d-1} \sin^{d-1}\varphi$$
$$< \frac{\partial}{\partial t}(t^d \cdot |B(\varphi, x)|).$$

Turning to (iii), we rewrite (6.14) in the form

$$|B(\varphi, x)| = d\kappa_d \int_0^{\sin\varphi} \frac{1}{\sqrt{1-r^2}} \cdot r^{d-1} dr; \tag{6.15}$$

thus the upper bound is a consequence of $1/\sqrt{1-r^2} \leq 1/\cos\varphi$ for $r \leq \sin\varphi$. However, if $(1-1/d)\sin\varphi \leq r \leq \sin\varphi$ then $\tan\varphi \leq \sqrt{d}$ yields

$$\frac{\sqrt{1-r^2}}{\cos\varphi} \leq \frac{\sqrt{1-(1-\frac{1}{d})^2\sin^2\varphi}}{\cos\varphi} < \sqrt{1+\frac{2}{d}\tan^2\varphi} \leq \sqrt{3}.$$

Since $(1-1/d)^d < 1/e$, we conclude that

$$|B(\varphi,x)| > d\kappa_d \int_{(1-\frac{1}{d})\sin\varphi}^{\sin\varphi} \frac{1}{\sqrt{1-r^2}} \cdot r^{d-1} dr$$

$$> \kappa_d \left(1-\frac{1}{e}\right) \cdot \frac{1}{\sqrt{3}\cos\varphi} \cdot \sin^d\varphi.$$

Therefore $(e-1)/(e\sqrt{3}) > 1/3$ yields (iii). Finally, (iv) follows by (iii), completing the proof of Lemma 6.8.6. $\square$

Proof of Lemma 6.8.5. Let $\eta = 1/(d\ln d)$, and let

$$N = \left\lfloor \frac{1}{\Omega((1-\eta)\varphi)} \cdot \ln\frac{\Omega((1-\eta)\varphi)}{\Omega(\eta\varphi)} \right\rfloor. \tag{6.16}$$

We distribute N centres $x_1, \ldots, x_N$ on S^d independently and with uniform distribution according to the probability measure on S^d. The expected probability measure of the part of S^d that is not covered by any of the balls $B((1-\eta)\varphi, x_i)$, $i = 1, \ldots, N$, is $(1 - \Omega((1-\eta)\varphi))^N$. Since $1 - t < e^{-t}$, we may choose the centres in a way that the probability measure of the part of S^d not covered by balls $B((1-\eta)\varphi, x_i)$, $i = 1, \ldots, N$, is at most $e^{-N\Omega((1-\eta)\varphi)}$.

Let $B(\eta\varphi, y_1), \ldots, B(\eta\varphi, y_M)$ be a maximal family of nonoverlapping spherical balls of radius $\eta\varphi$ that are contained in the uncovered part. In particular, M satisfies

$$M \leq \frac{(1 - \Omega((1-\eta)\varphi))^N}{\Omega(\eta\varphi)} < \frac{e^{-N\cdot\Omega((1-\eta)\varphi)}}{\Omega(\eta\varphi)}.$$

For any $z \in S^d$, the spherical ball $B(\eta\varphi, z)$ intersects either some $B(\eta\varphi, y_j)$ or some $B((1-\eta)\varphi, x_i)$; hence, we obtain a covering of S^d by $n = N + M$ spherical balls of radius φ, where

$$n \leq N + \frac{e^{-N\cdot\Omega((1-\eta)\varphi)}}{\Omega(\eta\varphi)}.$$

Differentiation suggests we choose N as in (6.16). Since readily $\Omega((1-\eta)\varphi) < 1/2$, we deduce that the density of the covering by the n spherical balls of radius φ is

$$n \cdot \Omega(\varphi) < \frac{\Omega(\varphi)}{\Omega((1-\eta)\varphi)} \cdot \left(\ln \frac{\Omega((1-\eta)\varphi)}{\Omega(\eta\varphi)} + \sqrt{e} \right).$$

Therefore, Lemma 6.8.6 and $\eta = 1/(d\ln d)$ yield Lemma 6.8.5. □

Proof of Theorem 6.8.1. In the following, c_0, c_1, c_2 always denote suitable positive absolute constants. If $\varphi \geq \arccos\sqrt{1/(d+1)}$ then the $2(d+1)$ spherical balls of radius φ centred at the vertices of a regular crosspolytope inscribed into B^{d+1} cover S^d. Therefore, having Lemma 6.8.5 at hand, it is sufficient to verify the following claim: If $d > 15$ and $1/\sqrt{d+1} \leq \cos\varphi \leq 1/\left[4^{1/4}(d+1)^{3/8}\right]$ then S^d can be covered by

$$c_0 \, \cos\varphi \cdot \frac{1}{\sin^d \varphi} \cdot d^{\frac{3}{2}} \ln 2(d+1)\cos^2\varphi \tag{6.17}$$

spherical balls of radius φ, and the density of the arrangement is at most

$$c_0 \, d \cdot \ln 2(d+1)\cos^2\varphi. \tag{6.18}$$

We define $m = \lceil 3(d+1)^2 \cos^4\varphi \rceil$ and $k = \lceil (d+1)/m \rceil$. The idea is to take pairwise orthogonal linear subspaces $L_1, \ldots, L_k$ that span $\mathbb{R}^{d+1}$, and $\dim L_i = m$ for $i = 1, \ldots, \lceil (d+1)/m \rceil$, and to cover each $L_i \cap S^d$ using Lemma 6.8.5. In the following, c_1 and c_2 denote positive absolute constants.

Let ψ be the acute angle satisfying $\cos\psi = \sqrt{k} \cdot \cos\varphi$. According to Lemma 6.8.5, $L_i \cap S^d$ can be covered by N_i spherical $(m-1)$-balls of radius ψ, where

$$N_i \leq c_1 \, \cos\psi \cdot \frac{1}{\sin^{\frac{m-1}{2}}\psi} \cdot m^{\frac{3}{2}} \ln m.$$

Since, for any $u \in S^d$, the length of the orthogonal projection of u into some L_i is at least $1/\sqrt{k}$, we obtain a covering of S^d by $n = \sum_1^k N_i$ spherical balls of radius φ, where

$$n \leq c_1 \cdot \left\lceil \frac{d+1}{m} \right\rceil^{\frac{3}{2}} \cdot \cos\varphi \cdot \frac{1}{\left(1 - \lceil \frac{d+1}{m} \rceil \cos^2\varphi\right)^{\frac{m-1}{2}}} \cdot m^{\frac{3}{2}} \ln m.$$

Let us observe that

$$m \leq 4(d+1)^2 \cos^4\varphi \leq \sqrt{d+1} \quad \text{and} \quad \text{hence} \quad \left\lceil \frac{d+1}{m} \right\rceil \leq \frac{5}{4} \cdot \frac{d+1}{m}, \tag{6.19}$$

which, in turn, yields $\lceil (d+1)/m \rceil \cos^2 \varphi < 1/2$. However, $1 - t > e^{-t-t^2}$ holds for $0 < t < 1/2$, and therefore

$$n < c_2 \, d^{\frac{3}{2}} \, \cos \varphi \cdot \frac{e^{\frac{m-1}{2} \lceil \frac{d+1}{m} \rceil^2 \cos^4 \varphi}}{e^{-\frac{m-1}{2} \lceil \frac{d+1}{m} \rceil \cos^2 \varphi}} \cdot \ln 4(d+1)^2 \cos^4 \varphi.$$

Here $\lceil (d+1)/m \rceil^2 (m-1)/2 \cos^4 \varphi \leq 25/96$ by (6.19). In addition, $m^2 \leq d+1$ and $\lceil (d+1)/m \rceil < (d+1)/m + 1$ yield $\lceil (d+1)/m \rceil (m-1)/2 < d$; thus,

$$e^{-\frac{m-1}{2} \lceil \frac{d+1}{m} \rceil \cos^2 \varphi} > (1 - \cos^2 \varphi)^d = \sin^d \varphi.$$

These inequalities complete the proof of the claims (6.17) and (6.18) and, in turn, the proof of Theorem 6.8.1.   □

Open Problem.

(i) Is the density of any covering of S^d by equal spherical balls of acute radius at least cd for some positive absolute constant c?

An affirmative answer would follow from Conjecture 6.7.3.

Comments. We note that Delone cells were first considered by B. N. Delone (see [Del1934]). Theorem 6.8.1 is due to K. Böröczky Jr. and G. Wintsche [BöW2003]. In addition, these authors construct a covering of S^d by spherical balls of any given acute radius φ such that no point of S^d is contained in more than $400 \, d \ln d$ balls.

We call an arrangement of spherical balls on S^d a k-fold covering if each point of S^d is contained in at least k of the balls. Various estimates for k-fold coverings are presented in Sh. I. Galiev [Gal1996].

6.9. The Simplex Bound for Hyperbolic Ball Packings

The only important estimate for packings of finitely many equal hyperbolic balls of dimension at least three is the simplex bound due to K. Böröczky [Bör1978] for packings on compact hyperbolic manifolds. We note that compact hyperbolic manifolds can be defined analogously as surfaces in Section A.5.4 (see J. G. Ratcliffe [Rat1994] for properties of hyperbolic spaces). Let us consider the $d+1$ balls of radius r centred at the vertices of a regular simplex of edge length $2r$ in H^d, and write $\sigma_{H^d}(r)$ to denote the ratio of the volume of the part of the simplex occupied by the $d+1$ balls to the volume of the simplex. The paper by K. Böröczky [Bör1978] actually verifies for any packing of balls of radius r in H^d that the density of each ball with respect to its Dirichlet–Voronoi cell is at most $\sigma_{H^d}(r)$. We deduce (see Corollary 4.1.3 in the planar case) the following:

Theorem 6.9.1 (Coxeter–Böröczky bound). *The density of any packing of embedded balls of radius r on a compact hyperbolic d-manifold X is at most $\sigma_{H^d}(r)$. Equality holds for $d \geq 3$ if and only if $d = 4$, ch $r = (\sqrt{5} + 1)/2$, and the centres are vertices of a facet-to-facet tiling of X by regular simplices of edge length 2r.*

The existence of the equality case follows from the existence of the tiling of H^4 by the corresponding regular simplices and from the Selberg Lemma (see Example A.5.10 for the analogous argument in the planar case). Each vertex in the tiling of H^4 is contained in 120 edges, and the other endpoints of these edges determine a regular 600-cells.

Theorem 6.9.1 was conjectured by H. S. M. Coxeter [Cox1963]. We do not provide the rather involved proof, which can be also found in Ch. Zong [Zon1999]. However, if $d = 3$ then the argument is simpler, and it is essentially the same as the argument for Theorem 6.4.1 (see also K. Böröczky and A. Florian [BöF1964]).

Similarly to the planar case (see Section 4.1), $\sigma_{H^d}(r)$ is increasing in r if $d = 3$ (see K. Böröczky and A. Florian [BöF1964]) and if d is large (see T. H. Marshall [Mar1999]). Moreover, T. H. Marshall [Mar1999] described the asymptotic behavior of $\sigma_{H^d}(r)$ when r is fixed and d tends to infinity. It sounds probably surprising but the simplex bound has profound applications in differential geometry; namely, it can be used to provide a lower bound on the volume of a hyperbolic d-manifold X. The idea is that it is easier to find an r close to being optimal such that a ball $B(r)$ of radius r is embedded into X; hence,

$$V(X) \geq \frac{1}{\sigma_{H^d}(r)} V(B(r)).$$

7

Congruent Convex Bodies

Given n and a convex body K, we consider arrangements of n congruent copies, namely, packings inside convex containers and coverings of compact convex sets. Concerning density, clusters are naturally related to periodic arrangements (see Theorem 7.1.1). In addition, we verify that an asymptotic density exists as n tends to infinity (see Theorem 7.2.2), and we characterize the case when this asymptotic density is one (see Lemma 7.2.3). Example 7.2.4 shows that the asymptotic structure of the optimal arrangement depends very much on K.

Concerning the mean i-dimensional projection for $i = 1, \ldots, d - 1$, we verify that the optimal convex hull of n nonoverlapping congruent copies of K is close to being a ball (see Theorem 7.2.2). In contrast, the compact convex set of maximal mean width covered by n congruent copies of K is close to being a segment (see Theorem 7.4.1).

7.1. Periodic and Finite Arrangements

Let K be a convex body in $\mathbb{R}^d$. If $\mathcal{K}$ is a periodic arrangement by congruent copies of K with respect to the lattice Λ (see Section A.13), and the number of equivalence classes is m, then the density of the arrangement is $m \cdot V(K)/\det \Lambda$. We define the *packing density* $\delta(K)$ to be supremum of the densities of periodic packings by congruent copies of K, and we let $\Delta(K) = V(K)/\delta(K)$. Moreover, the *covering density* $\vartheta(K)$ is the infimum of the densities of periodic coverings by congruent copies of K, and we let $\Theta(K) = V(K)/\vartheta(K)$. In particular, $\Delta(K)$ is the infimum of $V(T)/m$ over all tori T and integers m such that there exists a packing of m isometric copies of K in T, and $\Theta(K)$ is the supremum of $V(T)/m$ over all tori T and integers m such that some m isometric copies of K cover T. We have $\delta(K) > 4^{-d}$ according to the Rogers–Shephard bound [Theorem 9.5.2 (ii)] and $\vartheta(K) < 4d \ln d$ according to the Rogers bound [Theorem 9.5.2 (i)]. Let us show that periodic arrangements are closely related to clusters.

Theorem 7.1.1. *Given convex bodies K and D in $\mathbb{R}^d$ with $r(D) > R(K)$, let N be the maximal number of nonoverlapping congruent copies of K inside D, and let M be minimal number of congruent copies of K that cover D. Then*

(i) $\left(1 - \frac{R(K)}{r(D)}\right)^d \cdot \frac{V(D)}{\Delta(K)} \leq N \leq \left(1 + \frac{R(K)}{r(D)}\right)^d \cdot \frac{V(D)}{\Delta(K)}$;

(ii) $\left(1 - \frac{R(K)}{r(D)}\right)^d \cdot \frac{V(D)}{\Theta(K)} \leq M \leq \left(1 + \frac{R(K)}{r(D)}\right)^d \cdot \frac{V(D)}{\Theta(K)}$.

Proof. The argument is only presented for packings because the case of covering is completely analogous. We place K and D in a way that $K \subset R(K) B^d \subset r(D)D \subset D$, and we define

$$D^- = \left(1 - \frac{R(K)}{r(D)}\right) \cdot D \text{ and } D^+ = \left(1 + \frac{R(K)}{r(D)}\right) \cdot D.$$

Assuming that K' is a congruent copy of K with circumcentre c, if c lies outside of D^+ then K' avoids D, and if $c \in D^-$ then $K' \subset D$. Given a torus T, we write the same symbol for a convex body in $\mathbb{R}^d$ and its embedded image on T.

We start with the upper bound for $N \Delta(K)$. Let $T = \mathbb{R}^d/\Lambda$ be some torus such that D^+ can be isometrically embedded into T, and let $K_1, \ldots, K_m$ be a family of nonoverlapping isometric copies of K on T of maximal cardinality. If we write x_i to denote the circumcentre of K_i, (A.50) yields

$$\int_T \#\left((D^+ + x) \cap \{x_1, \ldots, x_m\}\right) dx = m \cdot V(D^+). \tag{7.1}$$

Thus some translate $D^+ + x$ contains at most $m \cdot V(D^+)/V(T)$ points out of $x_1, \ldots, x_m$, say, the points $x_{i_1}, \ldots, x_{i_k}$. Replacing $K_{i_1}, \ldots, K_{i_k}$ by the N nonoverlapping isometric copies of K contained in $D + x$, we obtain a packing of $m - k + N$ isometric copies of K in T. Therefore, $N \leq k$ follows from the maximality of m and, in turn, $k \leq m \cdot V(D^+)/V(T)$ yields

$$\left(1 + \frac{R(K)}{r(D)}\right)^d \cdot V(D) = V(D^+) \geq N \cdot \frac{1}{m} V(T) \geq N \cdot \Delta(K).$$

To verify the lower bound for $N \cdot \Delta(K)$, let $0 < \lambda < 1$ satisfy that $\lambda \cdot V(D^-)/\Delta(K) > \lceil V(D^-)/\Delta(K) \rceil - 1$. We deduce by the definition of $\Delta(K)$ that there exist a torus $T = \mathbb{R}^d/\Lambda$ and m nonoverlapping isometric copies $K_1, \ldots, K_m$ of K on T such that $V(T) < \lambda^{-1} \cdot m\Delta(K)$ and D can be isometrically embedded into T. We write x_i to denote the circumcentre of K_i, and we observe that the analogue of (7.1) provides a translate $D^- + x$ that contains at least $m \cdot V(D^-)/V(T)$ points out of $x_1, \ldots, x_m$. We may assume that these points are $x_1, \ldots, x_k$; hence, $k \geq \lambda \cdot V(D^-)/\Delta(K)$ and $K_1, \ldots, K_k$

are contained in $D + x$. Therefore, $N \geq k \geq V(D^-)/\Delta(K)$ by the definition of λ, completing the proof of Theorem 7.1.1. $\square$

Comments. Given a convex body K in $\mathbb{R}^d$, $\delta(K)$ $(\vartheta(K))$ is the maximal (minimal) density of any infinite packing in $\mathbb{R}^d$ (covering of $\mathbb{R}^d$) by congruent copies of K, namely, even if nonperiodic arrangements are allowed (see G. Fejes Tóth and W. Kuperberg [FTK1993]).

H. Groemer [Gro1963] proved a statement (the analogue of Corollary 9.4.3) resembling Theorem 7.1.1. G. Fejes Tóth and W. Kuperberg [FTK1995] prove that any strictly convex body K in $\mathbb{R}^d$ has an affine image K' such that no lattice convering by K' is a thinnest periodic covering by congruent copies of K'.

7.2. Density for Finite Packings and Coverings

Given a convex body K in $\mathbb{R}^d$, this section discusses the asymptotic density of optimal finite packings and coverings by congruent copies of K. Theorem 7.2.2 proves the existence of such densities, and Lemma 7.2.3 characterizes the case when the asymptotic densities are one. First, we construct economic finite packings and coverings.

Lemma 7.2.1. *For any n and convex body K in $\mathbb{R}^d$, there exist convex bodies D_n and $\tilde{D}_n$ with the following properties:*

  (i) $V(D_n) < n \cdot d \, V(K)$, and D_n contains n nonoverlapping translates of K;
  (ii) $V(\tilde{D}_n) > n \cdot \frac{1}{ed} \, V(K)$, and $\tilde{D}_n$ is covered by n translates of K.

Proof. We may assume that $o \in \partial K$, diam $K = 1$, and the segment ou is a diameter of K.

We define D_n to be the convex hull of $iu + K$ for $i = 1, \ldots, n$. Since $V(D_n) < n \cdot |K/u|$, and $V(K) \geq (1/d)|K/u|$, we deduce (i).

Turning to coverings, we observe that for any $y \in (1 - 1/d)(K/u)$, the line $\lin u + y$ intersects K in a segment of length at least $1/d$. Now the covering is given by the translates $(i/d)u + K$, $i = 1, \ldots, n$, and $\tilde{D}_n$ is the intersection of the union of these translates and the infinite cylinder $\lin u + (1 - 1/d)(K/u)$. Therefore, $V(K) < |K/u|$ and $V(\tilde{D}_n) > n/d \, (1 - 1/d)^{d-1}|K/u|$ yield (ii). $\square$

The main result of the section is

Theorem 7.2.2. *Given a convex body K in $\mathbb{R}^d$, there exist $\Delta_{\text{fin}}(K)$ and $\Theta_{\text{fin}}(K)$ satisfying the following properties:*

(i) *If D_n is a convex body of minimal volume containing n nonoverlapping congruent copies of K then*

$$\lim_{n\to\infty} \frac{V(D_n)}{n} = \Delta_{\text{fin}}(K);$$

(ii) *if $\widetilde{D}_n$ is a convex body of maximal volume that can be covered by n congruent copies of K then*

$$\lim_{n\to\infty} \frac{V(\widetilde{D}_n)}{n} = \Theta_{\text{fin}}(K).$$

In particular, we define the asymptotic density of finite packings and coverings using the notation of Theorem 7.2.2 as follows:

$$\delta_{\text{fin}}(K) = \frac{V(K)}{\Delta_{\text{fin}}(K)} = \lim_{n\to\infty} \frac{n\,V(K)}{V(D_n)};$$

$$\vartheta_{\text{fin}}(K) = \frac{V(K)}{\Theta_{\text{fin}}(K)} = \lim_{n\to\infty} \frac{n\,V(K)}{V(\widetilde{D}_n)}.$$

We deduce by Lemma 7.2.1 that

$$\frac{1}{d} \le \delta_{\text{fin}}(K) \le 1 \quad \text{and} \quad 1 \le \vartheta_{\text{fin}}(K) \le ed.$$

Since $\delta(B^d) < (d/2 + 1)\,2^{-0.5d}$ according to the Blichfeldt bound (Example 8.7), we observe that finite packings can be substantially denser than periodic packings.

Proof of Theorem 7.2.2. We present the argument in detail only for packings, and we sketch the necessary changes for coverings at the end. In the following $|\cdot|$ stands for the relative content of compact convex sets, and the implied constant in $O(\cdot)$ depends only on K. We define

$$\Delta_{\text{fin}}(K) = \liminf_{n\to\infty} \frac{V(D_n)}{n}.$$

Let D_{n_k} be a subsequence such that $V(D_{n_k})/n_k$ tends to $\Delta_{\text{fin}}(K)$. If $r(D_{n_k})$ is not bounded then $\Delta_{\text{fin}}(K) = \Delta(K)$ according to Theorem 7.1.1. However, Theorem 7.1.1 yields $\limsup_{n\to\infty} V(D_n)/n \le \Delta(K)$, and, in turn, we conclude (i) in this case. Therefore, we may assume that $r(D_{n_k})$ is bounded.

We recall that $r_m(D_{n_k})$ is the radius of a largest m-dimensional ball in D_{n_k}. Let $1 \le m \le d - 1$ be the smallest integer such that $r_{m+1}(D_{n_k})$ is bounded, and hence we may assume that $r_m(D_{n_k})$ tends to infinity with n_k. According to (A.20), there exist a constant R independent of n_k and an affine m-plane L_{n_k} for each D_{n_k} such that

$$D_{n_k} \subset L_{n_k} + R \cdot B^d. \tag{7.2}$$

For any small $\varepsilon > 0$, we verify that there exist an m-dimensional cube W contained in some $L = L_{n_k}$, a $(d-m)$-dimensional compact convex set $C \subset L^{\perp}$, and an integer N with the following properties: $W + C$ contains N nonoverlapping congruent copies of K and satisfies

$$\frac{V(W+C)}{N} \leq (1 + O(\varepsilon)) \cdot \Delta_{\text{fin}}(K). \tag{7.3}$$

Let the edge length of W be $2(a + R(K))$, where a is any number satisfying $(a + R(K))^m / a^m < 1 + \varepsilon$. To fix the right n_k, we may readily assume that L_{n_k} is a linear m-space, and

$$\frac{V(D_{n_k})}{n_k} < (1 + \varepsilon) \, \Delta_{\text{fin}}(K)$$

holds for any n_k. We write M_{n_k} to denote the orthogonal projection of D_{n_k} into L_{n_k}; hence, $r_m(M_{n_k})$ tends to infinity by (7.2). Thus, we fix $\tilde{n} = n_k$ in a way that

$$\frac{2a\sqrt{m}}{\varepsilon \, r_m(M_{\tilde{n}})} < \frac{\varepsilon}{2R}, \tag{7.4}$$

where $2a\sqrt{m}$ is an upper bound for the diameter of W. We simply write D, L, and M instead of $D_{\tilde{n}}$, $L_{\tilde{n}}$, and $M_{\tilde{n}}$, and we identify L with $\mathbb{R}^m$. In addition, we may assume (see (A.19)) that

$$-M \subset m \cdot M. \tag{7.5}$$

Let $K_1, \ldots, K_{\tilde{n}}$ be the $\tilde{n}$ nonoverlapping congruent copies of K in D, and let x_i be the circumcentre of K_i. For $W_0 = [-a, a)^m$, we consider the tiling $W_0 + 2a\,\mathbb{Z}^m$ of L, and we write $\alpha(y)$ to denote the number of x_is in $y + W_0 + L^{\perp}$.

Our plan is to cut off the "rim" of D. Thus, we consider

$$D(\varepsilon) = ((1 - 2\varepsilon)\, M + L^{\perp}) \cap D.$$

Let A be a section of D of maximal content by an affine $(n-m)$-plane parallel to $L^{\perp}$. Since (7.5) yields $|(x + L^{\perp}) \cap D| \geq |A|/(2m+2)^{d-m}$ for any $x \in M/2$, we deduce

$$V(D) - V(D(\varepsilon)) = O(\varepsilon) \cdot V(D) \tag{7.6}$$

by the Fubini theorem. Next we write Ω to denote the family of $y \in 2a\mathbb{Z}^m$ such that $y + 2W_0 \subset (1 - \varepsilon)\, M$. It follows by the condition (7.4) on $\tilde{n}$ that

$\cup_{y \in \Omega}(y + W_0)$ covers $(1 - 2\varepsilon)M$, and hence the number of x_i in $\Omega + W_0 + L^\perp$ is at least $(1 - O(\varepsilon))\,\tilde{n}$. Therefore, there exists a $y_0 \in \Omega$ satisfying

$$\frac{V\big((y_0 + W_0 + L^\perp) \cap D\big)}{\alpha(y_0)} < (1 + O(\varepsilon)) \cdot \frac{V(D)}{\tilde{n}}$$

$$< (1 + O(\varepsilon))\Delta_{\text{fin}}(K).$$

We define $\alpha(y_0) = N$ and $W = y_0 + [-(a + R(K)), a + R(K)]^m$, and define C to be the projection of $(W + L^\perp) \cap D$ into $L^\perp$. Then $W + C$ contains the corresponding N copies of K out of $K_1, \ldots, K_{\tilde{n}}$. Therefore, (7.3) will follow if, for any $x \in W$, we verify

$$|C| \leq (1 + O(\varepsilon)) \cdot |(x + L^\perp) \cap D|.$$

Since $(x + L^\perp) \cap D$ is contained in $x + C$, and $r_{d-m}(C) \geq r(K)$, it is sufficient to show that for any $u \in C$ there exists a $v \in (x + L^\perp) \cap D$ whose distance from $x + u$ is at most ε. Let $z \in W$ satisfy that $z + u \in D$. If $z = x$ then we are done; thus, let w' be the point where the half line zx intersects ∂M. This w' is the projection of some $w \in D$, and we define v to be the point of the segment connecting w and $z + u$ that lies in $x + L^\perp$. Now the estimates $d(x, z) \leq 2a\sqrt{m}$ and $d(w', z) > \varepsilon\, r_m(M)$ and the condition (7.4) on $\tilde{n}$ yield that $d(v, x + u) < \varepsilon$; hence, we conclude the claim (7.3).

Finally, for large n, some positive integer γ satisfies $\gamma^m/(\gamma - 1)^m < 1 + \varepsilon$ and $(\gamma - 1)^m N < n \leq \gamma^m N$. Since $C + \gamma\, W$ contains n congruent copies of K, (7.3) yields $V(D_n)/n \leq (1 + O(\varepsilon))\Delta_{\text{fin}}(K)$, completing the proof of Theorem 7.2.2 for packings.

For coverings we define

$$\Theta_{\text{fin}}(K) = \limsup_{n \to \infty} \frac{V(\widetilde{D}_n)}{n}.$$

Now the constant in $O(\cdot)$ depends on both R and K. Instead of (7.3), we require that $W + C$ is covered by N congruent copies of K, and

$$\frac{V(W + C)}{N} \geq (1 - O(\varepsilon)) \cdot \Theta_{\text{fin}}(K).$$

After fixing the y_0, we define $W = y_0 + [a - R(K), a - R(K)]^m$, and we define C to the intersection of the projections of the sections $(x + L^\perp) \cap D$ into $L^\perp$ for $x \in W$. In addition, let C_0 be the projection of $(W + L^\perp) \cap D$ into $L^\perp$. The same argument as before yields that the Hausdorff distance of C and C_0 is at most ε; hence, all we should show is that $r_{d-m}(C) \geq \varrho$ for some positive ϱ depending only on m, R, and K. It follows by the optimality of $\widetilde{D}$ that $V(W + C) \geq (N/2) \cdot \Theta_{\text{fin}}(K)$. Since readily $|W| = O(N)$, we deduce

$|C| > \nu$ for some positive ν depending only on R and K. Therefore, $R(C) \leq R$ and the Steinhagen inequality (Theorem A.8.2) yield $r_{d-m}(C) \geq \varrho$ for suitable ϱ. In turn, we conclude Theorem 7.2.2. $\qquad\square$

Because no practical way is known for determining $\delta_{\text{fin}}(K)$ or $\vartheta_{\text{fin}}(K)$, at least we provide a characterization for convex bodies such that either asymptotic density is one:

Lemma 7.2.3. *Either* $\delta_{\text{fin}}(K) = 1$ *or* $\vartheta_{\text{fin}}(K) = 1$ *hold for a convex body* K *in* $\mathbb{R}^d$ *if and only if there exist a linear m-plane L with $1 \leq m \leq d - 2$ or $m = d$ and a $(d - m)$-dimensional compact convex set C in $L^\perp$ such that $L + C$ can be tiled by congruent copies of K.*

Proof. We discuss only packings because coverings can be discussed in an analogous way. If L and C exist then readily $\delta_{\text{fin}}(K) = 1$; hence, let us assume $\delta_{\text{fin}}(K) = 1$.

The argument is similar to the one for Theorem 7.2.2. Let D_n be a convex body of minimal volume that contains n nonoverlapping copies of K. In addition, let $1 \leq m \leq d$ be the largest integer such that $r_m(D_n)$ is not bounded, and hence we may assume that $r_m(D_{n_k})$ tends to infinity with n_k for a subsequence D_{n_k}. Since $r_{m+1}(D_{n_k})$ is bounded, (A.20) yields the existence of an affine m-plane L_{n_k} for each D_{n_k} and a constant R such that

$$D_{n_k} \subset L_{n_k} + R \cdot B^d. \tag{7.7}$$

We may assume that L_{n_k} is a linear m-space, and $\{L_{n_k}\}$ tends to some linear m-space L.

Let β be a positive integer. For any small $\varepsilon > 0$, the proof of Theorem 7.2.2 yields the existence of some $n_k > 1/\varepsilon$, an m-cube $W(\beta, \varepsilon)$ in L_{n_k} with edge length 2β, some $(d - m)$-dimensional compact convex set $C(\beta, \varepsilon)$ in $L_{n_k}^\perp \cap R B^d$, and a family $\mathcal{K}(\beta, \varepsilon)$ of nonoverlapping congruent copies of K with the following properties: The members of $\mathcal{K}(\beta, \varepsilon)$ cover all $W(\beta, \varepsilon) + C(\beta, \varepsilon)$ but a part of volume $O(\varepsilon) \cdot V(W(\beta, \varepsilon) + C(\beta, \varepsilon))$ and, moreover, given a $K_j \in \mathcal{K}(\beta, \varepsilon)$, we have $K_j/L_{n_k} \subset C(\beta, \varepsilon)$, and K_j intersects $W(\beta, \varepsilon) + C(\beta, \varepsilon)$.

Now let ε tend to zero. According to the Blaschke selection theorem (Theorem A.3.1), we may assume that $C(\beta, \varepsilon)$ tends to some $(d - m)$-dimensional compact convex set C_β in $L^\perp \cap R B^d$, and $W(\beta, \varepsilon)$ tends to the cube $[-\beta, \beta]^m$ in L. Since the cardinality of $\mathcal{K}(\beta, \varepsilon)$ is bounded, we obtain a family $\mathcal{K}_\beta$ of nonoverlapping congruent copies of K such that $[-\beta, \beta]^m + C_\beta$ is covered by the members of $\mathcal{K}_\beta$; moreover, given a $K_j \in \mathcal{K}_\beta$, we have $K_j/L \subset C_\beta$, and K_j intersects $[-\beta, \beta]^m + C_\beta$.

Finally, let β tend to infinity; hence, we may assume that C_β tends to some $(d-m)$-dimensional compact convex set C in $L^\perp$. For any $i \geq 1$, we define a subsequence $\{\mathcal{K}^i_j\}$, $j = 0, 1, \ldots$, of $\{\mathcal{K}_\beta\}$: $\{\mathcal{K}^1_j\}$ is a subsequence such that the union of members $\{\mathcal{K}^1_j\}$ that intersect B^d tends to the union of a certain finite family $\widetilde{\mathcal{K}}_1$ of nonoverlapping congruent copies of K. We note that the elements of $\widetilde{\mathcal{K}}_1$ are contained in $L + C$ and cover $(L + C) \cap B^d$. Next we apply induction on i for $i \geq 2$; namely, $\{\mathcal{K}^i_j\}$ is a subsequence of $\{\mathcal{K}^{i-1}_j\}$ such that the union of the members of $\{\mathcal{K}^i_j\}$ intersecting $i \cdot B^d$ tends to the union of a certain finite family $\widetilde{\mathcal{K}}_i$ of nonoverlapping congruent copies of K. Then $\{\widetilde{\mathcal{K}}_i\}$ is an increasing family, and hence the congruent copies of K that are contained in some $\widetilde{\mathcal{K}}_i$ form a tiling of $L + C$.

Naturally, it may happen that $m = d - 1$ in this construction but in this case the whole space can be tiled by the strips $L + C$. $\qquad\square$

Example 7.2.4. *Given* $1 \leq m \leq d - 2$ *or* $m = d \geq 3$, *there exists a convex body* K *in* $\mathbb{R}^d$ *with the following properties: If* D_n *denotes the convex body of minimal volume that contains n nonoverlapping copies of* K *then, as n tends to infinity,* $r_m(K)$ *tends to infinity and* $r_{m+1}(K)$ *stays bounded. In addition, the analogous properties hold for coverings.*

Let P be the Dirichlet–Voronoi cell of a lattice Λ in $\mathbb{R}^m$ such that no two nonzero vectors of Λ are orthogonal. We define $K = P + B^{d-m}$, and ideas leading to Lemma 7.2.3 prove the required properties of K.

An argument similar to the proof of Lemma 7.2.3 yields the following:

Remark 7.2.5. Either $\delta(K) = 1$ or $\vartheta(K) = 1$ hold for a convex body K in $\mathbb{R}^d$ if and only if there exists a tiling of $\mathbb{R}^d$ by congruent copies of K. In this case K is a polytope.

Let us finally compare the asymptotic density and the optimal density for periodic arrangements. Clusters show (see Theorem 7.1.1) that $\delta_{\text{fin}}(K) \geq \delta(K)$ and $\vartheta_{\text{fin}}(K) \leq \vartheta(K)$. Readily, equality holds, say, if K is a parallelepiped. However, if K is a cylinder whose base is a $(d-1)$-ball in $\mathbb{R}^d$, $d \geq 3$, then readily $\delta(K) < 1 = \delta_{\text{fin}}(K)$ and $\vartheta(K) > 1 = \vartheta_{\text{fin}}(K)$.

Open Problem.

(i) Does there exists a convex body in $\mathbb{R}^d$ for $d \geq 3$ such that certain congruent copies of K tile $\mathbb{R}^d$ but none of these tilings is periodic?

 If such a K exists then there exists no densest periodic packing or densest periodic covering for K. P. Schmitt, J. H. Conway, and

L. Danzer constructed a rhomboid R in $\mathbb{R}^3$ such that certain images of R by orientation-preserving isometries tile $\mathbb{R}^3$ but there exists no periodic tiling of this kind (see M. Senechal [Sen1995] for a detailed description).

Comments. Lemma 7.2.1 and Example 7.2.4 are due to P. Gritzmann [Gri1984]. Remark 7.2.5 is proven by W. M. Schmidt [Sch1961].

7.3. Minimal Mean Projection for Packings

Let $1 \leq i \leq d - 1$. In this section we consider the minimum of the mean i-dimensional projection for a large number of nonoverlapping congruent copies of a given convex body in $\mathbb{R}^d$. The presentation is simpler if we do it in terms of the intrinsic i-volume.

Theorem 7.3.1. *Given a convex body K in $\mathbb{R}^d$ and $1 \leq i \leq d - 1$, let D_n be a convex body containing n nonoverlapping congruent copies of K such that $V_i(D_n)$ is minimal. Then $D_n / \sqrt[d]{n}$ tends to the ball of volume $\Delta(K)$ as n tends to infinity.*

Proof. During the argument, c_1, c_2, and c_3 denote suitable positive constants that depend only on K. Let B_n be the ball of minimal radius, which contains n nonoverlapping copies of K. We deduce by the minimality of $V_i(D_n)$ and by Theorem 7.1.1 that

$$V_i(D_n) \leq V_i(B_n);$$
$$V(B_n) \sim n \cdot \Delta(K). \tag{7.8}$$

The first step is to show that both $r(D_n)$ and $R(D_n)$ are of order $\sqrt[d]{n}$. According to the Steinhagen inequality (Theorem A.8.2), there exists a line l such that the projection of D_n into l is of length at most $2\sqrt{d}\, r(D_n)$. Let M be a section of D_n of maximal $(d-1)$-measure among hyperplane sections of D_n orthogonal to l; hence, the Fubini theorem yields that $|M| \geq V(D_n)/[2\sqrt{d}\, r(D_n)]$. Applying the isoperimetric inequality (A.37) to M, we conclude

$$V_i(D_n) > V_i(M) \geq c_1 \left(\frac{V(D_n)}{r(D_n)} \right)^{\frac{i}{d-1}} \geq c_1 \left(\frac{n V(K)}{r(D_n)} \right)^{\frac{i}{d-1}},$$

and thus $r(D_n) \geq c_2 \sqrt[d]{n}$ follows by (7.8). As n is the maximal number of nonoverlapping congruent copies of K in D_n, Theorem 7.1.1 yields

$$V(D_n) \sim n \cdot \Delta(K). \tag{7.9}$$

We also deduce $R(D_n) \leq c_3 \sqrt[d]{n}$ because D_n contains a segment of length $R(D_n)$, and hence any subsequence of $\{D_n / \sqrt[d]{n}\}$ contains a convergent subsequence according to the Blaschke selection theorem (Theorem A.3.1). Let C be the limit of such a subsequence, and let B be the ball of volume $\Delta(K)$. Then $V(C) = V(B)$ follows by (7.9), and $V_i(C) \leq V_i(B)$ follows by (7.8). Therefore, the isoperimetric inequality (A.37) and its equality case yield that C is congruent to B. In turn, we conclude Theorem 7.3.1.   □

Comments. The analogue of Theorem 7.3.1 for translates of a given convex body is proved independently by K. Böröczky Jr. [Bör1994] and Ch. Zong [Zon1995a]. Actually, the proof in [Bör1994] yields that $R(D_n)/r(D_n) \leq 1 + c \, n^{-2/d(d+3)}$ in Theorem 7.3.1, where c depends on K. The improved estimate is based on a stability version of the isoperimetric inequality (see H. Groemer [Gro1993] for a related survey).

7.4. Maximal Mean Width for Coverings

Given a convex body K in $\mathbb{R}^d$ and a large n, our main goal is to show that the compact convex set D of maximal mean width that can be covered by n congruent copies of K is close to being a segment. We measure the distance of D from segments in terms of the radius $r_2(D)$ of a largest circular disc contained in D; namely, we claim that if s is a diameter of D then

$$D \subset s + 3r_2(D) \cdot B^d. \tag{7.10}$$

To verify (7.10), let h be the maximal distance of points of D from s. Then D contains a triangle T such that s is a longest side, and the distance of the opposite vertex from s is h. Therefore, (7.10) follows by $hs/2 = A(T) \leq 3s \cdot r_2(T)/2$.

We have seen in the planar case (see Section 1.9) that it is essentially equivalent to cover only the perimeter. In higher dimensions the appropriate "one-dimensional" subset of a compact convex set D is the *one-skeleton* skel$D$; namely, the family of points x such that no circular disc centred at x is a subset of D. The one-skeleton is not empty because it contains the extremal points of D. It may not be closed, and it is not continuous on the space of compact convex sets.

Instead of the mean width, we present the results in terms of the first intrinsic volume because it is more convenient to work with.

Theorem 7.4.1. *For any n and convex body K in $\mathbb{R}^d$, there exists a compact convex set D_n such that $V_1(D_n)$ is maximal under the condition that the one-skeleton of D_n can be covered by n congruent copies of K. If $n > c^d$ for some* *positive absolute constant c then*

(i) $r_2(D_n) < \frac{48\sqrt{2\pi d}}{n} \cdot \operatorname{diam} K$;

(ii) $n \cdot \operatorname{diam} K \leq V_1(D_n) \leq \left(n + \frac{24\pi d}{n}\right) \cdot \operatorname{diam} K$.

Remark. We also verify that inequalities (i) and (ii) hold if not only the one-skeleton of D_n but all of D_n is covered by $n \geq 215\, d$ congruent copies of K.

Let us show that the estimates of Theorem 7.4.1 are essentially optimal. We need some estimates on the first intrinsic volume of the unit ball B^d; namely, (A.29) yields

$$\sqrt{\frac{2\pi d^2}{d+1}} < V_1(B^d) < \sqrt{2\pi\, d}. \tag{7.11}$$

Example 7.4.2.

(i) *There exists a convex body K in $\mathbb{R}^d$ such that the optimal D_n in Theorem 7.4.1 is a segment of length $n \operatorname{diam} K$ for large n.*

We consider an isosceles triangle whose equal angles are $\pi/6$, and we define K to be the rotated image around the longest side. Assuming that this longest side is one, inequality (7.15) in the proof of Theorem 7.4.1 can be replaced by

$$V_1(D_n) \leq n - \tilde{h} \cdot cn + \tilde{h} \cdot \sqrt{2\pi d}$$

for some positive absolute constant c. Therefore, $\tilde{h} = 0$ for large n.

(ii) $r_2(D_n) > \sqrt{d}/(50n)$ *and* $V_1(D_n) > 2n + d/(6n)$ *if $K = B^d$ and n is large.* Expression (7.10) and $18 V_1(B^d) < 50\sqrt{d}$ shows that it is sufficient to verify the inequality for $V_1(D_n)$. Let Z_n be the right cylinder whose base is a $(d-1)$-ball of radius $\varrho = V_1(B^{d-1})/(2n)$, and whose height is $2n(1 - 2\varrho^2/3)$. Then some n unit balls cover Z_n for large n, and

$$V_1(Z_n) = 2n \cdot \left(1 - \frac{2\varrho^2}{3}\right) + \varrho \cdot V_1(B^{d-1}) = 2n + \frac{1}{6n} \cdot V_1(B^{d-1})^2,$$

where $V_1(B^{d-1}) > \sqrt{d}$ according to (7.11).

The case when all of D_n is covered by the congruent copies of K in Theorem 7.4.1 is much easier to handle; therefore, we discuss it in Step I. After that we consider the problem when the one-skeleton of D_n is covered.

We establish in Step II that the extremal D_n actually exists. Then we show that $r_2(D_n)$ is small, where the most involved part of the argument is to verify that $r_2(D_n)$ stays bounded as n tends to infinity.

In the proof of Theorem 7.4.1, we assume that diam $K = 1$, and we write $K_1, \ldots, K_n$ to denote the n congruent copies of K that cover either D_n or the one-skeleton of D_n. We fix a diameter $\tilde{s}$ of D_n, write $\tilde{h}$ to denote the maximal distance of the points of D_n from $\tilde{s}$, and write $\tilde{T}$ to denote a triangle contained in D_n such that $\tilde{s}$ is a side and the opposite vertex is of distance $\tilde{h}$ from $\tilde{s}$. Since $V_1(B^d) < \sqrt{2\pi d}$ (see (7.11)) and D_n is contained in $\tilde{s} + \tilde{h} \cdot B^d$, we deduce

$$V_1(D_n) \leq \operatorname{diam} D_n + \tilde{h} \cdot \sqrt{2\pi d}. \tag{7.12}$$

However, the segment of length n can be readily covered by n congruent copies of K, and hence

$$V_1(D_n) \geq n.$$

In addition, we assume that $o \in \tilde{T}$ and write $\tilde{L}$ to denote the two-dimensional linear subspace spanned by $\tilde{T}$ and use $\tilde{C}$ to denote $D_n | \tilde{L}$, where $\cdot | \tilde{L}$ stands for the projection into $\tilde{L}$.

Step I: All of D_n Is Covered

Each $K_i | \tilde{L}$ is of diameter at most one; hence, its perimeter is at most π according to the Cauchy formula (A.34). In turn, the isometric inequality Theorem A.5.7 yields that the area of $K_i | \tilde{L}$ is at most $\pi/4$. As the projections $K_1 | \tilde{L}, \ldots, K_n | \tilde{L}$ cover $\tilde{T}$, we deduce that diam $D_n \leq \pi n/(2\tilde{h})$ and $\tilde{h} \leq \sqrt{\pi n/2}$. Substituting these estimates into (7.12) results in

$$V_1(D_n) \leq \frac{\pi}{2\tilde{h}} \cdot n + \pi \sqrt{d} \cdot \sqrt{n}.$$

However, $V_1(D_n) \geq n$, which, in turn, yields $\tilde{h} \leq 2$ for $n \geq 215\,d$.

Next we show that $\tilde{h}$ is very small based on the claim

$$\operatorname{diam} D_n \leq \left(1 - \frac{\tilde{h}^2}{48}\right) \cdot n. \tag{7.13}$$

We may assume that diam $D_n \geq (1 - 2^2/48) \cdot n \geq 11n/12$; hence, $\tilde{s}$ contains a segment s' of length at least $11n/24$ such that any line orthogonally intersecting s' intersects D_n in a segment of length at least $\tilde{h}/2$. We write σ and σ' to denote the two arcs of $\partial \tilde{C}$ connecting the endpoints of $\tilde{s}$. If K_i intersects σ then let p_i and q_i be the furthest points of $\sigma \cap K_i$, and otherwise let $p_i = q_i$ be any point of σ. We define the points $p_i', q_i' \in \sigma'$ analogously and claim that if the projection of K_i into aff $\tilde{s}$ is contained in s' then

$$d(p_i, q_i) + d(p'_i, q'_i) \leq 2 - \frac{\tilde{h}^2}{8}. \tag{7.14}$$

We may assume that p_i, q_i, p'_i, and q'_i are vertices of a quadrilateral Q. Since Q has an angle of at least $\pi/2$, and its diagonals are at most one, we deduce that either $p_i q_i$ or $p'_i q'_i$ is of length at most $\sqrt{1 - \tilde{h}^2/4}$. Hence, claim (7.14) follows. Now the number of K_i whose projection into aff s is contained in s' is at least $11n/24 - 2 \geq n/3$; therefore,

$$\begin{aligned}
\text{diam } D_n &\leq \frac{1}{2} \left[\sum_{i=1}^{n} d(p_i, q_i) + d(p'_i, q'_i) \right] \\
&\leq \frac{2n}{3} + \frac{n}{3} \cdot \left(1 - \frac{\tilde{h}^2}{16} \right) = \left(1 - \frac{\tilde{h}^2}{48} \right) \cdot n,
\end{aligned}$$

completing the proof of claim (7.13). In turn, (7.12) and (7.13) yield

$$V_1(D_n) \leq \left(1 - \frac{\tilde{h}^2}{48} \right) \cdot n + \tilde{h} \cdot \sqrt{2\pi d}. \tag{7.15}$$

Optimizing the upper bound in $\tilde{h}$ leads to Theorem 7.4.1 (ii), and (i) is a direct consequence of $r(D_n) \leq \tilde{h}$ and $V_1(D_n) \geq n$.

Step II: Existence of the Extremal Set When the One-Skeleton Is Covered

If L is a two-dimensional linear subspace and D is a compact convex set then $\partial(D|L)$ is contained in the projection of skel$D$ into L. We deduce

$$\partial(D_n|L) \subset K_1|L \cup \ldots \cup K_n|L, \tag{7.16}$$

which, in turn, yields

$$\text{diam } D_n \leq n. \tag{7.17}$$

Now the existence of the extremal D_n readily follows from (7.17), the Blascke selection theorem (Theorem A.3.1) and the following claim: Given a sequence $\{C_k\}$ of compact convex sets that tends to a compact convex set C, if $x \in$ skel$C$ and $\varepsilon > 0$ then

$$(x + \varepsilon B^d) \cap \text{skel} C_k \neq \emptyset \tag{7.18}$$

holds for large k. We suppose that the claim does not hold for certain $\varepsilon > 0$, and we seek a contradiction. We choose a hyperplane H in a way that x is an extremal point of $H \cap C$. Let x_k be the closest point of C_k to x, and let H_k be the hyperplane that contains x_k and is parallel to H. We may assume that $d(x, x_k) < \varepsilon/2$, and x_k is not the midpoint of any segment in $H_k \cap C_k$ that has length at least ε/d. Let us write x_k in the form $\sum_{i=0}^{m} \lambda_i y_i$ for extremal points $y_0, \ldots, y_m$ of $H_k \cap C_k$ in a way that the coefficients satisfy $\lambda_0 \geq \ldots \geq \lambda_m \geq 0$ and $\sum_{i=0}^{m} \lambda_i = 1$. The Carathéodory theorem says that $m \leq d$ can be assumed, and hence $\lambda_0 \geq 1/(d+1)$. Since $y_0 \in \mathrm{skel} C_k$, we have $d(x, y_0) \geq \varepsilon$, and thus $d(x_k, y_0) > \varepsilon/2$. In addition, the point $p = \sum_{i=1}^{m} \lambda_i y_i / (1 - \lambda_0)$ of $H_k \cap C_k$ satisfies $p - x_k = -\lambda_0 (y_0 - x_k)/(1 - \lambda_0)$; therefore, the segment $y_0 p$ contains a segment whose midpoint is x_k, and whose length is at least ε/d. This contradiction yields claim (7.18) and, in turn, the existence of D_n.

Step III: The Bound on $r_2(D_n)$

The main part of the argument is to show that $\tilde{h}$ is bounded. Here we heavily use the nontrivial fact that the one-skeleton is connected (see the comments at the end of this section for references).

First, we associate a geometric graph G to the covering via constructing a sequence of four graphs. Let G_1 be the graph on $K_1, \ldots, K_n$ as vertices such that a pair $\{K_i, K_j\}$ is an edge if and only if $i \neq j$, and K_i and K_j intersect. Since the one-skeleton of D_n is connected, G_1 is a connected graph. Let G_2 be a spanning tree of G_1, namely, a minimal connected graph on $K_1, \ldots, K_n$ such that each edge of G_2 is an edge of G_1. We number the $n-1$ edges of G_2 from one to $n-1$ and let $v_k \in K_i \cap K_j$ if $\{K_i, K_j\}$ is the kth edge. Coincidences may occur, and hence we assume that $v_1, \ldots, v_m, m \leq n-1$, are the different points out of $v_1, \ldots, v_{n-1}$. Next we define the geometric graph G_3 on $v_1, \ldots, v_m$ in a way that the segment $v_k v_l$ represents an edge if and only if $k \neq l$, and some K_i contains both v_k and v_l. Thus G_3 is connected as well, and let G be a spanning tree of G_3. We write σ to denote the union of the edges of G, which is a connected set consisting of segments such that any two segments intersect in at most one point. In addition, the total length $|\sigma|$ of σ is at most $n-2$, and

$$\mathrm{skel}\, D_n \subset \sigma + B^d.$$

We think about the first intrinsic volume as a mean perimeter of two-dimensional projections (see (A.36) and the preceding definitions). We recall that $\mu_{d,i}$ denotes the unique invariant measure on the Grassmannian $\mathrm{Gr}(d, i)$ of linear i-spaces in $\mathbb{R}^d$ such that $\mu_{d,i}(\mathrm{Gr}(d, i)) = 1$. Then

$$V_1(D_n) = \frac{V_1(B^d)}{\pi} \int_{\mathrm{Gr}(d,2)} V_1(D_n|L)\, d\mu_{d,2}(L)$$

and

$$|\sigma| = \frac{V_1(B^d)}{\pi} \int_{\mathrm{Gr}(d,2)} |\sigma|L| \, d\mu_{d,2}(L).$$

We note that if L is any linear two-space then

$$\partial(D_n|L) \subset \sigma|L + B^d.$$

Proposition 7.4.3. *Given a convex domain C in $\mathbb{R}^2$, if γ is a connected union of finitely many segments such that $\partial C \subset \gamma + B^2$ then*

(i) $V_1(C) \leq |\gamma| + \pi$;

(ii) $\frac{33}{32} V_1(C) \leq |\gamma| + \pi$ *if $r(C) \geq 19$.*

Proof. We may assume that any two segments of γ intersect in at most one point. If s is a segment and l is a linear subspace of dimension one then the length of $s|l$ is just the integral of the function $\#(s \cap (x + l^{\perp}))$ of x over l. Therefore, the Cauchy formula (A.34) yields

$$V_1(C) = \frac{1}{2} P(C) = \frac{\pi}{2} \int_{\mathrm{Gr}(2,1)} \int_{l^{\perp}} \frac{1}{2} \#(\partial C \cap (x + l)) \, dx \, d\mu_{2,1}(l);$$

$$|\gamma| = \frac{\pi}{2} \int_{\mathrm{Gr}(2,1)} \int_{l^{\perp}} \#(\gamma \cap (x + l)) \, dx \, d\mu_{2,1}(l),$$

where the integrals readily make sense. For any line l passing through the origin, let a and $\tilde{a}$ denote the lengths of $C|l$ and $C|l^{\perp}$, respectively. Since the projections of γ into l and into $l^{\perp}$ are of lengths at least $a - 2$ and $\tilde{a} - 2$, respectively, we deduce that

$$a + \tilde{a} \leq \int_{l^{\perp}} \#(\gamma \cap (x + l)) \, dx + \int_{l} \#(\gamma \cap (x + l^{\perp})) \, dx + 4,$$

which, in turn, yields (i). Therefore, let $r(C) > 19$, and we claim

$$\frac{33}{32}(a + \tilde{a}) \leq \int_{l^{\perp}} \#(\gamma \cap (x + l)) \, dx + \int_{l} \#(\gamma \cap (x + l^{\perp})) \, dx + 4. \quad (7.19)$$

We may assume that $a \geq \tilde{a}$ and that l is the first coordinate axis. For any integer k, we define the interval I_k on l to be the closed interval $[2k - 1, 2k + 1]$ if k is odd, and the open interval $(2k - 1, 2k + 1)$ if k is even. We say that a vertical line is *proper* if it does not go through either a point contained in at least two segments of γ or an endpoint of some segment of γ. In addition, an interval I_k is called *saturated*, if k is even and any vertical proper line intersecting I_k intersects γ in at least two points. If the number of saturated intervals is at least $a/32$ then (7.19) readily follows; hence, we assume that the number of saturated intervals is at most $a/32$.

Now there exists an interval s contained in $C\|l$ of length at least $a/2$ such that if a vertical line intersects s then it intersects C in a segment of length at least $r(C)$. Since there exist at least $a/8$ intervals I_k with even k that intersect s, we can find at least $a/16 + 1$ out of them that are not saturated. If I_k is one of these at least $a/16 + 1$ intervals then we associate a vertical proper line l_k to I_k, which intersects I_k, and which intersects γ only in one point. We call l_k a witness. Let l_{k_1} and l_{k_2} be two consecutive witnesses; hence, there exists an I_m with odd m that lies between l_{k_1} and l_{k_2}. Next we consider the points $p, q \in \partial C$ whose first coordinate is $2m$, and let $p', q' \in \gamma$ be the points whose distances from p and q, respectively, are at most one. Then the projection of the segment $p'q'$ into $l^\perp$ is of length at least $r(C) - 2 \geq 17$, and there is a polygonal path in γ that connects p' and q' and does not intersect l_{k_1} and l_{k_2}. Therefore,

$$\int_{l^\perp} \#(\gamma \cap (x + l))\, dx > 17 \cdot \frac{a}{16}.$$

In turn, we conclude (7.19) and Proposition 7.4.3 as well.   □

Since the diameter of the o-symmetric set $(D_n - D_n)/2$ is $|\tilde{s}|$, we deduce that $V_1(D_n) \leq |\tilde{s}|/2 \cdot V_1(B^d)$. We assume that $\tilde{h} \geq 60$, and hence $r_2(\tilde{T}) \geq 20$ according to (7.10). Let Σ be the family of linear two-spaces such that $\tilde{s}|L$ is of length at least $|\tilde{s}|/2$ and $r_2(D_n|L) \geq 19$. Then $\mu_{d,2}(\Sigma) > c_1^d$ for an absolute constant $c_1 > 0$, and if $L \in \Sigma$ then $V_1(D_n|L) \geq |\tilde{s}|/2 \geq V_1(D_n)/V_1(B^d)$. We deduce by Proposition 7.4.3 and by $V_1(B^d) < \sqrt{2\pi d}$ (see (7.11))

$$V_1(D_n) + \frac{c_1^d}{32\,\pi} \cdot V_1(D_n) < |\sigma| + V_1(B^d) \leq n - 2 + \sqrt{2\pi d}.$$

Thus, $V_1(D_n) \geq n$ yields that $n < c_2^d$ for some positive absolute constant c_2; or in other words, if $n \geq c_2^d$ then $\tilde{h} \leq 60$.

In the final part of the proof we use the notation set up in Step I for $\tilde{C}$. If $\tilde{h} \geq 2$ then there exist at least $n/3$ out of $K_i|\tilde{L}$ that do not intersect either σ or σ'; hence, these $K_i|\tilde{L}$ satisfy $d(p_i, q_i) + d(p'_i, q'_i) \leq 1$. In particular, diam $D_n \leq 5n/6$. However, we deduce by $V_1(D_n) \geq n$ and by (7.12) that diam $D_n \geq n - 60\sqrt{2\pi d}$, which leads to a contradiction if n is large. Therefore, $\tilde{h} < 2$. Now the argument used in Step I completes the proof of Theorem 7.4.1.   □

We close the section by discussing the maximum of $V_m(D)$ for some $2 \leq m \leq d - 1$, where D is covered by n congruent copies of a given convex body K. Besides that, we consider the problem when only the m-skeleton of D is covered, namely, the points, that are not centres of any

$(m + 1)$-dimensional balls that are contained in D. In particular, the 0-skeleton is formed by the extreme points. The m-skeleton of D is connected for $m \geq 1$, and its projection to an m-plane coincides with the projection of D. We note that the argument in Step II of the proof of Theorem 7.4.1 yields the following:

Lemma 7.4.4. *Given a convex body K in $\mathbb{R}^d$ and $1 \leq m \leq d - 1$, there exists a compact convex set of maximal intrinsic m-volume whose m-skeleton can be covered by n congruent copies of K.*

If $2 \leq m \leq d - 1$ then the asymptotic structure of the optimal covering with respect to the intrinsic m-volume seems to depend on the gauge body K, so no analogue of Theorem 7.4.1 can be expected. Yet we can make some simple observations:

Lemma 7.4.5. *Let K be a convex body in $\mathbb{R}^d$, and let $2 \leq m \leq d - 1$.*

(i) *If the m-skeleton of the compact convex set D is covered by n congruent copies of K then $V_m(D) \leq n \cdot V_m(K)$.*

(ii) *For any n, there exists a covering of some convex body D by n translates of K such that $V_m(D) \geq n \cdot V_m(K)/(2em)$.*

(iii) *If D_n is a convex body of maximal intrinsic m-volume that can be covered by n congruent copies of K then $r_{m+1}(D_n)$ stays bounded as n tends to infinity.*

Proof. (i) is a consequence of the facts that the intrinsic m-volume is proportional to the mean m-dimensional projection and that the projection of D into an m-dimensional linear subspace coincides with the projection of the m-skeleton of D.

To verify (ii), we may assume that diam $K = 1$ and that the segment ou is a diameter of K. Therefore, if $y \in [(m - 1)/m](K/u)$ then the line $y + \lin u$ intersects K in a segment of length at least $1/m$. We consider the translates $(i/m)u + K$ for $i = 1, \ldots, n$, and we define D to be the intersection of union of these translates and the infinite cylinder $\lin u + [(m - 1)/m](K/u)$. In particular,

$$V_m(K) < 2 V_{m-1}(K/u)$$

and

$$V_m(D) > \frac{n \cdot (m - 1)^{m-1}}{m \cdot m^{m-1}} V_{m-1}(K/u)$$

yield $V_m(D) \geq n \cdot V_m(K)/(2em)$.

Turning to (iii), we make the observation that D_n has an m-dimensional section whose content is at least $cV_m(D_n)$, where $c > 0$ depends only on m and d. It follows that $V_{m+1}(D_n) \geq [c/(m+1)]\, V_m(D_n)\, r_{m+1}(D_n)$, which, in turn, yields the boundedness of $r_{m+1}(D_n)$. $\square$

Comments. The one-skeleton of a compact convex set was defined by D. G. Larman and C. A. Rogers [LaR1970], who also proved the deep theorem that the one-skeleton is connected. The m-skeleton was introduced in D. G. Larman and C. A. Rogers [LaR1973].

If the one-skeleton of the compact convex set D is covered by n translates of a given convex body K then G. Fejes Tóth, P. Gritzmann, and J. M. Wills [FGW1984] prove that

$$V_1(D) \leq (n-1) \cdot \operatorname{diam} K + V_1(K).$$

Lemma 7.4.5 is due to P. Gritzmann [Gri1984].

Functions defined on the Grassmannians is a topic that belongs to integral geometry. Various aspects of this field are presented in the monographs by R. Gardner [Gar1995] and L. A. Santaló [San1976].

7.5. Related Problems

Sequences of Unrelated Convex Bodies

We consider packings and coverings of convex bodies of any size and shape in $\mathbb{R}^d$. In case packings, the topic is also known as the potato-sack problem because of its relation to the following question: What is the largest volume of a sack of potatoes such that, regardless of their shapes, the potatoes can be packed into a given sack? Naturally, we are also given an upper bound on the diameters of the potatoes.

Let $K_1, \ldots, K_n$ be a sequence of convex bodies of diameter at most one. We say that the sequence can be packed into a given cube if there exists a packing $K_1', \ldots, K_n'$ into the cube such that K_i' is congruent to K_i for each i, and the sequence permits a covering of a given cube if there exists a covering $K_1'', \ldots, K_n''$ of the cube such that K_i'' is congruent to K_i for each i. In addition, the packing or the covering is called *on-line* if an algorithm is given where we learn K_i only after K_j' or K_j'', respectively, have been determined for all $j < i$. What may sound surprising is that recent on-line methods prove to be almost as efficient as methods using the knowledge of every $K_1, \ldots, K_n$ from the beginning. The related results are discussed in more detail in the survey papers by H. Groemer [Gro1985] when all $K_1, \ldots, K_n$ are known ahead and by M. Lassak [Las1997] when on-line methods are applied.

In all of the proofs the convex bodies $K_1, \ldots, K_n$ are replaced by rectangular boxes. It is known that any convex body K contains a rectangular box of volume at least $(2/d^d) V(K)$ (see M. Lassak [Las1993], and H. Hadwiger [Had1955] without the factor two), and K is contained in a rectangular box of volume at most $d! V(K)$ (see H. Hadwiger [Had1955]). This is the reason why the coefficient $d^d/2$ occurs in inequalities for packings and $1/d!$ occurs in the inequalities for coverings.

For packings, H. Groemer [Gro1982] (see also E. Makai and J. Pach [MaP1983]) proved that if $s \geq 3$ and

$$\sum_{i=1}^{n} V(K_i) \leq \frac{1}{d!} \cdot \left((s-1)^d - \frac{s-1}{s-2} \cdot (s-1)^{n-2} - 1 \right)$$

then $K_1, \ldots, K_n$ can be packed into the cube of edge length s. The on-line version is due to M. Lassak and J. Zhang [LaZ1991]: If $s > 1$ and

$$\sum_{i=1}^{n} V(K_i) \leq \frac{1}{d!} (s-1)(\sqrt{s}-1)^{2(d-1)}$$

then $K_1, \ldots, K_n$ can be packed on-line into the cube of edge length s. For coverings, again H. Groemer [Gro1982] (see also E. Makai and J. Pach [MaP1983]) proved the best bounds known: If

$$\sum_{i=1}^{n} V(K_i) \geq \frac{d^d}{2} \cdot ((1+s)^d - 1)$$

for some $s > 0$ then $K_1, \ldots, K_n$ permits a covering of the cube of edge length s. The corresponding on-line results are more complicated: If

$$\sum_{i=1}^{n} V(K_i) \geq \frac{d^d}{2} \cdot 6^{d-1}(s+1)\left(s + \frac{1}{2} \right)^{d-1} - 3^{d-1}$$

then $K_1, \ldots, K_n$ permits an on-line covering of the cube of edge length s according to J. Januszewski and M. Lassak [JaL1994]. In addition, the right-hand side can be replaced by $(d^d/2) f(d, s)$, where $\lim_{s \to \infty} f(d, s)/s^d = 1$ (see J. Januszewski and M. Lassak [JaL1995b]).

Newton Number

The Newton number $N(K)$ of a convex body K is the maximal number of nonoverlapping congruent copies of K that touch K. C. J. A. Halberg, Jr., E. Levin, and E. G. Straus [HLS1959] prove that $N(K) \geq N(B^d)$ in $\mathbb{R}^d$.

8

Packings and Coverings by Unit Balls

We start this chapter by discussing the Newton number: the maximal number of nonoverlapping unit balls in $\mathbb{R}^d$ touching a given unit ball. The three-dimensional case was the subject of the famous debate between Isaac Newton and David Gregory, and it was probably the first finite packing problem in history.

In the later part of the chapter, optimality of a finite packing of n unit balls means that the volume or some mean projection of the convex hull is minimal. If the dimension d is reasonably large then the packing minimizing the volume of the convex hull is the sausage; namely, the centres are collinear. However, if some mean projection is considered then the convex hull of the balls in an optimal arrangement is essentially some ball for large n in any dimension. For the mean width, we also verify that, in the optimal packing of $d + 1$ balls, the centres are vertices of a regular simplex.

Concerning optimal coverings of compact convex sets by n unit balls in $\mathbb{R}^d$, mostly conjectures are known; namely, it is conjectured that the optimal coverings are sausage-like (see Section 8.6). However, sound density estimates will be provided when a larger ball is covered by unit balls.

In this chapter only a few proofs are provided because the arguments either use the linear programming bound (6.1) or are presented in Chapter 7 for packings and coverings by congruent copies of a given convex body.

8.1. The Newton Number

This section is a survey because many of the important results were proved using the linear programming bound (6.1). These arguments are thoroughly discussed by J. H. Conway and N. J. A. Sloane [CoS1999] and T. Ericson and V. Zinoviev [ErZ2001].

The notion of Newton number grew out of the old dispute between Isaac Newton and David Gregory in 1694. After observing that twelve solid unit three-balls can touch a given unit ball, they discussed the problem of whether the same can be done with thirteen unit balls. It turned out that there is no room for the thirteenth ball as Newton claimed. The first and still incomplete solutions were provided only in 1874 and 1875 by the physicists C. Bender [Ben1874], S. Günter [Gün1875], and R. Hoppe [Hop1874]. The first "rigorous proof" was due to K. Schütte and B. L. van der Waerden [ScW1953], which was simplified by J. Leech [Lee1956]. New arguments are provided by W. Y. Hsiang [Hsi2001], K. Böröczky [Bör2003], O. R. Musin [Mus?], and K. M. Anstreicher [Ant?] (see the comments to Section 4.5 concerning the main ideas of the proofs).

We write $N(B^d)$ to denote the *Newton number* in $\mathbb{R}^d$; namely, the maximal number of nonoverlapping unit balls that touch a given unit ball. This notion is also known as the *kissing number* or as the Hadwiger number of ellipsoids (see Section 9.6). We observe that $N(B^d)$ is the maximal number of nonoverlapping spherical balls of radius $\pi/6$ on S^{d-1}; hence, the corresponding estimates for spherical ball packings can be applied. Actually, the Newton number is known only in a few dimensions: Besides $N(B^2) = 6$, we have the following values:

$$N(B^3) = 12, \qquad N(B^4) = 24$$
$$N(B^8) = 240, \qquad N(B^{24}) = 196\,560 \tag{8.1}$$

The three-dimensional case is the subject of Theorem 4.5.1 (and see the aforementioned references). $N(B^4) = 24$ is due to O. R. Musin [Mus?]. The values $N(B^8) = 240$ and $N(B^{24}) = 196,560$ were determined independently by V. I. Levenšteĭn [Lev1979] and A. M. Odlyzko and N. J. A. Sloane [OdS1979]. We note that the (unique) optimal arrangement is provided by the lattice E_8 and the so-called Leech lattice, respectively, and the uniqueness was verified by E. Bannai and N. J. A. Sloane [BaS1981].

In large dimensions we have the estimates

$$\frac{2^d}{\sqrt{3}^d} < N(B^d) < 2^{0.401d+o(d)}, \tag{8.2}$$

where $2/\sqrt{3} = 2^{0.207...}$. The lower bound in (8.2) is due to C. E. Shannon [Sha1959] and J. M. Wyner [Wyn1965] (see also Theorem 6.3.1), and the upper bound was verified by G. A. Kabatjanskii and V. I. Levenšteĭn [KaL1978] using the linear programming bound (6.1).

Although the Newton number is known to be exponential in the dimension, it is still unclear what the order of the maximal number of unit balls that touch a given one in a lattice packing is. The best construction to this day based on the so-called Barnes–Wall lattices (see J. H. Conway and N. J. A. Sloane

[CoS1999]) provides $d^{\log_2 d - 2\log_2 \log_2 d}$ balls touching a given one in some lattice packing.

We note that Ch. Zong has introduced the notion of the *blocking number* of the unit ball, namely, the minimal number of nonoverlapping unit balls touching B^d such that no additional unit ball can touch B^d without overlapping at least one ball that is already there. This notion is thoroughly discussed in Ch. Zong [Zon1999].

8.2. Periodic Arrangements of Balls

We recall from Section 7.1 that the packing density $\delta(B^d)$ is the supremum of the densities of periodic packings of the unit ball, and the covering density $\vartheta(B^d)$ is the infimum of the densities of periodic coverings by the unit ball. In this section we collect some basic estimates about the packing and covering density of balls. The estimates use the Riemann zeta function $\zeta(d) = \sum_{i=1}^{\infty} 1/i^d$. In addition, let us consider the $d + 1$ unit balls centred at the vertices of a regular d-simplex of circumradius one, and define τ_d to be the ratio of the sum of the volumes of the sectors of the unit balls in the simplex to the volume of the simplex. We note that

$$\tau_d \sim \frac{d}{e\sqrt{e}} \text{ as } d \text{ tends to infinity}$$

according to H. S. M. Coxeter, L. Few, and C. A. Rogers [CFR1959] (see also C. A. Rogers [Rog1964]), and approximate values of τ_d/κ_d for $d = 3, \ldots, 24$ can be found in J. H. Conway and N. J. A. Sloane [CoS1999] (see their table 2.1 under the heading "Bound for the center density").

Theorem 8.2.1.

$$\vartheta(B^d) > \tau_d \qquad \textbf{(Coxeter–Few–Rogers bound)}; \qquad (8.3)$$

$$\vartheta(B^d) < d\ln d + d\ln\ln d + 5d \qquad \textbf{(Rogers bound)}; \qquad (8.4)$$

$$\delta(B^d) > 2(d-1)\zeta(d) \cdot 2^{-d} \qquad \textbf{(Ball–Minkowski bound)}; \qquad (8.5)$$

$$\delta(B^d) < 2^{-0.599d + o(d)} \qquad \textbf{(Kabatjanskii–Levenšteĭn bound)}. \qquad (8.6)$$

We do not prove most of these bounds in this monograph since the arguments are usually very involved and discussed thoroughly elsewhere. The Coxeter–Few–Rogers bound is due to H. S. M. Coxeter, L. Few, and C. A. Rogers [CFR1959] (see also C. A. Rogers [Rog1964] or Ch. Zong

[Zon1999]). The Rogers bound is actually verified for any convex body by C. A. Rogers [Rog1957] (see also C. A. Rogers [Rog1964]). This is the only bound in Theorem 8.2.1 whose proof we present in this book (see Theorem 9.5.2).

Concerning packings, the greedy algorithm yields that $\delta(B^d) > 2^{-d}$ (see Theorem 9.5.2). H. Minkowski showed the existence of a lattice packing of density at least $2\zeta(d)/2^d$, and the improved lower bound of (8.5) is due to K. Ball [Bal1992]. Finally, the Kabatjanskii–Levenšteĭn bound is verified in G. A. Kabatjanskii and V. I. Levenšteĭn [KaL1978] (see also the monographs by J. H. Conway and N. J. A. Sloane [CoS1999], T. Ericson and V. Zinoviev [ErZ2001], and Ch. Zong [Zon1999]) using the linear programming bound (6.1). We only verify a much simpler estimate due to H. F. Blichfeldt [Bli1929], which has the advantage that it is explicit in any dimension.

Example 8.2.2.

$$\delta(B^d) \leq \frac{d+2}{2} \cdot 2^{-0.5d} \quad \textbf{(Blichfeldt bound)}. \tag{8.7}$$

Let T be a d-dimensional torus such that any ball of radius $\sqrt{2}$ embeds isometrically into T, and let $B_1, \ldots, B_n$ form a packing of embedded unit balls on T. We write x_i to denote the centre of B_i, and $d(\cdot, \cdot)$ to denote distance on T. We define the density function $\sigma_i(x)$ for $i = 1, \ldots, n$ by

$$\sigma_i(x) = \begin{cases} 1 - \dfrac{1}{2} d(x, x_i)^2 & \text{if } d(x, x_i) \leq \sqrt{2}; \\ 0 & \text{if } d(x, x_i) \geq \sqrt{2}. \end{cases}$$

Given $x \in T$, let $x_1, \ldots, x_m$ be the centres whose distance from x is at most $\sqrt{2}$. Writing $x', x'_1, \ldots, x'_m$ to denote some corresponding inverse images in $\mathbb{R}^d$ such that all are contained in $x' + \sqrt{2}B^d$, we have

$$m \sum_{i=1}^{m} d(x, x_i)^2 = \left(mx' - \sum_{i=1}^{m} x'_i \right)^2 + \sum_{1 \leq i < j \leq m} (x'_i - x'_j)^2 \geq 2m(m-1).$$

We deduce $\sum_{i=1}^{n} \sigma_i(x) \leq 1$, which, in turn, yields with the help of polar coordinates (see A.30)

$$V(T) \geq n \int_T \sigma_1(x)\, dx = n d\kappa_d \int_0^{\sqrt{2}} \left(1 - \frac{1}{2} r^2 \right) r^{d-1} dr = n\kappa_d \frac{2}{d+2} \sqrt{2}^d.$$

In turn, we conclude the Blichfeldt bound.

The Blichfeldt bound was slightly improved by C. A. Rogers [Rog1958], whose bound has more geometric appeal (see also C. A. Rogers [Rog1964] and Ch. Zong [Zon1999]). Let us consider $d + 1$ unit balls centred at the vertices of a regular simplex of edge length two in $\mathbb{R}^d$. We define σ_d to be the ratio of the volume of the part of the simplex covered by the balls to the volume of the simplex, which satisfies

$$\delta(B^d) < \sigma_d \quad \textbf{(Rogers's simplex bound)}. \tag{8.8}$$

Approximate values of σ_d/κ_d for $4 \le d \le 24$ are contained in J. H. Conway and N. J. A. Sloane [CoS1999], (see their table 1.2 under the heading "Bound for the center density"). We note that $\sigma_d \sim (d/e)2^{-0.5d}$ as d tends to infinity (see C. A. Rogers [Rog1958, Rog1964], and Ch. Zong [Zon1999]).

We close the section by recalling the relation between periodic packings and clusters from Section 7.1. Given a convex body D in $\mathbb{R}^d$ satisfying $r(D) > 2$, we write N to denote the maximal number of nonoverlapping unit balls in D, and we write M to denote the minimal number of unit balls covering D. Then Theorem 7.1.1 yields that

$$\frac{N \cdot \kappa_d}{V(D)} = \left(1 + O\left(\frac{1}{r(D)}\right)\right) \cdot \delta(B^d); \tag{8.9}$$

$$\frac{M \cdot \kappa_d}{V(D)} = \left(1 + O\left(\frac{1}{r(D)}\right)\right) \cdot \vartheta(B^d), \tag{8.10}$$

where the implied constant in $O(\cdot)$ depends on d.

8.3. Finite Ball Packings of Maximal Density

Among packings of n unit balls in $\mathbb{R}^d$, we search for the one whose convex hull is of minimal volume, or, in other words, the one of maximal density. First, we review the known results, and the reader should consult the comments for references.

Two type of arrangements seem to compete to be the optimal one: Clusters and the so-called sausage arrangement, where the centres of the balls are aligned and consecutive balls touch each other. If the number of balls is large in $\mathbb{R}^d$ then the density of clusters is asymptotically $\delta(B^d)$ according to (8.9), and the density in the sausage arrangement is asymptotically $\kappa_d/(2\kappa_{d-1})$.

Let us discuss packings in $\mathbb{R}^3$ and $\mathbb{R}^4$ (see the comments for references). The so-called checkerboard lattice Λ_d in $\mathbb{R}^d$ is the family of points of the integer lattice $\mathbb{Z}^d$ whose coordinates add up to an even number (which is also called the face-centred cubic lattice in the three-dimensional case). Now

$\sqrt{2}\Lambda_d$ is a packing lattice for the unit ball whose determinant is $2^{\frac{d+2}{2}}$; hence,

$$\delta(B^d) \leq \frac{\kappa_d}{2^{\frac{d+2}{2}}} < \frac{\kappa_d}{2\kappa_{d-1}}$$

for $d = 3, 4$. In particular, if the number n of unit balls in a packing is high then clusters are denser than the sausage in $\mathbb{R}^3$ and $\mathbb{R}^4$. However, if $n \leq 4$ then the sausage packing is the optimal packing. In $\mathbb{R}^3$, some clusters are known to be denser than the sausage if $n \geq 58$ or $n = 56$, and it is conjectured that the sausage is optimal for the rest of the values of n. In $\mathbb{R}^4$, clusters denser than the sausage are known if $n \geq 375{,}769$. This phase transition in $\mathbb{R}^3$ and $\mathbb{R}^4$ was named the *sausage catastrophe* by J. M. Wills.

Turning to packings in $\mathbb{R}^d$ for $d \geq 5$, we note that $\kappa_d/(2\kappa_{d-1}) > \delta(B^d)$ in this case; namely, the sausage is denser than clusters for large n. This fact can be derived from the Rogers simplex bound (8.8) if $d = 5, 6$, and, moreover, from the Blichfeldt bound $\delta(B^d) \leq (d/2 + 1) \cdot 2^{-0.5d}$ (see (8.7)) and from the estimate $\kappa_d/(2\kappa_{d-1}) > \sqrt{\pi/(2d+2)}$ (see (A.29)) if $d \geq 7$. Similar considerations lead to the celebrated *sausage conjecture* of L. Fejes Tóth: *If $d \geq 5$ then the sausage arrangement minimizes the volume of the convex hull of n nonoverlapping unit d-balls for any n.*

As this book goes to press, the sausage conjecture has been verified for $d \geq 42$. The proofs are very involved, even if d is rather large. Therefore, a relatively simple proof is presented on the expense of the lower bound on d. We write S_n to denote the segment of length $2(n - 1)$; hence, the sausage $S_n + B^d$ holds n nonoverlapping unit balls and is of volume $2(n - 1)\kappa_{d-1} + \kappa_d$.

Theorem 8.3.1. *Given $d \geq 217{,}163$ and $n \geq 2$, if a convex body D contains n nonoverlapping unit balls then*

$$V(D) \geq V(S_n + B^d)$$

with equality if and only if D is congruent to $S_n + B^d$.

We may assume that D is the convex hull of the n unit balls. We write $x_1, \ldots, x_n$ to denote the centres of the unit balls and Ω_i to denote the set of points in $\mathbb{R}^d$, which are not closer to any x_j than to x_i. Thus Ω_i is a possibly unbounded convex polyhedral set that contains $x_i + B^d$, and we call $\Omega_i \cap D$ the Dirichlet–Voronoi cell of x_i. If the packing is the sausage packing then there exist two types of Dirichlet–Voronoi cells: The ones in the middle are cylinders of volume $2\kappa_{d-1}$, and the two cells at the end are of volume $\kappa_d + \kappa_{d-1}$. Therefore, Theorem 8.3.1 follows from the following claim:

If the convex hull C of $x_1, \ldots, x_n$ is not a segment then

$$V(\Omega_i \cap D) > \begin{cases} 2\kappa_{d-1} & \text{for all but at most two } \Omega_i; \\ \frac{1}{2}\kappa_d + \kappa_{d-1} & \text{otherwise.} \end{cases} \tag{8.11}$$

The proof of (8.11) occupies the rest of the section. First, we analyze Ω_i. We claim that if F is a $(d-2)$-face of Ω_i then the distance of aff F from x_i is at least $2/\sqrt{3}$. There exist two other centres x_j and x_k such that x_i, x_j, and x_k have the same distance ω from aff F, and aff F intersects aff $\{x_i, x_j, x_k\}$ at a point p that is of the same distance ω from x_i, x_j, x_k. Since any pair of centres are of distance at least two apart, and one of the angles $\angle x_i p x_j$, $\angle x_j p x_k$, and $\angle x_k p x_i$ is at most $2\pi/3$, we conclude that $\omega \geq 2/\sqrt{3}$, as claimed. Now a crucial observation is that if $\varrho < 2/\sqrt{3}$ and d is large then Ω_i contains not only $x_i + B^d$ but the greatest part of $x_i + \varrho\, B^d$ as well. More precisely, we consider the case $\varrho = 1/\cos 0.5 < 2/\sqrt{3}$, and we write $|\cdot|$ to denote the Hausdorff $(d-1)$-measure in $\mathbb{R}^d$.

Proposition 8.3.2. *For $m \geq 111$, let P be a possible unbounded polyhedron in $\mathbb{R}^m$ containing the unit ball B^m such that the affine hulls of the $(m-2)$-faces are of distance at least $2/\sqrt{3}$ from o, and let $\Sigma \subset S^{m-1}$ be the radial projection of the points $x \in P$ with $\|x\| \geq 1/\cos 0.5$. Then*

$$|\Sigma| > 0.99 \cdot |S^{m-1}|.$$

Proof. Let P' be the intersection of $(2/\sqrt{3})B^m$ and P; hence, it is sufficient to prove Proposition 8.3.2 for P' in place of P. We set $\varrho = 1/\cos 0.5$.

Now the boundary of P' consists of finite number of Euclidean $(m-1)$-balls and from points whose distance from o is $2/\sqrt{3}$. Therefore, it is sufficient to prove the following statement: Let M be an $(m-1)$-ball with center v such that $1 \leq \|v\| < \varrho$, the affine hull of M is orthogonal to v, and the radius of M is $\sqrt{(2/\sqrt{3})^2 - \|v\|^2}$. If Σ_1 denotes the radial projection of M into S^{m-1}, and Σ_0 denotes the radial projection of the points $x \in M$ with $\|x\| \leq \varrho$ into S^{m-1}, then

$$|\Sigma_1| \geq 100 \cdot |\Sigma_0|. \tag{8.12}$$

We observe that the projections of Σ_0 and Σ_1 into $v^\perp$ are $(m-1)$-balls of radius

$$\frac{\sqrt{\varrho - \|v\|^2}}{\varrho} \quad \text{and} \quad \frac{\sqrt{(2/\sqrt{3})^2 - \|v\|^2}}{(2/\sqrt{3})},$$

respectively, centred at the origin. If $x' \in v^\perp$ is the orthogonal projection of an $x \in \Sigma_1$ then the ratio of the surface area element at $x \in \Sigma_1$ to the area element at $x' \in v^\perp$ is $\langle x, v \rangle^{-1}$, which is an increasing function of $\|x'\|$. Therefore,

$$\frac{|\Sigma_1|}{|\Sigma_0|} > \left(\frac{\sqrt{(2/\sqrt{3})^2 - \|v\|^2}}{2/\sqrt{3}} \cdot \frac{\varrho}{\sqrt{\varrho^2 - \|v\|^2}} \right)^{m-1}.$$

Now the right-hand side is minimal if $\|v\| = 1$, and hence (8.12) follows by $m \geq 111$. $\qquad\square$

We fix an x_i, which we keep fixed until (8.18). We may assume

$$x_i = o, \quad \text{and let } u_l = \frac{x_l}{\|x_l\|} \quad \text{and} \quad \alpha_{kl} = \angle u_k o u_l \text{ for } k, l \neq i.$$

Since our plan is to compare $\Omega_i \cap D$ to the Dirichlet–Voronoi cells in the sausage packings, we search for cylinders in Ω_i:

Proposition 8.3.3. *Let us assume that for certain $0 < \beta < \pi/2$ and $1 \leq r < 1/\sin\beta$, there exists a unit vector u such that $\sin \angle u o u_l \leq \sin\beta$ for any $l \neq i$. If s is the segment connecting $\pm(1 - r\sin\beta) \cdot u$ then*

$$s + r \cdot (u^\perp \cap B^d) \subset \Omega_i.$$

Proof. Let x be a point of the cylinder. For any u_l, $|\langle u, u_l \rangle| = \cos\gamma_l$ holds for some $0 < \gamma_l \leq \beta$, and hence

$$|\langle x, u_l \rangle| \leq (1 - r\sin\beta)\cos\gamma_l + r\sin\gamma_l < 1 - r\sin\beta + r\sin\gamma_l \leq 1.$$

In turn, we deduce Proposition 8.3.3. $\qquad\square$

To prove (8.11), we will classify the Dirichlet–Voronoi cell $\Omega_i \cap D$ into three possibly types. After determining the type of $\Omega_i \cap D$, we will choose two characteristic vectors u_j and u_m in a way that $\sin \angle x_j o x_m$ is reasonably large. For each type, we define $L = \text{lin}\{u_j, u_m\}$, and we single out some pairwise nonoverlapping convex domains in L. Any of these domains will be denoted by Π^ν for some symbol ν, and we write Ω_i^ν to denote the intersection of the Dirichlet–Voronoi cell $\Omega_i \cap D$ and $\Pi^\nu + L^\perp$. In particular, let

$$\Pi^0 = \text{conv}\{o, u_j, u_m\},$$

which occurs in all types of Dirichlet–Voronoi cells. Readily, $\Pi^0 \subset C$, and $A(\Pi^0) = (1/2)\sin\alpha_{jm}$. We may apply Proposition 8.3.2 to $P = L^\perp \cap \Omega_i$ with $m = d - 2$. In this case, Σ is the set of all $x \in L^\perp \cap S^{d-1}$ such that $(1/\cos 0.5)x \in \Omega_i$ and satisfies $|\Sigma| \geq 0.99 \cdot |S^{d-3}|$. If $x \in \Sigma$ then $B^d \subset \Omega_i$

yields that $x + u_j \tan 0.25$ lies in Ω_i; hence, $y + tx \in \Omega_i$ for any $y \in \tan 0.25 \cdot \Pi^0$ and $0 \le t \le 1$. We conclude using $\kappa_{d-2}/\kappa_{d-1} > \sqrt{(d-1)/(2\pi)}$ (see (A.29)) and $d \ge 217{,}163$ that

$$V\left(\Omega_i^0\right) > \frac{\tan^2 0.25}{2} \cdot \sin \alpha_{jm} \cdot 0.99 \,\kappa_{d-2} > 6 \sin \alpha_{jm} \cdot \kappa_{d-1}. \quad (8.13)$$

We define $0 < \alpha \le \pi/2$ by the property that $\sin \alpha$ is the maximum of all $\sin \alpha_{kl}$, and we use α to distinguish three types of Dirichlet–Voronoi cells.

Type I: $\alpha \ge \pi/6$

Let u_j and u_m satisfy $\sin \alpha_{jm} = \sin \alpha$; hence (8.13) yields

$$V(\Omega_i \cap D) > 2\kappa_{d-1}. \quad (8.14)$$

Type II: $\alpha < \pi/6$ and Some Angle $\angle u_k o u_l$ Is Obtuse

We may choose u_j and u_m in a way such that $\pi - \alpha/2 \le \alpha_{jm} \le \pi - \alpha$. In addition to Π^0, we consider the unit squares Π^j and Π^m satisfying that one–one sides are the segments $o u_j$ and $o u_m$, respectively, and Π^j, Π^m, and Π^0 pairwise do not overlap. If s is the segment connecting the origin and $(1 - \sin \alpha) u_j$ then the cylinder $s + (u_j^\perp \cap B^d)$ is contained in Ω_i according to Proposition 8.3.3. Since $s \subset C$, we conclude that

$$V\left(\Omega_i^j\right),\ V\left(\Omega_i^m\right) \ge \frac{1 - \sin \alpha}{2} \cdot \kappa_{d-1}. \quad (8.15)$$

The next domain in L is Π^{jm}, which is the rectangle such that one side is $u_j u_m$, the sides orthogonal to $u_j u_m$ are of length one, and Π^{jm} does not overlap Π^0. We write u to denote the unit vector $(u_j - u_m)/\|u_j - u_m\|$ define $r = 1 + \sin(\alpha/2)$ and s_1 to be the segment connecting $\pm(1 - r \sin(3\alpha/2)) \cdot u$. Since $\angle u o u_j \le \alpha/2$, the angle of any u_l is at most $3\alpha/2$, either with u or with $-u$. In particular, Proposition 8.3.3 yields that the cylinder $Z = s_1 + r\,(u^\perp \cap B^d)$ is contained in Ω_i. If s_2 is the translate of s_1, that is, the intersection of Z and the segment $u_j u_m$, then the distance of s_1 and s_2 is at most $\sin(\alpha/2)$. Now the cylinder $s_2 + (u^\perp \cap B^d)$ is contained in $\Omega_i \cap D$ because $s_2 \subset C$; hence,

$$V\left(\Omega_i^{jm}\right) \ge \left(1 - r \sin \frac{3\alpha}{2}\right) \cdot \kappa_{d-1} > (1 - 2 \sin \alpha) \cdot \kappa_{d-1}. \quad (8.16)$$

We deduce $V(\Omega_i^j) + V(\Omega_i^m) + V(\Omega_i^{jm}) > (2 - 3 \sin \alpha)\,\kappa_{d-1}$, which, in turn, yields by $\sin \alpha_{jm} > (1/2) \sin \alpha$ and (8.13) that

$$V(\Omega_i \cap D) > V\left(\Omega_i^0\right) + V\left(\Omega_i^j\right) + V\left(\Omega_i^m\right) + V\left(\Omega_i^{jm}\right) > 2\kappa_{d-1}. \quad (8.17)$$

Type III: $\alpha < \pi/6$ and All Angles $\angle u_k o u_l$ Are Acute

We define u_j and u_m by the property $\angle u_j o u_m = \alpha$, and we separate three domains in L besides Π^0. Two of them are Π^j and Π^m, which are defined in the same way as for Type II cells, and still satisfy (8.15). The last domain Π^* is the family of points in $B^d \cap L$ whose angle with both u_j and u_m is at least $\pi/2$. Since $\alpha < 2 \sin \alpha$ and $\kappa_d/\kappa_{d-1} < \sqrt{2\pi/(d+1)}$ (see (A.29)), it follows that

$$V(\Omega_i^*) \geq \left(\frac{1}{2} - \frac{\alpha}{2\pi} \right) \cdot \kappa_d > \frac{\kappa_d}{2} - \sin \alpha \cdot \kappa_{d-1}.$$

Therefore, summing the estimates leads to

$$V(\Omega_i \cap D) > V\left(\Omega_i^0\right) + V\left(\Omega_i^j\right) + V\left(\Omega_i^m\right) + V\left(\Omega_i^*\right) > \frac{\kappa_d}{2} + \kappa_{d-1}.$$

$$(8.18)$$

Finally, there exist at most two Dirichlet–Voronoi cells of Type III because any triangle has an angle that is at least $\pi/6$. Thus (8.14), (8.17), and (8.18) yield (8.11) and, in turn, Theorem 8.3.1. ☐

Open Problems.

(i) Prove that the sausage arrangement is the optimal packing of five balls in $\mathbb{R}^3$ or in $\mathbb{R}^4$.

(ii) Prove that if D_n is the convex hull of n nonoverlapping unit balls in $\mathbb{R}^3$ such that $V(D_n)$ is minimal then $r(D_n)$ tends to infinity.

Possibly, $D_n/\sqrt[3]{n}$ tends to a certain truncated octahedron, as for finite lattice packings (see Example 10.12.2).

(iii) Prove that if the convex body D in $\mathbb{R}^3$ contains $n \geq 4$ nonoverlapping unit balls then $V(D) > n \cdot 4\sqrt{2}$.

Here we assume that the Kepler conjecture $\Delta(B^3) = 4\sqrt{2}$ holds, as T. Hales announced (see [Hal2000] and [Hal?]). The sausage of three balls is of volume less than $3 \cdot 4\sqrt{2}$. In contrast, the sausage of four balls is the optimal packing of four balls (see K. Böröczky Jr. [Bör1993]), and its volume is larger than $4 \cdot 4\sqrt{2}$.

(iv) Prove the sausage conjecture in $\mathbb{R}^d$ for some $5 \leq d \leq 41$.

Comments. The history of the sausage conjecture and related results are discussed in detail in U. Betke, M. Henk, and J. M. Wills [BHW?]. Clusters that are denser then the sausage were found by P. M. Gandini and J. M. Wills [GaW1992] when $n = 56$, $59 \leq n \leq 62$, and $n \geq 65$ in $\mathbb{R}^3$, by P. Scholl [Sch2000b] when $n = 58, 63, 64$ in $\mathbb{R}^3$, and by P. M. Gandini and A. Zucco [GaZ1992] when $n \geq 375{,}769$ in $\mathbb{R}^4$. The fact that the optimal four-ball packing is the sausage is proved by K. Böröczky Jr. [Bör1993] in $\mathbb{R}^3$ and by

U. Betke and P. Gritzmann [BeG1984] in $\mathbb{R}^4$ (see also K. Böröczky Jr. and J. M. Wills [BöW1997] for a different proof).

The sausage conjecture was stated in the note [FTL1975a] by L. Fejes Tóth, and considerable effort was made towards verifying it. To describe some partial results, let C_n be the convex hull of the centres of n nonoverlapping unit balls. The sausage conjecture is proved by U. Betke, P. Gritzmann, and J. M. Wills [BGW1982] if dim $C_n \leq 7/12(d-1)$ and by U. Betke and P. Gritzmann [BeG1984] if dim $C_n \leq 9$ and $d \geq$ dim $C_n + 1$. K. Böröczky Jr. and M. Henk [BöH1995] verified the conjecture for any $d \geq 5$ if $r(C_n)$ is not too small. The breakthrough was achieved in U. Betke, M. Henk, and J. M. Wills [BHW1994], where the case $d \geq 13,387$ is settled. A much finer analysis leads to the proof of the sausage conjecture for $d \geq 42$ by U. Betke and M. Henk [BeH1998].

We recall that, in the proof of Theorem 8.3.1, Ω_i denotes the Dirichlet–Voronoi cell in the space corresponding to the centre x_i. We note that Dirichlet–Voronoi cells were first considered by G. L. Dirichlet (see [Dir 1850]) and by G. F. Voronoi (see [Vor1908]). It was H. F. Blichfeldt who observed that the affine hull of any $(d-k)$–face of Ω_i is of distance at least $\sqrt{2k/(k+1)}$ away from x_i. It follows for any given $0 < \varrho < \sqrt{2}$ that the most part of $x_i + \varrho B^d$ is contained in Ω_i if the dimension d is large (see U. Betke, M. Henk, and J. M. Wills [BHW1995a]).

Given $k \geq 1$, the paper by G. Fejes Tóth, P. Gritzmann, and J. M. Wills [FGW1990] discusses finite k-fold packings of unit balls; namely, no point is contained in the interiors of $k + 1$ of the balls. If a convex body D in $\mathbb{R}^d$ contains a k-fold packing of n unit balls then [FGW1990] verifies $V(D) \geq c_k V(S_{[n/k]} + B^d)$, where c_k is a specific positive constant depending only on k. It is conjectured in [FGW1990] that if d is sufficiently large then even $c_k = 1$ works, a conjecture that holds according to Theorem 8.3.1 when $k = 1$.

We call a packing of unit balls in $\mathbb{R}^3$ *separable* if any two balls can be separated by a plane that avoids the interiors of all balls. G. Kertész [Ker1988] proves that if some rectangular box R in $\mathbb{R}^3$ contains a separable packing of n unit balls then $V(R) \geq 8n$, which estimate is readily optimal.

For packings of balls of different sizes, A. Pór [Pór?b] proves the following result, improving an earlier estimate of D.G. Larman [Lar1966b]. If a convex body D in $\mathbb{R}^d$ with $r(D) \geq 2$ contains nonoverlapping balls whose radii lie in the interval $[r, 1]$ then the density of the packing is at most $1 - cr^{196/(196+d)}$ for some positive c that depends on d. It is easy to see that for any $d \geq 3$, there exists a periodic packing of balls of radius $r = \sqrt{2d/(d+1)} - 1$ and 1 whose density is larger than $\delta(B^d)$. Assuming that $\delta(B^3) = \pi/\sqrt{12}$ as T. Hales (see [Hal2000] and [Hal?]) claims,

S. Vassallo and J. M. Wills [VaW1994] prove that even $r = 0.623$ works if $d = 3$.

8.4. Minimal Mean Projection for Ball Packings

For ball packings, the optimal arrangements with respect to the volume prefer the linear shape (see Theorem 8.3.1). It turns out that the situation is just the opposite with respect to the mean projections, or equivalently, with respect to the corresponding intrinsic volumes. For packings of large number of unit balls, Theorem 7.3.1 essentially solves the problem in the context of packings of congruent copies of a given convex body K. Since the proof is not simpler even if K is a ball, we only recall the result:

Theorem 8.4.1. *If $1 \leq i \leq d - 1$, and D_n is a convex body containing n nonoverlapping unit balls such that $V_i(D_n)$ is minimal, then*

$$\lim_{n \to \infty} \frac{D_n}{n^{1/d}} = B,$$

where B is a ball of volume $\kappa_d / \delta(B^d)$.

For the mean width (or first intrinsic volume) we are even able to determine the optimal packing of at most $d + 1$ unit balls:

Theorem 8.4.2. *If the mean width of the convex hull of $n \leq d + 1$ nonoverlapping unit balls in $\mathbb{R}^d$ is minimal then the centres are vertices of a regular $(n - 1)$-simplex of edge length two.*

It follows by the linearity of the first intrinsic volume that it is sufficient to consider the convex hull of the centres of the balls. Therefore, the basic idea of the proof for Theorem 8.4.2 is to verify that the first intrinsic volume of a simplex is a monotonic function of the edge lengths (see Corollary 8.4.5). We start with local considerations:

Proposition 8.4.3. *Let M be a d-simplex. If the length of one edge is increased while the other edge lengths stay unchanged then $V_1(M)$ strictly increases.*

Proof. We write $x_0, \ldots, x_d$ to denote the vertices of M. Let the vertices $x_1, \ldots, x_d$ stay fixed, and let x_0 arrive to the new position x_0' in a way that $d(x_0', x_1) > d(x_0, x_1)$, and $d(x_0', x_i) = d(x_0, x_i)$ for $i \geq 2$. We write H to denote the perpendicular bisector hyperplane of the segment $x_0 x_0'$ and use

X' to denote the reflected image of X through H for any object X. We observe that $x'_i = x_i$ for $i \geq 2$, and the simplex M has been deformed into the simplex $C = \text{conv}\{x'_0, x_1, \ldots, x_d\}$. As the first intrinsic volume is linear and invariant under reflection, we deduce that $V(C) > V(M)$ is equivalent to the inequality

$$V_1(C_0) > V_1(M_0), \tag{8.19}$$

where $C_0 = (C + C')/2$ and $M_0 = (M + M')/2$.

We present the proof in terms of $u = (x_1 + x_0)/2$, $v = (x_1 + x'_0)/2$, and

$$\sigma = \left\{ \frac{x_0 + x'_0}{2} \right\} \bigcup \bigcup_{i=2,\ldots,d} \left\{ x_i, \frac{x_0 + x_i}{2}, \frac{x_1 + x_i}{2} \right\}.$$

In particular, C_0 is the convex hull of u, σ, and their reflected images, and M_0 is the convex hull of v, σ, and their reflected images. Since $v - u = (x'_0 - x_0)/2$, the condition $d(x_1, x'_0) > d(x_1, x_0)$ yields that the points v and v' are contained in the segment $u'u$. We conclude that $M_0 \subset C_0$. However, u lies in C_0 but not in M_0, which, in turn, yields (8.19) by the strict monotonicity of the first intrinsic volume. $\qquad\square$

Next we deform two nondegenerate simplices into each other in a way that each edge length changes monotonically.

Proposition 8.4.4. *Given the d-simplices $M_0 = \text{conv}\{x_0, \ldots, x_d\}$ and $M_1 = \text{conv}\{y_0, \ldots, y_d\}$, let $d(x_i, x_j) \leq d(y_i, y_j)$ hold for any $i \neq j$. Then there exists a continuous one-parameter family $\{M_t\}$ of d-simplices for $t \in [0, 1]$ such that each edge is a monotonically increasing function of t.*

Proof. Let L_i, $i = 0, \ldots, d$, be a linear subspace of dimension i such that $L_i \subset L_j$ holds for $i < j$, and let L^+_{i-1} be one of the half spaces of L_i determined by L_{i-1}, $i = 1, \ldots, d$. We say that a d-simplex M with vertex order $v_0, \ldots, v_d$ is in proper position if $v_0 = o$ and $v_i \in L^+_{i-1}$ for $i \geq 1$. To such a simplex, we assign the positive definite matrix

$$\Omega(M) = [v_1, \ldots, v_d]^T \cdot [v_1, \ldots, v_d] = [\langle v_i, v_j \rangle]_{1 \leq i, j \leq d}.$$

We observe that

$$d(v_i, v_j)^2 = \langle v_i, v_i \rangle - 2\langle v_i, v_j \rangle + \langle v_j, v_j \rangle \quad \text{for} \quad 1 \leq i, j \leq d. \tag{8.20}$$

Let Ω be a positive definite matrix. We claim that there exists a unique simplex M in proper position such that $\Omega(M) = \Omega$. According to the principal axis theorem, there exists an orthogonal matrix Q such that $D = Q^T \Omega Q$ is a diagonal matrix whose entries in the diagonal are positive. Taking the square roots of the entries of D we obtain a diagonal matrix A, and hence $B = AQ^T$ satisfies that $\Omega = B^T B$. Now there exists an orthogonal matrix Q_0 such that the columns of $Q_0 B$ form the nonzero vertices of some simplex M in proper position, and readily $\Omega = \Omega(Q_0 B)$. Finally, the uniqueness of M follows by (8.20).

We may assume that M_0 and M_1 are in proper position with the natural ordering of the vertices. Now the matrix

$$\Omega_t = (1-t) \cdot \Omega(M_0) + t \cdot \Omega(M_1)$$

is readily positive definite, and we write M_t for the simplex in proper position satisfying $\Omega_t = \Omega(M_t)$. Then M_t is a continuous function of t because of its uniqueness. We write $v_0(t), \ldots, v_d(t)$ to denote the vertices of M_t; hence, (8.20) yields that

$$d(v_i(t), v_j(t))^2 = (1-t) \cdot d(x_i, x_j)^2 + t \cdot d(y_i, y_j)^2$$

for any $0 \le i, j \le d$. In turn, we conclude Proposition 8.4.4. $\qquad\square$

Proof of Theorem 8.4.2. We present the argument when $n = d + 1$ because the case $n \le d$ can be handled in an analogous way. Let $y_0, \ldots, y_d$ be the centres of $d + 1$ nonoverlapping unit balls, where $d(y_0, y_1) > 2$ can be assumed, and let M_1 denote the convex hull of the centres. Now the convex hull of the balls is $M_1 + B^d$, and its first intrinsic volume is $V_1(M_1) + V_1(B^d)$. Therefore, Theorem 8.4.2 follows if $V_1(M_1) > V_1(T^d)$, where T^d is the regular simplex of edge length two.

Let M_0 be a d-simplex of vertices $x_0, \ldots, x_d$ such that all edges but $x_0 x_1$ are of length two, and $d(y_0, y_1) > d(x_0, x_1) > 2$. Then Proposition 8.4.3 yields $V_1(M_0) > V_1(T^d)$; hence, it is sufficient to verify

$$V_1(M_1) \ge V_1(M_0). \tag{8.21}$$

We observe that M_1 can be assumed to be a d-simplex. Then Proposition 8.4.4 provides a family $\{M_t\}$, $t \in [0, 1]$, of d-simplices connecting M_0 to M_1 such that writing $v_0(t), \ldots, v_d(t)$ to denote the vertices of M_t, we have that the edge length $d(v_i(t), v_j(t))$ increases for any $i \ne j$. Now there exist $0 = t_0 < \ldots < t_m = 1$ with the following property: If d_{ij}, $0 \le i, j \le d$, satisfy that $d(v_i(t_k), v_j(t_k)) \le d_{ij} \le d(v_i(t_{k+1}), v_j(t_{k+1}))$ then there exists a d-simplex M such that the distance of the ith and jth vertex is d_{ij}. Among the simplices corresponding to a pair (t_k, t_{k+1}), the minimum of the first intrinsic volume is attained by M_{t_k} according to Proposition 8.4.3. Therefore, $V_1(M_{t_k})$ is increasing in k, which yields (8.21) and, in turn, Theorem 8.4.2. $\qquad\square$

The proof of Theorem 8.4.2 leads to the following stronger statement:

Corollary 8.4.5. *If the sets* $\{x_0, \ldots, x_n\}$ *and* $\{y_0, \ldots, y_n\}$ *satisfy that* $d(x_i, x_j) \leq d(y_i, y_j)$ *for any* $0 \leq i < j \leq n$ *then*

$$V_1(\text{conv}\{x_0, \ldots, x_n\}) \leq V_1(\text{conv}\{y_0, \ldots, y_n\}). \qquad (8.22)$$

If either convex hull is n *dimensional and* $d(x_i, x_j) < d(y_i, y_j)$ *holds for at least one pair* i, j *then strict inequality holds in* (8.22).

Open Problems.

(i) Prove that among $d + 2$ nonoverlapping unit balls in $\mathbb{R}^d$, $d \geq 4$, if the mean width of the convex hull is minimal then the convex hull of the centres is a bipyramid over some $(d - 1)$-simplex, and all edges of the bipyramid are of length two.

The case $d = 3$ is due to K. Böröczky and K. Böröczky Jr. [BöB1996].

(ii) Prove that if the mean width of the convex hull of $2d$ nonoverlapping unit balls is minimal in $\mathbb{R}^d$ for $d \geq 3$ then the convex hull of the centres is a regular crosspolytope.

Comments. Theorem 8.4.1 is proved independently by K. Böröczky Jr. [Bör1994] and by Ch. Zong [Zon1995a].

Theorem 8.4.2 and Corollary 8.4.5 are proved by V. Capoyleas and J. Pach [CaP1991] without the characterization of the equality case, which is due to K. Böröczky Jr. [Bör1997].

Concerning packings of few balls with respect to other mean projections, the only known result is that if the surface area of the convex hull of four nonoverlapping unit balls in $\mathbb{R}^3$ is minimal then the convex hull of the centres is a regular tetrahedron (see K. Böröczky Jr. [Bör1997]).

Let us discuss the minimal diameter of n nonoverlapping unit balls in $\mathbb{R}^d$. If $2 \leq n \leq d + 1$ then the minimal diameter is readily four, and the centres in the optimal arrangement are vertices of a regular $(n - 1)$ – simplex of edge length two. If $n = d + 2$ then K. Schütte [Sch1963b] proves that the centres in the optimal arrangement are vertices of a $\lfloor d/2 \rfloor$- and a $\lceil d/2 \rceil$-dimensional regular simplex of edge length two. A simple proof of this result is provided by I. Bárány [Bár1994]. For large n, it follows via Theorem 8.4.1 that the convex hull of the centres in an optimal packing is asymptotically some large ball.

8.5. Packings and Coverings with Respect to a Larger Ball

Given $R > 1$, we verify three simple statements concerning the maximal number $\nu(d, R)$ of nonoverlapping unit balls contained in $R\,B^d$ and the minimal number $\bar{\nu}(d, R)$ of unit balls covering $R\,B^d$. If R is large then the density of the optimal packing is asymptotically $\delta(B^d)$ (see (8.9)), and the density of the optimal covering is asymptotically $\vartheta(B^d)$ (see (8.10)). Therefore, the interesting case is when R is not too large. In this case both problems are intimately connected to arrangements of equal spherical balls on S^{d-1}; hence, we write $M(d-1, \varphi)$ to denote the maximal number of nonoverlapping spherical balls of radius φ contained in S^{d-1}, and we use $\widetilde{M}(d-1, \varphi)$ to denote the minimal number of spherical balls of radius φ covering S^{d-1}.

For packings we readily have $v(d, R) = 1$ if $R < 2$. In addition, $M(d-1, \pi/6)$ is the so-called Newton number: the maximal number of nonoverlapping unit balls in $\mathbb{R}^d$ that touch a given unit ball.

Lemma 8.5.1.

$$v(d, R) = M\left(d-1, \arcsin\frac{1}{R-1}\right) \quad \text{if } 2 \leq R < 3;$$
$$v(d, 3) = M(d-1, \pi/6) + 1.$$

Proof. Some $M(d-1, \arcsin(1/(R-1)))$ nonoverlapping unit balls whose centres lie on $(R-1)S^{d-1}$ shows that $v(d, R) \geq M(d-1, \arcsin(1/(R-1)))$, and adding B^d if $R = 3$ yields $v(d, 3) \geq M(d-1, \pi/6) + 1$. Next we consider a packing $x_1 + B^d, \ldots, x_n + B^d$ in RB^d for $n \geq 2$ and $2 \leq R \leq 3$. If some $x_i = o$ then $R = 3$, and any $x_j \neq x_i$ is contained in $2S^{d-1}$. If $x_i \neq o$ for all x_i then let x_i' be the radial projection of x_i into $(R-1)S^{d-1}$. For any $x_j \neq x_i$, the diameter – the longest side of the triangle $ox_i'x_j'$ – is at least two, and therefore $\angle x_i'ox_j' \geq \arcsin(1/(R-1))$. In turn, we conclude Lemma 8.5.1. $\qquad\square$

It follows by Theorem 6.2.1 that $v(d, 1 + \sqrt{2}) = 2d$, and $v(d, R) \leq d + 1$ if $R < 1 + \sqrt{2}$. In addition, Theorems 6.1.1 and 6.2.1 yield that the optimal packing of $n = 2, \ldots, d + 1$ or of $2d$ unit balls is determined by a $(n-1)$-dimensional regular simplex, or by a regular crosspolytope, respectively. In the three-space the value of $v(3, R)$ for $1 + \sqrt{2} < R \leq 3$ follows either by the known solution of the Tammes problem for $7 \leq n \leq 12$ circular discs on S^2 (see Section 4.4) or by the fact that $M(2, \pi/6) = 12$ (see Theorem 4.5.1). In addition, $v(4, 3) = 25$, $v(8, 3) = 241$ and $v(24, 3) = 196,561$ (see the introduction of Chapter 6). For other values of d and R, Theorem 6.3.1 and the Kabatjanskii–Levenšteĭn bound (6.7) yield

$$v(d, R) > \left(\frac{(R-1)^2}{2\sqrt{R^2 - 2R}}\right)^d \quad \text{if } 1 + \sqrt{2} < R \leq 3;$$
$$v(d, R) < 23\, d^{\frac{3}{2}}(R-1)^d 2^{-0.5d} \quad \text{if } 1 + \sqrt{2} < R < 3;$$
$$v(d, 3) < 2^{0.401d + o(d)}.$$

Turning to coverings for $R > 1$, we let $\widetilde{M}(d - 1, \arcsin(1/R))$ unit balls cover $R\,S^{d-1}$ in a way that each ball intersects $R\,S^{d-1}$ in a $(d-2)$-dimensional unit ball. If $R \leq \sqrt{2}$ then the balls cover RB^d, and if $R \leq 2$ then the balls together with B^d cover RB^d. We deduce the following:

Lemma 8.5.2. *If* $1 < R \leq \sqrt{2}$ *then*

$$\tilde{v}(d, R) = \widetilde{M}\left(d - 1, \arcsin \frac{1}{R}\right),$$

and if $\sqrt{2} < R \leq 2$ *then*

$$\widetilde{M}\left(d - 1, \arcsin \frac{1}{R}\right) \leq \tilde{v}(d, R) \leq \widetilde{M}\left(d - 1, \arcsin \frac{1}{R}\right) + 1.$$

Remark. $\tilde{v}(d, R) = \widetilde{M}(d - 1, \arcsin(1/R)) + 1$ if $\sqrt{2} < R \leq 2$ and there is a unique optimal covering of S^{d-1} by spherical balls of radius $\arcsin(1/R)$.

If n unit balls cover a larger ball then readily $n \geq d + 1$. Assuming $n \leq d + 2$, the centres of the n unit balls that cover a ball of maximal radius are vertices of some highly symmetric polytope, which is a regular simplex if $n = d + 1$ (see Theorem 6.5.1), and the convex hull of a regular $\lfloor (d + 1)/2 \rfloor$-dimensional simplex and an orthogonal regular $\lceil (d + 1)/2 \rceil$-dimensional simplex if $n = d + 2$ (see Theorem 6.6.1). For six and seven unit balls in $\mathbb{R}^3$, the solutions are provided by a certain regular octahedron and by a certain bipyramid over the regular pentagon, respectively (see Section 5.4). When $n = 11, 13, 15$ in $\mathbb{R}^3$ then some $n - 1$ of the unit balls in an optimal covering induce an optimal covering of S^2 by $n - 1$ equal spherical circular discs (see Section 5.4), and the nth unit ball covers a neighbourhood of the origin. For eight balls in $\mathbb{R}^4$, the solution is provided by a certain regular crosspolytope (see Theorem 6.7.1).

When the exact arrangement is not known, we use economic coverings of S^d. If $\cos \varphi \leq 1/\sqrt{d}$ then the regular crosspolytope inscribed into B^d shows that $\widetilde{M}(d - 1, \varphi) \leq 2d$, and if $\cos \varphi \geq 1/\sqrt{d}$ then Theorem 6.8.1 yields an absolute constant c satisfying

$$\widetilde{M}(d - 1, \varphi) \leq c \, \cos \varphi \cdot \sin^{-(d-1)} \varphi \cdot d^{\frac{3}{2}} \ln 2d \cos^2 \varphi. \tag{8.23}$$

In particular, Lemma 8.5.2 leads to

$$\tilde{v}(d, R) \leq 2d \quad \text{if } 1 < R \leq \sqrt{\frac{d}{d-1}}; \tag{8.24}$$

$$\tilde{v}(d, R) \leq R^d \cdot cd^{\frac{3}{2}} \sqrt{R - 1} \ln 3d(R - 1) \quad \text{if } \sqrt{\frac{d}{d-1}} \leq R \leq 2, \tag{8.25}$$

where c is an absolute constant. Next we find economic coverings of balls whose radius is larger than two:

Theorem 8.5.3. *There exists an absolute constant c such that*

$$\tilde{\nu}(d, R) \leq \frac{cd^{\frac{3}{2}} \ln d}{R} \cdot R^d \quad \text{if } 2 \leq R \leq \sqrt{d}; \tag{8.26}$$
$$\tilde{\nu}(d, R) \leq cd \ln d \cdot R^d \quad \text{if } R \geq \sqrt{d}. \tag{8.27}$$

Proof. If $2 \leq R \leq \sqrt{d}$ then the argument is based on the following observation for any $r > 1$: Let us cover $r S^{d-1}$ with $\widetilde{M}(d-1, \arcsin 1/r)$ unit balls in a way that each ball intersects $r S^{d-1}$ in a $(d-2)$-sphere of radius one. Then the shell between $r S^{d-1}$ and $(r - 1/r) S^{d-1}$ is covered by the balls. Writing m to denote the maximal integer such that $R - m/R > 1$, we deduce by (8.23) that

$$\tilde{\nu}(d, R) \leq 1 + \sum_{i=0}^{m} \widetilde{M}\left(d - 1, \arcsin \frac{1}{R - \frac{i}{R}}\right)$$
$$< c_1 d^{\frac{3}{2}} \ln d \sum_{i=0}^{m} \left(R - \frac{i}{R}\right)^{d-1},$$

where c_1 is an absolute constant. Therefore, (8.26) is a consequence of

$$\sum_{i=0}^{m} \left(R - \frac{i}{R}\right)^{d-1} < R^{d-1} \sum_{i=0}^{m} e^{-\frac{(d-1)i}{R^2}} < \frac{1}{1 - e^{-\frac{d-1}{R^2}}} \cdot R^{d-1} < 3R^{d-1}.$$

If $R \geq \sqrt{d}$ then let $\varrho = 1 - 1/d$. Given $r \geq 1$, let us cover $r S^{d-1}$ with $\widetilde{M}(d - 1, \arcsin \frac{\varrho}{r})$ balls of radius ϱ in such a way that each ball intersects $r S^{d-1}$ in a $(d - 2)$-sphere of radius ϱ. Since $\sqrt{1 - \varrho^2} > 1/\sqrt{d}$, the shell between $r S^{d-1}$ and $(r - 1/\sqrt{d}) S^{d-1}$ is covered by the balls. If we write k to denote the maximal integer such that $R - k/\sqrt{d} \geq 1$, (8.23) yields

$$\tilde{\nu}(d, R) \leq 1 + \sum_{i=0}^{k} \widetilde{M}\left(d - 1, \arcsin \frac{1 - \frac{1}{d}}{R - \frac{i}{\sqrt{d}}}\right)$$
$$< c_2 d^{\frac{3}{2}} \ln d \sum_{i=0}^{k} \left(R - \frac{i}{\sqrt{d}}\right)^{d-1},$$

where c_2 is an absolute constant. We conclude (8.27) by

$$\sum_{i=0}^{k} \left(R - \frac{i}{\sqrt{d}}\right)^{d-1} < \sqrt{d} \int_{0}^{R + \frac{1}{\sqrt{d}}} t^{d-1} dt = \frac{1}{\sqrt{d}} \left(R + \frac{1}{\sqrt{d}}\right)^d < \frac{e}{\sqrt{d}} R^d.$$

$\square$

Let us observe that the upper bounds of (8.24)–(8.27) change continuously as R increases (up to an absolute constant factor). The estimate (8.27) is close to being optimal because

$$\tilde{\nu}(d, R) \geq c_0 d \, R^d \quad \text{if } R \geq d \text{ or } R = \sqrt{d}, \tag{8.28}$$

where c_0 is a positive absolute constant. Here the case $R \geq d$ follows from the Coxeter–Few–Rogers bound (8.3) on the covering density and from Theorem 7.1.1. If $R = \sqrt{d}$ then we simply use the fact that the unit balls covering $R B^d$ cover $R S^{d-1}$ as well; therefore, Lemma 6.8.4 yields (8.28).

Comments. All the results about packings except for the results about $\nu(d, 3)$ are due to N. M. Blachman [Bla1963] (see also W. Kuperberg [Kup?]).

Theorem 8.5.3 is proved by J.-L. Verger-Gaugry [Ver?], who improves on an ealier estimate by C. A. Rogers [Rog1963].

Given a convex polytope C in $\mathbb{R}^d$ and a number n, let us discuss the maximal radius of n equal nonoverlapping balls that are contained in C and the minimal radius of n equal balls that cover C. If n is large then the density of the optimal packing is asymptotically $\delta(B^d)$ (see (8.9)), and the density of the optimal covering is asymptotically $\vartheta(B^d)$ (see (8.10)). Therefore, the interesting case is when the number n is not too large.

We start with packings of n equal balls into the unit cube in $\mathbb{R}^d$. In $\mathbb{R}^3$ the optimal packing is determined by J. Schaer if $n \leq 10$, $n \neq 7$. He solves the cases $n = 5$, $n = 6$, and $n = 10$ in [Sch1966b], [Sch1966c], and [Sch1994], respectively, and the remaining cases in [Sch1966a], and he describes the conjecturally optimal packing of seven balls in [Sch1966a]. In addition, dense packings of eleven to twenty-seven balls in the unit cube are provided by M. Goldberg [Gol1971]. The results in higher dimensions are mostly due to G. Kuperberg and W. Kuperberg [KuK?]: If $n = 2$ then the centres in the extremal arrangement are readily contained in a diameter. In $\mathbb{R}^4$ the optimal radius is the same for $n = 6, 7, 8$, and it is also the same (smaller) value for $n = 16$ and $n = 17$. When the dimension is five then only the case $n = 16$ has been solved besides the case $n = 2$. In even higher dimensions the known solutions are tied with the existence of certain Hadamard matrices: These cases include the one when there exists a Hadamard matrix of rank n and d is multiple of $n - 1$ (the only case that was settled earlier, namely, by S. A. Chepanov, S. S. Ryshkov, and N. N. Yakovlev [CRY1991]), and the case when there exists a Hadamard matrix of rank $2n$ and d is multiple of $2n - 2$. Finally, if there exists a Hadamard matrix of rank d then the optimal radii coincide for $n = d + 2, \ldots, 2d$.

The problem of the maximal number $m(d)$ of nonoverlapping balls of radius $1/4$ in the unit cube has received much attention because $m(d)$ is the maximal number of points in the unit cube whose pairwise distances are all at least one. It was observed by A. Meir (unpublished) that $m(2) = 4, m(3) = 8$, $m(4) = 17$ and

$$\log m(d) = (1/2 + o(1)) d \log d \quad \text{as } d \to \infty.$$

Published proofs of these results are due to V. Bálint and V. Bálint, Jr. [BáB2003].

For covering a three-dimensional unit cube by n equal balls, G. Kuperberg and W. Kuperberg [KuK?] have found the optimal solution if $n = 2, 3, 4, 8$. In addition, they have determined the optimal covering by four balls in $\mathbb{R}^4$. Finally, if $d \geq 4$ and $n = 2$ then the solution is due to N. P. Dolbilin and P. I. Sharygin (personal communication).

Next we consider packings of n equal balls into the d-dimensional regular crosspolytope O^d for $d \geq 3$. The solution is simple if $n = 2$, and the centres in an optimal packing of three balls form a suitable regular triangle. Now if at least four equal balls of certain radius can be packed into O^d then the same can be done even with $2d$ balls of the same radius; in turn, the centres of the $2d$ equal balls in the optimal packing are vertices of a homothetic copy of O^d. Finally, in the optimal packing of $2d + 1$ balls, one ball is centred at the origin, and the other $2d$ are situated near the vertices of O^d. In the three-space these results were proved by G. Golser [Gol1977] if $n = 3, 4, 5, 6$ and by K. Bezdek [Bez1987] if $n = 7$, who also provided alternative proofs if $n \leq 6$. The higher dimensional case was settled by K. Böröczky Jr. and G. Wintsche [BöW2000].

Much less is known about coverings of the d-dimensional regular crosspolytope by n equal balls, but the cases $n = 2, d, 2d$ are solved by K. Böröczky, Jr. and G. Wintsche [BöW?].

Finally, K. Bezdek [Bez1987] determined the maximal radius of n equal nonoverlapping balls in a regular tetrahedron if $n \leq 5$ or $n = 8, 9, 10$.

8.6. Optimal Finite Coverings by Unit Balls

Concerning finite coverings by unit balls, beautiful conjectures are circulated but very little has been proved. Part of the problem is that, given a finite family of unit balls, it is not clear how to find the convex body of maximal volume covered by these balls. Similarly, as in case of packings, it is plausible that optimal finite coverings by balls are either clusters or some sort of sausages, where a sausage is a finite arrangement of unit balls such that the centres are aligned and equally spaced.

First, we discuss the density of coverings of convex bodies by n unit balls in $\mathbb{R}^d$ if n is large. For clusters the density is asymptotically $\vartheta(B^d)$ (see (8.10)). In contrast, asymptotically optimal sausages are obtained by placing the consecutive centres $2/\sqrt{d}$ apart (see G. Fejes Tóth, P. Gritzmann, and J. M. Wills [FGW1984]). These balls cover a right cylinder whose height is $2n/\sqrt{d}$, and whose base is a $(d-1)$-ball of radius $\sqrt{(d-1)/d}$, and the asymptotic density is (see (A.29))

$$\frac{\kappa_d \sqrt{d}}{2\kappa_{d-1}(1 - \frac{1}{d})^{\frac{d-1}{2}}} \sim \sqrt{\frac{\pi e}{2}}.$$

Since $\vartheta(B^d) > \tau_d$ and $\tau_d \sim d/(e\sqrt{e})$ according to the Coxeter–Few–Rogers bound (see Theorem 8.2.1), it follows that sausages are more efficient packings in high dimensions. Checking the known estimates on $\vartheta(B^d)/\kappa_d$ for $d = 3, \ldots, 24$ (see J. H. Conway and N. J. A. Sloane [CoS1999], table 2.1 under the heading "center density") shows that clusters are more efficient than sausages if $d = 3, 4$, and sausages take over starting from dimension five. Similar considerations lead J. M. Wills [Wil1983] to formulate the following:

Conjecture 8.6.1 (Sausage conjecture for coverings). *Let $d \geq 5$ and $n \geq 2$. If D_n is a convex body of maximal volume that can be covered by n unit balls in $\mathbb{R}^d$ then the centres of the corresponding n unit balls are aligned and equally spaced.*

We note that G. Fejes Tóth, P. Gritzmann, and J. M. Wills [FGW1984]) verify that, in $\mathbb{R}^d$, the density of any finite covering by unit balls is at least $1 + 2^{-(1.21d+6.04)}$.

Now we turn to extremal coverings with respect to a mean projection; more precisely, we formulate the corresponding conjecture in terms of the intrinsic volumes $V_i(\cdot), i = 1, \ldots, d-1$. Actually, not only are coverings of a compact convex set D of interest but coverings of the i-*skeleton* of D as well. We recall that the i-skeleton of D consists of the points of D that are not centres of any $(i+1)$-dimensional ball contained in D (see Section 7.4). G. Fejes Tóth, P. Gritzmann, and J. M. Wills [FGW1984] propose the following:

Conjecture 8.6.2 (Sausage–skin conjecture). *Let $1 \leq i \leq d-1$ and $n \geq 2$. If D_n is a compact convex set of maximal intrinsic i-volume whose i-skeleton can be covered by n unit balls then the centres of the corresponding n unit balls are aligned and equally spaced.*

Remark. The analogous conjecture is also proposed when the whole D_n is covered by the n unit balls.

We note that the extremal compact convex set D_n actually exists according to Lemma 7.4.4. Strong evidence for the sausage-skin conjecture is only available for the first intrinsic volume. If $i = 1$ and $d = 2$ then the conjecture is verified by G. Fejes Tóth, P. Gritzmann and J. M. Wills [FGW1984] (see also Theorem 5.7.1). In higher dimensions the "thickness" of a compact convex set D can be conviniently measured by the maximal radius $r_2(D)$ of circular discs in D. We note that if s is a diameter of D then

$$D \subset s + 3r_2(D) \cdot B^d$$

according to (7.10). Now Theorem 7.4.1 and Example 7.4.2 yield

Theorem 8.6.3. *If n is large, and D_n is a compact convex set of maximal mean width whose one-skeleton can be covered by n unit balls in $\mathbb{R}^d$, then*

(i) $\frac{\sqrt{d}}{50n} < r_2(D_n) < \frac{98\sqrt{2\pi d}}{n}$;
(ii) $2n + \frac{d}{6n} \leq V_1(D_n) \leq 2n + \frac{48\pi d}{n}$.

Remark. (i) and (ii) also hold if not only the one-skeleton of D_n but the whole D_n is covered.

If $2 \leq i \leq d - 1$ then only some basic estimates are known concerning the sausage-skin conjecture. G. Fejes Tóth, P. Gritzmann, and J. M. Wills [FGW1984] observed that the asymptotically optimal sausages are obtained by placing the consecutive centres $2/\sqrt{i}$ apart. These balls cover a right cylinder Z_n whose height is $2n/\sqrt{i}$, and whose base is a $(d - 1)$-ball of radius $\sqrt{(i - 1)/i}$; hence,

$$V_i(Z_n) > n \cdot \sqrt{\frac{2}{\pi e}} \sqrt{\frac{i}{d}} \, V_i(B^d).$$

In addition, [FGW1984] verifies that if the i-skeleton of a compact convex set D is covered by n unit balls then

$$V_i(D_n) < n \cdot \left(1 + 2^{-(1.21i + 6.04)}\right)^{-1} V_i(B^d).$$

For thorough references about related problems, see P. Gritzmann and J. M. Wills [GrW1993b] and U. Betke, M. Henk, and J. M. Wills [BHW?].

8.7. Comments on Ball Packings and on Shapes in Nature

Let us introduce two quantities associated with a packing of n unit balls with centres $x_1, \ldots, x_n$ in $\mathbb{R}^3$. The *second moment* is $\sum_i (x_i - c)^2 = 1/n \sum_{i<j} (x_i - x_j)^2$, where $c = (x_1 + \ldots + x_n)/n$ is the centroid of the centres. J. H. Conway, T. D. Duff, R. H. Hardin, and N. J. A. Sloane [CDH1995] constructed packings of $n \leq 32$ unit balls with low second moment, and they conjecture that their arrangements minimize the second moment (see the homepage [Slo] of N. J. A. Sloane for additional arrangements). Given $\varrho > 0$, the *parametric density* of the packing is $nV(B^3)/V(C_n + \varrho B^3)$, where C_n is the convex hull of $x_1, \ldots, x_n$. Parametric density was introduced by J. M. Wills [Wil1993] (see also U. Betke, M. Henk, and J. M. Wills [BHW1994]), and a whole theory is centred around it (see Chapter 10 for results and history). When n is small and ϱ is relatively large then P. Scholl, A. Schürmann, and J. M. Wills [SSW2003] constructed packings with high density.

Investigating the iridescence of opals, V. N. Manoharan, M. T. Elsesser, and D. J. Pine [MEP2003] managed to separate n colloidal microspheres of the common plastic polystyrene for $n = 2, \ldots, 15$. It turns out that for small n the structure of the tiny balls coincides with the conjecturally optimal arrangement of n unit balls with respect to either the second moment or the parametric density.

If n is large then the convex hull of the centres in an optimal packing with respect to the second moment is close to some ball. However, parametric density can be used to model crystals even if n is large: Assuming that the centres are points of a given lattice and ϱ is relatively large, the optimal C_n is asymptotically homothetic to a polytopal Wulff-shape. In many cases this asymptotic shape coincides with the shape of crystals with related parameters (see Section 10.11).

9

Translative Arrangements

This chapter covers quite a number of topics because finite translative arrangements have been rather intensively investigated. Actually, one more topic is parametric density, which is the subject of Chapter 10.

Given a convex body K in $\mathbb{R}^d$, Section 9.1 introduces the associated Minkowski space, namely, the normed space induced by $K_0 = (K - K)/2$. The main body of the chapter starts with density-type problems. We characterize K when some translates of K tile a certain convex body (see Theorem 9.2.1). Then, appealing to methods in Chapter 7, we investigate the asymptotic behavior of optimal packings and coverings of a large number of translates of K in Sections 9.3 and 9.4. We also describe the classical economic periodic packings and coverings by translates of K (see Theorem 9.5.2).

The next topic is the so-called Hadwiger number $H(K)$: the maximal number of nonoverlapping translates of K that touch K. Theorem 9.6.1 says that $\lambda^d < H(K) \leq 3^d - 1$, where $\lambda > 1$ is an absolute constant. The lower bound $H(K) \geq d^2 + d$ is verified as well (see Theorem 9.7.1), which is optimal if $d = 2, 3$. For positive α, Sections 9.8 to 9.10 discuss a natural generalization of the Hadwiger number, that is, the maximal number $H_\alpha(K)$ of nonoverlapping translates of αK touching K. We note that $H_\infty(K)$ is related to antipodal sets and equilateral sets (see Section 9.11).

A family of nonoverlapping translates of K that do not overlap K is called a *translational cloud* if every half line emanating from K intersects the interior of at least one translate. Theorem 9.12.1 provides economic clouds of size $2^{d^2+o(d^2)}$ if K is centrally symmetric and of size $4^{d^2+o(d^2)}$ in general. However, if K is a ball then at least $2^{0.599d^2+o(d^2)}$ copies are needed to form a cloud according to Corollary 9.13.2. Moreover, Section 9.14 discusses sparse clouds, that is, clouds when the translates are far apart.

The final topic is the minimal number of smaller homothetic copies of K that cover K, which is at most 2^d according to a celebrated conjecture of H. Hadwiger and I. Z. Gohberg (see Section 9.15). We show that at least the order of the conjecture is right; more precisely, the minimal number of smaller homothetic copies is at most $5d \ln d \cdot 2^d$ if K is centrally symmetric, and it is at most $5d \ln d \cdot 4^d$ in the general case.

9.1. Minkowski Geometry and the Busemann Surface Area

In this section $|\cdot|$ stands for the Hausdorff $(d-1)$-measure. Given an o-symmetric convex body K in $\mathbb{R}^d$, the associated norm is

$$\|x\|_K = \min\{\lambda \geq 0 : x \in \lambda K\}.$$

In the related *Minkowski geometry*, the distance of x and y is $\|x - y\|_K$. To investigate packings of a not necessarily centrally symmetric convex body C, let $C_0 = (C - C)/2$. Since $x + C$ and $y + C$ overlap if and only if $x + C_0$ and $y + C_0$ overlap, $x + C$ and $y + C$ form a packing if and only if $\|y - x\|_{C_0} \geq 2$.

Next we characterize parallelepipeds that are common extremal bodies in Chapter 9:

Lemma 9.1.1. *If C is a convex body in $\mathbb{R}^d$ and $C_0 = (C - C)/2$ is a parallelepiped then C is a parallelepiped as well.*

Proof. C can be translated in a way that the facets of C_0 touch C, and hence both C and $-C$ are contained in C_0. Therefore, $C = C_0$ follows from $C_0 = (C - C)/2$. □

Given an o-symmetric convex body K in $\mathbb{R}^d$, the *Busemann surface area* of a polytope P with facets $\{F_i\}$ is

$$S_K(P) = \sum_i \frac{|F_i|}{|K \cap \mathrm{lin}\,(F_i - F_i)|}. \tag{9.1}$$

Moreover, the Busemann surface area $S_K(C)$ of any convex body C is the supremum of $S_K(P)$ for inscribed polytopes P. Assuming the presence of an auxiliary Euclidean metric, and choosing a unit exterior normal $u(x)$ at any $x \in \partial C$, we have that

$$S_K(C) = \int_{\partial C} \frac{1}{|u(x)^\perp \cap K|}\, dx. \tag{9.2}$$

In particular, the Busemann surface area of any Jordan measurable subset σ

of ∂C can be defined by the formula

$$S_K(\sigma) = \int_\sigma \frac{1}{\left|u(x)^\perp \cap K\right|} \, dx.$$

If K is the unit Euclidean ball then $\kappa_{d-1} S_K(C)$ coincides with the usual notion of the surface area. Actually, many books call $\kappa_{d-1} S_K(C)$ the Busemann notion of surface area, but we drop the factor κ_{d-1} to simplify various formulae.

According to the celebrated Busemann intersection inequality (which we do not verify), there exists an o-symmetric convex body I^*K satisfying

$$S_K(C) = V(C, I^*K; 1). \tag{9.3}$$

Therefore, $S_K(\cdot)$ is continuous and monotonic.

Now let $C = K$. Since K is o-symmetric, the Brunn–Minkowski inequality (A.44) yields that, among the sections of K orthogonal to a given unit vector u, the area of the central section is maximal. Therefore,

$$\frac{2}{d} h_K(u) \cdot \left|u^\perp \cap K\right| \le V(K) \le 2 h_K(u) \cdot \left|u^\perp \cap K\right|,$$

and equality holds in the upper bound if and only if K is a cylinder whose base is parallel to $u^\perp$. Given a Jordan measurable subset σ of ∂K, let $\text{Cone}(\sigma)$ be the union of all segments ox for $x \in \sigma$. Since

$$V(\text{Cone}(\sigma)) = \frac{1}{d} \int_\sigma h_K(u(x)) \, dx,$$

we deduce that

$$\frac{V(K)}{2d} \cdot S_K(\sigma) \le V(\text{Cone}(\sigma)) \le \frac{V(K)}{2} \cdot S_K(\sigma). \tag{9.4}$$

In particular, taking σ to be the whole ∂K, we conclude

$$2 \le S_K(K) \le 2d, \tag{9.5}$$

where $S_K(K) = 2d$ if and only if K is a parallelepiped.

Comments. The main references for this section are the books by R. Gardner [Gar1995] and A. C. Thompson [Tho1996].

9.2. Finite Translative Tilings

We characterize convex bodies such that there exists a translative finite packing or covering of density one.

Theorem 9.2.1. *Certain $n \geq 2$ translates of a convex body K in $\mathbb{R}^d$ tile some convex body if and only if K is of the form $s + M$, where s is a segment and M is a $(d - 1)$-dimensional compact convex set.*

Remark. We actually verify that if the union of the $n \geq 2$ nonoverlapping translates $x_1 + K, \ldots, x_n + K$ is a convex body C then K is of the form $P + M$, where P is an m-dimensional parallelepiped for some $1 \leq m \leq d$, and M is a $(d - m)$-dimensional compact convex set. In addition, aff$\{x_1, \ldots, x_n\}$ is of dimension m, and $x_1 + P, \ldots, x_n + P$ is an $(m - 1)$-face to $(m - 1)$-face tiling of its union.

Proof. If K is of the form $s + M$ then $ns + M$ is the union of n translates. To prove the converse statement, we prove the remark by induction on the dimension d, where the case $d = 1$ is obvious. Since the part of $\partial(x_1 + K)$ contained in int C is covered by the intersections of $x_1 + K$ with all other $x_i + K$, we deduce that ∂K contains a $(d - 1)$-dimensional convex set. In particular, let u be a (Euclidean) unit vector such that the supporting hyperplane of equation $\langle x, u \rangle = h_K(u)$ intersects K in a $(d - 1)$-dimensional convex set K_0. We define $w = h_K(u) + h_K(-u)$ and write K_α and C_α to denote the intersections of K and C with the hyperplanes of equations $\langle x, u \rangle = h_K(u) - \alpha$ and $\langle x, u \rangle = h_C(u) - \alpha$, respectively.

Now C_0 is tiled, say, by $x_1 + K_0, \ldots, x_k + K_0$. It follows by the induction hypothesis that $K_0 = P_0 + M_0$, where P_0 is an m_0-dimensional parallelepiped and M_0 is a $(d - 1 - m_0)$-dimensional compact convex set, and $m_0 = 0$ if $k = 1$ and $m_0 \geq 1$ if $k \geq 2$. In addition, aff $\{x_1, \ldots, x_k\}$ is of dimension m_0, and $x_1 + P_0, \ldots, x_k + P_0$ is a tiling of its union determined by the edge vectors of P_0. If $k < n$ then we consider the x_i maximizing $\langle x_j, u \rangle$ for $j > k$, and the convexity of C yields that $\langle x_i, u \rangle \leq \langle x_1, u \rangle - w$. Thus, for both $k < n$ and $k = n$, C_α is tiled by $x_1 + K_\alpha, \ldots, x_k + K_\alpha$ for $0 \leq \alpha < w$; hence, $K_\alpha = P_0 + M_\alpha$ for some $(d - 1 - m_0)$-dimensional compact convex set M_α. For any $0 < \alpha < w$ and $0 < \lambda < 1$, readily,

$$\lambda M_\alpha + (1 - \lambda)M_{w-\alpha} \subset M_{\lambda\alpha+(1-\lambda)(w-\alpha)}.$$

We deduce that $K = P_0 + M'$, where M' is a $(d - m_0)$-dimensional compact convex set; namely, M' is the closure of the union of the sets M_α, $0 \leq \alpha < w$.

Therefore, we are done if $k = n \geq 2$; namely, $P_0 = P$ and $M = M'$ in this case. Let $k < n$, and hence C_w, which is tiled by $x_1 + K_w, \ldots, x_k + K_w$, is also tiled say by $x_{k+1} + K_0, \ldots, x_{k+l} + K_0, l \geq 1$. It follows that C_{2w} is tiled by $x_{k+1} + K_w, \ldots, x_{k+l} + K_w$. Since C_w is in the middle between C_0 and

C_{2w}, the Brunn–Minkowski theorem (A.44) yields

$$2 \cdot k^{\frac{1}{d-m_0}} \cdot |K_w|^{\frac{1}{d-m_0}} \geq k^{\frac{1}{d-m_0}} \cdot |K_0|^{\frac{1}{d-m_0}} + l^{\frac{1}{d-m_0}} \cdot |K_w|^{\frac{1}{d-m_0}}. \qquad (9.6)$$

In turn, (9.6) is equivalent to $(k^{1/(d-m_0)} - l^{1/(d-m_0)})^2 \leq 0$ because $k\,|K_w| = l\,|K_0|$. Thus $k = l$, and we deduce by the equality condition for the Brunn–Minkowski theorem that C_0, C_w, and C_{2w} are translates, and hence C_α is a translate of C_0 for any $0 \leq \alpha \leq 2w$. In particular, all K_α are of the same $(d-1)$-content for any $0 \leq \alpha \leq w$, which, in turn, yields again by the equality condition for the Brunn–Minkowski theorem that $K = s + K_0$ for some segment s. Therefore, we may choose $P = P_0 + s$ and $M = M_0$, completing the proof of Theorem 9.2.1. $\qquad\square$

Given convex bodies C and K in $\mathbb{R}^d$ such that int$C$ contains a translate of K, and some translates of K tile C, the remark after Theorem 9.2.1 yields that K is a parallelepiped. In turn, we deduce the following:

Corollary 9.2.2. *Let $\nu > 1$, and let K be a convex body in $\mathbb{R}^d$. If νK can be tiled by translates of K then K is a parallelepiped, the tiling is facet to facet, and ν is an integer.*

Comments. Theorem 9.2.1 is due to P. Gritzmann [Gri1984], and Corollary 9.2.2 was proved earlier by H. Groemer [Gro1961].

9.3. Optimal Arrangements of a Large Number of Translates

We discuss the optimal packings or coverings by n translates of a given convex body K for large n. In this section we do not provide certain proofs, mostly because essentially the same arguments are presented in Chapter 7 for arrangements of congruent copies of K. For example, the proof of Theorem 7.2.2 yields the following:

Theorem 9.3.1. *Given a convex body K in $\mathbb{R}^d$, let D_n be a convex body of minimal volume containing n nonoverlapping translates of K, and let $\widetilde{D}_n$ be a convex body of maximal volume that can be covered by n translates of K. Then there exist $\delta_{\mathrm{fin},T}(K)$ and $\vartheta_{\mathrm{fin},T}(K)$ satisfying*

$$\lim_{n \to \infty} \frac{nV(K)}{V(D_n)} = \delta_{\mathrm{fin},T}(K);$$

$$\lim_{n \to \infty} \frac{nV(K)}{V(\widetilde{D}_n)} = \vartheta_{\mathrm{fin},T}(K).$$

We say that a convex body K in $\mathbb{R}^d$ is a *paralleletop* if K is a centrally symmetric polytope such that each facet is also centrally symmetric, and for each $(d - 2)$-face F, either four or six facets of K have a face that is a translate of F. The following is known (see the comments for references):

Lemma 9.3.2. *Some translates of a convex body K tile $\mathbb{R}^d$ if and only if K is a paralleletop.*

In the rest of the section we use the notation $M|L$ to denote the projection of a compact convex set M into some linear space L, and we use $|M|$ to denote the relative content of M.

Lemma 9.3.3. *Given a convex body K in $\mathbb{R}^d$, either $\delta_{\mathrm{fin},T}(K) = 1$ or $\vartheta_{\mathrm{fin},T}(K) = 1$ holds if and only if $K = P + C$, where P is an m-dimensional paralleletop for some $1 \leq m \leq d - 2$ or $m = d$, and C is a $(d - m)$-dimensional compact convex set.*

Proof. It follows via the proof of Lemma 7.2.3 that there exist a linear m-plane L with either $1 \leq m \leq d - 2$ or $m = d$ and a $(d - m)$-dimensional compact convex set M in $L^{\perp}$ such that $L + M$ can be tiled by translates of K. Let x be an exposed point of M. We place K in such a way that x is a common boundary point of $K|L^{\perp}$ and M, and they share the same exterior normals at x. Then x is an exposed point of $K|L^{\perp}$ as well; hence, L is tiled by the corresponding translates of $P = K \cap L$, which is a paralleletop in L according to Lemma 9.3.2. It also follows that if the projection of a tile into $L^{\perp}$ intersects the relative interior of $K|L^{\perp}$ then the two projections coincide. In particular, any nonempty intersection of K by an affine m-plane parallel to L is of m-content $|P|$, which, in turn, yields Lemma 9.3.3 by the equality case of the Brunn–Minkowski inequality (A.44). $\qquad\square$

We now show that if C and the paralleletop P in Lemma 9.3.3 are sufficiently general then the optimal packing or covering by n translates of K is truly m dimensional for large n. We measure the size of a compact convex set M using $r_m(M)$, which is the radius of the largest m-dimensional Euclidean ball contained in M.

Example 9.3.4. Given $1 \leq m \leq d - 2$ and $m = d$, let $K = P + C$ be a convex body in $\mathbb{R}^d$ such that C is a strictly convex body in $L^{\perp}$ for some linear m-space L; moreover, P is a paralleletop in L, and P cannot be written as the Minkowki sum of two lower dimensional compact convex sets.

(i) If D_n is a convex body of minimal volume that contains n nonoverlapping translates of K then $r_m(D_n)$ tends to infinity, and $D_n|L^\perp$ is a translate of C for large n.

(ii) If $\widetilde{D}_n$ is a convex body of maximal volume that can be covered by n translates of K then $r_m(\widetilde{D}_n)$ tends to infinity, and $\widetilde{D}_n|L^\perp$ is a translate of C for large n.

In particular, the convex hulls of the translation vectors in the optimal arrangements are m dimensional.

We note that a suitable P can be obtained as the Dirichlet–Voronoi cell of an m-lattice Λ in $\mathbb{R}^m$ such that no two nonzero vectors are orthogonal (e.g., the basis satisfies that the altogether m^2 coordinates are algebraically independent). We consider only the case of packings because the argument for coverings is quite analogous.

It is sufficient to verify (i) for any subsequence $D_{n'}$. Since $\delta_{\text{fin},T}(K) = 1$, we may assume that $D_{n'}$ is a subsequence such that the construction of the tiling of Lemma 7.2.3 applies. More precisely, the packings in $D_{n'}$ lead to a tiling of $L' + M$ by translates of K, where L' is a linear m'-plane with $1 \leq m' \leq d - 2$ or $m' = d$, and M is a $(d - m')$-dimensional compact convex set in $L'^\perp$. If $m' = d$ then $m = d$, and we are done. Thus let $m' \leq d - 2$, and let x be an exposed point of M'. We place K in such a way that x is a common boundary point of $K|L'^\perp$ and M', and they share the same exterior normals at x. We deduce that, just as for Lemma 9.3.3, $K = P' + C'$, where P' is a parallelotop in L', and C' is a $(d - m')$-dimensional compact convex set. Thus the conditions on P and C yield that $P' = P$ and $C' = C$. In addition, if the projection of a tile into $L^\perp$ intersects the relative interior of $K|L^\perp$ then the two projections coincide. As C is strictly convex, it also follows that $M = C$. Now for large n', $D_{n'}|L^\perp$ is contained in a translate of $3C/2$, and hence the projections of the tiles into L form a packing. Therefore, the Fubini theorem yields that $V(D_{n'}) \geq |D_{n'}|L| \cdot |C|$, and if equality holds then $D_{n'}|L^\perp$ is a translate of C.

Comments. Lemma 9.3.2 is proved independently by B. A. Venkov [Ven1954] and by P. McMullen [McM1980]. In addition, Example 9.3.4 is due to P. Gritzmann [Gri1984].

9.4. Periodic Arrangements and Clusters

Given a convex body K in $\mathbb{R}^d$, we link economic periodic arrangements to economic clusters. If $\mathcal{K}$ is a periodic arrangement of translates of K with respect to the lattice Λ, and the number of equivalence classes is m, then the density of the arrangement is $m \cdot V(K)/\det \Lambda$. We define the *translative packing density* $\delta_T(K)$ to be the supremum of the densities of periodic translative packings of K, and we let $\Delta_T(K) = V(K)/\delta(K)$. In addition, the *translative covering density* $\vartheta_T(K)$ is the infimum of the densities of periodic coverings by translates of K, and we let $\Theta_T(K) = V(K)/\vartheta(K)$. We note that $\Delta_T(K)$ is the infimum of $V(T)/m$ over all tori T and integers m such that there exists a packing of m translates of K in T, and $\Theta_T(K)$ is the supremum of $V(T)/m$ over all tori T and integers m such that there exists a covering of T by m translates of K. We recall that if D and K are convex bodies and int D contains a translate of K then $D \ominus K$ is the family of x such that $x + K \subset D$, and $D \ominus K$ is a convex body.

Theorem 9.4.1. *Let K and D be convex bodies such that* int D *contains a translate of K. If N denotes the maximal number of nonoverlapping translates of K inside D, and M denotes the minimal number of translates of K covering D, then*

(i) $V(D \ominus K) \le N \cdot \Delta_T(K) \le V(D - K)$;
(ii) $V(D \ominus K) \le M \cdot \Theta_T(K) \le V(D - K)$.

Proof. The proof is presented only for packings because the case for coverings is completely analogous. Given a torus T, we write the same symbol for a convex body in $\mathbb{R}^d$ and its embedded image on T.

We start with the upper bound for $N\Delta_T(K)$. Let $T = \mathbb{R}^d/\Lambda$ be a torus such that $D - K$ can be embedded into T, and let $x_1 + K, \ldots, x_m + K$ be a family of nonoverlapping translates of K of maximal cardinality on T. Since (A.50) yields

$$\int_T \#((D - K) + x) \cap \{x_1, \ldots, x_m\}) \, dx = m \cdot V(D - K), \qquad (9.7)$$

there exists a translate $(D - K) + x$ that contains $k \le mV(D - K)/V(T)$ points out of $x_1, \ldots, x_m$, say, $x_{i_1}, \ldots, x_{i_k}$. Replacing $x_{i_1} + K, \ldots, x_{i_k} + K$ by the N nonoverlapping translates of K contained in $x + D$, we obtain a packing of $m - k + N$ translates of K in T. It follows by the maximality of m that $N \le k$, and therefore

$$V(D - K) \ge N \cdot \frac{V(T)}{m} \ge N \cdot \Delta_T(K).$$

To verify the lower bound for $N \cdot \Delta_T(K)$, let $0 < \lambda < 1$ satisfy that

$$\lambda \cdot \frac{V(D \ominus K)}{\Delta_T(K)} > \left\lceil \frac{V(D \ominus K)}{\Delta_T(K)} \right\rceil - 1.$$

We deduce by the definition of $\Delta_T(K)$ that there exist a torus $T = \mathbb{R}^d/\Lambda$ and m nonoverlapping translates $x_1 + K, \ldots, x_m + K$ of K in T such that $V(T) < \lambda^{-1} \cdot m \Delta_T(K)$, and D can be embedded into T. Now (9.7) for $D \ominus K$ yields a translate $(D \ominus K) + x$ that contains at least $m \cdot V(D \ominus K)/V(T)$ points out of $x_1, \ldots, x_m$. We may assume that these points are $x_1, \ldots, x_k$, and hence $k \geq \lambda \cdot V(D \ominus K)/\Delta_T(K)$ and $x_1 + K, \ldots, x_k + K$ are contained in $x + D$. Therefore, $N \geq k \geq V(D \ominus K)/\Delta_T(K)$ by the definition of λ, completing the proof of Theorem 9.4.1. $\qquad\square$

In the proof of Theorem 9.4.1, we may readily replace $D - K$ with the union of $x_1 + (K - K), \ldots, x_m + (K - K)$. We deduce the following:

Corollary 9.4.2. *If $x_1 + K, \ldots, x_n + K$ are nonoverlapping translates of a convex body K in $\mathbb{R}^d$ then the union of $x_1 + (K - K), \ldots, x_n + (K - K)$ is of volume at least $n \Delta_T(K)$.*

Finally, we consider packings and coverings with respect to Jordan measurable sets:

Corollary 9.4.3. *Given a convex body K and a Jordan measurable set σ with $\operatorname{int} \sigma \neq \emptyset$ in $\mathbb{R}^d$, if $N_\sigma(R)$ is the maximal number of nonoverlapping translates of K that are contained in $R\sigma$, and $M_\sigma(R)$ is the minimal number of translates of K that cover $R\sigma$, then, as R tends to infinity,*

$$N_\sigma(R) \sim \frac{V(\sigma)}{\Delta_T(K)} \cdot R^d \quad \text{and} \quad M_\sigma(R) \sim \frac{V(\sigma)}{\Theta_T(K)} \cdot R^d.$$

Proof. The sets $\Omega_R^+ = R \cdot \operatorname{cl} \sigma$ and $\Omega_R^- = R \cdot \operatorname{int} \sigma$ satisfy that

$$\Omega_R^+ - K = \cup\{x - K : x \in \Omega_R^+\} \quad \text{and} \quad \Omega_R^- \ominus K = \{x : x + K \subset \Omega_R^-\}$$

are compact and open, respectively, and hence are measurable. We deduce by the proof of Theorem 9.4.1 that if R is large then

$$\frac{V(\Omega_R^- \ominus K)}{\Delta_T(K)} \leq N_\sigma(R) \leq \frac{V(\Omega_R^+ - K)}{\Delta_T(K)} \quad \text{and}$$

$$\frac{V(\Omega_R^- \ominus K)}{\Theta_T(K)} \leq M_\sigma(R) \leq \frac{V(\Omega_R^+ - K)}{\Theta_T(K)};$$

thus Corollary 9.4.3 follows by $V(\Omega_R^+ - K) \sim V(\Omega_R^- \ominus K) \sim V(R\sigma)$ (see (A.21)). $\qquad\square$

Comments. Theorem 9.4.1 is due to U. Betke, M. Henk, and J. M. Wills; namely, [BHW1994] discusses the case of packings, and [BHW1995b] discusses the case of coverings (in a somewhat different setting). Corollary 9.4.2 is proved by K. Bezdek and P. Brass [BeB2003], and Corollary 9.4.3 is proved by H. Groemer [Gro1963].

Given a convex body K, $\delta_T(K)$ and $\vartheta_T(K)$ are the maximal and minimal density of any infinite packing and covering, respectively, by translates of K, namely, even if nonperiodic arrangements are allowed (see G. Fejes Tóth and W. Kuperberg [FTK1993], or G. Fejes Tóth [FTG1997]).

9.5. Economic Periodic Packings and Coverings by Translates

Given a convex body K, we show that there exist effective packings and coverings by translates of K. We note that combining Lemma 9.3.2 and the translative version of Lemma 7.2.5 yields the following:

Lemma 9.5.1. $\delta_T(K) = 1$ or $\vartheta_T(K) = 1$ holds for a convex body K in $\mathbb{R}^d$ if and only if K is a paralleletop.

In general we have the following estimates:

Theorem 9.5.2. If K is a convex body in $\mathbb{R}^d$, $d \geq 3$, then

   (i) $\vartheta_T(K) < 4d \ln d$ **(Rogers bound)**;

   (ii) $\delta_T(K) \geq \frac{V(K)}{V(K-K)} > 4^{-d}$ **(Rogers–Shephard bound)**;

   (iii) $\delta_T(K) \geq 2^{-d}$ if $K = -K$ **(Minkowski bound)**.

Remarks. The proofs actually yield stronger statements: If Λ is a lattice such that $\Lambda + (K - K)$ is a packing then there exist translative periodic packing and covering with respect to Λ that witness the preceding bounds. We note that the bounds in Theorem 9.5.2 are of the right order: If K is a ball then $\delta_T(K) \leq 2^{-0.599d+o(d)}$ and $\vartheta_T(K) \geq c\,d$ for some absolute constant c (see Section 8.2).

Proof. Let Λ be a lattice in $\mathbb{R}^d$ such that $\Lambda + (K - K)$ is a packing, and let T^d denote the quotient $\mathbb{R}^d / \Lambda$. We write the same symbol for a convex body in $\mathbb{R}^d$ and its embedded image on T^d.

For packings, the argument is a direct greedy algorithm: We place nonoverlapping translates $x_i + K$, $i = 1, \ldots, n$, on the torus T^d in a way that any translate of K in the torus intersects at least one of $x_1 + K, \ldots, x_n + K$.

Since the translates $\{x_i + K - K\}$ cover T^d, the density of the packing $x_1 + K, \ldots, x_n + K$ on T^d is at least $V(K)/V(K - K)$. Now if K is centrally symmetric then we conclude that $\delta_T(K) \geq 2^{-d}$. In general, the Rogers–Shephard inequality $V(K - K) < 4^d V(K)$ (see (A.45) and (A.46)) yields $\delta_T(K) > 4^{-d}$.

Turning to coverings, we assume that $-K \subset d\, K$ (see (A.19)) and $V(T^d) = 1$. Moreover, let $\eta = 1/(d + 1)$, and let $K_\eta = (1 - \eta)K$.

The first part of the argument is probabilistic. We place points $x_1, \ldots, x_N$ on T^d independently and with uniform distribution with respect to the Lebesgue measure, where

$$N = \left\lfloor d \cdot (1 - \eta)^{-d} V(K)^{-1} \ln \frac{1 - \eta}{\eta} \right\rfloor. \tag{9.8}$$

The part of T^d not covered by $x_1 + K_\eta, \ldots, x_N + K_\eta$ is of expected volume $(1 - V(K_\eta))^N$; hence, there exist translates $x_1 + K_\eta, \ldots, x_N + K_\eta$ such that the volume of the uncovered part is at most $(1 - V(K_\eta))^N$.

Next we place as many nonoverlapping translates of $-\eta\, K$ into the uncovered part of T^d as possible, say translates by $y_1, \ldots, y_M$. If $x \in T^d$ then $x - \eta\, K$ intersects either some $x_i + K_\eta$ or some $y_j - \eta\, K$. Thus, $x \in x_i + K$ in the first case, whereas $\eta = 1/(d + 1)$ and $-K \subset d\, K$ yield that $x \in y_j + K$ in the second case. Therefore, we have obtained a covering of the torus T^d by $n = N + M$ translates of K.

Since $1 - t < e^{-t}$ holds for positive $t < 1$, we deduce that

$$M \leq \frac{(1 - (1 - \eta)^d V(K))^N}{\eta^d V(K)} < \frac{e^{-N \cdot (1 - \eta)^d V(K)}}{\eta^d V(K)},$$

and hence the density of the covering is

$$n \cdot V(K) < N \cdot V(K) + \frac{e^{-N \cdot (1 - \eta)^d V(K)}}{\eta^d}.$$

Differentiating the right-hand side with respect to N suggests choosing N as in (9.8). It follows by $V(K - K) \leq 1$, and by the Brunn–Minkowski inequality $V(K - K) \geq 2^d V(K)$ (see (A.45)), that $V(K) \leq 1/2^d$, and hence

$$n \cdot V(K) < N_0 \cdot V(K) + \frac{e^{-(N_0 - 1) \cdot (1 - \eta)^d V(K)}}{\eta^d}$$

$$< \left(1 - \frac{1}{d + 1} \right)^{-d} \left(d \ln d + e^{\frac{1}{2d}} \right) < 4d \ln d.$$

Therefore, the proof of Theorem 9.5.2 is now complete. $\qquad\square$

Open Problems.

(i) Does there exist a $\lambda > 1/2$ such that $\delta(K) > \lambda^d$ holds for large d and for any centrally symmetric convex body K in $\mathbb{R}^d$?

(ii) Does there exist a $\lambda_0 > 1/4$ such that $\delta(K) > \lambda_0^d$ holds for large d and for any convex body K in $\mathbb{R}^d$?

(iii) Can one replace the Rogers bound in Theorem 9.5.2 (i) with $cd \ln \ln d$ or even with cd for some absolute constant c?

Comments. Let K be a convex body in $\mathbb{R}^d$. The density bounds of Theorem 9.5.2 for packings are essentially due to H. Minkowski. More precisely, he was the one observing that there exists a packing whose density is at least $V(K)/V(K-K)$. A nontrivial improvement is the celebrated Minkowski–Hlawka theorem; namely, there exists a packing lattice Λ such that

$$\det \Lambda \leq \frac{1}{2\zeta(d)} \cdot V(K-K),$$

where $\zeta(d) = \sum_{i=1}^{\infty} 1/i^d$ is the Riemann zeta function. This theorem was proved by H. Minkowski for balls and by E. Hlawka for general convex bodies (see, e.g., P. M. Gruber and C. G. Lekkerkerker [GrL1987]). The current record is due to W. Schmidt [Sch1963a]; namely, if d is large then even

$$\det \Lambda \leq \frac{1}{d \ln \sqrt{2} - c} \cdot \frac{1}{\zeta(d)} \cdot V(K-K)$$

can be assumed, where c is an absolute constant.

The bound of Theorem 9.5.2 for coverings was obtained by C. A. Rogers [Rog1957] (see also C. A. Rogers [Rog1964]); more precisely, he proved the bound $\vartheta_T(K) \leq d \ln d + d \ln \ln d + 5d$ by choosing $\eta = 1/[(d+1)\ln d]$. In addition, P. Erdős and C. A. Rogers [ErR1962] verified the Rogers bound in a very strong form: Given a convex body K in $\mathbb{R}^d$, there exists a periodic covering by translates of K such that no point is covered more than $cd \ln d$ times, where c is an absolute constant.

9.6. The Hadwiger Number

Given a convex body C in $\mathbb{R}^d$, the *Hadwiger number* $H(C)$ is the maximal number of nonoverlapping translates of C that touch C, which is finite because such translates of C are contained in $C - C$. Let us summarize the main properties: The Hadwiger number is affine invariant; namely, if φ is an affine transformation then $H(C) = H(\varphi(C))$. Moreover, $H(C)$ coincides with the Hadwiger number of the o-symmetric body $C_0 = (C-C)/2$ because $x + C$

and $y + C$ overlap if and only if $x + C_0$ and $y + C_0$ do. In turn, the Hadwiger number of C_0 can be expressed in terms of the norm $\| \cdot \|_{C_0}$; namely, it is the maximal cardinality of a set $\{x_i\}$ on ∂C_0 such that $\|x_i - x_j\|_{C_0} \geq 1$ holds for any $i \neq j$.

We have seen in Section 2.9 that the Hadwiger number attains its maximum for parallelograms in the planar case. Now we prove the corresponding upper bound in higher dimensions, and we show that the Hadwiger number is exponential in the dimension.

Theorem 9.6.1. *If C is a convex body in $\mathbb{R}^d$ then*

$$\lambda^d \leq H(C) \leq 3^d - 1,$$

where $\lambda > 1$ is an absolute constant, and equality holds in the upper bound if and only if C is a parallelepiped.

Unfortunately, no explicit λ is known in Theorem 9.6.1 because its proof uses the so-called QS-theorem (see Lemma 9.6.2) whose known proofs do not yield an explicit $c(\varepsilon)$ for a given $\varepsilon > 0$.

Proof of the upper bound of Theorem 9.6.1. Writing $C_0 = (C - C)/2$, we have that any translate of C_0 that touches C_0 is contained in $3C_0$, which, in turn, yields that $H(C) = H(C_0) \leq 3^d - 1$. If $H(C_0) = 3^d - 1$ then C_0 is a parallelepiped according to Corollary 9.2.2, and hence C is a parallelepiped as well by Lemma 9.1.1. $\qquad\square$

To prove the lower bound in Theorem 9.6.1, we take a cd-dimensional projection of a certain section of C so that the projection is almost an ellipsoid, where c is an absolute constant. Therefore, we can use the fact the Hadwiger number of ellipsoids is exponential in the dimension. We say that some o-symmetric compact convex sets M and N are ε-close if $(1 + \varepsilon)^{-1}N \subset M \subset (1 + \varepsilon)N$, and we quote the celebrated QS-theorem (where QS stands for quotient and section) as

Lemma 9.6.2. *Let $\varepsilon > 0$. If K is an o-symmetric convex body in $\mathbb{R}^d$ then there exist linear spaces $L \subset G$ such that $\dim L \geq c(\varepsilon) \cdot d$, and the orthogonal projection of $G \cap K$ into L is ε-close to some ellipsoid in L, where $c(\varepsilon) > 0$ depends only on ε.*

We do not prove the QS-theorem because it belongs to the very heart of the local theory of Banach spaces (see the comments for references).

Proof of the lower bound of Theorem 9.6.1. We may assume that C is o-symmetric. Observing that $\pi/6 < 2\pi/11 < \pi/4$, we choose $\varepsilon > 0$ satisfying $2\sin(2\pi/11) > (1 + \varepsilon)^2$. We apply the QS-theorem to C and ε, and we use the notation of Lemma 9.6.2. Let M be the orthogonal projection of $G \cap C$ into L, and let E be the ellipsoid in L that is centred at o and satisfies $(1 + \varepsilon)^{-1} E \subset M \subset (1 + \varepsilon) E$. Since the Hadwiger number is affine invariant, we may assume that E is a Euclidean unit ball in L. Let $u_1, \ldots, u_n$ be a subset of ∂E of maximal cardinality such that the angle between any two u_i and u_j is at least $4\pi/11$. Since the dimension of E is at least $c(\varepsilon)d$, we deduce $n > (\sin(4\pi/11))^{-c(\varepsilon)d}$ by Theorem 6.3.1. We define $v_i \in \partial M$ by the property that $v_i = t_i u_i$ for some $t_i > 0$, $i = 1 \ldots, n$, and hence the Euclidean distance between different v_i and v_j is at least $[2/(1 + \varepsilon)]\sin(2\pi/11)$, which, in turn, yields that $\|v_i - v_j\|_M > 2$. Therefore, the sets $M + 2v_i$, $i = 1, \ldots, n$, do not overlap, and touch M. Now let $w_i \in \partial(G \cap C)$ be a point whose projection onto L is v_i. It follows that the sets $C + 2w_i$, $i = 1, \ldots, n$, do not overlap nor touch C; thus we may choose $\lambda = (\sin(4\pi/11))^{-c(\varepsilon)}$. $\qquad\square$

Open Problems.

   (i) Does there exist an absolute constant $\lambda < 3$ such that $H(C) < \lambda^{d+o(d)}$ holds for any strictly convex body C in $\mathbb{R}^d$?

   We note that I. Talata [Tal?c] constructs a strictly convex body C satisfying $H(C) > \sqrt{7}^d/4$.

   (ii) Determine the Hadwiger number of the crosspolytope in $\mathbb{R}^4$.

   (iii) Determine the Hadwiger number of a simplex in $\mathbb{R}^4$.

   (iv) Does there exist a convex body in $\mathbb{R}^d$ whose Hadwiger number is $3^d - 2$? This problem is due to L. Fejes Tóth.

Comments. H. Hadwiger [Had1957b] posed the problem of finding the maximal number $H(C)$ of nonoverlapping translates of a convex body C that touch C. Following L. Fejes Tóth, this maximal number is called the Hadwiger number. The naturalness of this notion is shown by the fact that various fundamental results were rediscovered several times.

The fact that the Hadwiger number in $\mathbb{R}^d$ is at most $3^d - 1$ was pointed out already in H. Hadwiger [Had1957b]. It was verified by H. Groemer [Gro1961] that parallelepipeds are the only extremal bodies. The exponential lower bound of Theorem 9.6.1 is due to J. Bourgain (see Z. Füredi and P. A. Loeb [FüL1994]) and was rediscovered independently by I. Talata [Tal1998a]. The QS-theorem (see Lemma 9.6.2) is due to V. Milman [Mil1985].

Even the range of the values of the Hadwiger number has attracted much attention. Let $h(d)$ be the minimal value of the Hadwiger number in $\mathbb{R}^d$, which is at least $d^2 + d$ (see Theorem 9.7.1), and exponential in d (see Theorem 9.6.1). A. Pór and I. Talata [PóT?] showed that given any two consecutive integers m and $m + 1$ satisfying $h(d) \le m, m + 1 \le 3^d - 1$, one of them is the Hadwiger number of some convex body in $\mathbb{R}^d$. In $\mathbb{R}^3$, I. Talata [Tal?a] actually showed that any even number that is at least twelve and at most twenty-six occurs as a Hadwiger number. Contrary to a conjecture in B. Grünbaum [Grü1961], I. Talata [Tal?b] exhibits a three-dimensional convex body whose Hadwiger number is odd, namely, seventeen. This phenomenon is surprising in light of the fact that it is sufficient to consider o-symmetric convex bodies.

Finally, we discuss some specific convex bodies. Ch. Zong [Zon1997a] and I. Talata [Tal?c] provide estimates and formulae for the Hadwiger number of the Cartesian product of two convex compact sets. The Hadwiger number of ellipsoids coincides with the so-called Newton number or kissing number of unit balls, an intensively investigated notion discussed thoroughly in Section 8.1. Next there are numerous three-polytopes whose Hadwiger number have been determined. I. Talata [Tal1999c] and D. Larman and C. Zong [LaZ1999] verified independently that the Hadwiger number of the octahedron is eighteen. The same number eighteen occurs for the tetrahedron, according to I. Talata [Tal1999b] who also verified that the optimal arrangement is unique. Actually, Ch. Zong [Zon1996b] showed earlier that the Hadwiger number of a tetrahedron is either eighteen or nineteen, and hence any tetrahedron has at most eighteen neighbours in a lattice packing. This last fact, in turn, can be proved directly in a rather simple way (see I. Talata [Tal1999c]). It is interesting to note that in the densest lattice packing of the octahedron, or of the tetrahedron, each translate is in contact only with fourteen other translates (see P. Gruber and K. Lekkerkerker [GrL1987]). Moreover, D. Larman and C. Zong [LaZ1999] proved that the Hadwiger number is eighteen for the rhombic dodecahedron and for the elongated octahedron. In high dimensions D. Larman and C. Zong [LaZ1999] give nontrivial bounds for the Hadwiger number of the L_p-balls, $p \ge 1$, namely, the set of points $(x_1, \ldots, x_d) \in \mathbb{R}^d$ such that $\sum_{i=1}^{p} |x_i|^p \le 1$. Finally, I. Talata [Tal2000b] provides an explicit lower bound exponential in d for the Hadwiger number of a d-simplex.

9.7. Lower Bound for the Hadwiger Number in Low Dimensions

Since the exponential lower bound in Theorem 9.6.1 is not explicit, we provide a lower bound for the Hadwiger number that is exact in dimensions two and three.

Theorem 9.7.1. *If C is a convex body in $\mathbb{R}^d$ then*

$$H(C) \geq d^2 + d.$$

Remark. The lower bound of Theorem 9.7.1 is optimal if $d = 2, 3$, as circular discs and three-balls (see (8.1)) show. However, most probably better lower bounds exist if $d \geq 4$.

Proof. We may assume that C is o-symmetric. Let Λ be a packing lattice of minimal determinant for $C/2$, or equivalently, let the origin o be the only lattice point in int$C$ and det Λ be minimal under this condition. Writing $\pm p_1, \ldots, \pm p_N$ to denote the family lattice points on ∂C, we claim that

$$N \geq \frac{d^2 + d}{2}. \tag{9.9}$$

We suppose that $N < (d^2 + d)/2$, and we seek a contradiction. For $i = 1, \ldots, N$, let u_i be an exterior unit normal to C at p_i, and let $A = [a_{kl}]$ be a nonzero $d \times d$ matrix such that

$$\langle A p_i, u_i \rangle = 0 \quad \text{and} \quad a_{kl} = a_{lk} \tag{9.10}$$

hold for $i = 1, \ldots, N$ and for $k < l$. This is possible, because A has d^2 entries, and (9.10) poses $N + \binom{d}{2} < d^2$ homogeneous linear conditions on the entries.

Given $t \in \mathbb{R}$, we define $\Lambda' = (I + tA)\Lambda$ and $p_i' = (I + tA)p_i$ for $i = 1, \ldots, N$, where I is the $d \times d$ identity matrix. Then $p_i' \notin \text{int}C$ by $p_i' - p_i = tAp_i$ and by (9.10). If $|t|$ is small then $\Lambda' \cap \text{int}C = \{o\}$ because the points of Λ different from o, $\pm p_1, \ldots, \pm p_N$ lie outside a small neighbourhood of C. We deduce by the minimality of det Λ that det $\Lambda' \geq$ det Λ, and hence $f(t) \geq 1$ holds for the function

$$f(t) = \det(I + tA) = 1 + C_1 \cdot t + \ldots C_d \cdot t^d$$

if $|t|$ is small. In particular,

$$C_1 = \sum_k a_{kk} = 0 \quad \text{and} \quad C_2 = \sum_{k<l} a_{kk}a_{ll} - a_{kl}a_{lk} \geq 0.$$

Now $a_{kl} = a_{lk}$ yields that

$$-\sum_{k,l} a_{kl}^2 = 2C_2 - C_1^2 \geq 0.$$

This contradicts the condition that A has some nonzero entry, and hence $N \geq (d^2 + d)/2$. In turn, we conclude Theorem 9.7.1. $\quad\square$

Comments. Theorem 9.7.1 and its proof are due to H. P. F. Swinnerton–Dyer [Swi1953]. The proof shows that if C is an o-symmetric convex body in $\mathbb{R}^d$, and Λ is a packing lattice of minimal determinant, then at least $d^2 + d$ translates of C by vectors in Λ touch C. This statement is close to being optimal because if C is a typical convex body then C is touched by at most $2d^2$ translates of C by vectors in Λ according to P. M. Gruber [Gru1986].

9.8. The Generalized Hadwiger Number

Given a convex body C in $\mathbb{R}^d$ and an $\alpha > 0$, we present two equivalent definitions of the *generalized Hadwiger number* $H_\alpha(C)$:

- $H_\alpha(C)$ is the maximal number of nonoverlapping translates of αC touching C;
- $H_\alpha(C)$ is the maximal number of points $x_1, \ldots, x_n \in \partial(C - \alpha C)$ such that $\|x_i - x_j\|_{C_0} \geq 2\alpha$ for any $i \neq j$, where $C_0 = (C - C)/2$.

We note that if $\alpha \neq 1$ then $H_\alpha(C)$ and $H_\alpha(C_0)$ may differ (see Example 2.10.5). In this section we generalize the upper bound $3^d - 1$ of Theorem 9.6.1.

Theorem 9.8.1. *If C is a convex body in $\mathbb{R}^d$ and $\alpha > 0$ then*

$$H_\alpha(C) \leq \frac{(1 + 2\alpha)^d - 1}{\alpha^d},$$

with equality if and only if C is a parallelepiped and $1/\alpha$ is an integer.

Proof. We define $C_0 = (C - C)/2$ and

$$M = \left(1 - \frac{\alpha}{2}\right) \cdot C - \frac{\alpha}{2} \cdot C.$$

Let $x_1, \ldots, x_n \in \partial(C - \alpha C)$ satisfy that $\|x_i - x_j\|_{C_0} \geq 2\alpha$ for $i \neq j$. In particular, the sets $x_i + \alpha C_0$ do not overlap, and each of them touches M and is contained in $M + 2\alpha C_0$. We deduce $V(M, C_0; i) \leq V(C_0, C; i)$ for $i = 1, \ldots, d$ by $C_0 = (M - M)/2$ and by the generalized Brunn–Minkowski inequality (A.47); therefore,

$$n \leq \frac{V(M + 2\alpha C_0) - V(M)}{\alpha^d V(C_0)} \leq \frac{V(C_0 + 2\alpha C_0) - V(C_0)}{\alpha^d V(C_0)} = \frac{(1 + 2\alpha)^d - 1}{\alpha^d}.$$

Assuming that equality holds in Theorem 9.8.1, C_0 and the translates $x_i + \alpha C_0$, $i = 1, \ldots, n$ together tile $C_0 + 2\alpha C_0$. Since C_0 lies in the interior and it can be separated by hyperplanes from the translates of αC_0, we deduce that C_0 is a polytope.

If F is a facet of C_0 then the facet $(1 + 2\alpha)F$ of $(1 + 2\alpha)C_0$ is tiled by some translates of αF; hence, Corollary 9.2.2 yields that F is a $(d-1)$-parallelepiped and $1/\alpha$ is an integer. It also follows that there exists a translate of αF in the tiling of $(1 + 2\alpha)F$ that lies in the relative interior of $(1 + 2\alpha)F$. Now the interior of the corresponding translate of αC_0 can be strictly separated from the other translates of αC_0 intersecting $(1 + 2\alpha)F$, and hence each pair of opposite $(d-2)$-faces of F is contained in some pair of parallel supporting hyperplanes of C_0. Thus, C_0 is a parallelepiped, which, in turn, yields that C is a parallelepiped by Lemma 9.1.1. $\qquad\qquad\qquad\qquad\qquad\square$

Comments. The generalized Hadwiger number was first considered for polytopes by L. Fejes Tóth in [FTL1970] and [FTL1975a]. Theorem 9.8.1 is proved by H. Groemer [Gro1961] and later independently by V. Boju and L. Funar [BoF1993] for centrally symmetric bodies and by K. Böröczky Jr., D. Larman, S. Sezgin, and C. Zong [BLS2000] in the general case.

9.9. Lower Bound for the Generalized Hadwiger Number

We fix $\alpha > 0$ and seek the true order of $H_\alpha(C)$ if C is a convex body in $\mathbb{R}^d$ and d is large. We have just seen in Theorem 9.8.1 that $H_\alpha(C)$ is at most exponential in d. To have an idea what lower bound to expect, let us consider the case when C is a ball. If $\alpha = \sqrt{2} + 1$ then $H_\alpha(C) = 2d$ according to Theorem 6.2.1, whereas if $\alpha < \sqrt{2} + 1$ then $H_\alpha(C)$ is exponential in d according to Theorem 6.3.1. Now we show that the second phenomenon holds not only for balls but for any centrally symmetric convex body.

Theorem 9.9.1. *For $\alpha > 0$, there exists $\lambda > 1$ depending on α such that if C is a convex body in $\mathbb{R}^d$ then*

(i) $H_\alpha(C) > \lambda^d$ in case $\alpha < \sqrt{2}$;
(ii) $H_\alpha(C) > \lambda^d$ in case C is o-symmetric and $\alpha < \sqrt{2} + 1$.

Proof. The argument is a variation of the proof of the lower bound for the Hadwiger number in Theorem 9.6.1. Let $K = (C - C)/2$. In what follows we use the notation of Lemma 9.6.2 (the QS-theorem) for K and for certain ε that we specify later. We write m to denote $\dim L > c(\varepsilon)\, d$ and use C' and K' to denote the projections of $C \cap G$ and $K \cap G$ into L, respectively. In addition, we may assume by affine invariance that

$$\frac{1}{1 + \varepsilon}\, B^m \subset L \cap K' \subset (1 + \varepsilon)\, B^m,$$

where B^m is the unit ball in L.

If C is o-symmetric and $\alpha < 1 + \sqrt{2}$ then let $\varepsilon > 0$ satisfy that $(1 + \varepsilon)^2 2\alpha/(1 + \alpha) < \sqrt{2}$. According to Theorem 6.3.1, we may choose $k \geq \lambda_0^m$ points $x_1, \ldots, x_k \in \partial B^m$ so that the Euclidean distance between any two is at least $(1 + \varepsilon)^2 2\alpha/(1 + \alpha)$, where $\lambda_0 > 1$ depends only on α. Let $y_i = t_i x_i$ be contained in $\partial(C' - \alpha C')$ for $t > 0$; hence, $\|y_i - y_j\|_{K'} \geq 2\alpha$ for any $i \neq j$. We write z_i to denote a point of $G \cap \partial(C - \alpha C)$ whose projection into L is y_i, $i = 1, \ldots, k$, which satisfy $\|z_i - z_j\|_K \geq 2\alpha$ for any $i \neq j$. In particular, $\lambda = \lambda_0^{c(\varepsilon)}$ works for (ii).

Turning to (i), we may assume that $1 < \alpha < \sqrt{2}$ and $2K \subset C - \alpha C$. Now let $\varepsilon > 0$ satisfy $(1 + \varepsilon)^2 \alpha < \sqrt{2}$. We choose $k \geq \lambda_1^m$ points $x_1, \ldots, x_k \in \partial B^m$ so that the Euclidean distance between any two is at least $(1 + \varepsilon)^2 \alpha$, where $\lambda_1 > 1$ depends only on α (see Theorem 6.3.1). Let again $y_i = t_i x_i$ be contained in $\partial(C' - \alpha C')$ for $t_i > 0$. Since $[2/(1 + \varepsilon)]B^m \subset C' - \alpha C'$, it follows that the Euclidean distance between any two y_i is at least $(1 + \varepsilon)2\alpha$; hence, $\|y_i - y_j\|_{K'} \geq 2\alpha$ for $i \neq j$. Finally, we finish the argument as before, completing the proof Theorem 9.9.1. $\qquad\square$

Comments. For centrally symmetric convex bodies, Theorem 9.9.1 is proved independently by J. Bourgain (see Z. Füredi and P. A. Loeb [FüL1994]) and by I. Talata [Tal1998a]. The case of Theorem 9.9.1 about bodies that are not centrally symmetric is a new result.

It is rather annoying that, in Theorem 9.9.1, we do not have the exponential lower bound for any convex body C and for any $\alpha < \sqrt{2} + 1$. Our argument breaks down because if C is not centrally symmetric then it is not clear whether the generalized Hadwiger number of any section is at most the generalized Hadwiger number of C. Actually, the QS-theorem (Lemma 9.6.2) does hold for non-centrally symmetric bodies as well. First, M. Rudelson [Rud2000] verified the QS-theorem for a fixed (unknown) positive ε_0 in place of ε. Given any $\varepsilon > 0$, repeated projections lead to the analogue of Lemma 9.6.2, using, say, the method of Y. Gordon, O. Guédon, and M. Meyer [GGM1998].

9.10. The Asymptotics of the Generalized Hadwiger Number

Given a convex body C in $\mathbb{R}^d$, our aim is to determine the asymptotics of $H_\alpha(C)$ as α tends to zero. First, we recall some basic facts. Writing $|\cdot|$ to denote the relative content of compact convex sets, if M is a $(d-1)$-dimensional o-symmetric compact convex set in $\mathbb{R}^{d-1}$ then we have

$$|M| \leq \Delta_T(M) \leq 2^{d-1}|M| \tag{9.11}$$

according to Theorem 9.5.2. In addition, let σ be a Jordan measurable subset of $\mathbb{R}^{d-1}$. If $m_\sigma(\alpha)$ denotes the maximal number of points contained in σ

whose mutual distances are at least 2α with respect to $\|\cdot\|_M$, where α tends to zero, then Corollary 9.4.3 yields

$$m_\sigma(\alpha) \sim \frac{|\sigma|}{\Delta_T(M)} \cdot \frac{1}{\alpha^{d-1}}. \tag{9.12}$$

It is instructive to discuss the asymptotics of $H_\alpha(P)$ for a polytope P in $\mathbb{R}^d$. Let $P_0 = (P - P)/2$, and we write $F_1, \ldots, F_k$ to denote the facets of P and u_i to denote the exterior unit normal for F_i. If $x + \alpha P$ and $y + \alpha P$ touch P in relative interior points of some F_i then $x - y \in u_i^\perp$, and naturally $\|x - y\|_{u_i^\perp \cap P_0} \geq 2\alpha$. Now, for small α, the number of translates of αC around the $(d-2)$-faces of C is $O(\alpha^{-(d-2)})$, where the implied constant in $O(\cdot)$ depends on P. Therefore, (9.12) yields

$$H_\alpha(P) \sim \sum_{i=1}^{k} \frac{|F_i|}{\Delta_T\left(u_i^\perp \cap P_0\right)} \cdot \frac{1}{\alpha^{d-1}} \tag{9.13}$$

as α tends to zero. We generalize this formula to any convex body:

Theorem 9.10.1. *If C is a convex body in $\mathbb{R}^d$ and $C_0 = (C - C)/2$ then*

$$H_\alpha(C) \sim \int_{\partial C} \frac{1}{\Delta_T\left(u(x)^\perp \cap C_0\right)} \, dx \cdot \frac{1}{\alpha^{d-1}}$$

holds as α tends to zero, where $u(x)$ is a unit exterior normal at $x \in \partial C$.

To prove Theorem 9.10.1, we attempt to imitate the proof of (9.13); hence, our goal is to subdivide the boundary of C into sufficiently flat patches. The basic tool is the Vitali covering theorem:

Proposition 9.10.2. *Let σ be an open Jordan measurable subset of ∂K for a convex body K in $\mathbb{R}^d$, and let $\{U_j\}$ be an open cover of σ on ∂K. Given $\varepsilon > 0$, there exist finite family $\{\sigma_i\}$ of Jordan measurable and pairwise disjoint subsets of σ such that any σ_i is closed and is contained in some U_j; moreover, $\sigma \setminus \bigcup_i \sigma_i$ is of $(d-1)$-measure at most ε.*

Proof. Using radial projection, the problem can be transferred to the boundary of some d-cube; hence, we may assume that $\sigma \subset \mathbb{R}^{d-1}$.

We place a countable family $\mathcal{F}$ of pairwise disjoint closed balls into σ such that their union covers all σ but a part of measure at most $\varepsilon/2$, and any ball from $\mathcal{F}$ is contained in some U_j. This is done step by step. For the first step, we place any ball into $\sigma \cap U_j$ for some U_j. For the kth step, $k \geq 2$, we choose an $r_k > 0$ such that, writing Σ_k to denote the set of points whose distance from the union of the boundaries of the balls already in place is at most r_k, we have $\mathcal{H}^{d-1}(\Sigma_k) < \varepsilon/2^k$, and $\mathcal{H}^{d-1}(\Sigma_k)$ is less than half of the measure of the uncovered part of σ. We colour Σ_k red, and we place a maximal family of pairwise disjoint balls with radius r_k into the yet uncovered part of σ so that each ball is contained in some U_j.

Now the part of σ that is neither painted red nor covered by $\mathcal{F}$ has measure zero. Since $\sum_{k\geq 2} \varepsilon/2^k = \varepsilon/2$, a finite subfamily of $\mathcal{F}$ can be chosen with the required properties.   □

Next we state a somewhat technical observation. If σ is a Jordan-measurable subset of the boundary of a convex body C then we write $\mathcal{N}^+(\sigma, t)$ to denote the family of points x in $\mathbb{R}^d$ such that the closest point y of C to x lies in σ and $d(x, y) \leq t$.

Proposition 9.10.3. *If C is a convex body in $\mathbb{R}^d$, and σ is a Jordan-measurable subset of ∂C then*

$$\mathcal{H}^d(\mathcal{N}^+(\sigma, t)) \leq |\sigma| \cdot t + \sum_{i=2}^{d} \kappa_i \, V_{d-i}(C) \cdot t^i.$$

Proof. For any polytope $P \subset C$, we write σ_P to denote the family of closest points of P to the points of σ. Since the closest point map does not increase the distance (see (A.3)), we deduce that $|\sigma_P| \leq |\sigma|$ (see (A.31)). Given any $\varepsilon > 0$, if P is suitably close to C then $\mathcal{N}^+(\sigma, t) \subset \mathcal{N}^+(\sigma_P, t + \varepsilon)$ and $V_{d-i}(C) < V_{d-i}(P) + \varepsilon$ for any $i = 2, \ldots, d$. In particular, we may assume that C is a polytope; hence, the points of $\mathcal{N}^+(\sigma, t)$ such that the closest point of C lies in a $(d-2)$-face are of d-measure at most $\sum_{i=2}^{d} \kappa_i \, V_{d-i}(C) \cdot t^i$ (see Section A.11). In turn, we conclude Proposition 9.10.3.   □

Proof of Theorem 9.10.1. It is sufficient to prove that, given a convex body C, for any $0 < \varepsilon < \varepsilon_0$ there is $\alpha_0 > 0$ such that if $0 < \alpha < \alpha_0$ then

$$H_\alpha(C) = (1 + O(\varepsilon)) \int_{\partial C} \frac{1}{\Delta_T(u(x)^\perp \cap C_0)} \, dx \cdot \frac{1}{\alpha^{d-1}}, \tag{9.14}$$

where $\varepsilon_0 > 0$ and both ε_0 and the implied constant in $O(\cdot)$ depend on C.

Since both $H_\alpha(C)$ and the integral expression (9.14) are affine invariant, we may assume by the John theorem (A.8.3) that

$$B^d \subset C \subset d \, B^d. \tag{9.15}$$

In particular, if $x \in \partial C$ then

$$\langle u(x), x \rangle \geq \frac{\|x\|}{d}. \tag{9.16}$$

The proof of (9.14) is based on constructing almost flat patches on the boundary of C; namely, we will construct supporting hyperplanes $H_1, \ldots, H_k$ of C and open Jordan measurable subsets $\sigma_1, \ldots, \sigma_k$ of ∂C whose closures are pairwise disjoint and satisfy the following properties:

(i) $\mathcal{H}^{d-1}(\partial C \backslash (\sigma_1 \cup \ldots \sigma_k)) < \varepsilon$;

(ii) if $x \in \sigma_i$ and H_i touches at x_i then

$$\Delta_T(u(x)^\perp \cap C_0) = (1 + O(\varepsilon)) \cdot \Delta_T(u(x_i)^\perp \cap C_0);$$

(iii) if, for $\alpha < \alpha_0$, the radial projections of $x, y \in \partial(C - \alpha C)$ into ∂C land in σ_i then their radial projections x', y' into H_i satisfy

$$(1 + \varepsilon)^{-3} \cdot \|x - y\|_{C_0} < \|x' - y'\|_{C_0} < (1 + \varepsilon)^3 \cdot \|x - y\|_{C_0};$$

(iv) if $x \in \sigma_i$ then the radial projection x' of x into H_i satisfies

$$\|x'\|_{C_0} < (1 + \varepsilon) \cdot \|x\|_{C_0}.$$

Before constructing the objects in (i), ..., (iv), let us show that these properties yield (9.14). We write F_i to denote the radial projection of σ_i into H_i. We deduce by (9.11) and (9.15) that $c_1 < \Delta_T(u^\perp \cap C_0) < c_2$ for any $u \in S^{d-1}$, where c_1 and c_2 are positive constants depending on d. Therefore,

$$\int_{\partial C} \frac{1}{\Delta_T(u(x)^\perp \cap C_0)} \, dx = (1 + O(\varepsilon)) \cdot \sum_{i=1}^k \frac{1}{\Delta_T(u_i^\perp \cap C_0)} \cdot |F_i|. \tag{9.17}$$

First, we prove that

$$H_\alpha(C) \geq (1 - O(\varepsilon)) \cdot \int_{\partial C} \frac{1}{\Delta_T(u(x)^\perp \cap C_0)} \, dx \cdot \frac{1}{\alpha^{d-1}}. \tag{9.18}$$

For small α, let $x_1', \ldots, x_{m_i(\alpha)}' \in F_i$ satisfy that $\|x_t' - x_s'\|_{C_0} \geq 2(1 + \varepsilon)^2$ holds for $t \neq s$, and $m_i(\alpha)$ is maximal with this property. Then

$$m_i(\alpha) \geq (1 - \varepsilon) \cdot \frac{|F_i'|}{\Delta_T(u_i^\perp \cap C_0)} \cdot \frac{1}{\alpha^{d-1}}$$

holds according to (9.12). We write x_t to denote the radial projection of x_t' into $\partial(C - \alpha C)$, and hence $\|x_t - x_s\|_{C_0} \geq 2\alpha$ for any $t \neq s$ by (iii). Therefore, (9.17) yields (9.18).

Next, we verify

$$H_\alpha(C) \leq (1 + O(\varepsilon)) \cdot \int_{\partial C} \frac{1}{\Delta_T(u(x)^\perp \cap C_0)} \, dx \cdot \frac{1}{\alpha^{d-1}}. \quad (9.19)$$

We write σ for the complement of $\sigma_1 \cup \ldots \cup \sigma_k$ on ∂C. Since σ is Jordan measurable, there exists an open neighbourhood σ_0 of σ in ∂C whose $(d-1)$-measure is at most 3ε. Now let $x_1, \ldots, x_m \in \partial(C - \alpha C)$ satisfy $\|x_t - x_s\|_{C_0} \geq 2\alpha$ for any $t \neq s$. If the radial projection x'_t of x_t into ∂C lands in σ then $d(x_t, x'_t) \leq d^2\alpha$ according to (9.15) and (9.16), and hence $x_t + \alpha C_0$ lies in $\mathcal{N}^+(\sigma_0, 2d^2\alpha)$ for small α. In particular, Proposition 9.10.3 yields that if α is small then the number $m_0(\alpha)$ of points x_t whose radial projection into ∂C lands in σ is

$$m_0(\alpha) \leq \frac{V\left(\mathcal{N}^+(\sigma_0, 2d^2\alpha)\right)}{V(\alpha C_0)} \leq O(\varepsilon) \cdot \frac{1}{\alpha^{d-1}}. \quad (9.20)$$

If the radial projections of x_s and x_t into ∂C both land in σ_i with $i \geq 1$ then their radial projections x'_s and x'_t, respectively, into F_i satisfy $\|x'_s - x'_t\|_{C_0} \geq (1 + \varepsilon)^{-3} 2\alpha$ according to (iii). Therefore, (9.12) yields for the number $m_i(\alpha)$ of such x_t that

$$m_i(\alpha) \leq (1 + O(\varepsilon)) \cdot \frac{|F_i|}{\Delta_T(u_i^\perp \cap C_0)} \cdot \frac{1}{\alpha^{d-1}}.$$

We conclude (9.19) by (9.17) and (9.20).

Finally, we construct the objects in (i)–(iv). For any smooth $x \in \partial C$, we write $H(x)$ to denote the tangent hyperlane at x, and we choose a convex cone $K(x)$, $x \in \text{int } K(x)$, with the following properties: For a suitable positive h and for $y \in \partial C \cap K(x)$,

- $\|u(x) - u(y)\| \leq h/2$;
- $\|y'\| \leq (1 + \varepsilon) \cdot \|y\|$ holds for the radial projection y' of y into $H(x)$;
- $\langle y, u(x) \rangle \geq \|y\|/(2d)$.

The last property is possible because of (9.16). The first condition on h is that whenever $\|u - v\| \leq h$ holds for unit vectors u, v then

$$(1 + \varepsilon)^{-1} \cdot \delta_T(u^\perp \cap C_0) \leq \delta_T(v^\perp \cap C_0) \leq (1 + \varepsilon) \cdot \delta_T(u^\perp \cap C_0). \quad (9.21)$$

The second condition is somewhat more complicated and is related to (iii): If the unit vector u, the point $z \in u^\perp$, and the vector w satisfy that

$$|\langle u, w \rangle| \leq h \cdot \|w\| \quad \text{and} \quad |\langle u, z - w \rangle| \geq \frac{\|z - w\|}{3d}$$

then

$$(1 + \varepsilon)^{-1} \cdot \|w\|_{C_0} \leq \|z\|_{C_0} \leq (1 + \varepsilon) \cdot \|w\|_{C_0}. \tag{9.22}$$

Now the union of int $K(x) \cap \partial C$ covers all ∂C but a set of $(d - 1)$-measure zero as x runs through the smooth points of ∂C (see Section A.10). Therefore, the Vitali covering theorem (Proposition 9.10.2) yields smooth points $x_1, \ldots, x_k$, and pairwise disjoint open Jordan-measurable subsets $\sigma_1, \ldots, \sigma_k$ of ∂C such that the closure of σ_i lies in int $K(x_i)$, and the complement of $\sigma_1 \cup \ldots \cup \sigma_k$ is of $(d - 1)$-measure at most ε. Therefore (i), (ii), and (iv) readily hold, and all we have to check is (iii). For small α, let the radial projections of $x, y \in \partial(C - \alpha C)$ into ∂C land in σ_i. We observe that pos $\{x, y\}$ intersects $\partial(C - \alpha C)$ in a convex arc, and let p be the farthest point of this arc from the segment xy. Then there exists an exterior unit normal v to $C - \alpha C$ at p that is orthogonal to $x - y$. Since $\|u(x_i) - v\| < h$, we deduce that

$$\langle u(x_i), x - y \rangle| \leq h \cdot \|x - y\|.$$

However, $|\langle u(x_i), y \rangle| \geq \|y\|/(3d)$. Thus, (9.22) yields that the radial projection y'' of y into the hyperplane that is parallel to $H(x_i)$ and passes through x satisfies

$$(1 + \varepsilon)^{-1} \cdot \|x - y\|_{C_0} \leq \|x - y''\|_{C_0} \leq (1 + \varepsilon) \cdot \|x - y\|_{C_0}.$$

In turn, we conclude (iii) and hence Theorem 9.10.1 as well. $\square$

Let us investigate the expression

$$\Omega(C) = \int_{\partial C} \frac{1}{\Delta_T(u(x)^\perp \cap C_0)} \, dx$$

occurring in Theorem 9.10.1. The term $\Omega(C)$ is closely related to the Busemann surface area with respect to $C_0 = (C - C)/2$ (see Section 9.1); namely,

$$\Omega(C) \leq S_{C_0}(C) = \int_{\partial C} \frac{1}{|u(x)^\perp \cap C_0|} \, dx.$$

Since $S_{C_0}(C) = V(C, I^*C_0; 1)$ for some o-symmetric convex body I^*C_0 (see (9.3)), the generalized Brunn–Minkowski inequality (A.47) yields that $S_{C_0}(C) \leq S_{C_0}(C_0)$. However, $S_{C_0}(C_0)$ is maximal if and only if C_0 is a parallelepiped (see (9.5)), which is equivalent to C being a parallelepiped according to Lemma 9.1.1. We conclude the following:

Lemma 9.10.4. *If C is a convex body in $\mathbb{R}^d$ then $\Omega(C) \leq 2d$, and equality holds if and only if C is a parallelepiped.*

Therefore, in accordance with Theorem 9.8.1, Theorem 9.10.1 yields that if P is a parallelepiped and C is any convex body not a parallelepiped in $\mathbb{R}^d$ then $H_\alpha(C) < H_\alpha(P)$ holds for small α.

Open Problems.

(i) Given the volume of the convex body C, is the Busemann surface area $S_{C_0}(C)$ with $C_0 = (C - C)/2$ minimal when C is a simplex?

The answer is yes in the planar case (see Lemma 2.1.7), but, for example, affine regular hexagons are also minimal domains.

(ii) Assuming that C is an o-symmetric convex body, determine the minimum of $S_C(C)$.

In the planar case, the minimum is obtained by the affine regular hexagons (see Lemma 2.1.7), but not even a safe conjecture is known in higher dimensions.

Comments. Theorem 9.10.1 for polytopes was proved in the papers [FTL1970] and [FTL1975b] by L. Fejes Tóth. Following a suggestion by T. Gowers, the general case was verified in K. Böröczky Jr., D. Larman, S. Sezgin, and C. Zong [BLS2000]. Proposition 9.10.2 is a version of the Vitali Covering Theorem (see, e.g., K. J. Falconer [Fal1986]).

9.11. Pairwise Touching Translates and Antipodal Points

In this section we consider four closely related concepts in $\mathbb{R}^d$, where C is a convex body:

- $H_\infty(C)$ is the maximal number of nonoverlapping translates of C that have a common point;
- the *Petty number* of C is the maximal number of pairwise touching translates of C;
- a finite set is called an *equilateral set* with respect to a given norm if the pairwise distances coincide;
- the family $x_1, \ldots, x_n$ of points, $n \geq 2$, is called *antipodal* if, for any x_i and x_j, $i \neq j$, there exists a strip containing $x_1, \ldots, x_n$ such that x_i and x_j lie on different hyperplanes on the boundary of the strip.

We observe that $H_\infty(C)$ is a natural extension of the notion of a generalized Hadwiger number. Now $H_\infty(C)$ is at most the Petty number of C, which is readily finite. Writing C_0 to denote $(C - C)/2$, we observe that the translates $x_1 + C, \ldots, x_n + C$ are pairwise touching if and only if each of the pairwise distances among $x_1, \ldots, x_n$ is two with respect to $\| \cdot \|_{C_0}$. Therefore, the

Petty number of C is the maximal cardinality of equilateral sets with respect to $\| \cdot \|_{C_0}$. Next, it is easy to see that any equilateral set is antipodal. However, we claim that any antipodal set is equilateral with respect to some norm: If the affine hull of an antipodal set is lower dimensional, and one adds a point outside of the affine hull to the family, then the extended family stays antipodal. Thus, let $x_1, \ldots, x_n$ be antipodal and span $\mathbb{R}^d$. Then the convex hull P of $x_1, \ldots, x_n$ satisfies that $P - x_1, \ldots, P - x_n$ do not overlap, and each translate contains the origin; hence, $x_1, \ldots, x_n$ is equilateral with respect to $\| \cdot \|_{P_0}$, where $P_0 = (P - P)/2$. In summary, we have the following equivalences:

pairwise touching translates $\leftrightarrow$ equilateral sets $\leftrightarrow$ antipodal sets.

Theorem 9.11.1. *For a convex body C in $\mathbb{R}^d$, any arrangement of pairwise touching translates contains at most 2^d translates. Equality holds if and only if C is a parallelepiped and the translation vectors are vertices of a translate of C.*

Remark. The previous discussion shows that $H_\infty(C) \leq 2^d$, and the cardinality of any equilateral set or antipodal set in $\mathbb{R}^d$ is at most 2^d with the obvious cases of equality.

Proof. Let $C_0 = (C - C)/2$. We write n to denote the number of translates, and we observe that the translates of C_0 by the same set of vectors are pairwise touching as well. The diameter of the union of these n translates of C_0 is four with respect to $\| \cdot \|_{C_0}$; hence, the same holds for the convex hull K of the translates, which, in turn, yields that $K_0 = (K - K)/2$ is contained in $2C_0$. Since $V(K_0) \geq V(K)$ according to the Brunn–Minkowski inequality (see (A.45)), we deduce $V(K) \leq V(2C_0)$; therefore, $n \leq 2^d$. If $n = 2^d$ then K is a translate of $2C_0$ by the equality case of the Brunn–Minkowski inequality, and the 2^d translates of C_0 tile K. It follows via Corollary 9.2.2 that C_0 is a parallelepiped; therefore, C is a parallelepiped as well, according to Lemma 9.1.1. $\square$

Let us observe that $H_\infty(C) = 2^d$ when C is a parallelepiped in $\mathbb{R}^d$ but $H_\infty(C) = 2$ if C is a smooth convex body. It is a nontrivial question whether $H_\infty(C)$ is necessarily very small if C is strictly convex. The related antipodal sets are the strictly antipodal ones: We say that $x_1, \ldots, x_n$ are *strictly antipodal* if, for any pair x_i, x_j, $i \neq j$, there exist parallel hyperplanes H_{ij} and H_{ji} such that $x_i \in H_{ij}, x_j \in H_{ji}$, and all other points lie strictly between H_{ij} and H_{ji}. In addition, we say that a norm satisfies the strict triangle inequality if whenever two vectors are not parallel then the strict triangle inequality holds for them. The following concepts are equivalent:

- pairwise touching translates of a strictly convex body,
- equilateral sets with respect to norms with strict triangle inequality, and
- strictly antipodal sets.

These equivalences are rather trivial, and so we only show how a strictly antipodal set $x_1, \ldots, x_n$ induces nonoverlapping translates of some strictly convex body that meet at the origin. For any ordered pair (i, j), $i \neq j$, we write B_{ij} to denote a ball that contains $x_1, \ldots, x_n$, and H_{ij} touches B_{ij} at x_i (where H_{ij} is defined as before). Then the intersection C of all B_{ij} is a strictly convex body, and $C - x_1, \ldots, C - x_n$ touch each other at the origin.

Lemma 9.11.2. *There exists an o-symmetric smooth and strictly convex body C in $\mathbb{R}^d$ whose Petty number is at least $\sqrt[3]{3^d}/3$.*

Remark. We note that $\sqrt[3]{3} = 1.44\ldots$. Naturally, the centres of the translates of C form an equilateral set with respect to C of cardinality at least $\sqrt[3]{3^d}/3$. The proof provides a strictly antipodal set of cardinality at least $\sqrt[3]{3^d}/3$, and hence a strictly convex body K with $H_\infty(K) \geq \sqrt[3]{3^d}/3$.

Proof. Let $k = \lfloor d/3 \rfloor$. The plan is to construct a strictly antipodal set of cardinality 3^k, and then to build the suitable C.

First, we construct three pairwise disjoint open circular arcs $\sigma_1, \sigma_2, \sigma_3$ in $\mathbb{R}^3$ with the following property: If x is a point of one of the arcs, and y is a point of another arc, then there exist parallel planes H_{xy} and H_{yx} such that $x \in H_{xy}$, $y \in H_{yx}$, and all other points of the arcs lie strictly between H_{xy} and H_{yx}. To achieve this, let e_1, e_2, and e_3 be edges of some three-cube such that neither two intersect and neither two are parallel. We write σ_i' to denote the shorter circular arc connecting the two endpoints of e_i whose centre is the centre of the cube, and we write y_i to denote the midpoint of σ_i'. Given $i \neq j$, we choose both $H_{y_i y_j}$ and $H_{y_j y_i}$ so that they are parallel to the tangent lines of σ_i' at y_i and σ_j' at y_j; hence, all points of $\sigma_1', \sigma_2', \sigma_3'$ except y_i and y_j lie strictly between $H_{y_i y_j}$ and $H_{y_j y_i}$. We conclude by continuity that σ_i, $i = 1, 2, 3$, can be chosen to be a small open subarc of σ_i' that contains y_i.

Next we write $\mathbb{R}^{3k}$ in the form $L_1 \oplus \ldots \oplus L_k$, where each L_m is isometric to $\mathbb{R}^3$, and place the images $\sigma_{m1}, \sigma_{m2}, \sigma_{m3}$ of the three arcs into L_m. For any k-tuple $\varphi = (i_1, \ldots, i_k) \in \{1, 2, 3\}^k$, we choose a point x_φ such that its orthogonal projection $x_{\varphi m}$ into L_m lands on $\sigma_{m i_m}$, and $x_{\varphi m} \neq x_{\psi m}$ if $\varphi \neq \psi$ and $m = 1 \ldots, k$. Now one of the coordinates of any two φ, $\psi \in \{1, 2, 3\}^k$ are different; hence, the family $\{x_\varphi\}_{\varphi \in \{1,2,3\}^k}$ is strictly antipodal by the definition of $\sigma_1, \sigma_2, \sigma_3$ and is of cardinality $3^k > 3^{d/3}/3$.

Finally, we construct the convex body C in $\mathbb{R}^d$. The strict antipodality yields some $\varrho > 0$ and parallel hyperplanes $H_{\varphi\psi}$ and $H_{\psi\varphi}$ such that $(x_\varphi - x_\psi)/2 \in H_{\varphi\psi}$, $(x_\psi - x_\varphi)/2 \in H_{\psi\varphi}$, and the balls of radius 2ϱ around all other differences $(x_\theta - x_\omega)/2$ lie strictly between $H_{\varphi\psi}$ and $H_{\psi\varphi}$. Let $B'_{\varphi\psi}$ be a Euclidean ball such that $H_{\varphi\psi}$ touches $B'_{\varphi\psi}$ at $(x_\varphi - x_\psi)/2$ and $B'_{\varphi\psi}$ contains all the balls $(x_\theta - x_\omega)/2 + 2\varrho\, B^d$ for $(\theta, \omega) \neq (\varphi, \psi)$. We may assume that all balls $B'_{\varphi\psi}$ are of the same radius r. We write $B''_{\varphi\psi}$ to denote the ball that is of radius $r - \varrho$ and has the same centre as $B_{\varphi\psi}$, and let C'' be the intersection of all $B''_{\varphi\psi}$. Therefore, $C = C'' + \varrho\, B^d$ is a smooth o-symmetric strictly convex body, and the translates $x_\varphi + C$, $\varphi \in \{1, 2, 3\}^k$, pairwise touch each other. $\qquad\square$

Open Problems.
(i) Do strictly antipodal families in $\mathbb{R}^d$ of cardinality $2^{d-o(d)}$ exist?
(ii) Does there exist an absolute constant $\lambda < 2$ with the following property: If, for a set $X \subset \mathbb{R}^d$, all the triangles whose vertices lie in X have only acute angles then $\#X \leq \lambda^d$?

     Naturally, such an X is strictly antipodal. We note that P. Erdős and Z. Füredi [ErF1983] constructs a family X of cardinality at least $(2/\sqrt{3})^{d-1}$.

Comments. The book by L. M. Blumenthal [Blu1953] discusses equilateral sets in $\mathbb{R}^d$ and verifies that their cardinality is at most 2^d. Moreover, C. M. Petty introduced the problem of pairwise touching translates in [Pet1971]. He verified that the Petty number of K is at most 2^d with equality if and only if K is a parallelepiped.

Z. Füredi, J. C. Lagarias, and F. Morgan [FLM1991] constructed the first example of a strictly convex body with differentiable boundary whose Petty number is exponential, namely, at least 1.02^d. We note that slight but technical modification of the proof of Lemma 9.11.2 yields that the boundary of the convex body C can be assumed to be differentiable of any order. For specific convex bodies, the Petty numbers of ellipsoids and parallelepipeds are readily $d + 1$ and 2^d, respectively. It is conjectured that the Petty number of the crosspolytope is $2d$, and J. Koolen, M. Laurent, and A. Schrijver [KLS2000] verified this conjecture for $d \leq 4$.

Theorem 9.11.1 for $H_\infty(\cdot)$ is due to A. Bezdek, K. Kuperberg, and W. Kuperberg [BKK1995]. The paper by J. Koolen, M. Laurent, and A. Schrijver [KLS2000] shows that $H_\infty(C) = 2d$ if C is a d-dimensional crosspolytope.

Lemma 9.11.2 is due to I. Talata [Tal?d], who even verifies that $\sqrt[3]{3}^d/3$ can be replaced by $\sqrt[4]{5}^d/4$. B. Grünbaum [Grü1963] proved that the maximal cardinality of a strictly antipodal set in $\mathbb{R}^3$ is five.

Following I. Talata [Tal1999c], a polytope P in $\mathbb{R}^d$ is called *edge-antipodal* if for any edge e of P, there exist parallel supporting hyperplanes to P at the endpoints of e. I. Talata [Tal1999c] exhibits a four-dimensional edge-antipodal polytope whose vertices do not form an antipodal set. Improving on an estimate of B. Csikós [Csi2003], K. Bezdek, T. Bisztriczky and K. Böröczky [BBB?] prove that the maximal number of vertices of an edge-antipodal polytope in $\mathbb{R}^3$ is eight. Moreover, A. Pór [Pór?a] shows that for any $d \geq 4$, there exist c_d depending only on d such that the number of vertices of any edge-antipodal polytope in $\mathbb{R}^d$ is at most c_d.

Given different $x_1, \ldots, x_n \in \mathbb{R}^d$, we say that x_i and x_j for $i \neq j$ form an *antipodal pair* if there exist parallel hyperplanes H_i and H_j such that $x_i \in H_i$, $x_j \in H_j$ and the points $x_1, \ldots, x_n$ lie in the strip bounded by H_i and H_j. Moreover, x_i and x_j form a *strictly antipodal pair* if each x_k for $k \neq i, j$ lies strictly between H_i and H_j. E. Makai, Jr. and H. Martini [MaM1991] discuss the minimal and the maximal number of antipodal pairs and of strictly antipodal pairs among n points in $\mathbb{R}^d$. We note that some auxiliary statements to [MaM1991] are proved in E. Makai, Jr. and H. Martini [MaM1993].

9.12. Economic Translational Clouds

Given a convex body K, we call a family of nonoverlapping translates of K that do not overlap K a *translational cloud* if every half line emanating from K intersects the interior of at least one translate. In this section we construct a cloud of small cardinality:

Theorem 9.12.1. *If K is a convex body in $\mathbb{R}^d$, $d \geq 3$, then there exists a cloud of $4^{d^2+o(d^2)}$ translates, and if in addition K is centrally symmetric then already some $2^{d^2+o(d^2)}$ translates form a cloud.*

Remark. If K is a ball then one needs at least $2^{0.599d^2+o(d)}$ balls to form a cloud (see Corollary 9.13.2).

To verify Theorem 9.12.1, we prove a much stronger statement; namely, a periodic packing is constructed such that the complement does not contain any long segment.

Lemma 9.12.2. *If K is a convex body in $\mathbb{R}^d$, $d \geq 3$, then there exists a periodic packing of translates of K such that any segment of length*

$$6{,}000d^2 \frac{V(K - K)}{V(K)}$$

with respect to $\| \cdot \|_{K_0}$, where $K_0 = (K - K)/2$, intersects the interior of some translate.

Let us show that the periodic packing in Lemma 9.12.2 cannot be replaced by some lattice packing for balls; namely, lattice packings of balls always contain huge unbounded holes.

Remark 9.12.3. Given a lattice packing $\Lambda + B^d$ for large d, there exists an affine plane G of dimension at least $d/(200\log_2 d)$ such that $G + (1/2)\, B^d$ avoids all balls in the packing.

Replacing Λ by $\lambda_1 \Lambda$ for $0 < \lambda_1 \leq 1$ if necessary, we may assume that Λ contains a vector u_1 of length two. Now replacing Λ by $T_2\Lambda$ if necessary, we may assume that Λ contains two independent vectors of length two, where T_2 is a linear transformation satisfying $T_2 u_1 = u_1$ and $T_2 v = \lambda_2 v$ for some $0 < \lambda_2 \leq 1$ and for all $v \in u_1^\perp$. After d steps, we may assume that Λ contains d independent vectors $u_1, \ldots, u_d$ of length two.

We define $k = \lceil 0.005\, d/\log_2 d \rceil$ and claim that if d is large then

$$1.5^{d-k}\kappa_{d-k} < \frac{\det \Lambda}{2^k}. \tag{9.23}$$

Now $\det \Lambda > 2^{0.599d + o(d)}\kappa_d$ according to the Kabatjanskii–Levenšteĭn bound (8.6) where $\log_2 2^{0.599d}/1.5^d > 0.00503\, d$. In addition, $\log_2 \kappa_{d-k}/\kappa_d = (k/2 + o(1))\log_2 d$ follows by the Stirling formula (A.27), which, in turn, yields (9.23) for large d.

The linear subspace L spanned by $u_1, \ldots, u_k$ satisfies $\det(L \cap \Lambda) \leq 2^k$; hence, $\det \Lambda_0 \geq \det \Lambda/2^k$ holds for the projection Λ_0 of Λ into $L^\perp$. Therefore, (9.23) yields that the projection of $\Lambda + 1.5B^d$ into $L^\perp$ does not cover $L^\perp$; or, in other words, there exists an affine k-plane G parallel to L that is disjoint from $\Lambda + 1.5B^d$.

Before verifying Lemma 9.12.2, we show how it leads to our main result.

Proof of Theorem 9.12.1 based on Lemma 9.12.2. Let $K_0 = (K - K)/2$ and let $R = 6000d^2 V(K - K)/V(K)$. We consider the packing $\{x_i + K\}$ of Lemma 9.12.2, where we assume that $x_1, \ldots, x_N$ are the x_is contained in $x_1 + (R + 3) K_0$. It follows by Lemma 9.12.2 that any ray emanating from a point of $x_1 + K$ is intersected by the interior of some $x_j + K$, $2 \leq j \leq N$. Since $V(K - K) \geq 2^d V(K)$ according to the Brunn–Minkowski inequality (A.44), and $x_2 + K_0, \ldots, x_N + K_0$ are nonoverlapping translates in $x_1 + (R + 3) K_0$, we have just constructed a cloud of cardinality at most

$$\frac{(R + 3)^d V(K_0)}{V(K)} < \left(\frac{V(K - K)}{V(K)} \right)^{d + o(d)}. \tag{9.24}$$

We are readily done if K is centrally symmetric. In general, the Rogers–Shephard bound yields $V(K - K) < 4^d V(K)$ (see (A.45) and (A.46)); hence, we conclude Theorem 9.12.1. $\square$

The proof of Lemma 9.12.2 will be probabilistic: We consider a large torus T^d, and we throw uniform random translates of K into T^d one by one. By keeping those that are disjoint from all earlier translates we obtain our periodic packing. We note that this method is not greedy, as our rule excludes a translate from the packing if it intersects some earlier translates even though all those earlier translates may have been excluded themselves. This suboptimal rule is necessary to obtain *independence* between the configurations occurring in two regions that are far from each other. We will show that the collection of the translates that are kept satisfies the requirement of Theorem 9.12.2 with high probability. To find a suitable torus, we will need some reasonably dense lattice packing:

Proposition 9.12.4. *If C is an o-symmetric convex body in $\mathbb{R}^d$ then there exists a packing lattice Λ for C satisfying* $\det \Lambda \leq 3^d V(C)$.

Proof. Readily, there exists a packing lattice Λ for C such that Λ is not a proper sublattice of any packing lattice. We define $\mu > 0$ to be minimal with the property that $\Lambda + \mu C$ covers $\mathbb{R}^d$. Thus, there exists $y \in \mathbb{R}^d$ such that $\|y\|_C = \mu$, and any $x \in \Lambda$ satisfies $\|x - y\|_C \geq \mu$. Since $\Lambda + \mu C$ is a covering, $\|x_0 - 3y\|_C \leq \mu$ holds for some $x_0 \in \Lambda$, or, in other words, $\|x_0/3 - y\|_K \leq \mu/3$. Let Λ' be the lattice generated by Λ and $x_0/3$. As Λ' is not a packing lattice for C, there exists some $x \in \Lambda$ such that $x - x_0/3 \in C$. Therefore,

$$2 \geq \|x - \frac{x_0}{3}\|_C \geq \|x - y\|_C - \|\frac{x_0}{3} - y\|_C \geq \frac{2}{3} \mu,$$

and hence $\mu \leq 3$, which, in turn, yields that $\det \Lambda \leq 3^d V(C)$. $\square$

Starting the actual proof of Lemma 9.12.2, we define $c = 6,000$ and $C = c\, d^2 4^d (K - K)$. Since $V(K - K) < 4^d V(K)$ according to the Rogers–Shephard inequality (see (A.45) and (A.46)), Proposition 9.12.4 provides a packing lattice Λ for C satisfying

$$\det \Lambda \leq 3^d V \left(c\, d^2 4^d (K - K) \right) < c^d 2^{6d^2} V(K). \tag{9.25}$$

In this proof, distance on the torus $T^d = \mathbb{R}^d / \Lambda$ is induced by the norm $\| \cdot \|_{K_0}$, not by the Euclidean norm. The condition on Λ yields that if $\|x - y\|_{K_0} \leq 2c\, d^2 4^d$ then the distance of the images of x and y on T^d is still $\|x - y\|_{K_0}$. We write the same symbol for a convex body in $\mathbb{R}^d$ and its embedded image on T^d.

We assume that $V(T^d) = 1$, and we choose an integer M satisfying

$$(1 - V(K - K))^M < \frac{1}{2}.$$

Let us throw points $x_1, \ldots, x_M$ into T^d independently with uniform distribution with respect to the Lebesgue measure. We colour an x_i red if $\|x_j - x_i\|_{K_0} > 2$ holds for any $j < i$, or, in other words, if $x_i + K$ is disjoint from any $x_j + K$ for $j < i$. For a measurable $A \subset T^d$, we write $P(A)$ to denote the probability that A contains no red point.

Proposition 9.12.5. *Let $A, B \subset T^d$ be measurable sets satisfying that $\|x - y\|_{K_0} < 2$ for any $x, y \in B$, and there exist translates $y_i + B \subset A$, $i = 1, \ldots, N$, with $\|y_i - y_j\|_{K_0} \geq 6$ for $i \neq j$. Then*

$$P(A) \leq \left(1 - \frac{V(B)}{2V(K - K)} \right)^N.$$

Proof. First we calculate $P(B)$. Since the probability that x_i lands in B and is coloured red is

$$P_i = V(B) \cdot (1 - V(K - K))^{i-1},$$

and at most one $x_i \in B$ is coloured red, we deduce that

$$P(B) = 1 - \sum_{i=1}^{M} P_i = 1 - \frac{V(B)(1 - (1 - V(K - K))^M)}{V(K - K)}$$
$$< 1 - \frac{V(B)}{2V(K - K)}.$$

Now the sets $y_i + B - (K - K)$ are disjoint; therefore, the events that $y_i + B$ contains no red point, $i = 1, \ldots, N$, are independent. Each of these events have probability $P(B)$; hence, Proposition 9.12.5 follows. $\qquad\square$

Our plan is to approximate segments in T^d by segments chosen from a suitable finite family. According to the Rogers bound Theorem 9.5.2 (i), there exists a covering of T^d by translates of K/d whose density is at most $4d \ln d$, and we write Θ to denote the family of translation vectors. Now we define Σ to be the family of segments in T^d whose length is between $c\, d^2 \cdot V(K - K)/V(K) - 1$ and $c\, d^2 \cdot V(K - K)/V(K) + 1$, and whose endpoints are chosen from Θ. Therefore (compare (9.25)),

$$\#\Sigma < (\#\Theta)^2 \leq \left(4d \ln d \cdot d^d\, \frac{\det \Lambda}{V(K)} \right)^2 \leq c^{2d} \cdot 2^{14d^2}.$$

For any $s \in \Sigma$, Proposition 9.12.5 can be applied to $A_s = s - (1 - 2/d)K$, to $B = -(1 - 2/d)K$, and to $N = \lceil (c/7)\, d^2 V(K - K)/V(K) \rceil$. We deduce that the probability P_0 that there exists an $s \in S$ such that $s - (1 - 2/n)K$ contains no red point satisfies

$$P_0 \leq \#\Sigma \cdot \left(1 - \frac{V(B)}{2V(K - K)} \right)^N$$

$$\leq c^{2d} \cdot 2^{14d^2} \left(1 - \left(1 - \frac{2}{d} \right)^d \frac{V(K)}{2V(K - K)} \right)^N.$$

Here $(1 - 2/d)^d/2 \geq 1/54$ and $1 - t < e^{-t}$ yield that

$$\ln P_0 < d^2 \cdot \left(\frac{2}{d} \ln c + 14 \ln 2 - \frac{1}{7 \cdot 54} c \right),$$

which is negative for $c = 6{,}000$. Therefore, $P_0 < 1$, and hence there exists a sequence $x_1, \ldots, x_M$ such that, for any $s \in S$, the set $s - (1 - 2/d)K$ contains a red point. We write $r_1, \ldots, r_m$ to denote the red points.

Finally, we verify that $\Lambda + \{r_1 + K, \ldots, r_m + K\}$ is a suitable periodic translative packing for Lemma 9.12.2. Let us consider a segment $\sigma = aa'$ of length $c\, d^2 V(K - K)/V(K)$. If we embed a and a' into T^d, the images of a and a' lie in $z + (1/d)K$ and in $z' + (1/d)K$ for some $z, z' \in Z$, respectively. Thus the segment $s = zz'$ is an element of S, and s is a subset of $\sigma - (1/d)K$. We have seen that there exists some red point $r_i \in s - (1 - 2/d)K$. Therefore, r_i lies in $\sigma - (1 - 1/d)K$, or, in other words, σ intersects the interior of $r_i + K$, completing the proof of Lemma 9.12.2. □

Open Problems.

    (i) Does there exist a translative cloud of cardinality $2^{d^2 + o(d^2)}$ for any convex body?

    (ii) Does there exist an absolute constant $c < 2$ with the following property? There exist a packing of unit balls in $\mathbb{R}^d$ and a direction such that any segment that is parallel to the given direction and is of length of at least $c^{d+o(d)}$ intersects the interior of some ball in the packing.

Such a packing would improve the lower bound $\delta(B^d) \geq 2^{-d}$ into $c^{-d+o(d)}$.

Comments. The original problem of Hornich (see L. Fejes Tóth [FTL1959]) was more restrictive than our formulation: It asked for the smallest number of unit three-balls that do not overlap each other and the unit ball centred at the origin, nor block any ray emanating from the origin. L. Fejes Tóth [FTL1959] proved that at least nineteen unit balls are needed, a lower bound that was improved subsequently to twenty-four by A. Heppes [Hep1967] and to the current record thirty by G. Csóka [Csó1977]. However, L. Danzer [Dan1960] constructed a restricted cloud of cardinality forty-two.

For translates of a given convex body K in $\mathbb{R}^d$, K. Böröczky and V. Soltan [BöS1996] were the first to construct a finite cloud. The first quantitative upper bound on the cardinality of some cloud, namely, a bound of order d^{d^2}, was due to Ch. Zong [Zon1997b]. It was improved to c^{d^2} for some specific absolute constant c independently by I. Talata [Tal2000a] and by I. Bárány and I. Leader (see Ch. Zong [Zon1999b]). Finally, K. Böröczky Jr. and G. Tardos [BöT?] proved Theorem 9.12.1.

Observing that any packing of unit balls in $\mathbb{R}^d$ avoids a segment of length $2^{0.599d+o(d)}$, Ch. Zong conjectured that there exists a packing by translates of any given convex body K in $\mathbb{R}^d$ such that the length of the longest segment in the complement is at most c^d with respect to the norm $\|\cdot\|_{K_0}$, where $K_0 = (K - K)/2$ and c is an absolute constant. M. Henk and Ch. Zong [HeZ2000] and G. Harcos and J. Pach (unpublished) prove the existence of a packing satisfying that the lengths of the segments in the complement are bounded. The close to being optimal bound of Lemma 9.12.2 is due to K. Böröczky Jr. and G. Tardos [BöT?]. Given $\varrho \geq 2$, M. Henk and Ch. Zong [HeZ2000] construct a packing $\{x_i + K\}$ such that $\|x_i - x_j\|_{K_0} \geq \varrho$, $i \neq j$, and the length of any segment in the complement is bounded, and K. Böröczky Jr. and G. Tardos [BöT?] prove the existence of such packing satisfying that the length of any segment in the complement is at most $c_1 \varrho^d \log \varrho$. However, the complement of any such "sparse packing" contains a segment of length $c_2 \varrho^d$ (where c_1 and c_2 are positive constants depending on the dimension d). We note that an analogous argument to the one for Lemma 9.12.2 is applied by P. Frankl, J. Pach, and V. Röld [FPR1984].

Given any lattice packing of unit three-balls, A. Heppes [Hep1960] proved that the complement contains an infinite open circular cylinder of radius $3\sqrt{2}/4 - 1$, which is the optimal bound as shown by the face-centred cubic lattice. In addition, J. Horváth and S. S. Ryskov [HoR1975] show that if

$d \geq 4$ then any lattice packing of unit balls in $\mathbb{R}^d$ avoids an infinite open circular cylinder of radius 0.3 (see also Ch. Zong [Zon1999]). Next, M. Henk [Hen?] proves that if d is large then any lattice packing of unit balls avoids an affine plane of dimension $d/\log_2 d$ (see Remark 9.12.3). However, M. Henk, G. Ziegler, and Ch. Zong [HZZ2002] construct a lattice packing of unit balls in $\mathbb{R}^d$ such that any affine plane of dimension cd intersects the interior of some of the balls, where $0 < c < 1$ is an absolute constant.

The simple proof of Proposition 9.12.4 is due to C. A. Rogers [Rog1950]. Proposition 9.12.4 is not optimal, but all the better bounds have substantially more complicated proofs (see the comments to Section 9.5).

9.13. Clouds Require Plentiful Balls

In this section we show that the order of the upper bound in Theorem 9.12.1 is optimal for ball packings.

Theorem 9.13.1. *At least $\delta(B^d)^{-d+o(d)}$ unit d-balls are needed to form a cloud.*

Substituting the Kabatjanskii–Levenšteĭn bound $\delta(B^d) \leq 2^{-0.599 \cdot d + o(d)}$ (see Section 8.2) into Theorem 9.13.1 leads to

Corollary 9.13.2. *At least $2^{0.599 \cdot d^2 + o(d^2)}$ unit balls are needed to form a cloud in $\mathbb{R}^d$.*

Proof of Theorem 9.13.1. In this proof $o(d)$ always denotes some function of only d such that $o(d)/d$ tends to zero as d tends to infinity. We write $|\cdot|$ to denote the Hausdorff $(d-1)$-measure.

The main idea is to show that if the radial projections of nonoverlapping unit balls cover S^{d-1} then one needs at least $\delta(B^d)^{-d+o(d)}$ unit balls. We recall (see Theorem 6.3.1) that the maximal number $M(d-1, \varphi)$ of nonoverlapping spherical balls of radius φ satisfies

$$M(d-1, \varphi) < \frac{1}{\sin^{d-1}\varphi} \cdot 2^{-0.5d+o(d)}. \tag{9.26}$$

We will apply (9.26) to unit balls that are reasonably close to the origin. Let the unit balls $B_1, \ldots, B_n$ form a cloud around the unit ball B^d. We write x_i to denote the centre of B_i, and we assume that $\|x_i\|$ is increasing with i. The radial projection of B_i onto S^{d-1} is a spherical ball B_i' whose centre is $y_i = x_i/\|x_i\|$, and its radius φ_i satisfies $\sin\varphi_i = 1/\|x_i\|$. In turn, the orthogonal projection of B_i' into the tangent hyperplane at y_i to S^{d-1} is a Euclidean $(d-1)$-ball of radius $\sin \varphi_i$. Since the angle of the exterior normals at y_i and at any point y of B_i' is at most $\pi/6$, we deduce that

$$|B_i'| \le \frac{2}{\sqrt{3}} \sin^{d-1} \varphi_i \cdot \kappa_{d-1} < \frac{|S^{d-1}|}{\|x_i\|^{d-1}}. \tag{9.27}$$

We start with the first $N = \lfloor 2^{0.1d^2} \rfloor$ balls. If $k/d \le \|x_i\|, \|x_j\| < (k+1)/d$ for $k \ge 2d$ then $\|x_i - x_j\| \ge 2$ yields that the angle γ_{ij} of x_i and x_j satisfies

$$\sin \frac{\gamma_{ij}}{2} = \sqrt{\frac{1 - \cos \gamma_{ij}}{2}} = \sqrt{\frac{\|x_i - x_j\|^2 - (\|x_i\| - \|x_j\|)^2}{4 \cdot \|x_i\| \cdot \|x_j\|}} \ge \frac{d\sqrt{1 - \frac{1}{4d^2}}}{k+1}.$$

It follows by $k \ge 2d$ and by (9.26) that the number of centres x_i with $k/d \le \|x_i\| < (k+1)/d$ is at most

$$\left(1 - \frac{1}{4d^2}\right)^{-\frac{d-1}{2}} \cdot \frac{(k+1)^{d-1}}{d^{d-1}} \cdot 2^{-0.5d+o(d)} = \frac{k^{d-1}}{d^{d-1}} \cdot 2^{-0.5d+o(d)}.$$

Given i, writing m_i to denote the integer satisfying $m_i/d \le \|x_i\| < (m_i + 1)/d$, we deduce

$$i \le \sum_{k=2d}^{m_i} \frac{k^{d-1}}{d^{d-1}} \cdot 2^{-0.5d+o(d)} \le d \int_0^{\|x_i\|+\frac{1}{d}} t^{d-1} dt \cdot 2^{-0.5d+o(d)}$$

$$\le \|x_i\|^d \cdot 2^{-0.5d+o(d)}. \tag{9.28}$$

Combining (9.27) and (9.28) shows that the total surface area of the radial projections of the first $N = \lfloor 2^{0.1d^2} \rfloor$ balls is at most

$$\sum_{i=1}^N \frac{2^{-0.5d+o(d)}}{i^{\frac{d-1}{d}}} \cdot |S^{d-1}| < \int_0^{2^{0.1d^2}} \frac{2^{-0.5d+o(d)}}{s^{\frac{d-1}{d}}} ds \cdot |S^{d-1}|$$

$$= 2^{-0.4 \cdot d+o(d)} \cdot |S^{d-1}|. \tag{9.29}$$

In particular, the surface area covered by the spherical balls B_i', $i > N$, is $(1 - o(1)) |S^{d-1}|$.

If $i > N$ then $\|x_i\| > 2^{0.1d+o(d)}$ holds according to (9.28). Now the ball $(\|x_i\| + 1) B^d$ hosts i nonoverlapping unit balls; moreover, Theorem 9.4.1 implies that the density of these i balls inside $(\|x_i\| + 2) B^d$ is at most $\delta(B^d)$; hence,

$$i \le (\|x_i\| + 2)^d \cdot \delta(B^d) = (1 + o(1)) \cdot \|x_i\|^d \cdot \delta(B^d). \tag{9.30}$$

Therefore, combining (9.27) and (9.30) leads to

$$(1 - o(1)) \cdot |S^{d-1}| < (1 + o(1)) \sum_{i=N+1}^n \frac{1}{i^{\frac{d-1}{d}}} \cdot \delta(B^d)^{\frac{d-1}{d}} |S^{d-1}|$$

$$< (1 + o(1)) \int_0^n \frac{1}{s^{\frac{d-1}{d}}} ds \cdot \delta(B^d)^{\frac{d-1}{d}} |S^{d-1}|.$$

Since the integral in the last expression is $dn^{1/d}$, the Blichfeldt bound $\delta(B^d) \le 2^{-0.5d+o(d)}$ (see (8.7)) yields that $n \ge \delta(B^d)^{-d+o(d)}$, completing the proof of Theorem 9.13.1. □

Open Problem.
  (i) Does there exist a $\lambda > 1$ such that the cardinality of any cloud for any convex body in $\mathbb{R}^d$ is at least λ^d?
      It is essential that the cloud is formed by the interiors of the translate. If the closures are permitted then even $2d$ translates of a d-cube can shadow the d-cube, where $2d$ is the minimal cardinality for any convex body according to K. Böröczky and V. Soltan [BöS1996].
      However, if K is the Dirichlet–Voronoi cell of the origin for a d-lattice Λ, where no two nonzero vectors of Λ are orthogonal, then there exists a cloud of cardinality at most 5^d. This cloud is formed by the other Dirichlet–Voronoi cells that are contained in $5K$.

Comments. Corollary 9.13.2 is due to I. Talata [Tal2000a]. A somewhat weaker bound of the same order was verified by I. Bárány and I. Leader (see Ch. Zong [Zon1999]).

9.14. Sparse Clouds

We discuss clouds where the distance between any two translates is large. Let K be a convex body in $\mathbb{R}^d$, and let $\varrho \ge 2$. Writing $K_0 = (K - K)/2$, we define Cloud(K, ϱ) to be the minimal N such that there exists a cloud $x_1 + K, \ldots, x_N + K$ for K such that $\|x_i\|_{K_0} \ge \varrho$ for any i, and $\|x_i - x_j\|_{K_0} \ge \varrho$ for $i \ne j$.

Theorem 9.14.1. *Given $\varrho \ge 2$, and a convex body K in $\mathbb{R}^d$, we have*

$$2^{-5d^2} \cdot \varrho^{d^2-d} \le \text{Cloud}(K, \varrho) \le 2^{10d^2} \cdot \varrho^{d^2-d}.$$

The upper bound of Theorem 9.14.1 is of the optimal order even if $\varrho = 2$ (see Corollary 9.13.2).

Proof. We assume that $o \in \text{int}K$, and we define $K_0 = (K - K)/2$. Length will always be measured in terms of $\| \cdot \|_{K_0}$. A fundamental tool of the argument is the notion $S_{K_0}(\cdot)$ of Busemann surface area (see Section 9.1), which satisfies (see (9.5))

$$2 \le S_{K_0}(K_0) \le 2d. \tag{9.31}$$

To verify the lower bound in Theorem 9.14.1, let us consider a cloud $x_1 + K, \ldots, x_N + K$ for K such that $\|x_i\|_{K_0} \geq \varrho$ for any i, and $\|x_i - x_j\|_{K_0} \geq \varrho$ for $i \neq j$. The argument is based on the fact that the radial projections of the translates into ∂K_0 cover ∂K_0.

We assume that $r_i = \|x_i\|_{K_0}$ is an increasing function of i. Since the translates $x_1 + (\varrho/2)K_0, \ldots, x_i + (\varrho/2)K_0$ are contained in $(r_i + \varrho/2)K_0$, we have

$$i \leq \frac{\left(r_i + \frac{\varrho}{2}\right)^d}{\left(\frac{\varrho}{2}\right)^d} \leq \frac{3^d}{\varrho^d} \cdot r_i^d. \tag{9.32}$$

Let $\tilde{x}_i$ be the radial projection of x_i into ∂K_0; thus, the radial projection σ_i of $x_i + K$ coincides with the radial projection of $\tilde{x}_i + (1/r_i)K$ into ∂K_0. Given some $x \in \tilde{x}_i + (1/r_i)K$, we write $\tilde{x}$ to denote the radial projection of x into ∂K. Since $\|x - \tilde{x}_i\|_{K_0} \leq 2/r_i$, we deduce $\|\tilde{x} - x\|_{K_0} \leq 2/r_i$ by the triangle inequality. Therefore, $\|\tilde{x} - \tilde{x}_i\|_{K_0} \leq 4/r_i$, or, in other words, $\tilde{x} \in \tilde{x}_i + (4/r_i)K_0$. In turn, (9.31) and (9.32) yield

$$S_{K_0}(\sigma_i) \leq S_{K_0}\left(\frac{4}{r_i}K_0\right) \leq 2d\left(\frac{4}{r_i}\right)^{d-1} \leq 2d \cdot \frac{12^{d-1}}{\varrho^{d-1}} \cdot \frac{1}{i^{\frac{d-1}{d}}}.$$

Now applying again (9.31) leads to

$$2 \leq \sum_{i=1}^{N} S_{K_0}(\sigma_i) \leq \sum_{i=1}^{N} \frac{2^{4d}}{\varrho^{d-1}} \frac{1}{i^{\frac{d-1}{d}}} \leq \frac{2^{4d}}{\varrho^{d-1}} \int_0^N \frac{1}{t^{\frac{d-1}{d}}}\,dt < \frac{2^{5d}}{\varrho^{d-1}} N^{\frac{1}{d}},$$

hence, we conclude the lower bound in Theorem 9.14.1.

The proof of the upper bound proceeds as follows: Let $R = 2^{10d}\varrho^d$. This choice of R allows us to find a covering $x_1 + \mathrm{int}K, \ldots, x_N + \mathrm{int}K$ of the boundary of RK_0, and $y_i = t_i x_i$ with $1/2 \leq t_i \leq 1$ for $i = 1, \ldots, N$ such that $\|y_i - y_j\|_{K_0} \geq \varrho$ holds for $i \neq j$. Since the origin lies in $\mathrm{int}\,K$, the translates $y_1 + K, \ldots, y_N + K$ form a suitable cloud.

Let Λ be a packing lattice for $3R\,K_0$, and we may assume that $V(T^d) = 1$ holds for $T^d = \mathbb{R}^d/\Lambda$. In addition, let $x_1 + K/2, \ldots, x_n + K/2$ be a thinnest covering of T^d by translates of $K/2$; hence, $n \leq 4d \ln d/V(K/2)$ according to the Rogers bound (see Theorem 9.5.2 (i) and the remarks). Given a compact subset A of $2R\,K_0$, we claim that

$$\#A \cap \{x_1, \ldots, x_n\} \leq \frac{2^{3d}}{V(K)}\,V(A + 2K_0). \tag{9.33}$$

Let $A + 2K_0 = B$. The formula (see (A.50))

$$\int_{T^d} \#((B + x) \cap \{x_1, \ldots, x_n\})\,dx = n \cdot V(B)$$

provides a translate $B + x$ with the property

$$\#((B + x) \cap \{x_1, \ldots, x_n\}) \leq \frac{4d \ln d}{V(\frac{1}{2} K)} \, V(B) < \frac{2^{3d}}{V(K)} \, V(B).$$

Since if $y + K/2$ intersects A then $y \in B$, we deduce that A can be covered by $2^{3d} V(B)/V(K)$ translates of $K/2$. Therefore, the minimality of n yields the claim (9.33).

In the calculations that follow, we frequently use the inequality

$$(R + 3)^d - (R - 3)^d < (R + 3)^d \cdot \frac{6d}{R + 3} < 2^{2d} \cdot R^{d-1}. \tag{9.34}$$

Let $x_1 + K/2, \ldots, x_N + K/2$ be the translates that intersect the boundary of RK_0. We observe that if $i \leq N$ then x_i is contained in $(R + 1)K_0$, and not in $(R - 1)K_0$. Therefore, applying (9.34) in the estimate (9.33) leads to

$$N < \frac{2^{3d}}{V(K)} \cdot 2^{2d} R^{d-1} \cdot V(K_0) < 2^{6d} R^{d-1} < 2^{10d^2} \varrho^{d^2 - d} \tag{9.35}$$

as $V(K_0) < 2^d V(K)$ according to the Rogers–Shephard inequality (see (A.45) and (A.46)).

Our next goal is to find $t_i \in [1/2, 1]$ for $i = 1, \ldots, N$ in a way that the homothetic copies $t_i x_i + (\varrho/2)K_0$ form a packing. We write x' to denote the radial projection of any $x \neq o$ into $\partial(RK_0)$.

For fixed i, let us count the number m_i of x_j such that $j \leq N$ and $t x_j + \varrho K_0$ intersects the ray ox_i for some $t \in [1/2, 1]$. Given such an x_j, $x_j + 2\varrho K_0$ intersects the ray ox_i in some point p. We deduce $\|p - x_i'\|_{K_0} \leq 2\varrho + 2$ by the triangle inequality; hence, $\|x_j - x_i'\|_{K_0} \leq 5\varrho$. In particular,

$$m_i \leq \#A_i \cap \{x_1, \ldots, x_n\}, \tag{9.36}$$

where A_i is the intersection of sets $x_i' + 5\varrho K_0 (R + 1)K_0$, and the complement of $(R - 1)K_0$. Let σ_i be the intersection of $\partial(RK_0)$ and $x_i' + 10\varrho K_0$. Since $A_i + 2K_0$ is contained in $(R + 3)K_0$ and in the complement of $(R - 3)K_0$, we deduce that the radial projection of $A_i + 2K_0$ into $\partial(RK_0)$ lies in σ_i. According to (9.4), the union $\mathrm{Cone}(\sigma_i)$ of the segments ox for $x \in \sigma_i$ satisfies

$$V(\mathrm{Cone}(\sigma_i)) \leq R \cdot \frac{V(K_0)}{2} \cdot S_{K_0}(\sigma_i).$$

We conclude that

$$V(A + 2K_0) \leq \frac{(R + 3)^d - (R - 3)^d}{R^d} \cdot V(\mathrm{Cone}(\sigma_i))$$

$$< 2^{3d} \cdot \frac{V(K)}{2} \cdot S_{K_0}(10\varrho K_0),$$

and thus (9.31) and (9.33) yield

$$m_i \leq d \cdot 2^{6d} 10^{d-1} \cdot \varrho^{d-1}.$$

It follows that the part of the ray ox_i that is covered by m_i translates of ϱK_0 is of length at most $m_i 2\varrho < 2^{10d} \varrho^d / 3 < R/3$. Therefore, we are able to choose $y_i = t_i x_i$, $t_i \in [1/2, 1]$ for $i = 1, \ldots, N$, by induction on i so that $\|y_i - y_j\|_{K_0} > \varrho$ holds for $i \neq j$. Since any ray emanating from K intersects some $x_i + K/2$, it also intersects $y_i + \text{int} K$, and hence (9.35) completes the proof of Theorem 9.14.1. $\qquad\qquad\qquad\qquad\qquad\square$

Comments. The existence of sparse clouds is established by M. Henk and Ch. Zong [HeZ2000], and a slightly weaker form of Theorem 9.14.1 is proved by K. Böröczky Jr. and G. Tardos [BöT?].

9.15. The Hadwiger–Gohberg Number

Given a convex body K in $\mathbb{R}^d$, we define its Hadwiger–Gohberg number as the minimal number of smaller homothetic copies of K that cover K. We may readily assume that the smaller homothetic copies are of the same size. If K is a parallelepiped then any smaller homothetic copy contains at most one vertex; therefore, the Hadwiger–Gohberg number is 2^d. We verify that not many more than 2^d smaller homothetic copies are sufficient for any centrally symmetric convex body:

Theorem 9.15.1. *The Hadwiger–Gohberg number of any convex body K in $\mathbb{R}^d$ is at most $5d \ln d \cdot 4^d$, and at most $5d \ln d \cdot 2^d$ if K is centrally symmetric.*

Proof. Let $\lambda = \sqrt[d]{4/5}$. Given any torus T, we write the same symbol for a convex body in $\mathbb{R}^d$ and its embedded image on T. According to the Rogers bound (Theorem 9.5.2 (i)), there exist a torus $T^d = \mathbb{R}^d / \Lambda$ and a covering $x_1 + \lambda K, \ldots, x_n + \lambda K$ of the torus with a density at most $4d \ln d$. We may assume that Λ is a packing lattice for $2(K - K)$ and that $V(T^d) = 1$; hence, $n \leq 4d \ln d / V(\lambda K)$.

Let $A = K - \lambda K$, and thus if $x_i + \lambda K$ intersects $x + K$ for some $x \in T^d$ then $x_i \in x + A$. Since (see (A.50))

$$\int_{T^d} \#((A + x) \cap \{x_1, \ldots, x_n\}) \, dx = n \cdot V(A),$$

there exists a translate $A + x$ with the property

$$\#((x + A) \cap \{x_1, \ldots, x_n\}) \leq \frac{4d \ln d}{V(\lambda K)} \cdot V(A) < \frac{5d \ln d}{V(K)} \cdot V(K - K).$$

We are readily done if K is centrally symmetric, whereas in the general case $V(K - K) < 4^d V(K)$ holds according to the Rogers–Shephard inequality (see (A.45) and (A.46)). $\qquad\square$

Given a convex body K, we finally state two equivalent formulations of the Hadwiger–Gohberg number:

Illumination: We say that a point $x \notin K$ illuminates a $y \in \partial K$ if the ray xy intersects $\operatorname{int} K$ but the segment xy does not intersect $\operatorname{int} K$. We ask for the minimal number of outside points that illuminate ∂K.

Separation of the faces of the polar: We assume that $o \in \operatorname{int} K$. The polar K^* of K is the set of y such that $\langle x, y \rangle \leq 1$ holds for any $x \in K$. Now K^* is a convex body containing o in its interior, and the intersection of K^* and some supporting hyperplane is called a face of K^*. We ask for the minimal number of open half spaces whose boundary contains the origin such that any face of K^* is contained in one of the half spaces.

Open Problems.

(i) Is the Hadwiger–Gohberg number at most 2^d for any d-dimensional convex body? (This is the celebrated Hadwiger–Gohberg conjecture.)

(ii) Let $f(d)$ be the minimal number such that any set of diameter 1 can be partitioned into $f(d)$ subsets of smaller diameter. What is the true order of $f(d)$?

It was a celebrated conjecture of K. Borsuk [Bor1933] that $f(d) = d + 1$. After various results supporting the conjecture, J. Kahn and G. Kalai [KaK1993] proved that $f(d) > 1.1^{\sqrt{d}}$. However, Theorem 9.15.1 shows that the $f(d)$ is at most exponential. For detailed account of the Borsuk conjecture, see the monographs by V. G. Boltyanskii, H. Martini, and P. Soltan [BMS1997] and P. Brass, W. Moser, and J. Pach [BMP?].

Comments. The first occurrence of the Hadwiger–Gohberg covering problem is the paper [Lev1955] by F. W. Levi, who essentially solved the planar case, and the same result appeared independently in I. Z. Gohberg and A. S. Markus [GoM1960]. The conjecture for any dimensions was posed independently by H. Hadwiger (see [Had1957b]) and by I. Z. Gohberg (see I. Z.

Gohberg and A. S. Markus [GoM1960]). The version based on illumination is due to H. Hadwiger [Had1960] and V. G. Boltyanskii [Bol1960], whereas the formulation using polar bodies was observed by K. Bezdek [Bez1991]. For detailed information about the Hadwiger–Gohberg number in $\mathbb{R}^d$, we refer to the books by V. G. Boltyanskii, H. Martini, and P. Soltan [BMS1997] and P. Brass, W. Moser, and J. Pach [BMP?]. We note that Theorem 9.15.1 has been observed by various people, and the conjecture has been verified for certain classes of convex bodies: for centrally symmetric convex bodies in $\mathbb{R}^3$, for three-polytopes with affine symmetry, for zonoids, and, more generally, for belt bodies, as well as for bodies of constant width and for dual cyclic polytopes. We are aware of only one extra fact compared to references in V. G. Boltyanskii, H. Martini, and P. Soltan [BMS1997]; namely, improving on the bound of K. Bezdek and T. Bisztriczky [BeB1997], I. Talata [Tal1999a] determines the Hadwiger–Gohberg number of dual cyclic polytopes if d is even, and he provides some close to being optimal bounds if d is odd (with all these bounds being polynomial in d).

9.16. Related Problems

Relatives of the Hadwiger Number

Let C be a convex body in $\mathbb{R}^d$. The first notion is the *Hadwiger number for lattice packings*; namely, $H_L(C)$ denotes the maximal number of neighbours in a lattice packing of C. Readily, $H_L(C) \leq 3^d - 1$ with equality if and only if C is a parallelepiped. If C is strictly convex then $H_L(C) \leq 2^{d+1} - 2$ according to a result by H. Minkowski (see P. M. Gruber and C. G. Lekkerkerker [GrL1987]). However, modifying a cube leads to a strictly convex C satisfying $H_L(C) \geq 2^d$. Moreover the proof of the Swinnerton–Dyer bound (Theorem 9.7.1) yields $H_L(C) \geq d^2 + d$ for any convex body C, say, $H_L(C) \geq 12$, in the three dimensional case. Surprisingly, the true order of $H_L(C)$ is not known even for balls. The record for large d is $H_L(B^d) \geq d^{\log_2 d - 2\log_2 \log_2 d}$, which is obtained using the so-called Barnes–Wall lattices (see J. H. Conway and N. J. A. Sloane [CoS1999]).

Readily, $H(C) \geq H_L(C)$, and $H(C) = H_L(C)$ holds when $d = 2$. If $d \geq 3$, first C. Zong [Zon1994] gave examples with $H(C) \geq H_L(C) + 2$, and the bodies constructed by I. Talata [Tal1998b] satisfy $H(C) \geq H_L(C) + 2^{d-1}$. In addition, I. Talata [Tal?c] constructs a strictly convex body C with the property $H(C) > \sqrt{7}^d/4$; hence $H(C)/H_L(C) \geq (\sqrt{7}/2)^d/8$.

K. Bezdek and P. Brass [BeB2003] discuss the *one-sided Hadwiger number* $H^+(C)$. This is the maximal number of nonoverlapping translates of C that touch C and are contained in a supporting half space of C. K. Bezdek and P. Brass [BeB2003] prove that $H^+(C) \leq 2 \cdot 3^{d-1} - 1$, and equality holds if and only if C is a parallelepiped. We note that the QS-theorem of V. Milman yields an exponential lower bound for $H^+(C)$ as well. Earlier G. Fejes Tóth [FTG1981] verified that the one-sided Hadwiger number of the three-ball is nine.

It was observed by J. Pach and P. K. Agarwal [PaA1995] that among n nonoverlapping translates of a given convex body K, the number of touching pairs is at most $H(K)n/2$. This upper bound was improved to $H(K)n/2 - \gamma \cdot n^{(d-1)/d}$ by K. Bezdek [Bez2002], where γ denotes a positive constant that depends on K. Finally, we mention the notion of *blocking number* introduced by Ch. Zong in [Zon1995b]. For a convex body C, the blocking number is the minimal number of nonoverlapping translates of C that touch C, and any other translate of C that touch C overlaps at least one of the given translates. We do not discuss this notion because the book by Ch. Zong [Zon1999] provides a detailed account.

Minimizing Some Mixed Volume

Given convex bodies K and C, let D_n to denote the convex hull of n translates of K such that $V(D_n, C; 1)$ is minimal. Then A. Schürmann [Sch?] proves that $D_n / \sqrt[d]{n}$ tends to the homothetic copy C' of C of volume $\Delta_T(K)$.

Homothetic Cubes

Let us consider a finite family of smaller homothetic copies of the unit cube $[0, 1]^d$ in $\mathbb{R}^d$. The main question is whether the given cubes can be translated into $[0, 1]^d$ in a way that either they form a packing or cover $[0, 1]^d$. Recently, on-line methods have also been considered; namely, we learn the ith cube only after the first $i - 1$ cubes are already in place. The results that follow exhibit how surprisingly efficient on-line methods are compared to optimal off-line methods.

J. Meir and L. Moser [MeM1968] prove the existence of a packing by the translates of the small cubes into $[0, 1]^d$ if the total volume of the small cubes is at most $1/2^{d-1}$. This bound is optimal, which is shown by two cubes of edge length $1/2 + \varepsilon$ for small $\varepsilon > 0$. If $d \geq 5$ then J. Januszewski and M. Lassak [JaL1997] even construct on-line packings under the very same condition. If $d = 3, 4$ then J. Januszewski and M. Lassak [JaL1997] provided on-line packings under the assumption that the total volume of the small cubes is at most $3/2^{d+1}$.

For coverings, J. Meir and L. Moser [MeM1968] stated that there exists a covering of $[0, 1]^d$ by the translates of the small cubes if the total volume of the small cubes is at least $2^d - 1$, and this conjecture was verified by H. Groemer [Gro1982] and by A. Bezdek and K. Bezdek [BeB1984]. This bound is optimal, which is shown by $2^d - 1$ cubes of edge length $1 - \varepsilon$ for small $\varepsilon > 0$. Again there exist very efficient on line methods: M. Lassak [Las2002] constructs on-line covering of $[0, 1]^d$ if the total volume of the small cubes is at least $2^d + 5/3 + 5/(3 \cdot 2^d)$.

Snakes of Cubes

A family $W_1, \ldots, W_n$ of unit cubes in $\mathbb{R}^d$ is a facet-to-facet snake if W_i and W_j intersect in a common facet when $|i - j| = 1$ and are disjoint otherwise. The snake is called limited if it is not the proper subset of any other snake, and L. Szabó and Z. Ujváry–Menyhárt [SzU2001] prove that the minimal cardinality of a limited facet-to-facet snake is $8d - 1$ for all $d \geq 3$.

10

Parametric Density

We consider the density of finite packings and coverings by translates of a given convex body K in $\mathbb{R}^d$, where the density depends also on a fixed real parameter. First, we discuss packings for which the parameter ϱ is positive. Given the nonoverlapping translates $x_1 + K, \ldots, x_n + K$, we write C_n to denote the convex hull of $x_1, \ldots, x_n$, and we define the *parametric density of the packing* to be

$$\frac{n \cdot V(K)}{V(C_n + \varrho K)}. \tag{10.1}$$

Our aim is to find the maximal parametric density over all packings of n translates of K for given ϱ, or, in other words, to find the minimum of $V(C_n + \varrho K)$. We observe that if $\varrho = 1$ then it is equivalent to find the minimal volume of convex bodies that contain n nonoverlapping translates of K. Parametric density for packings was introduced by J. M. Wills [Wil1993], and the fundamental properties were established in U. Betke, M. Henk, and J. M. Wills [BHW1994].

In general the optimal C_n is known only for very few choices of K, ϱ, and n. Say if $\varrho < 1/(32d)$ then the optimal C_n is a segment for any n and K (see Theorem 10.1.1), and the optimal arrangement is called a *sausage packing*.

We know much more about the asymptotic behavior of the optimal packings if K and ϱ are fixed and n tends to infinity. First, there exists an asymptotic parametric density $\delta(K, \varrho)$ (see Theorem 10.4.1). We define the sausage radius $\varrho_s(K)$ to be the supremum of all ϱ such that $\delta(K, \varrho)$ coincides with the asymptotic density of the optimal linear arrangement (hence $\varrho_s(K) \geq 1/(32d)$). We note that if $\varrho < \varrho_s(K)$ and K is smooth then any optimal packing is some sausage for large n (see Theorem 10.6.1). However, the critical radius $\varrho_c(K)$ is the minimal ϱ such that $\delta(K, \varrho)$ coincides with the translative packing density $\delta_T(K)$. In particular, the inradius of the optimal C_n tends to infinity with n if $\varrho > \varrho_c(K)$, and it stays bounded if $\varrho < \varrho_c(K)$. We note that $\varrho_c(K) \leq d$, and even $\varrho_c(K) \leq 2$ if K is centrally symmetric (see Section 10.5). Readily,

287

$\varrho_c(K) \geq \varrho_s(K)$, and in any dimension at least three, we exhibit a convex body
K satisfying $\varrho_c(K) > \varrho_s(K)$ (see Theorem 10.7.1).

For coverings, we discuss parametric density for any real parameter θ. If
$\theta < 0$ then we use the family $D \ominus M$ of all points x such that $x + M \subset L$
for some convex bodies D and M (see Section A.2). When writing $D \ominus M$
we always assume that D contains a translate of M. Given $\theta \in \mathbb{R}$ and convex
compact set D, we define

$$
D \oplus \theta \cdot K = \begin{cases} D + \theta K & \text{if } \theta \geq 0; \\ D \ominus |\theta| K & \text{if } \theta \leq 0. \end{cases}
$$

Now if the translates $x_1 + K, \ldots, x_n + K$ cover a compact convex set D then
the *parametric density of the covering* is

$$
\frac{n \cdot V(K)}{V(D \oplus \theta \cdot K)}. \tag{10.2}
$$

Given K, θ, and n, our aim is to find the compact convex set D covered by n
translates of K such that the parametric density is minimal, or equivalently,
$V(D \oplus \theta \cdot K)$ is maximal. Parametric density for coverings was investigated
first by U. Betke, M. Henk, and J. M. Wills [BHW1995b] in a somewhat
different context, namely, when the compact convex set D covered by the
translates $x_1 + K, \ldots, x_n + K$ is the convex hull of $x_1, \ldots, x_n$.

We have results only in the case when the real parameter θ is fixed and n
tends to infinity. In particular, we verify that there exists an asymptotic density
$\vartheta(K, \theta)$ (see Theorem 10.4.1). We define the critical radius $\theta_c(K)$ to be the
maximal θ such that $\vartheta(K, \theta)$ coincides with the translative covering density
$\vartheta_T(K)$. Writing D_n to denote the optimal compact convex set covered by n
translates of K, we have that the inradius of D_n tends to infinity with n if
$\theta < \theta_c(K)$ and stays bounded if $\theta > \theta_c(K)$ (see Section 10.5). In addition
if $\theta > d^2$ then there exists a segment s such that $D_n \subset s + 6(K - K)$ (see
Theorem 10.2.1). We note that the critical radius satisfies $-1 \leq \theta_c(K) \leq$
$10 \ln d / d$.

After presenting the general theory, we exhibit some most probably typical
properties by discussing some examples in Sections 10.7 to 10.11. Among
other specific problems, we survey results about parametric density for finite
lattice packings. In this case, given an o-symmetric K and reasonably large
ϱ, if n tends to infinity then there exists an asymptotic polytopal shape of the
optimal lattice packings of n translates; this shape resembles the well-known
Wulff-shape from crystallography.

Concerning notation, in this chapter we frequently write $|M|$ to denote
relative content of a compact convex set M and $M|L$ to denote the projection

of M into the affine subspace L; moreover, M/L denotes the projection of M into $L^\perp$.

10.1. Sausage Packings

Let K be a convex body in $\mathbb{R}^d$ for $d \geq 3$. We call a packing $x_1 + K, \ldots, x_n + K$ a *sausage packing* if $x_1, \ldots, x_n$ are collinear. Our aim is to show that, given a small $\varrho > 0$ and an integer n, any packing of n translates of K with maximal parametric density is some sausage.

Let us discuss sausage packings. We define a *cylinder* Z to be a convex body of the form $Z = A + s$, where s is a segment and A is a $(d-1)$-dimensional compact convex set. Given a line l, now we assume that $Z/l = K/l$, and Z contains K; moreover, s is a secant of K of maximal length parallel to l. We claim that in this case

$$V(K) \leq V(Z) \leq d \cdot V(K).\qquad(10.3)$$

We write x' to the image of an object X by the Steiner symmetrization with respect to $l^\perp$. Since K' contains a bypiramid that is the convex hull of K/l and s', the Fubini theorem yields $V(K) \geq |s| \cdot |K/L|/d = V(Z)$ and, in turn, (10.3). We also observe that $(n-1)s$ contains $x_1, \ldots, x_n$ such that $x_1 + K, \ldots, x_n + K$ is a packing, and

$$V((n-1)s + \varrho K) = (n-1)V(Z) \cdot \varrho^{d-1} + V(K) \cdot \varrho^d$$

for any $\varrho > 0$. We say that the packing $x_1 + K, \ldots, x_n + K$ is associated to the cylinder Z. In particular, given any $\varrho > 0$, a sausage packing of n translates of K is of maximal density (among sausage packings) if and only if it is associated to some cylinder Z containing K of minimal volume. These packings we call *optimal sausage packings*. We write $Z(K)$ to denote one of the cylinders of minimal volume containing K and write S_n to denote the convex hull of the translation vectors in the associated sausage packing of n translates of K; therefore, $Z(K)$ satisfies (10.3), and

$$V(S_n + \varrho K) = (n-1)V(Z(K)) \cdot \varrho^{d-1} + V(K) \cdot \varrho^d.\qquad(10.4)$$

Theorem 10.1.1. *Given a convex body K in $\mathbb{R}^d$, and a positive $\varrho < 1/(32d)$, if $C_n = \mathrm{conv}\{x_1, \ldots, x_n\}$ for a packing $x_1 + K, \ldots, x_n + K$ then*

$$V(C_n + \varrho K) \geq (n-1)V(Z(K)) \cdot \varrho^{d-1} + V(K) \cdot \varrho^d,$$

with equality if and only if the packing is an optimal sausage.

Proof. Let $K_0 = (K - K)/2$, and let the distance $d_{K_0}(x, L)$ of x from an affine subspace L be the minimum of $\|x - y\|_{K_0}$ for $y \in L$. We reindex $x_1, \ldots, x_n$ and define ω_k, $k = 1, \ldots, n$, in a way that

(a) $x_1 = -x_2$ and $2\omega_1 = \|x_1 - x_2\|_{K_0} = \max_{i,j} \|x_i - x_j\|_{K_0}$;

(b) if $k \geq 2$ and $L_k = \mathrm{aff}\{x_1, \ldots, x_k\}$ then

$$\omega_k = d_{K_0}(x_{k+1}, L_k) = \max_i d_{K_0}(x_i, L_k).$$

We prove Theorem 10.1.1 by contradiction; hence, assume that $\omega_2 > 0$. We observe that $\omega_1 \geq 1$, and we define m to be the minimal index such that $\omega_k \geq 1/2$ if $k \leq m$, and $\omega_{m+1} < 1/2$.

We recall that (A.42) yields that if an i-dimensional compact convex set M is contained in C_n then

$$\binom{d}{i} V(C_n, K; i) \geq \binom{d}{i} V(M, K; i) = |M| \cdot |K/\mathrm{aff}\, M|. \tag{10.5}$$

To compare volume estimates to $V(Z(K))$, we verify that if L is a linear i-space then

$$V(Z(K)) \leq i \cdot |K_0 \cap L| \cdot |K/L|. \tag{10.6}$$

Choosing a line $l \subset L$, we let $Z = s + A$ be a cylinder containing K such that $Z/l = K/l$, A is a $(d - 1)$-dimensional compact convex set, and s is a secant of K parallel to l. We may assume that the origin is the midpoint of s, and hence s is a common secant of K_0 and the cylinder $Z_0 = (Z - Z)/2$. If M is an i-dimensional section of Z parallel to L then $(M - M)/2 \subset L \cap Z_0$; thus, the Brunn–Minkowski inequality (see (A.45)) yields that $|M| \leq |L \cap Z_0|$. Since $|L \cap Z_0| \leq i \cdot |L \cap K_0|$ according to (10.3), we conclude (10.6) by $Z/L = K/L$ and the Fubini theorem.

Case I: $m = 1$

The definition of the mixed volumes and (10.4) yield

$$V(C_n + \varrho K) - V(S_n + \varrho K) \geq \sum_{i=d-2}^{d-1} \binom{d}{i} V(C_n, K; i) \cdot \varrho^i$$
$$-(n - 1)V(Z(K)) \cdot \varrho^{d-1}. \tag{10.7}$$

Let $l = \mathrm{lin}\{x_2\}$ and $L = \mathrm{lin}\{x_1, x_2, x_3\}$. Since $|(x_i + K_0) \cap l|$ is at least $(1 - \omega_2)|K_0 \cap l|$ for each x_i, we deduce that

$$d\, V(C_n, K; d - 1) \geq |l \cap C_n| \cdot |K/l| \geq (n - 1)(1 - \omega_2) \cdot V(Z(K)).$$

Next the triangle $T = x_1 x_2 x_3$ is contained in C_n and satisfies

$$|T| \geq \frac{\omega_2}{4} \frac{|l \cap C_n|}{|l \cap K_0|} \cdot |K_0 \cap L| \geq \frac{\omega_2(1 - \omega_2)}{4} \cdot (n - 1)|K_0 \cap L|,$$

which, in turn, yields the estimates (compare (10.5) and (10.6))

$$\binom{d}{d - 2} V(C_n, K; d - 2) \geq |T| \cdot |K/L|$$

$$\geq \frac{\omega_2(1 - \omega_2)}{8} \cdot (n - 1)V(Z(K)).$$

However, the conditions $0 < \omega_2 < 1/2$ and $\varrho \leq 1/16$ imply that

$$\frac{\omega_2(1 - \omega_2)}{8} \cdot \varrho^{d-2} + (1 - \omega_2) \cdot \varrho^{d-1} - \varrho^{d-1} > 0; \qquad (10.8)$$

hence, $V(C_n + \varrho K) > V(S_n + \varrho K)$ holds by (10.7), contradicting the minimality of $V(C_n + \varrho K)$.

Case II: $m \geq 2$

We define $M = \mathrm{conv}\{x_1, \ldots, x_{m+1}\}$ and $L = \mathrm{lin}\, M$. Property (b) yields that M is contained in an m-parallelepiped P such that, for suitable pairwise nonparallel $(m - 1)$-faces $F_1, \ldots, F_m$, F_1 touches $\omega_1 K_0$ at x_2, and F_k touches $\mathrm{lin}\{x_2, \ldots, x_k\} + \omega_k K_0$ at x_{k+1} for $k \geq 2$. It follows that $|P| = 2^{m-1} m! \cdot |M|$. We write $\tilde{x}_i$ to denote a closest point of L to x_i with respect to $\|\cdot\|_{K_0}$, and hence $\tilde{x}_1 + (K_0 \cap L)/2, \ldots, \tilde{x}_n + (K_0 \cap L)/2$ form a packing because $\omega_{m+1} < 1/2$. We deduce by the definition of ω_k, $k = 1, \ldots, m$, the existence of a certain m-parallelepiped $\tilde{P}$ that contains each $\tilde{x}_i + (K_0 \cap L)/2$ and whose content is

$$|\tilde{P}| = \prod_{i=1}^{m} \frac{\omega_i + \frac{1}{2}}{\omega_i} \cdot |P| \leq 3 \cdot 2^{m-2} |P| = 3 \cdot 2^{2m-3} m! \cdot |M|.$$

It follows that $|M| \geq n \cdot |K_0 \cap L|/(2^{3m-3} 3 m!)$; hence, first applying (10.5) and afterwards (10.6) leads to

$$V(C_n + \varrho K) > |M| \cdot |K/L| \cdot \varrho^{d-m}$$

$$\geq \frac{1}{2^{3m-3} 3 m! m} \cdot n V(Z(K)) \cdot \varrho^{d-m}.$$

Now the Stirling formula (see A.27) and the condition $\varrho < 1/(32d)$ yield the inequality $V(C_n + \varrho K) > n V(Z(K)) \varrho^{d-1}$, contradicting the minimality of $V(C_n + \varrho K)$ and completing the proof of Theorem 10.1.1. $\qquad \square$

Open Problem.

(i) Does Theorem 10.1.1 hold if $\varrho > \gamma$ for some absolute constant γ?

Comments. A somewhat weaker version of Theorem 10.1.1 is due to U. Betke, M. Henk, and J. M. Wills [BHW1995a]; namely, their upper bound on ϱ is of order $d^{-3/2}$ if K is centrally symmetric and of order d^{-2} for any K.

10.2. Near-Sausage Coverings

For coverings we verify that if the parameter θ is large then the optimal covering is not far from being a sausage.

Theorem 10.2.1. *Let K be a convex body in $\mathbb{R}^d$, $d \geq 3$, and let $\theta \geq d^2$. If, for given $n > 72 d^2$, D_n is a compact convex set such that it is covered by n translates of K, and $V(D_n + \theta K)$ is maximal, then there exists a segment $s_n \subset D_n$ satisfying*

$$D_n \subset s_n + 6 \cdot (K - K).$$

To verify Theorem 10.2.1, we will need the following simple estimate:

Proposition 10.2.2. *If $0 \leq i \leq d$ and the compact convex set D is covered by n translates of the convex body K in $\mathbb{R}^d$ then*

$$V(D, K; i) \leq n \cdot V(K).$$

Proof. We may assume that K is strictly convex and D is a polytope and, in addition, that the interiors of some translates $K_1, \ldots, K_n$ of K cover D. We use the definition (A.39) of the mixed volume, and for any K_i, consider the sum of

$$|F \cap K_i| \cdot |N(F) \cap (K / \text{aff } F)|$$

over all $(d - i)$-faces F of D that intersect int K_i. Since this sum is at most $V(D \cap K_i, K; i) \leq V(K)$, we conclude Proposition 10.2.2. $\square$

Proof of Theorem 10.2.1. We may assume that D_n is at least two dimensional. We write K_0 to denote $(K - K)/2$ and measure distances with respect to $\| \cdot \|_{K_0}$. Let $p, q \in D_n$ maximize the distances of pairs of points from D_n, and let $w \in D_n$ be a point of maximal distance from the line pq, where we write ω to denote the maximal distance. We prove Theorem 10.2.1 (ii) by contradiction; therefore, we suppose that $\omega \geq 12$. We write L to denote the affine two plane pqw, and we use T to denote the triangle pqw, that satisfies

$$|T| \geq \|p - q\|_{K_0} \cdot \frac{\omega}{8} \cdot |L \cap K_0|.$$

According to the Brunn–Minkowski inequality (see (A.45)), the area of any section of K parallel to L is at most $|L \cap K_0|$. Since T is covered by at most n translates of K, we deduce that its area is at most $n\,|L \cap K_0|$, which, in turn, yields

$$\|p - q\|_{K_0} \leq \frac{8n}{\omega}. \tag{10.9}$$

Next, we write s to denote a segment of maximal length that is contained in K and parallel to $p - q$. Since the segment $(n - 1)s$ can be covered by n translates of K, we have

$$V(D_n + \theta K) \geq (n - 1) \cdot d\, V(s, K; d - 1) \cdot \theta^{d-1} + V(K) \cdot \theta^d.$$

However,

$$
\begin{aligned}
V(D_n, K; d - 1) &\leq V\left(\frac{\|p - q\|_{K_0}}{2} \cdot s + \omega K, K; d - 1 \right) \\
&\leq \frac{4n}{\omega} \cdot V(s, K; d - 1) + \omega \cdot V(K),
\end{aligned}
$$

and hence

$$
\begin{aligned}
\sum_{i=0}^{d-2} \binom{d}{i} \frac{V(D_n, K; i)}{n V(K)} \cdot \theta^i &\geq \left(1 - \frac{1}{n} - \frac{4}{\omega} \right) \\
&\times \frac{d\, V(s, K; d - 1)}{V(K)} \cdot \theta^{d-1} - \frac{d\,\omega}{n} \cdot \theta^{d-1}.
\end{aligned}
$$

We deduce by $d\, V(s, K; d - 1) = |s| \cdot |K/\text{aff}\, s| \geq V(K)$ and by Proposition 10.2.2 that

$$\sum_{i=0}^{d-2} \binom{d}{i} \theta^i \geq \left(1 - \frac{1}{n} - \frac{4}{\omega} - \frac{d\,\omega}{n} \right) \cdot \theta^{d-1}. \tag{10.10}$$

Now $\omega \leq \sqrt{8n}$ because of (10.9) and $\omega \leq \|p - q\|_{K_0}$, and the function $f(\omega) = 4/\omega + d\,\omega/n + 1/n$ of ω is convex. It follows by $n > 72\,d^2$ that $f(\omega)$ is less than $5/12$ if $\omega = 12$ or $\omega = \sqrt{8n}$; therefore, the right-hand side of (10.10) is greater than $(7/12)\,\theta^{d-1}$. However, $\theta \geq d^2$ and $\binom{d}{i} \leq d^{d-i}/(d - i)!$

yield

$$\sum_{i=0}^{d-2} \binom{d}{i} \theta^i < \frac{d^2}{2} \theta^{d-2} + \frac{d^3}{6} \theta^{d-3} + \sum_{i=0}^{d-4} \frac{d^{d-i}}{24} \theta^i < \frac{7}{12} \cdot \theta^{d-1},$$

a contradiction that verifies Theorem 10.2.1.   □

It is a natural question to ask how close the optimal D_n is to being a segment if θ and n is large. Given a strictly convex body K and $\varepsilon > 0$, the proof of Theorem 10.2.1 provides $\theta_\varepsilon(K)$ such that if $\theta > \theta_\varepsilon(K)$ and n is large then $D_n \subset l + \varepsilon \cdot K$ for some line l. In addition, if K is a cube then D_n is actually a segment if θ and n are large (see Theorem 10.10.1). However, if K is a regular hexagon then, given any θ, the optimal D_n contains a translate of $K/3$ for large n (see Theorem 3.4.8).

10.3. Cluster-like Optimal Packings and Coverings

Given a convex body K in $\mathbb{R}^d$, the aim of this section is to investigate the optimal arrangements when the parameter ϱ is large for packings and the parameter θ is small for coverings. We show that cluster-like arrangements are optimal in both cases, and as the number of translates tends to infinity, the optimal parametric density tends to the translative packing density $\delta_T(K)$ and to the translative covering density $\vartheta_T(K)$, respectively (see Section 9.4 for definitions of $\delta_T(K)$ and $\vartheta_T(K)$). In particular, we need good density estimates for clusters: Given a convex body D with $r_K(D) \geq 1$, we write N to denote the maximal number of nonoverlapping translates of K inside D, and write M to denote the minimimal number of translates of K that cover D. If the write $\Delta_T(K) = V(K)/\delta_T(K)$ and $\Theta_T(K) = V(K)/\vartheta_T(K)$, Theorem 9.4.1 says

$$V(D - K) \geq \frac{N \cdot \Delta_T(K)}{M \cdot \Theta_T(K)} \geq V(D \ominus K). \tag{10.11}$$

In Theorem 10.3.1, C_n denotes the convex hull of some $x_1, \ldots, x_n$ such that $x_1 + K, \ldots, x_n + K$ is a packing, and D_n denotes some compact convex set such that D_n is covered by n translates of K.

Theorem 10.3.1. *Let K be a convex body in $\mathbb{R}^d$.*

(i) *Given $\varrho > 2$ in case K is centrally symmetric, and $\varrho > d + 1$ otherwise, if $V(C_n + \varrho K)$ is minimal then*

$$\lim_{n \to \infty} \frac{nV(K)}{V(C_n + \varrho K)} = \delta_T(K),$$

and for large n, possibly after translation, we have

$$\frac{1}{c_\varrho} \sqrt[d]{\frac{n}{\delta_T(K)}} \cdot K \subset C_n \subset c_\varrho \sqrt[d]{\frac{n}{\delta_T(K)}} \cdot K, \qquad (10.12)$$

where c_ϱ depends on ϱ and d, and $\lim_{\varrho \to \infty} c_\varrho = 1$.
(ii) Given $\theta < -1$, if $V(D_n \ominus |\theta|K)$ is maximal then

$$\lim_{n \to \infty} \frac{n\, V(K)}{V(D_n \ominus |\theta|K)} = \vartheta_T(K),$$

and for large n, possibly after translation, we have

$$\frac{1}{c_\theta} \sqrt[d]{\frac{n}{\vartheta_T(K)}} \cdot K \subset D_n \subset c_\theta \sqrt[d]{\frac{n}{\vartheta_T(K)}} \cdot K, \qquad (10.13)$$

where c_θ depends on θ and d, and $\lim_{\theta \to -\infty} c_\theta = 1$.

Proof. The statements of Theorem 10.3.1 about the asymptotics of the parametric density follow directly from (10.11), (10.12), and (10.13). With respect to the shape of the optimal arrangement, we present the proof for packings of centrally symmetric bodies in detail, and we discuss the necessary changes for the other cases at the end. Let $M_n = C_n + 2K$. Since $\sqrt[d]{n\Delta_T(K)/V(K)} \cdot K$ contains the centres of n nonoverlapping translates of K according to (10.11), the optimality of C_n yields for large n that

$$V(M_n + (\varrho - 2)K) \leq V\left(\sqrt[d]{\frac{n\Delta_T(K)}{V(K)}} \cdot K + \varrho K\right)$$
$$< n\Delta_T(K) + d(\varrho + 1)(n\Delta_T(K))^{\frac{d-1}{d}} V(K)^{\frac{1}{d}}.$$

However, $V(M_n) \geq n\Delta_T(K)$ by (10.11); hence,

$$V(M_n + (\varrho - 2)K) > n\Delta_T(K) + d(\varrho - 2) \cdot V(M_n, K; 1).$$

Comparing the two inequalities leads to

$$V(M_n, K; 1) < \frac{\varrho + 1}{\varrho - 2} \cdot V(M_n)^{\frac{d-1}{d}} V(K)^{\frac{1}{d}}. \qquad (10.14)$$

Let $M_n' = M_n / \sqrt[d]{n}$. Since $V(M_n) \sim n\Delta(K)$ and $V(M_n, K; 1) \geq V(M_n)/(dr_K(M_n))$ according to (A.48), we deduce for large n that $V(M_n') \sim \Delta(K)$ and

$$r_K(M_n') > \sqrt[d]{\frac{\Delta(K)}{V(K)}} \cdot \frac{\varrho - 2}{2d(\varrho + 1)}.$$

In turn, we conclude the existence of c_ϱ.

To understand the asymptotic behavior of c_ϱ, we assume that $\varrho \geq 3$. It follows that, possibly after translation, $c_1 K \subset M'_n \subset c_2 K$ hold for large n, where c_1 and c_2 depend only on K. Let $\mathcal{D}_\varrho$ denote the family of all convex bodies D satisfying (10.14) in place of M_n and $c_1 K \subset D \subset c_2 K$. Combining the Blaschke selection theorem (Theorem A.3.1) and the Minkowski inequality (A.43) yields that $R_K(D)/r_K(D) \leq \gamma(\varrho)$ holds for any $D \in \mathcal{D}_\varrho$, where $\lim_{\varrho \to \infty} \gamma(\varrho) = 1$. Therefore, we may choose c_ϱ so that $\lim_{\varrho \to \infty} c_\varrho = 1$. Finally, (10.11) and (10.12) imply that the asymptotic density is $\delta_T(K)$.

If K is any convex body then $-K \subset d K$ can be assumed according to (A.19), and hence $K - K \subset (d + 1)K$. Therefore, the same argument as before works with $M_n = C_n + (\varrho + 1)K$.

For coverings let $M_n = D_n \ominus |\theta| K$. Since $\left(\sqrt[d]{n\Theta_T(K)/V(K)} - 1 \right) \cdot K$ can be covered by n translates of K according to (10.11), it follows for large n that

$$V(M_n) \geq V\left(\left(\sqrt[d]{\frac{n\Theta_T(K)}{V(K)}} - 1 - |\theta| \right) \cdot K \right)$$
$$> n\Theta_T(K) - d\left(|\theta| + 1\right) \cdot (n\Theta_T(K))^{\frac{d-1}{d}} V(K)^{\frac{1}{d}}.$$

However, $V(M_n + (|\theta| - 1)K) \leq n\Theta_T(K)$ holds again by (10.11); therefore, if n is large then

$$V(M_n, K; 1) < \frac{|\theta| + 1}{|\theta| - 1} (n\Theta_T(K))^{\frac{d-1}{d}} V(K)^{\frac{1}{d}} < \frac{|\theta| + 2}{|\theta| - 1} V(M_n)^{\frac{d-1}{d}} V(K)^{\frac{1}{d}}.$$

Now we conclude Theorem 10.3.1. $\qquad\qquad\qquad\qquad\qquad\qquad\qquad \square$

Comments. The density estimates of Theorem 10.3.1 are due to U. Betke, M. Henk, and J. M. Wills; namely, [BHW1994] is concerned with packings, and [BHW1995b] is concerned with coverings (in a somewhat different setting). The shape of the optimal packings is discussed in K. Böröczky Jr. and U. Schnell [BöS1998a], who verify that $c_\varrho \leq 1 + \gamma \cdot \varrho^{-1/(d+1)}$ for large ϱ, where γ depends only on d. A similar argument yields $c_\theta \leq 1 + \gamma \cdot |\theta|^{-1/(d+1)}$ for coverings if θ is close to $-\infty$.

10.4. Asymptotic Density for Packings and Coverings

Let K be a convex body in $\mathbb{R}^d$. We show that, as n tends to infinity, the extremal density of n translates of K tend to some asymptotic density for both packings and coverings. We write C_n to denote the convex hull of $x_1, \ldots, x_n$ such that $x_1 + K, \ldots, x_n + K$ is a packing and $V(C_n + \varrho K)$ is minimal, and we write

D_n to denote a compact convex set such that D_n is covered by n translates of K and $V(D_n \oplus \theta \cdot K)$ is maximal. Then the candidates for the *asymptotic parametric density* are

$$\delta(K, \varrho) = \limsup_{n \to \infty} \frac{nV(K)}{V(C_n + \varrho K)} \quad \text{and} \quad \vartheta(K, \theta) = \liminf_{n \to \infty} \frac{nV(K)}{V(D_n \oplus \theta \cdot K)}.$$

For various applications, it is convenient to define $\Delta(K, \varrho) = V(K)/\delta(K, \varrho)$ for packings and $\Theta(K, \theta) = V(K)/\vartheta(K, \theta)$ for coverings, quantities that measure the average volume corresponding to one translate. Readily, we see that $\delta(K, \varrho)$ is monotonically decreasing and $\vartheta(K, \theta)$ is a monotonically increasing function of ϱ. Since $\delta(K, \varrho_2) \leq (\varrho_2/\varrho_1)^d \cdot \delta(K, \varrho_1)$ holds for $\varrho_2 > \varrho_1$, we deduce that $\delta(K, \varrho)$ is continuous in both variables. A similar argument yields that $\Delta(K, \varrho)$, $\vartheta(K, \theta)$, and $\Theta(K, \theta)$ are continuous as well.

Theorem 10.4.1. *Let K be a convex body in $\mathbb{R}^d$.*

(i) Let $\varrho > 0$. If, for given n, $x_1 + K, \ldots, x_n + K$ is a packing such that $V(C_n + \varrho K)$ is minimal for $C_n = \mathrm{conv}\{x_1, \ldots, x_n\}$ then

$$\lim_{n \to \infty} \frac{nV(K)}{V(C_n + \varrho K)} = \delta(K, \varrho).$$

(ii) Let $\theta \in \mathbb{R}$. If, for given n, D_n is a compact convex set covered by n translates of K such that $V(D_n \oplus \theta \cdot K)$ is maximal then

$$\lim_{n \to \infty} \frac{nV(K)}{V(D_n \oplus \theta \cdot K)} = \vartheta(K, \theta).$$

It is relatively easy to estimate the density of clusters; therefore, the following statement is at the heart of the proof of Theorem 10.4.1. We recall that if M is a compact convex set and $1 \leq k \leq d$ then $r_k(M)$ denotes the radius of the largest k-dimensional ball contained in M.

Lemma 10.4.2. *Using the notation of Theorem 10.4.1, let us assume that for an $R > 0$, for $1 \leq k \leq d - 1$, and for a subsequence $\{C_{n'}\}$, $r_k(C_{n'})$ tends to infinity and $r_{k+1}(C_{n'}) < R$. Then, for any $\varepsilon > 0$, there exist a k-dimensional parallelepiped Q contained in a linear k-plane L, a $(d - k)$-dimensional compact convex set A in $L^\perp$ with $R(A) \leq 2(k + 1)R$, and a packing of N translates of K in $Q + A + (K/L)$ such that*

$$V(Q + A + \varrho(K/L)) < N \cdot (1 + \varepsilon)\,\Delta(K, \varrho).$$

In addition, we have the analogous statement for coverings.

Before verifying Lemma 10.4.2, we show how it leads to Theorem 10.4.1.

Proof of Theorem 10.4.1 based on Lemma 10.4.2. We present the proof of Theorem 10.4.1 only for packings, when it is equivalent to verify

$$\limsup_{n \to \infty} \frac{V(C_n + \varrho K)}{n} \le \Delta(K, \varrho). \tag{10.15}$$

The definition of $\Delta(K, \varrho)$ provides a subsequence $\{C_{n'}\}$ satisfying

$$\lim_{n' \to \infty} \frac{V(C_{n'} + \varrho K)}{n'} = \Delta(K, \varrho).$$

If $r(C_{n'})$ is bounded then we may also assume that $r_k(C_{n'})$ tends to infinity, and $r_{k+1}(C_{n'})$ is bounded for some $1 \le k \le d - 1$. Using the notation of Lemma 10.4.2, we conclude (10.15) via the tiling of L by translates of Q.

Therefore, we may assume that $r(C_{n'})$ tends to infinity. In this case (10.11) yields that $V(C_{n'} + \varrho K) \sim n' \Delta_T(K)$, and hence $\Delta(K, \varrho) = \Delta_T(K)$. Now we let λ_n be minimal with the property $\lambda_n K + K$ contains n nonoverlapping translates of K. Since $V(\lambda_n K + \varrho K) \sim n \Delta_T(K)$ by (10.11), we conclude (10.15) and, in turn, Theorem 10.4.1. □

The rest of the section is devoted to the proof of Lemma 10.4.2. We present the argument for packings and sketch the necessary changes for coverings at the end.

It follows by (A.20) that $C_{n'} \subset L_{n'} + (k + 1) R B^d$, where $L_{n'}$ is an affine k-plane. Possibly taking a subsequence further, we obtain a linear k-plane L such that the linear k-planes $L_{n'} - L_{n'}$ tend to L. Now $r_k(C_{n'}|L)$ tends to infinity, and we may assume according to (A.19) that

$$-C_{n'}|L \subset k C_{n'}|L.$$

We need some auxiliary notions. We define $K' = K|L^{\perp}$ and $K_0 = (K - K)/2$, and we observe that there exists some $\Xi_n \subset C_n$ of cardinality n such that $\|x - y\|_{K_0} \ge 2$ for different $x, y \in \Xi_n$. In addition, let $P \subset L$ be any k-parallelepiped that contains $K|L$, and we choose some $\lambda > 0$ such that

$$\frac{(\lambda + 1)^k}{\lambda^k} < \sqrt{1 + \varepsilon}.$$

We write Λ to denote a k-lattice in L such that $\Lambda + \lambda P$ tile L. For any $y \in \Lambda \cap (C_{n'}|L)$, we define

$$A(y) = \left((y + \lambda P + L^{\perp}) \cap C_{n'} \right) | L^{\perp},$$

and we observe that $R(A(y)) \le 2(k + 1)R$ for large n'. In addition, if $x \in$

$C_{n'}|L$ then $B(x) \subset L^\perp$ is defined by

$$x + B(x) = (x + L^\perp) \cap C_{n'},$$

and hence $A(y)$ is the union of all $B(x)$ with $x \in y + \lambda P$. Naturally, $A(y)$ and $B(x)$ depend on n' as well. Now we show that the greatest part of the boundary of $C_{n'}$ is asymptotically parallel to L.

Proposition 10.4.3. *There exists some $\mu(n')$ such that $\mu(n')r_k(C_{n'}|L)$ tends to infinity, $\mu(n')$ tends to zero, and if n' is large and $y + \lambda P$ intersects $(1 - \mu(n'))C_{n'}|L$ for some $y \in \Lambda$ then*

$$V(\lambda P + A(y) + \varrho K') < (1 + \mu(n')) \cdot V((y + \lambda P + L^\perp) \cap (C_{n'} + \varrho K)).$$

Proof. It is sufficient to prove that, for any fixed $0 < \mu < \varrho$, if n' is large and $y + \lambda P$ intersects $(1 - \mu)C_{n'}|L$ for some $y \in \Lambda$ then

$$V(\lambda P + A(y) + (\varrho - \mu)K') < V((y + \lambda P + L^\perp) \cap (C_{n'} + \varrho K)). \quad (10.16)$$

In turn, (10.16) will follow if

$$A(y) \subset B(z) + \mu K' \quad (10.17)$$

for any $z \in y + \lambda P - (K|L)$. We may assume that n' is large enough to ensure that if $y + \lambda P$ intersects $(1 - \mu)C_{n'}|L$ for some $y \in \Lambda$ then $y + (2/\mu)(\lambda P - (K|L))$ is contained in $C_{n'}|L$, and the projection of

$$\left(y + \frac{8R}{\mu}(\lambda P - (K|L)) + L^\perp\right) \cap C_{n'}$$

into $L^\perp$ is contained in a translate of $2RK'$.

Let $x_1' \in A(y)$ be the projection of some $x_1 \in (y + \lambda P + L^\perp) \cap C_{n'}$ into $L^\perp$. We write x_1'' to denote the projection of x_1 into L and observe that $x_1'' + (4R/\mu)(z - x_1'')$ is the projection of some $x_2 \in C_{n'}$ into L. Now $x_3 = x_1 + [\mu/(4R)](x_2 - x_1)$ lies in $C_{n'}$, its projection into L is z, and $x_1' \in x_3' + \mu K'$ because $x_1' \in x_2' + 4RK'$. We conclude (10.17) and, in turn, Proposition 10.4.3.   □

To prove Lemma 10.4.2, next we show that the effect of the rim of C_n can be neglected; namely, we claim that

$$V\left((C_{n'} + \varrho K) \cap ((1 - 2\mu(n'))C_{n'}|L + L^\perp)\right)$$
$$> (1 - O(\mu(n'))) \cdot V(C_{n'} + \varrho K), \quad (10.18)$$

where the implied constant in $O(\cdot)$ depends only on d and k. We choose an

$x_0 \in (C_{n'} + \varrho K)|L$ such that the $(d - k)$-volume of

$$\widetilde{B} = (x_0 + L^\perp) \cap (C_{n'} + \varrho K)$$

is maximal. It follows for large n that the volume of the part of $C_{n'} + \varrho K$ whose projection into L does not intersect $(1 - 2\mu(n'))C_{n'}|L$ is at most

$$(1 - (1 - 3\mu(n'))^k) \cdot V_k((C_{n'} + \varrho K)|L) \cdot V_{d-k}(\widetilde{B}).$$

However, if $x \in (1/2)(C_{n'} + \varrho K)|L + (1/2)x_0$ then

$$V_{d-k}\left((x + L^\perp) \cap (C_{n'} + \varrho K)\right) \geq \frac{V_{d-k}(\widetilde{B})}{2^{d-k}}.$$

Therefore,

$$V(C_{n'} + \varrho K) > \frac{V_k((C_{n'} + \varrho K)|L)}{2^d} \cdot V_{d-k}(\widetilde{B}),$$

and we conclude the claim (10.18).

Let $\Omega_{n'}$ denote the family of all $y \in \Lambda$ such that $y + \lambda P$ intersects $(1 - \mu(n'))C_{n'}|L$. It follows by (10.18) that

$$\frac{V((C_{n'} + \varrho K) \cap (\Omega_{n'} + \lambda P + L^\perp))}{\#(\Xi_{n'} \cap (\Omega_{n'} + \lambda P + L^\perp) \cap C_{n'})} \leq (1 + O(\mu(n'))) \cdot \Delta(K, \varrho),$$

and hence there exists $y_{n'} \in \Omega_{n'}$ such that

$$\frac{V((C_n + \varrho K) \cap (y_{n'} + \lambda P + L^\perp))}{\#(\Xi_{n'} \cap (y_{n'} + \lambda P + L^\perp) \cap C_{n'})} \leq (1 + O(\mu(n'))) \cdot \Delta(K, \varrho).$$

Therefore, fixing some large n', Lemma 10.4.2 holds if we choose

$$Q = y_{n'} + (\lambda + 1)P \quad \text{and} \quad A = A(y_{n'}).$$

Turning to coverings, we use the analogous argument for D_n in place of C_n, only we need some obvious changes. The only nontrivial change is that $A(y)$ is defined to be the intersection of all $B(x)$ for $x \in y + \lambda P$. Therefore, the proof of Lemma 10.4.2 is now complete. $\qquad\square$

10.5. The Critical Radius for Packings and Coverings

Let K be a convex body in $\mathbb{R}^d$. According to Theorem 10.3.1, the asymptotic parametric densities are $\delta(K, \varrho) = \delta_T(K)$ if $\varrho > d + 1$ and $\vartheta(K, \theta) = \vartheta_T(K)$ if $\theta < -1$. Since $\delta(K, \varrho)$ is monotonically increasing and $\vartheta(K, \theta)$ is monotonically decreasing, and both are continuous, it is natural to define the *critical radii* $\varrho_c(K)$ *and* $\theta_c(K)$ *to be*

$$\varrho_c(K) = \min\{\varrho : \delta(K, \varrho) = \delta_T(K)\},$$
$$\theta_c(K) = \max\{\theta : \vartheta(K, \theta) = \vartheta_T(K)\}.$$

Theorem 10.5.1. *Let K be a convex body in $\mathbb{R}^d$.*

(i) Let $\varrho > 0$. If, for given n, $x_1 + K, \ldots, x_n + K$ is a packing such that $V(C_n + \varrho K)$ is minimal for $C_n = \operatorname{conv}\{x_1, \ldots, x_n\}$ then, as n tends to infinity,

$$r_K(C_n) \text{ tends to infinity } \quad \text{if } \varrho > \varrho_c(K);$$
$$r_K(C_n) \text{ stays bounded } \quad \text{if } \varrho < \varrho_c(K).$$

(ii) Let $\theta \in \mathbb{R}$. If, for given n, D_n is a compact convex set that is covered by n translates of K and $V(D_n \oplus \theta \cdot K)$ is maximal then, as n tends to infinity,

$$r_K(D_n) \text{ tends to infinity } \quad \text{if } \theta < \theta_c(K);$$
$$r_K(D_n) \text{ stays bounded } \quad \text{if } \theta > \theta_c(K).$$

Proof. We simply write ϱ_c and θ_c to denote $\varrho_c(K)$ and $\theta_c(K)$, respectively. We present the argument for packings in detail, and we sketch the necessary changes for coverings at the end.

Let $\varrho < \varrho_c$, and hence $\delta(K, \varrho) > \delta_T(K)$. We suppose that there exists a subsequence $\{C_{n'}\}$ such that $r_K(C_{n'})$ tends to infinity, and we seek a contradiction. Since $K - K \subset (d + 1)K$ can be assumed according to (A.19), it follows by (10.11) that

$$\left(\frac{r_K(C_{n'}) + \varrho}{r_K(C_{n'}) + d + 1}\right)^d \cdot n' \Delta_T(K) \leq V(C_{n'} + \varrho K)$$
$$\leq \left(\frac{r_K(C_{n'}) + \varrho}{r_K(C_{n'})}\right)^d \cdot n' \Delta_T(K).$$

Therefore, $\delta(K, \varrho) = \lim_{n' \to \infty} n' V(K)/V(C_{n'} + \varrho K) = \delta_T(K)$. This contradiction verifies that $r_K(C_n)$ is bounded.

If $\varrho > \varrho_c$ then we consider $M_n = C_n + \varrho_c K$. Now $\sqrt[d]{n \Delta_T(K)/V(K)} \cdot K$ contains $x_1, \ldots, x_n$ such that $x_1 + K, \ldots, x_n + K$ is a packing according to (10.11); thus, we deduce by the optimality of C_n that

$$V(M_n + (\varrho - \varrho_c)K) \leq V\left(\sqrt[d]{\frac{n \Delta_T(K)}{V(K)}} \cdot K + \varrho K\right) < n \Delta_T(K) + v(n) \cdot n,$$

where $\lim_{n \to \infty} v(n) = 0$. However, the definition of ϱ_c yields that $V(M_n) =$

$n\Delta_T(K) + \mu(n) \cdot n$, where $\lim_{n\to\infty} \mu(n) = 0$, and hence

$$V(M_n + (\varrho - \varrho_c)K) > n\Delta_T(K) + \mu(n) \cdot n + d(\varrho - \varrho_c) \cdot V(M_n, K; 1).$$

Therefore,

$$V(M_n, K; 1) \leq \frac{|\mu(n)| + |\nu(n)|}{d(\varrho - \varrho_c)} \cdot n.$$

Now $V(M_n) \sim n\Delta_T(K)$, whereas $V(M_n) \leq d\, V(M_n, K; 1) \cdot r_K(M_n)$ according to (A.48); thus, $r_K(M_n)$ tends to infinity.

For coverings, the case $\theta > \theta_c$ is completely analogous to the case $\varrho < \varrho_c$ for packings. If $\theta < \theta_c$ then we define $M_n = D_n \oplus \theta \cdot K$. The proof is based on the fact that

$$V(M_n + (\theta_c - \theta)K) \leq V(D_n \oplus \theta_c \cdot K) \sim n\Theta_T(K)$$

and uses ideas similar to those used for packings.   □

For the magnitude of the critical radii, we have the followings:

Lemma 10.5.2. *If K is a convex body in $\mathbb{R}^d$ then*

$$1 - \frac{\ln d}{d-1} \leq \varrho_c(K) \leq \begin{cases} 2 & \text{when } K \text{ is centrally symmetric,} \\ d & \text{in general;} \end{cases}$$

$$-1 \leq \theta_c(K) \leq \frac{10\ln d}{d}.$$

Remarks. The lower bound for $\varrho_c(K)$ and the upper bound for $\theta_c(K)$ are essentially optimal for large d because if K is a parallelepiped then $\varrho_c(K) = 1$ and $\theta_c(K) = 0$. In addition, if d is large then $\varrho_c(B^d) > 1.5$ according to (10.28). Finally, if $d \geq 3$ and $K = [0, 1]^{d-2} + B^2$ then $\theta_c(K) < -0.09$. This follows by $\vartheta_T(K) = \vartheta(B^2) = 2\pi/\sqrt{27}$ (see Corollary 5.1.2) and by the inequality $\vartheta(K, \theta) \leq (1 + \theta)^{-2}$ that is induced for $\theta > -1$ by the tiling of $[0, m]^{d-2} + B^2$ by m^{d-2} translates of K.

Proof. Theorem 10.3.1 verifies $\theta_c(K) \geq -1$ for any convex body and $\varrho_c(K) \leq 2$ if K is centrally symmetric. In addition, Theorem 10.3.1 yields that $\varrho_c(K) \leq d + 1$ holds for any convex body, but we slightly improve on this bound.

We suppose that $\varrho_c(K) > d$, and we seek a contradiction. Let ϱ satisfy $d < \varrho < \varrho_c(K)$, and, for given n, let C_n be the convex hull of $x_1, \ldots, x_n$, where $x_1 + K, \ldots, x_n + K$ is a packing such that $V(C_n + \varrho K)$ is minimal. We choose an $\varepsilon > 0$ with the property $(1 + \varepsilon)^2 \Delta(K, \varrho) < \Delta_T(K)$. Since $r(C_n)$ is bounded according to Theorem 10.5.1, Lemma 10.4.2 yields an

m-dimensional parallelepiped Q for some $1 \leq m \leq d - 1$ and a compact convex set $A \subset L^{\perp}$ for $L = \text{aff} Q$ such that $Q + A + (K/L)$ contains some N nonoverlapping translates of K, and

$$V(Q + A + \varrho(K/L)) \leq (1 + \varepsilon) \Delta(K, \varrho) \cdot N < \frac{\Delta_T(K)}{1 + \varepsilon} \cdot N. \quad (10.19)$$

We may assume that $|(Q + K - K)|L| < (1 + \varepsilon)|Q|$ holds in addition to (10.19), since we may replace Q by kQ for suitable integer k if necessary. However, $-(K/L) \subset (d - m)(K/L)$ can be assumed according to (A.19), and hence (10.11) yields that

$$V(Q + A + \varrho(K/L)) \geq \frac{1}{1 + \varepsilon} V(Q + A + K - K) \geq \frac{\Delta_T(K)}{1 + \varepsilon} \cdot N.$$

This contradicts (10.19); therefore, $\varrho_c(K) \leq d$.

For the lower bound for $\varrho_c(K)$, sausage packings (see Section 10.1) show that

$$V(K) \leq \Delta_T(K) \leq \varrho_c(K)^{d-1} V(Z(K)) \leq d\varrho_c(K)^{d-1} V(K),$$

and hence $\varrho_c(K) \geq d^{-1/(d-1)} \geq 1 - \ln d/(d - 1)$.

Turning to the upper bound for $\theta_c = \theta_c(K)$, we let l be a line such that $s = l \cap K$ has maximal length among the sections of K parallel to l. To obtain a rough estimate, we observe that $n \cdot s$ can be covered by n translates of K, and $V(ns + \theta_c K) > n \cdot \theta_c^{d-1} V(K)$. We deduce by the Rogers bound $\vartheta(K, \theta_c) = \vartheta_T(K) \leq 4d \ln d$ (see Theorem 9.5.2) that $\theta_c \leq (4d \ln d)^{1/(d-1)}$. This settles the case $d = 3$ and shows that $\theta_c < d - 1$ if $d \geq 4$.

When $d \geq 4$, let M be the family of $x \in \mathbb{R}^d/l$ such that the length of $(x + l) \cap K$ is at least $(1 + \theta_c)/d$ times the length of s; therefore, M contains a translate of $(1 - (1 + \theta_c)/d) \cdot (K/l)$. Since suitable n translates of K cover

$$D_n = (n - \gamma) \frac{1 + \theta_c}{d} \cdot s + \left(1 - \frac{1 + \theta_c}{d}\right) \cdot (K/l)$$

for some constant γ, we deduce by $\vartheta(K, \theta_c) \leq 4d \ln d$ that

$$\frac{1}{d} \cdot \left(\frac{d - 1}{d}\right)^{d-1} \cdot (1 + \theta_c)^d \leq 4d \ln d.$$

Here $[(d - 1)/d]^{d-1} > 1/e$; hence, simple estimates yield Lemma 10.5.2.

$\square$

Open Problem.

(i) Is $\varrho_c(K) \leq 4$ for any convex body K? Or at least $\varrho_c(K) \leq \gamma$ for some absolute constant γ?

Comments. The bound $\varrho_c(K) \leq 2$ if K is centrally symmetric is due to U. Betke, M. Henk, and J. M. Wills [BHW1994], and they prove $\varrho_c(K) \leq d + 1$ in general. The bound $\theta_c(K) \geq -1$ is in essence a result of U. Betke, M. Henk, and J. M. Wills [BHW1995b]. In addition, K. Böröczky Jr. and U. Schnell [BöS1998a] proved for packings that the inradius for the optimal packing tends to infinity if $\varrho > \varrho_c(K)$.

10.6. The Sausage Radius for Packings

Given a convex body K in $\mathbb{R}^d$, we introduce a natural threshold $\varrho_s(K)$ such that any optimal packing of given large number of translates is a sausage if the parameter $\varrho < \varrho_s(K)$ and is at least two dimensional if $\varrho > \varrho_s(K)$. We write K_0 to denote $(K - K)/2$, and we measure distances with respect to $\| \cdot \|_{K_0}$.

We recall from Section 10.1 that $Z(K)$ is a cylinder of minimal volume containing K, and writing S_n to denote the segment of length $2(n - 1)$ containing the translation vectors of the associated packing of n translates of K, we have

$$V(S_n + \varrho K) = (n - 1)V(Z(K)) \cdot \varrho^{d-1} + V(K) \cdot \varrho^d \qquad (10.20)$$

for any $\varrho > 0$. In this section C_n always denotes the convex hull of some $x_1, \ldots, x_n$, where $x_1 + K, \ldots, x_n + K$ is a packing. In particular, the packing is called an *optimal sausage packing* if C_n is a segment and $V(C_n + \varrho K) = V(S_n + \varrho K)$ for any $\varrho > 0$.

According to Theorem 10.1.1, if $\varrho < 1/(32d)$ then the optimal packings are sausages, and their density is asymptotically $V(K)/[\varrho^{d-1}V(Z(K))]$. Since $\delta(K, \varrho)$, and hence $\Delta(K, \varrho)$ is continuous, we may define the *sausage radius* $\varrho_s(K)$ by

$$\varrho_s(K) = \max\{\varrho : \ \Delta(K, \varrho) = \varrho^{d-1}V(Z(K))\}.$$

We note that $1/(32d) \leq \varrho_s(K) \leq 4$ (see Lemma 10.6.2 later in this section).

Theorem 10.6.1. *Given a smooth convex body K in $\mathbb{R}^d$, let $\varrho < \varrho_s(K)$. If $x_1 + K, \ldots, x_n + K$ is a packing for large n then $C_n = \mathrm{conv}\{x_1, \ldots, x_n\}$ satisfies*

$$V(C_n + \varrho K) \geq (n - 1)V(Z(K)) \cdot \varrho^{d-1} + V(K) \cdot \varrho^d,$$

with equality if and only if the packing is an optimal sausage.

Remarks. Let K be any convex body in $\mathbb{R}^d$. If $\varrho > \varrho_s(K)$ and n is large then $r_2(C_n) > \gamma$ holds for the optimal C_n, where γ is a positive constant depending on K and ϱ. However, if $\varrho < \varrho_s(K)$ and $V(C_n + \varrho K)$ is minimal for given n then

(i) $\lim\limits_{n\to\infty} \dfrac{V(C_n + \varrho K)}{n} = V(Z(K)) \cdot \varrho^{d-1}$;

(ii) $C_n \subset l_n + \mu(n) \cdot K$, where l_n is a line, $\lim_{n\to\infty} \mu(n) = 0$, and $\mu(n)$ depends on n, ϱ, and K;

(iii) C_n is segment provided that $\varrho < 1/4$ and n is large.

Proof of Theorem 10.6.1 and its remarks. We first prove the remarks. The case $\varrho > \varrho_s(K)$ of the remarks follows from the inequality $\Delta(K, \varrho) < \varrho^{d-1} V(Z(K))$; therefore, let $\varrho < \varrho_s(K)$.

Our main tool is the auxiliary function (compare (A.38) and (10.20))

$$f(t, C_n) = V(S_n + t\,K) - V(C_n + t\,K) \tag{10.21}$$

$$= t^{d-1}\left((n-1)\,V(Z(K)) - \sum_{i=0}^{d-1} \binom{d}{i} V(C_n, K; i)\, \frac{1}{t^{d-i-1}} \right)$$

for $t \ge 0$. It follows by the definition of the sausage radius that

$$f(\varrho_s(K), C_n) \le \nu(n) \cdot n, \tag{10.22}$$

where $\lim_{n\to\infty} \nu(n) = 0$ holds for the nonnegative $\nu(n)$ that depends only on n and K. Since $f(\varrho, C_n) \ge 0$ by the optimality of C_n, the function $f(t, C_n)$ is increasing in t for $t \ge \varrho$; therefore, $f(\varrho, C_n) \le \nu(n) \cdot n$ as well. In turn, we conclude (i).

The rest of the statements are proved by contradiction; therefore, we suppose that C_n is at least two dimensional. We assume that $-K \subset d\,K$ (see (A.19)), write K_0 to denote $(K - K)/2$, and measure distances always with respect to the norm $\|\cdot\|_{K_0}$. Let $l = \text{aff}\{x_1, x_2\}$, where x_1 and x_2 maximize the distances of points in C_n, and let x_3 maximize the distances of points in C_n from l. We may assume that $o \in l$. We write ω to denote the distance of x_3 and l, and we define $L = \text{aff}\{x_1, x_2, x_3\}$. Moreover, let Π be the parallelogram in L that is circumscribed around $L \cap K_0$ and has two sides that are parallel to l with the other two sides touching at the endpoints of $l \cap K_0$. In particular, the area of the triangle $T = x_1 x_2 x_3$ is

$$A(T) = \frac{\omega}{8} \cdot \|x_1 - x_2\|_{K_0} \cdot |\Pi|.$$

In addition, any two-section of K parallel to L is contained in a translate of Π, and thus $V(Z(K)) \le |\Pi| \cdot |K/L|$. We deduce by (A.42) that

$$\binom{d}{2} V(C_n, K; d-2) \ge |T| \cdot |K/L| \ge \frac{\omega}{8} \cdot \|x_1 - x_2\|_{K_0} V(Z(K)).$$

$$\tag{10.23}$$

After these preparations, the proof of (ii) also uses the already introduced function $f(t, C_n)$. Now $f(\varrho, C_n) \geq 0$ yields that $f(t, C_n) \geq 0$ for $t \geq \varrho$; hence,

$$t \cdot \frac{\partial}{\partial t} f(t, C_n) = (d - 1) f(t, C_n) + \sum_{i=0}^{d-2} (d - 1 - i) \binom{d}{i} V(C_n, K; i) \cdot t^i$$

$$\geq \sum_{i=0}^{d-2} \binom{d}{i} V(C_n, K; i) \cdot \varrho^i.$$

This inequality and (10.22) lead to

$$\sum_{i=0}^{d-2} \binom{d}{i} V(C_n, K; i) \cdot \varrho^i \leq \frac{\varrho_s(K)}{\varrho_s(K) - \varrho} \cdot v(n) n, \qquad (10.24)$$

and we deduce by $f(\varrho, C_n) \leq v(n) \cdot n$ that if n is large then

$$d V(C_n, K; d - 1) \geq \frac{V(Z(K))}{2} \cdot n. \qquad (10.25)$$

We will introduce positive constants $c_1, \ldots, c_4$ that depend only on K, d, and ϱ. First, (10.24) yields $\binom{d}{2} V(C_n, K; d - 2) \leq c_1 v(n) \cdot n$, and hence $\|x_1 - x_2\|_{K_0} \leq c_2 n v(n) / \omega$ follows by (10.23). Since $C_n - C_n$ is contained in $\|x_1 - x_2\|_{K_0} \cdot K_0$, it follows that $V(C_n, K; d - 1) \leq c_3 n v(n) / \omega$. Therefore, (10.25) yields $\omega \leq c_4 v(n)$, and hence we may choose $\mu(n) = c_4 d \cdot v(n)$ in (ii). For (iii), if n is large and the line l intersects ωK_0 then the length of $l \cap K_0$ is at least $2(1 - \omega)$; hence, $\|x_1 - x_2\|_{K_0} \geq 2(1 - \omega)(n - 1)$. Given $\varrho < 1/4$, if n is large then $\omega \leq c_4 v(n)$ leads to

$$1 - \frac{\omega(1 - \omega)}{4\varrho} - (1 - \omega) < 0.$$

Therefore $d V(C_n, K; 1) \geq (1 - \omega)(n - 1) V(Z(K))$ and (10.23) yield $f(\varrho, C_n) < 0$. This contradiction verifies (iii).

Finally, let the boundary of K be smooth. If n is large then the ω in (10.23) becomes arbitrarily small according to (ii). In particular, for any $\varepsilon > 0$, if n is large and the line l intersects ωK_0 then the length of $l \cap K_0$ is at least $2(1 - \varepsilon\omega)$, and hence $\|x_1 - x_2\|_{K_0}$ is at least $2(1 - \varepsilon\omega)(n - 1)$. We choose $\varepsilon > 1$ satisfying $\varepsilon + 4\varrho\varepsilon < 1$; hence,

$$1 - \frac{\omega \cdot (1 - \varepsilon\omega)}{4\varrho} - (1 - \varepsilon\omega) < 0.$$

Therefore, we conclude Theorem 10.6.1. □

Concerning the value of the sausage radius, we have the following:

Lemma 10.6.2. *If K is a convex body in $\mathbb{R}^d$ then*

$$\frac{1}{32d} \leq \varrho_s(K) \leq \begin{cases} 4 & \text{in general;} \\ 2 & \text{if } K \text{ is centrally symmetric.} \end{cases}$$

Proof. Here the lower bound is a consequence of Theorem 10.1.1. For the upper bounds, clusters show (see (10.11))

$$V(Z(K)) \cdot \varrho_s(K)^{d-1} \leq \Delta_T(K) \leq \Delta_T(Z(K)).$$

Now $Z(K) = s + M$ for a segment s and $(d-1)$-dimensional compact convex set M. Since $\Delta_T(M) \leq |M - M|$ holds according to Theorem 9.5.2 (ii), we deduce that $\varrho_s(K)^{d-1} \leq |M - M|/|M|$. Thus, the centrally symmetric case readily follows. If K is any convex body then we use the Rogers–Shephard inequality (see (A.45)): $|M - M|/|M| \leq \binom{2(d-1)}{(d-1)} < 4^{d-1}$.   □

Open Problem.

(i) Can $\varrho_s(K)$ be bounded below by a positive absolute constant γ for any d and convex body K in $\mathbb{R}^d$?

The existence of γ is known for centrally symmetric bodies that are almost ellipsoids (see M. Henk [Hen1995]).

Comments. The paper by U. Betke, M. Henk, and J. M. Wills [BHW1995a] provides a lower bound of order d^{-2} for the sausage radius for general K (and of order $d^{-3/2}$ if K is centrally symmetric).

10.7. The Critical and the Sausage Radius May Not Be Equal

According to Corollary 3.2.2, the sausage radius and the critical radius coincide for planar packings. In this section we construct convex bodies in any dimension $d \geq 3$ such that the critical and sausage radius differ.

Theorem 10.7.1.

(i) If K is a regular octahedron in $\mathbb{R}^3$ then $\varrho_s(K) \leq 3/4 < 8/9 \leq \varrho_c(K)$,
(ii) If $d \geq 4$ and K is the convex body in $\mathbb{R}^d$ that is the direct sum of B^{d-2} and a regular hexagon then $\varrho_s(K) < 1 < \varrho_c(K)$.

Proof. We start with (ii). The tiling of the plane by regular hexagons shows that $\delta(K, 1) = 1$; thus, $\delta_T(K) = \delta_T(B^{d-2}) < 1$ leads to $\varrho_c(K) > 1$. However, K is not a cylinder, and hence $\Delta(K, 1) < V(Z(K))$, which, in turn, yields that $\varrho_c(K) < 1$.

Next let K be the convex hull of $\pm e_1, \pm e_2, \pm e_3$ where e_1, e_2, e_3 form an orthonormal system in $\mathbb{R}^3$. We claim that

$$V(Z(K)) = 2. \tag{10.26}$$

For $u \in \partial K$, we define

$$\sigma(K, u) = 2\|u\| \cdot |K/u|,$$

hence, $V(Z(K))$ is the minimum of $\sigma(K, u)$ over $u \in \partial K$. We may assume that u is contained in $F = \text{conv}\{e_1, e_2, e_3\}$ and that e_1 is the closest vertex of F to u. In addition, let m_i be the midpoint of the side $e_j e_k$, where $\{i, j, k\} = \{1, 2, 3\}$.

First, we assume that $u \in \text{conv}\{e_1, m_2, m_3\}$; hence, K/u is the projection of the square $\text{conv}\{\pm e_2, \pm e_3\}$ into $u^\perp$. Since the area of the square is two, we deduce that

$$\sigma(K, u) = 4\|u\| \cdot \cos \angle e_1 o u,$$

and thus $\sigma(K, u)$ is minimal if and only if $u \in m_2 m_3$. In particular, we may assume that u is contained in the triangle $m_1 m_2 m_3$. In this case let $v = (e_1 + e_2 + e_3)/3$, and let T denote the triangle $e_1 e_2 e_3$. Now K/u is the projection of the (possibly degenerate) hexagon

$$\text{conv}\{(T - u) \cup (u - T)\},$$

whose area is $2|T|$ (see Lemma 2.2.5). Therefore,

$$\sigma(K, u) = 4|T| \cdot \|u\| \cdot \cos \angle vou = \sigma(K, v) = 2,$$

completing the proof of (10.26).

Now we prove Theorem 10.7.1 (i) by comparing the densities of the one-, two-, and three-dimensional packings when n is large. The two-dimensional arrangement is defined by the linear plane L that is parallel to the face $e_1 e_2 e_3$. It follows that $L \cap K$ is a regular hexagon H with area $3\sqrt{3}/4$, and the distance between L and the face $e_1 e_2 e_3$ is $1/\sqrt{3}$. Thus, if λ_n is minimal with the property that $\lambda_n H$ contains $x_1, \ldots, x_n$ such that $x_1 + K, \ldots, x_n + K$ form a packing, then

$$V(\lambda_n H + \varrho K) \sim n \cdot \frac{3}{2}\varrho,$$

which, in turn, yields for any $\varrho > 0$ that

$$\Delta(K, \varrho) \leq \frac{3}{2}\varrho. \tag{10.27}$$

Since $\Delta(K, \varrho_s(K)) = 2\varrho_s(K)^2$ according to Proposition 10.26, we deduce $\varrho_s(K)) \leq 3/4$. However, $\Delta(K, \varrho_c(K)) \geq V(K) = 4/3$ yields $\varrho_c(K) \geq 8/9$, and hence the proof of Theorem 10.7.1 is complete. $\square$

Comments . The at least four-dimensional example here is due to P. Gritzmann [Gri1984], and the estimates for the regular octahedron are proved by V. Arhelger, U. Betke, and K. Böröczky Jr. [ABB2001].

10.8. Ball Packings

For packings of $n \leq d + 1$ of unit balls in $\mathbb{R}^d$, K. Böröczky Jr. and J. M. Wills [BöW1997] established the existence of some $\varrho_{n,d}$ with the following property: Given ϱ, let C_n be the convex hull of n nonoverlapping unit balls in $\mathbb{R}^d$ such that $V(C_n + \varrho B^d)$ is minimal. Then C_n is a regular $(n - 1)$-simplex if $\varrho > \varrho_{n,d}$, and C_n is a segment if $n \leq 4$ and $\varrho < \varrho_{n,d}$. We note that some possibly optimal configurations for larger ϱ in $\mathbb{R}^3$ are exhibited by P. Scholl, A. Schürmann, and J. M. Wills [SSW2003].

With respect to the critical radius, we have

$$2^{-0.599+o(1)} \leq \varrho_c(B^d) \leq 2. \tag{10.28}$$

Here the upper bound is stated in Lemma 10.5.2. For the lower bound, the sausage arrangement shows that $2\kappa_{d-1}\varrho_c(B^d)^{d-1} \geq \Delta(B^d)$. Since $\delta(B^d) \leq 2^{-0.599 d + o(d)}$ according to the Kabatjanskii–Levenštein bound (8.6), $\varrho_c(B^d) \geq 2^{-0.599+o(1)}$ follows by $\kappa_d/\kappa_{d-1} < \sqrt{2\pi/d}$ (see (A.29)).

Next U. Betke, M. Henk, and J. M. Wills [BHW1995a] prove

$$\varrho_s(B^d) \geq 2^{0.5+o(1)}. \tag{10.29}$$

In particular, the gap between $\varrho_s(B^d)$ and $\varrho_c(B^d)$ is rather small. Actually, U. Betke, M. Henk and J. M. Wills [BHW1994] propose

Conjecture 10.8.1 (Strong sausage conjecture).

$$\varrho_s(B^d) = \varrho_c(B^d).$$

The conjecture holds for $d = 2$ (actually for any convex domains, not only for circular discs; see Corollary 3.2.2). For higher dimensions (see U. Betke, M. Henk, and J. M. Wills [BHW?] for thorough discussion), the conjecture has been verified for finite lattice packings in $\mathbb{R}^3$ (see V. Arhelger, U. Betke,

310   Parametric Density

and K. Böröczky Jr. [ABB2001] or Example 10.12.2) and for packings based on lattice zonotopes in any dimension (see K. Böröczky Jr. and J. M. Wills [BöW1997]). Further evidence is provided by U. Schnell and J. M. Wills [ScW2000]; namely, the optimal arrangement is never two dimensional for $d \geq 3$ and $n \geq 4$.

10.9. Packings of Parallelepipeds

Parallelepipeds are the only convex bodies in higher dimensional spaces such that an essentially complete description of the optimal packings with respect to the parametric density is available:

Theorem 10.9.1. *If $s_1, \ldots, s_d$ are segments and $P = \sum s_i$ is a parallelepiped in $\mathbb{R}^d$ then $\varrho_s(P) = \varrho_c(P) = 1$. In addition, let C_n be the union the centres of n nonoverlapping translates of P such that $V(C_n + \varrho K)$ is minimal. Then*

(i) $C_n = (n - 1)s_i$ for some i if $\varrho < 1$;
(ii) $C_n = (k - 1)P$ if $\varrho > 1$ and $n = k^d$ for some positive integer k;
(iii) $\lim_{n \to \infty} n^{-\frac{1}{d}} C_n = P$ if $\varrho > 1$.

Proof. We may assume that $P = [0, 1]^d$. If $\varrho = 1$ then both sausages and clusters occur as optimal packings; therefore, $\varrho_s(P) = \varrho_c(P) = 1$. Since, for any compact convex set C, $dV(C, P; 1)$ is the sum of the projections of C into the coordinate axes, we deduce (i) by induction on n. Now if $\varrho > 1$ then

$$V(C_n + \varrho P) = V(Q_n + (\varrho - 1)P)$$

for $Q_n = C_n + P$, and hence (ii) follows by $V(Q_n) \geq n$ and by the Brunn–Minkowski inequality (A.44).

Let us verify (iii). For large n, we define $m = \lfloor n^{\frac{1}{d}} \rfloor$. Let α be the minimal integer such that $n \leq m^d + \alpha\, m^{d-1}$, and let β be the maximal integer such that

$$n \leq m^d + \alpha \cdot m^{d-1} - \beta(\alpha + \beta) \cdot m^{d-2} = m^{d-2}(m + \alpha + \beta) \cdot (m - \beta).$$

Therefore, the rectangular box R whose two edges are of length $m + \alpha + \beta$ and $m - \beta$, and whose other edges are of length m, contains n nonoverlapping translates of the unit cube. Since $\alpha \leq d + 1$ and $\beta < \sqrt{m}$, we deduce for large n that

$$V(Q_n) + d(\varrho - 1)V(Q_n, P; 1) \leq V(R + (\varrho - 1)P)$$
$$\leq m^d + d(\varrho - 1)m^{d-1} + O\left(m^{d-1.5}\right).$$

In particular, $V(Q_n) \geq n$ yields that

$$V(Q_n, P; 1) \leq \left(1 + O\left(n^{-2d}\right)\right) \cdot V(Q_n)^{\frac{d-1}{d}} V(P)^{\frac{1}{d}};$$

hence, (iii) follows by the Minkowski inequality (A.43). □

Comments. Theorem 10.9.1 is observed by K. Böröczky Jr. and U. Schnell [BöS1998a].

10.10. Coverings by Parallelepipeds

For coverings by parallelepipeds, the optimal compact convex set is eventually a segment if θ is large.

Theorem 10.10.1. *If P is a parallelepiped then $\theta_c(P) = 0$. In addition, let $\theta \in \mathbb{R}$ and, for any n, let D_n be a compact convex set, that is covered by n translates of P and satisfies that $V(D_n + \theta P)$ is maximal.*

(i) If $\theta < 0$ then $\lim_{n\to\infty} n^{-\frac{1}{d}} D_n = P$;
(ii) if $\theta > 300\,d^2$ and n is large then D_n is a segment.

Proof. We may assume that P is the unit cube $[-1, 1]^d$. Since both sausages and clusters occur as optimal coverings when $\theta = 0$, we deduce that $\theta_c(P) = 0$. The case $\theta < 0$ can be handled similarly to the case $\varrho > 1$ for packings in the previous section.

Therefore, let $\theta > 300\,d^2$. The main idea is to slice the covering of D_n orthogonally to a suitable diagonal of P and to show that if D_n is not a segment parallel to that diagonal then D_n is "too short" in the direction of the diagonal. We write H_t to denote the hyperplane consisting of the points whose coordinates add up to t and, for a compact convex set M, we write $\sigma(M)$ to denote the maximal t such that H_t intersects M. In particular, if M is a unit cube then $\sigma(M) - 2d$ is the minimal t such that H_t intersects M.

We always measure distances and lengths with respect to $\|\cdot\|_P$, and we assume that n is sufficiently large. Let $s_n \subset D_n$ be a segment of maximal length, and we write ω_n to denote the maximal distance of points of D_n from s_n and use p_n to denote the point of D_n realizing ω_n. It follows by Theorem 10.2.1 that $\omega_n \leq 12$. We may assume that s_n connects the origin to a point whose coordinates are all positive. In particular, $d\,V(P, s_n; 1) = \sigma(s_n)$, and hence (A.40) implies

$$V(D_n + \theta P) \leq V(s_n + (\omega_n + \theta)\,P) = (\omega_n + \theta)^d + \sigma(s_n) \cdot (\omega_n + \theta)^{d-1}.$$

However, $V(D_n + \theta P) \geq \theta^d + 2dn \cdot \theta^{d-1}$ because n times a diagonal of P can be covered by n translates of P. Therefore, $\omega_n \leq 12$ and $\theta \geq 300\, d^2$ yield that

$$2d \cdot n \leq 2d \cdot \omega_n + \sigma(s_n) \cdot \left(1 + \frac{\omega_n}{200\, d}\right). \qquad (10.30)$$

In particular, $\sigma(s_n) > (2d - 1)n$ if n is large. Since D_n is covered by n translates of P, we deduce that $-n < \sigma(p_n) \leq 2d\, n$. Thus, for any t satisfying $(1/2)\, dn \leq t \leq (3/2)\, dn$, the intersection $H_t \cap D_n$ contains a segment of length at least $\omega/12$.

Let us index the n translates $P_1, \ldots, P_n$ of P covering D_n in a way that $\sigma(P_i) \geq \sigma(P_j)$ if $i > j$. We observe that $\sigma(P_{i+1}) - \sigma(P_i) \leq 2d$ holds for any $i = 1, \ldots, n - 1$, and we claim that if $(1/2)\, dn \leq \sigma(P_i) \leq (3/2)\, dn$ then

$$\sigma(P_{i+2}) - \sigma(P_i) \leq 4d - \frac{\omega_n}{12}.$$

The claim simply follows from the fact that otherwise the diameter of $H_{\sigma(P_i)} \cap P_{i+i}$ is smaller than $\omega_n/12$. Therefore, $\sigma(s_n) \leq 2dn - n\, \omega_n/75$, which, in turn, yields that $\omega_n = 0$ for large n by (10.30). $\qquad \square$

10.11. Finite Lattice Packings and the Wulff-Shape

Given an o-symmetric convex body K in $\mathbb{R}^d$, a *finite lattice packing* means a finite translative packing where the centres are points of a packing lattice for K. Let $\varrho > 0$. For any n, we consider the finite lattice packing of n translates of K minimizing $V(C_n + \varrho K)$, where C_n is the convex hull of the centres. In this section we discuss the phenomenon that if the parameter ϱ is not too small then C_n is asymptotically homothetic to a certain polytope as n tends to infinity. We call the polytope a Wulff-shape because it is intimately related to the well-known Wulff-shape from crystallography. After describing the results about densest finite lattice packings, we will survey the rich theory that surrounds the subject in the comments. All the arguments leading to the results in this section can be found in the paper by U. Betke and K. Böröczky Jr. [BeB?] and in its references and in the monograph by U. Betke, M. Henk, and J. M. Wills [BHW?]. Actually, planar lattice packings are discussed in more detail in Section 3.4.2.

First, we consider packings with respect to a fixed packing lattice Λ. Let $\varrho > 0$ and, for any n, let C_n be the convex hull of n points of Λ such that $V(C_n + \varrho K)$ is minimal. Then there exists a *critical radius* $\varrho_c(K, \Lambda)$ such that

$$\lim_{n\to\infty} \frac{V(C_n + \varrho K)}{n} \begin{cases} < \det \Lambda & \text{if } \varrho < \varrho_c(K, \Lambda); \\ = \det \Lambda & \text{if } \varrho \geq \varrho_c(K, \Lambda). \end{cases}$$

Writing $\mu(K, \Lambda)$ to denote *covering radius*, namely, the minimal μ such that $\Lambda + \mu K$ covers $\mathbb{R}^d$, we have

$$\max_{\substack{u \in S^{d-1} \\ \dim(u^{\perp} \cap \Lambda) = d-1}} \left\{ \frac{\det \Lambda}{2h_K(u) \cdot \det(u^{\perp} \cap \Lambda)} \right\} \leq \varrho_c(K, \Lambda) \leq \mu(K, \Lambda), \quad (10.31)$$

where $h_K(\cdot)$ is the support function. Given $\varrho > \varrho_c(K, \Lambda)$, we define the *Wulff-shape* by

$$\widetilde{W}_{K,\Lambda,\varrho} = \bigcap_{\substack{u \in S^{d-1} \\ \dim(u^{\perp} \cap \Lambda) = d-1}} \left\{ x : \langle u, x \rangle \leq \varrho h_K(u) - \frac{\det \Lambda}{2\det(u^{\perp} \cap \Lambda)} \right\}. \quad (10.32)$$

Actually, $\widetilde{W}_{K,\Lambda,\varrho}$ is a polytope. Since the asymptotic density of optimal clusters is $V(K)/\det \Lambda$, we renormalize the Wulff-shape by letting

$$W_{K,\Lambda,\varrho} = \sqrt[d]{\frac{\det \Lambda}{V\left(\widetilde{W}_{K,\Lambda,\varrho}\right)}} \cdot \widetilde{W}_{K,\Lambda,\varrho}. \quad (10.33)$$

Theorem 10.11.1. *Let K be an o-symmetric convex body and let Λ be lattice in $\mathbb{R}^d$. If, for given $\varrho > \varrho_c(K, \Lambda)$, C_n is the convex hull of n points of Λ such that $V(C_n + \varrho K)$ is minimal then*

$$\lim_{n\to\infty} \frac{1}{\sqrt[d]{n}} \cdot C_n = W_{K,\Lambda,\varrho}.$$

The way how the Wulff-shape is related to finite packings with respect to Λ is transparent in the planar case (see Section 3.4.2 for a more detailed discussion): Let P_n denote some lattice polygon containing n lattice points of Λ, and hence

$$A(P) + \frac{\det \Lambda}{2} \cdot \#(\Lambda \cap \partial P) + \det \Lambda = n \det \Lambda$$

according to the Pick formula (see Lemma 3.4.3). For any edge e of P_n, we write u_e to denote the exterior unit normal. Therefore, minimizing $A(P_n + \varrho K)$ for given n and ϱ is equivalent to minimizing

$$\mathbf{E}(P_n) = \sum_{e \text{ edge of } P_n} \left(\varrho h_K(u_e) - \frac{\det \Lambda}{2\det(u_e^{\perp} \cap \Lambda)} \right) \cdot |e|.$$

The interesting case is when the area of P_n is essentially $n \det \Lambda$; hence, we obtain an isoperimetric type problem. Finally, the Minkowski inequality for mixed areas yields that the solution is asymptotically homothetic to the Wulff-shape $W_{K,\Lambda,\varrho}$.

Unfortunately, if the dimension is at least three then no direct analogue of the Pick theorem holds, and a more tedious argument is needed. Nonetheless, the basic idea is very similar: The first step is to show that $r(C_n) \geq c_1 \sqrt[d]{n}$ and $R(C_n) \leq c_2 \sqrt[d]{n}$ for some positive constants c_1 and c_2 depending on K, Λ, and ϱ (see V. Arhelger, U. Betke, and K. Böröczky Jr. [ABB2001]). Therefore, we may consider a subsequence $C_{n'}$ such that $C_{n'}/\sqrt[d]{n'}$ tends to some convex body M.

Now our problem translates into the following one (see U. Betke and K. Böröczky Jr. [BeB?]): Given the volume of a convex body M ($\det \Lambda$ in our case), minimize certain surface energy $\mathbf{E}_\varphi(M)$. We first present the definition of $\mathbf{E}_\varphi(M)$ when M is a polytope with facets $F_1, \ldots, F_k$ such that aff F_i is parallel to some lattice hyperplane for each i. Writing $u_1, \ldots, u_k$ to denote the corresponding exterior unit normals, we have

$$\mathbf{E}_\varphi(M) = \sum_{i=1}^{k} \left(\varrho h_K(u) - \frac{\det \Lambda}{2 \det(u^\perp \cap \Lambda)} \right) \cdot |F_i|. \tag{10.34}$$

In general, the "free energy" φ is defined for any $u \in S^{d-1}$ by

$$\varphi(u) = \begin{cases} \varrho h_K(u) - \frac{\det \Lambda}{2 \det(u^\perp \cap \Lambda)} & \text{if } \dim(u^\perp \cap \Lambda) = d - 1; \\ \varrho \cdot h_K(u) & \text{otherwise.} \end{cases} \tag{10.35}$$

We choose an exterior unit normal u_x at each $x \in \partial M$, and we define the surface energy of M to be

$$\mathbf{E}_\varphi(M) = \int_{\partial M} \varphi(u_x) \, dx.$$

Finally, $M = W_{K,\Lambda,\varrho}$ follows via the celebrated Wulff theorem.

Next we consider finite lattice packings of the o-symmetric convex body K in $\mathbb{R}^d$ where the packing lattice may vary. We write $\Delta_L(K)$ to denote the minimal determinant of packing lattices for K. Let $\varrho > 0$ and, for any n, let C_n be the convex hull of n points of a packing lattice for K such that $V(C_n + \varrho K)$ is minimal. Then there exists $\varrho_{c,L}(K)$ satisfying

$$\lim_{n \to \infty} \frac{V(C_n + \varrho K)}{n} \begin{cases} < \Delta_L(K) & \text{if } \varrho < \varrho_{c,L}(K); \\ = \Delta_L(K) & \text{if } \varrho \geq \varrho_{c,L}(K). \end{cases}$$

It is known (see the proof of Lemma 10.5.2 for the lower bound and M. Henk [Hen1995] or U. Betke, M. Henk, and J. M. Wills [BHW?] for the upper bound) that

$$1 - \frac{\ln d}{d-1} < \varrho_{c,L}(K) \leq 3.$$

Theorem 10.11.2. *Let K be an o-symmetric convex body in $\mathbb{R}^d$, and let $\varrho > \varrho_{c,L}(K)$. If C_n is the convex hull of n points of a packing lattice such that $V(C_n + \varrho K)$ is minimal then the circumradius of $C_n / \sqrt[d]{n}$ stays bounded as n tends to infinity, and any convergent subsequence $C_{n'} / \sqrt[d]{n'}$ satisfies*

$$\lim_{n' \to \infty} \frac{1}{\sqrt[d]{n'}} \cdot C_{n'} = W_{K,\Lambda,\varrho},$$

where Λ is a packing lattice of minimal determinant.

Remarks. Since the circumradius of $C_n / \sqrt[d]{n}$ is bounded, any subsequence of $C_n / \sqrt[d]{n}$ contains a convergent subsequence according to the Blaschke selection theorem (Theorem A.3.1).

We note that Theorem 10.11.2 has a nicer form if there exists a unique packing lattice Λ of minimal determinant for K up to the affine symmetries of K. In this case $\lim_{n \to \infty} C_n / \sqrt[d]{n} = W_{K,\Lambda,\varrho}$.

Based on an observation of H. Minkowski, the packing lattice of minimal determinant has been determined for various three-dimensional convex bodies (see P. M. Gruber and C. G. Lekkerkerker [GrL1987] and U. Betke and M. Henk [BeH2000]). The packing lattice of minimal determinant for B^d is known if either $d \leq 8$ (see [GrL1987]) or Ch. Zong [Zon1999]) or $d = 24$ (see H. Cohn and A. Kumar [CoK?]). Concerning the densest lattice packing of n unit balls for given $\varrho > \varrho_{c,L}(B^d)$ in any dimension. U. Betke and A. Schürmann [BeS?] verified that the centres are always part of some packing lattice of minimal determinant if n is large. However, A. Schürmann [Sch2000d] showed for a certain o-symmetric octagon and ϱ that the centres in the densest lattice packing of n translates are not part of the packing lattice of minimal determinant if n is large.

Comments. The subject of this section is related to the extensively studied world of crystals. General references are U. Betke, M. Henk, and J. M. Wills [BHW?] and K. Böröczky Jr., U. Schnell, and J. M. Wills [BSW2000].

Very Brief History of Geometrical Crystallography

This section owes much to the a nice survey by M. Senechal [Sen1990]. The shape of crystals has attracted the attention of various scientists since the ancient Greeks, and investigating the shape quickly led to the question about the structure. From our point of view, the first important treatise is Johann Kepler's *Strena Seu de Nive Sexangula* (A New Year's Gift or the Six Cornered Snowflake), which he presented to a friend in 1611. Kepler represents crystal structures as lattice packing of spheres and attempts to explain (without success) why snowflakes always have six corners, never five or seven. Towards understanding the form, a cornerstone was the discovery by Romé de l'Isle and his students (1783) of the law of the constancy of interfacial angles for crystals of given species. Romé de l'Isle did not give an explanation, but one was provided somewhat later by the French abbé René-Just Haüy. Haüy also assumes an underlying periodic structure, and he proposes that crystals grow by adding layers to the previous stage starting from a basic polytopal form. He states (implicitly) that the new layers are always parallel to some lattice plane of the underlying lattice.

A successful mathematical model for crystal growth was proposed by W. Gibbs [Gib1878]. For each possible face direction, he introduced a certain free energy, which depends on the temperature and on the direction of the face. As W. Gibbs [Gib1878] suggested, given the volume, the equilibrium shape of a crystal minimizes the total free energy of the surface, namely, the integral of the free energy on the surface. A necessary condition for the *equilibrium shape* was described by G. Wulff [Wul1901]: *If the crystal is at equilibrium then the distance of any face from the origin is proportional to the free energy of that face.*

Mathematical proofs for the Wulff theorem were given somewhat later, for example by the 1914 Nobel prize winner Max von Laue [Lau1943]. The first argument that actually constructs the equilibrium shape if the free energy is known for any direction is due to A. Dinghas [Din1943]. We note that around the time when G. Wulff stated his theorem, H. Minkowski developed his theory of mixed volumes. In particular, he proved the Minkowski inequality (A.43), which, in turn, yields directly the Wulff theorem (see say U. Betke and K. Böröczky Jr. [BeB?]).

Now if a face is parallel to some reasonable dense two-dimensional sublattice of the underlying lattice then the free energy coincides with (10.35) for certain crystals. Therefore, finite lattice packings model crystal growth reasonable well in these cases. We note that the free energy depends continuously on the direction; therefore, it may coincide with (10.35) only in the "dense directions." Fortunately, these directions are the essential ones for crystal growth in many cases.

Finite Lattice Packings and Crystallography

A systematic study about possible shapes that can be associated with dense finite packings of the unit ball B^d with respect to a given lattice Λ was started by J. M. Wills in [Wil1996] and [Wil1997]. Let $\varrho > \varrho_0$, where ϱ_0 depends on Λ. Given a lattice polytope P, we assign the finite ball packing to any dilate mP, $m \in \mathbb{N}$, where the centres are the lattice points in mP. Then the asymptotic parametric density is $V(B^d)/\det\Lambda$, and J. M. Wills proposed to measure the optimality of P by the *density deviation*

$$\lim_{m\to\infty} V(mP)^{1/d} \cdot \left(\frac{V(B^d)}{\det\Lambda} - \frac{\#(\Lambda \cap mP) \cdot V(B^d)}{V(mP + \varrho B^d)} \right).$$

In particular, let $\lambda_{n,P}$ be the minimal integer such that $\lambda_{n,P}P$ contains at least n lattice points. Now if the density deviation for a lattice polytope P is less than for another lattice polytope Q, then $\lambda_{n,P}P$ provides a denser arrangement than $\lambda_{n,Q}Q$. If the facets of P are $F_1, \ldots, F_k$ with exterior unit normals $u_1, \ldots, u_k$ then the *density deviation* is (compare (10.34))

$$\frac{V(B^d)}{\det\Lambda} \cdot V(P)^{\frac{1-d}{d}} \cdot \sum_{i=1}^{k} \left(\varrho - \frac{\det\Lambda}{2\det(\Lambda \cap u_i^\perp)} \right) \cdot |F_i|. \tag{10.36}$$

Learning about this result in a sunny café in Cortona in 1995, P. M. Gruber noted that the density deviation resembles the Gibbs–Curie energy, and he pointed out a possible connection to the Wulff theorem.

Progress accelerated after this observation: J. M. Wills [Wil1996,Wil1997] showed that the Wulff-shape (10.32) is always a polytope if the gauge body is a ball, and U. Schnell extended this result to any gauge body (an argument published later in [Sch2000a]). In the meantime, the asymptotic structure was better and better understood when ϱ is larger than the critical radius: K. Böröczky Jr. and U. Schnell [BöS1998a] verified that the inradius of the optimal convex hull tends to infinity, and V. Arhelger, U. Betke, and K. Böröczky Jr. [ABB2001] showed that the ratio of the circum- and the inradius stays bounded. Finally, U. Betke and K. Böröczky Jr. [BeB?] proved that the Wulff-shape is the optimal asymptotic shape (see Theorems 10.11.1 and 10.11.2). We note that, for balls, M. Henk [Hen1995] verified $\varrho_{c,L}(B^d) < \sqrt{21/4}$ (see also U. Betke, M. Henk, and J. M. Wills [BHW?]).

Finite Periodic Packings

The underlying set for a crystal may not be a lattice but rather a set that is periodic with respect to some lattice. This occurs, for example, when at least two kinds of atoms build the crystal.

Let Γ be the periodic set $\{v_1, \ldots, v_k\} + \Lambda$ in $\mathbb{R}^d$, where Λ is a lattice. Before defining the Wulff-shape, we consider some basic properties of Γ. First, if P is any polytope then

$$\#(\Gamma \cap \lambda P) \sim \frac{k \cdot V(P)}{\det \Lambda} \cdot \lambda^d$$

as λ tends to infinity. Now if λ is large and the facets $F_1, \ldots, F_m$ of P have exterior normals $v_1, \ldots, v_m$ that are primitive vectors of the dual lattice Λ^* then (see U. Betke, M. Henk, and J. M. Wills [BHW?])

$$\#(\Gamma \cap \lambda P) = \frac{k \cdot V(P)}{\det \Lambda} \cdot \lambda^d + O\left(\lambda^{d-2}\right)$$
$$+ \sum_{i=1}^{m} \sum_{j=1}^{k} \left(\frac{1}{2} - \{\lambda h_P(v_i) - \langle v_i, w_j \rangle\}\right) \cdot \frac{|F_i|}{\|v_i\|} \cdot \lambda^{d-1},$$

where $\{t\}$ is the fractional part $t - \lfloor t \rfloor$. Therefore, for any primitive $v \in \Lambda^*$, we define

$$\tau_\Gamma(v) = \max_{0 \le t < 1} \sum_{j=1}^{k} \left(\frac{1}{2} - \{t - \langle v, w_j \rangle\}\right).$$

In particular, given an o-symmetric convex body K and $\varrho > \varrho_\Gamma$, the corresponding *Wulff-shape* is

$$\widetilde{W}_{K,\Gamma,\varrho} = \bigcap_{v \in \Lambda^* \text{ prim.}} \{x : \langle v, x \rangle \le \varrho \cdot h_K(v) - \tau_\Gamma(v)\}. \qquad (10.37)$$

We will use the normalization $W_{K,\Gamma,\varrho}$ of $\widetilde{W}_{K,\Gamma,\varrho}$ that is of volume $\det \Lambda / k$. The definition of the Wulff-shape for periodic packings is due to U. Schnell [Sch2000a], who also verified that $W_{K,\Gamma,\varrho}$ is a polytope. We observe that, unlike in the lattice case, $W_{K,\Gamma,\varrho}$ may not be o-symmetric.

Now let C_n denote the convex hull of some n points of Γ. Similarly to lattice packings, one can define the critical radius $\varrho_c(K, \Gamma)$ to be the minimal ϱ such that, for any choice of C_n,

$$\liminf_{n \to \infty} \frac{V(C_n + \varrho K)}{n} \ge \frac{\det \Lambda}{k}.$$

If $\mu(K, \Gamma)$ denotes the minimal μ such that $\Gamma + \mu K$ covers $\mathbb{R}^d$ then $\varrho_c(K, \Gamma) \le \mu(K, \Gamma)$. We have the following analogue of Theorem 10.11.1:

Let K be an o-symmetric convex body in $\mathbb{R}^d$, let Γ be a periodic set, and let $\varrho > \varrho_c(K, \Gamma)$. If $V(C_n + \varrho K)$ is minimal for any n then

$$\lim_{n \to \infty} \frac{1}{\sqrt[d]{n}} \cdot C_n = W(K, \Gamma, \varrho).$$

This statement can be proved similarly to Theorem 10.11.1, only one needs a more delicate argument (see U. Betke, M. Henk, and J. M. Wills [BHW?]).

The notion of *density deviation* can be extended to periodic packings, and this approach has the advantage that different convex bodies $K_1, \ldots, K_k$ might be assigned to the equivalences classes corresponding to $v_1, \ldots, v_k$. For a thorough discussion, see U. Schnell [Sch1999, Sch2000a] and J. M. Wills [Wil2000].

These papers also exhibit examples showing that the Wulff-shape provided by optimal packings coincides with the experimental shape of the corresponding crystal in many cases.

We note that the possible symmetries of a periodic set are rather limited. In the planar case, only three-fold, four-fold, or six-fold rotations are possible (see, e.g., the beautiful book by B. Grünbaum and G. C. Shephard [GrS1987]).

Finite Ball Packings and Quasi-periodic Sets

The discovery of quasi-crystals in 1984 was based on the fact that their structure exhibits some long-range order, whereas they have symmetries forbidden for crystals (see C. Janot [Jan1992]). Ever since, the so-called quasi-periodic sets have been extensively investigated. We now discuss finite packings of unit balls where the centres are part of a given quasi-periodic set. There are very few results in this direction; therefore, we rather propose a possible theory. One result will be reported at the end about the classical planar Penrose tiling, which had been described by R. Penrose [Pen1974] ten years before the discovery of quasi-crystals. The Penrose tiling exhibits five-fold rotational symmetry and a rather strong long-range order.

We recall that the definition of density deviation for a given lattice polytope used dilates by integers. To define "finite quasi-periodic packings," we consider special quasi-periodic sets that possess a natural notion of expansion. Our discrete set will be the family of vertices of a tiling $\mathcal{T}$ of $\mathbb{R}^d$ by translates of finitely many polytopes, and we assume that the intersection of any two tiles is a common face. A finite union of tiles in $\mathcal{T}$ is called a *patch*. To inflate the tiles, we use a linear map $\varphi : \mathbb{R}^d \to \mathbb{R}^d$, which is called *expansive* if it is diagonalizable over $\mathbb{C}$, and each eigenvalue is of absolute value larger than one. We say that $\mathcal{T}$ is a *self-affine* if the following conditions are satisfied:

- *Repetition:* For any patch, there exists an $R > 0$ such that each ball of radius R contains a translate of the patch that is part of $\mathcal{T}$,
- *Inflation:* There exists an expansive linear map φ such that the φ image of any tile is filled by a patch,
- *Deflation:* If the tiles T_1 and T_2 are translates then the patches corresponding to $\varphi(T_1)$ and $\varphi(T_2)$ are translates as well.

The term repetitive is used for example, in M. Senechal [Sen1995], and the same property is called local isomorphism in B. Solomyak [Sol1997]. The finiteness conditions imply that some power of φ acts by multiplication by a $\eta > 1$. The dilation by η is an inflation, but it may not be a deflation. The strength of long-range order of a self-affine tiling is shown by the fact that it has uniform patch frequency (see say B. Solomyak [Sol1997]). According to this property, for any given patch Π, there exists a $\delta > 0$ such that the number of translates of Π that are patches and are contained in a "large set" S is asymptotically $\delta V(S)$.

We observe that the family of all translations that leave $\mathcal{T}$ invariant form a d_0-dimensional lattice Λ for some $d_0 \leq d$ and $\varphi\Lambda \subset \Lambda$. Therefore, $\mathcal{T}$ has no translational symmetry in general, say, if η is irrational. In the planar case this happens if φ acts by multiplication by a nonreal complex number whose rotational component is m-fold with $m \neq 3, 4, 6$.

Now the underlying set Γ for our finite packings is the family of vertices of a given self-affine tiling $\mathcal{T}$, where we assume that the distance between any two vertices is at least two. It follows by the uniform patch frequency that there exists a $\Delta(\Gamma)$ associated to Γ such that if P is any patch then

$$\lim_{k \to \infty} \frac{V(\varphi^k P)}{\#(\Gamma \cap \varphi^k P)} = \Delta(\Gamma). \tag{10.38}$$

The missing brick of our theory is a more precise asymptotic estimate for $\#(\Gamma \cap \varphi^k P)$ that is analogous to the Pick formula (Lemma 3.4.3). We conjecture that if P is a patch then

$$\#(\Gamma \cap \varphi^k P) = \frac{V(\varphi^k P)}{\Delta(\Gamma)} + \frac{1}{2}\#\left(\Gamma \cap \partial(\varphi^k P)\right) + o\left|\partial(\varphi^k P)\right|. \tag{10.39}$$

If F is a facet of some tile then we define

$$\Delta_F = \liminf_{k \to \infty} \frac{|\varphi^k F|}{\#(\Gamma \cap \varphi^k F)}.$$

To any patch P, we assign the finite packing of unit balls where the balls are centred at the points of Γ in P. For $\varrho > 0$, the parametric density of the packing is

$$\frac{\#(\Gamma \cap P) \cdot V(B^d)}{V(P + \varrho B^d)}.$$

Now if $\varrho > \varrho_0$ for suitable $\varrho_0 > 0$ then this parametric density is always less then $V(B^d)/\Delta(\Gamma)$. Therefore, we may define the *density deviation* to be

$$\liminf_{k \to \infty} V(\varphi^k P)^{1/d} \cdot \left(\frac{V(B^d)}{\Delta(\Gamma)} - \frac{\#(\Gamma \cap \varphi^k P) \cdot V(B^d)}{V(\varphi^k P + \varrho B^d)} \right).$$

The conjecture (10.39) would yield that the density deviation is

$$\frac{V(B^d)}{\Delta(\Gamma)} \cdot V(P)^{\frac{1-d}{d}} \sum_{F \subset \partial P} \left(\varrho - \frac{\Delta(\Gamma)}{2\Delta(\Gamma_F)} \right) \cdot |F|, \qquad (10.40)$$

where F always denote a facet of some tile. Therefore, denoting the exterior unit normal to the facet F by u_F, we define the *Wulff-shape* to be

$$W(\Gamma, \varrho) = \bigcap_F \left\{ x : \langle x, u_F \rangle \le \varrho - \frac{\Delta(\Gamma)}{2\Delta(\Gamma_F)} \right\}.$$

In particular, to have dense finite packings, the shape of P has to be close to the shape of $W(\Gamma, \varrho)$.

As mentioned at the beginning, the Penrose tiling is the only case when an analogous theory has been worked out (see B. Grünbaum and G. C. Shephard [GrS1987] or M. Senechal [Sen1995] for nice reviews about the Penrose tiling). The paper by K. Böröczky Jr. and U. Schnell [BöS1999] considered the version of the Penrose tiling using two rhombi, a large one with angles $2\pi/5$ and $3\pi/5$, and a small one with angles $\pi/5$ and $4\pi/5$. We note that to obtain a self-affine tiling, one should consider the tiling by suitable triangles that are halves of the rhombi; namely, one triangle has side lengths $2, 2\tau$, and 2τ (where τ is the golden ratio $(\sqrt{5} + 1)/2$), and the other triangle has side lengths $2\tau^2, 2\tau$, and 2τ. The expansive map φ is the multiplication by the complex number $\tau\, e^{3\pi i/5}$. Now K. Böröczky Jr. and U. Schnell [BöS1999] showed that if P is the union of finitely many rhombi then (10.40) does hold. The minimal choice for the "critical radius" ϱ_0 is $0.27, \ldots$ and the resulting Wulff-shape is a regular decagon if $\varrho_0 < \varrho < 6\varrho_0$.

Many interesting examples and thorough discussions about quasi-periodic tilings with various symmetry groups can be found in the papers by M. Baake, J. Hermisson, and P. A. B. Pleasants [BHP1997] and M. Baake, J. Hermisson, and C. Richard [BHR1997]. In addition to the Penrose tiling, possible candidates for Wulff-shapes with icosahedral symmetry were discussed in

K. Böröczky Jr. and U. Schnell [BöS1998b]. The proposed Wulff-shapes correspond surprisingly well to the shape of real quasi-crystals.

Finally, J. C. Lagarias and P. A. B. Pleasant [LaP2003] introduced the notion of *linearly repetitive* tilings, namely, when there exists a constant c such that if a patch is contained in a ball radius R then each ball of radius cR contains a translate of the patch that is part of the tiling. We still assume that the tiling is face to face, and there exist only finitely many tiles up to translation. Such a tiling is periodic if and only if cR can be replaced by $R + c$. Many constructions of self-affine tilings in the literature produce linearly repetitive tilings, like the Penrose tiling (see B. Grünbaum and G. C. Shephard [GrS1987], p. 563). Linearly repetitive tilings have uniform patch frequency (see J. C. Lagarias and P. A. B. Pleasant [LaP2003]). Therefore the theory here can be possibly applied to linearly repetitive tilings with an inflation factor $\eta > 1$.

10.12. Finite Lattice Packings in Three Space

In Chapter 3 we described rather thoroughly the asymptotic structure of planar optimal finite translative packings with respect to parametric density. More precisely, Corollary 3.2.2 says that the optimal packing is eventually a sausage if the parameter ϱ is less than the critical radius, and it is cluster if ϱ is larger than the critical radius. We now provide an essentially complete description of finite lattice packings in $\mathbb{R}^3$ when the gauge body is centrally symmetric. The material is covered mostly by V. Arhelger, U. Betke, and K. Böröczky Jr. [ABB2001], and the role of the Wulff-shape is discussed in U. Betke and K. Böröczky Jr. [BeB?]. In addition, U. Betke, M. Henk, and J. M. Wills [BHW?] is a thorough reference.

Let K be an o-symmetric convex body in $\mathbb{R}^3$, and let $\varrho > 0$. First, we discuss the parametric density of certain special types of arrangements. For a three-dimensional cluster, the asymptotic density is $V(K)/\Delta_L(K)$. To understand two-dimensional clusters, we recall some facts from Chapter 2: Given an o-symmetric convex domain M, $H(M)$ denotes a circumscribed o-symmetric hexagon of minimal area, and hence $A(H(M))$ is the minimal determinant of packing lattices of M. We define

$$\Xi(K) = \min_{u \in S^2} 2h_K(u) \cdot H(u^{\perp} \cap K);$$

thus, the asymptotic parametric density of optimal two-dimensional clusters is $V(K)/[\varrho \, \Xi(K)]$.

Finally, when discussing sausage packings, we write $\|x\|$ to denote the Euclidean norm of $x \in \mathbb{R}^3$, use K/x to denote the projection of K into $x^\perp$, and recall (see Section 10.1) that

$$V(Z(K)) = \min_{x \in \partial K} 2\|x\| \cdot |K/x|.$$

Then $V(K)/[\varrho^2 V(Z(K))]$ is the asymptotic parametric density of optimal sausage packings. We define

$$\varrho_{s,L}(K) = \min\left\{ \frac{\Xi(K)}{V(Z(K))}, \sqrt{\frac{\Delta_L(K)}{V(Z(K))}} \right\};$$

$$\varrho_{c,L}(K) = \max\left\{ \frac{\Delta_L(K)}{\Xi(K)}, \sqrt{\frac{\Delta_L(K)}{V(Z(K))}} \right\}.$$

Theorem 10.12.1. *Let K be an o-symmetric convex body in $\mathbb{R}^3$ and, for given $\varrho > 0$, let C_n be the convex hull of n points of a packing lattice such that $V(C_n + \varrho K)$ is minimal.*

(i) If $\varrho < \varrho_{s,L}(K)$ then C_n is a segment for large n.
(ii) If $\varrho_{s,L}(K) < \varrho < \varrho_{c,L}(K)$ then C_n is planar for large n, and $c_1\sqrt{n} < r_2(C_n) < c_2\sqrt{n}$.
(iii) If $\varrho > \varrho_{c,L}(K)$ then $c_3\sqrt[3]{n} < r_3(C_n) < c_4\sqrt[3]{n}$ for large n, and any convergent subsequence of $C_n/\sqrt[3]{n}$ tends to the polytope $W_{K,\Lambda,\varrho}$ for some packing lattice Λ of minimal determinant.

Here c_1, c_2, c_3, c_4 are positive constants depending on K and ϱ.

Remark. The analogous statements hold for packings with respect to a fixed lattice.

We note that the planar phase in Theorem 10.12.1 does occur when K is an octahedron; namely,

$$\varrho_{s,L}(K) = \frac{3}{4} \quad \text{and} \quad \varrho_{c,L}(K) = \frac{76}{81}$$

(see V. Arhelger [Arh1998] for more about packings of octahedra).

Finally, we discuss ball packings. The packing lattice of minimal determinant for B^3 is the face centred cubic lattice Λ_3 according to Gauß [Gau1840]. We recall the strong sausage conjecture 10.8.1, which says that the sausage and the critical radius of unit balls coincide. According to Example 10.12.2

(i), the strong sausage conjecture holds at least for finite lattice packings in $\mathbb{R}^3$.

Example 10.12.2.

(i) $\varrho_{s,L}(B^3) = \varrho_{c,L}(B^3) = \sqrt{2\sqrt{2}/\pi} = 0.9488\ldots$

(ii) *If* $\sqrt{2\sqrt{2}/\pi} < \varrho \leq (\sqrt{3}+1)/(2\sqrt{2})$ *then* $W_{B^3, \Lambda_3, \varrho}$ *is a regular octahedron of volume* $4\sqrt{2}$.

(iii) *If* D_n *is a convex body of minimal volume in* $\mathbb{R}^3$ *that contains n unit balls whose centres lie in some packing lattice for the unit ball then*

$$\lim_{n\to\infty} \frac{1}{\sqrt[3]{n}} D_n = W(B^3, \Lambda_3, 1).$$

In addition, $W(B^3, \Lambda_3, 1)$ *is obtained by truncating a regular octahedron of edge length* $2.2905\ldots$ *by cutting off pyramids of edge length* $0.1797\ldots$ *at the vertices. Thus* $W(B^3, \Lambda_3, 1)$ *is bounded by six squares of side length* $0.1797\ldots$ *and by eight congruent (but not regular) hexagons. Let us observe that the regular octahedron of volume* $4\sqrt{2}$ *is of edge length* $\sqrt[3]{12} = 2.2894\ldots$.

Appendix
Background

In this chapter we review some topics that are needed in the main part of the text. The book uses tools from various branches of mathematics related to convexity; hence, to give the reader a chance to follow the presentation without constantly checking references, this introductory part is more extensive than usual. We present only a few proofs where the exact statement we need is not so easy to access. The chapter is organized in a way that most of the material needed in Part 1 is contained in Sections A.1–A.6.

Before going into details we define the central notions of this book. A set K is called *convex* if it contains any segment whose endpoints lie in K. In addition, K is a *convex body* if K is compact and its interior is nonempty. Planar convex bodies are also known as *convex domains*. A family $\{K_n\}$ of convex bodies is called a *packing* if the interiors of any two K_i and $K_j, i \neq j$, are disjoint, or, in other words, if K_i and K_j do not overlap. Next, $\{K_n\}$ is a *covering* of a set X if the union of the convex bodies contains X. Finally, $\{K_n\}$ is a *tiling* of X if each K_n is contained in X, and $\{K_n\}$ is both a packing and a covering of X.

A.1. Some General Notions

As usual $\mathbb{N}$, $\mathbb{Z}$, and $\mathbb{R}$ denote the family of natural numbers, integers, and real numbers, respectively. If x is a real number then $\lfloor x \rfloor$ denotes the largest integer not larger than x, and $\lceil x \rceil$ denotes the smallest integer not smaller than x. As usual $0! = 1$, and $n!$ is the product of the first n integers for a positive integer n. In addition $\binom{n}{k} = n!/[k!(n-k)!]$. Given real numbers $u_i, v_i, i = 1, \ldots, n$, the Cauchy–Schwartz inequality says that

$$|u_1 v_1 + \cdots + u_n v_n| \leq \sqrt{u_1^2 + \cdots + u_n^2} \cdot \sqrt{v_1^2 + \cdots + v_n^2}. \quad (A.1)$$

Let f and g be real functions on certain objects. We write $f = O(g)$ if there exists a constant c such that $|f| \leq c\,g$. The constant c may be an

325

absolute constant, or it may depend on certain parameters. We always specify what the implied constant depends on. If $f(t)$ and $g(t)$ are positive real functions of some real parameter t that tends to a then we write $f \sim g$ if $\lim_{t \to a} f(t)/g(t) = 1$. Here we allow $a = \infty$.

If C is a subset of $\mathbb{R}^d$ then its *characteristic function* is

$$\chi_C(x) = \begin{cases} 1 & \text{if } x \in C; \\ 0 & \text{if } x \in \mathbb{R}^d \backslash C. \end{cases}$$

In addition we write int$C$, cl$C$, and bd$C$ to denote the *interior*, the *closure*, and the *boundary* of C, respectively. For an introduction into topology from a geometric view point, see M. A. Armstrong [Arm1983].

The rest of the section is concerned with the basic notions of linear algebra. We write o to denote the *origin* of $\mathbb{R}^d$. For any set $S \subset \mathbb{R}^d$, the *linear hull* lin S is the smallest linear subspace containing S. A real function $\| \cdot \|$ on $\mathbb{R}^d$ is a *norm* if $\|u\| > 0$ for $u \neq o$, $\|\lambda u\| = |\lambda| \cdot \|u\|$ for $\lambda \in \mathbb{R}$, and $\|u\| + \|v\| \geq \|u + v\|$. The last property is commonly known as the *triangle inequality*. The corresponding distance between two points x and y is defined to be $d(x, y) = \|x - y\|$, which is a *metric* on $\mathbb{R}^d$.

We always assume tacitly the existence of a Euclidean structure on $\mathbb{R}^d$, and we write $\langle \cdot, \cdot \rangle$ to denote the scalar product. Actually, $\|u\|$ will always denote the *Euclidean norm* $\sqrt{\langle u, u \rangle}$ unless otherwise stated, and it satisfies $|\langle u, v \rangle| \leq \|u\| \cdot \|v\|$ according to the Cauchy–Schwartz inequality (A.1). In addition, the *angle* α between the nonzero vectors u and v is defined by the conditions $0 \leq \alpha \leq \pi$ and $\cos \alpha = \langle u, v \rangle / (\|u\| \cdot \|v\|)$. Given different $x, y, z \in \mathbb{R}^d$, we write $\angle xyz$ to denote the angle of $x - y$ and $z - y$.

We write diam$X$ to denote the *diameter* of a compact set X, namely, the maximum of the distances between two points of X. Given a point p and an $r > 0$, the family of points whose distance from p is at most r is called the *ball* of centre p and of radius r (or *circular disc* if $d = 2$), which is called a *unit ball* when $r = 1$ (or *unit disc* if $d = 2$). In addition, the boundary of a ball is called a *sphere* and that of a circular disc is a *circle*. Moreover, we write B^d and S^{d-1} to denote the unit ball centred at the origin in $\mathbb{R}^d$ and its boundary, respectively.

Given an orthonormal basis of $\mathbb{R}^d$, we identify linear maps and their matrices. An *orthogonal transformation* is a linear map $\mathbb{R}^d \to \mathbb{R}^d$ satisfying $\langle u, v \rangle = \langle Qu, Qv \rangle$ for any u, v. This property is equivalent to saying that the inverse of Q is its transpose; namely, $Q^{-1} = Q^T$. We say that the $d \times d$ matrix $A = [\alpha_{ij}]$ is positive semidefinite if A is symmetric (namely, $A^T = A$), and $u^T A u \geq 0$ holds for any u. Moreover A is positive definite if in addition $u^T A u > 0$ holds for any $u \neq 0$. The simplest positive definite matrices are the diagonal ones where $\alpha_{ij} = 0$ for $i \neq j$ and $\alpha_{ii} > 0$. A fundamental property of positive definite matrices is expressed by the *principal axis theorem*: For any positive definite matrix A, there exists an orthogonal matrix Q such that $Q^T A Q$ is diagonal.

If L is a linear subspace of dimension k and $x \in \mathbb{R}^d$ then $G = x + L$ is called an *affine subspace* of dimension k, or simply a k-*plane*. An affine subspace of dimension one or $d - 1$ is also known as a *line* or a *hyperplane*, respectively. We say that a vector u is orthogonal to the affine subspace G if $\langle u, x - y \rangle = 0$ for any $x, y \in G$. If u is nonzero and G is a hyperplane then there exists a number α such that G is the set of points x satisfying $\langle u, x \rangle = \alpha$. In this case the family of points x satisfying $\langle u, x \rangle \geq \alpha$ is called a (closed) *half space*. Actually, G determines two half spaces, where the other one is of equation $\langle u, x \rangle \leq \alpha$. Frequently, either of the two half spaces is called a *side* of G.

We say that two affine subspaces G_1 and G_2 are *parallel* if there exists an x such that $x + G_1$ either is contained in G_2 or contains G_2. If G is an affine subspace in $\mathbb{R}^d$ of dimension m, $1 \leq m \leq d - 1$, then the family of vectors that are orthogonal (normal) to G is a $(d - m)$-dimensional linear subspace, and this subspace is the *orthogonal complement* $G^\perp$ of G. If u is a nonzero vector then we write $u^\perp$ to denote the linear $(d - 1)$-subspace orthogonal to u. The map that sends any $x \in \mathbb{R}^d$ to the closest point $(x + G^\perp) \cap G$ of G is the *orthogonal projection* onto G. Given a set X and an affine subspace G, we write $X|G$ to denote the orthogonal projection of X into G and let X/G to denote the orthogonal projection of X into $G^\perp$. In addition, if u is a nonzero vector then X/u denotes the orthogonal projection of X into $u^\perp$.

For any set S, the *affine hull* aff$S$ of S is the minimal affine subspace containing S. We say that a family of k points is *affinely independent* if their affine hull is of dimension $k - 1$. Next, if $\{G_n\}$ is a sequence of affine m-planes and G is an affine m-plane then we say that $\{G_n\}$ tends to G if for any $x \in G$ there exists a sequence of $x_n \in G_n$ such that $\{x_n\}$ tends to x. Actually, it is sufficient to require this property for some $m + 1$ affinely independent points of G. It also follows that, if $\{G_n\}$ is a sequence of affine m-planes that intersect a compact set, then $\{G_n\}$ has a convergent subsequence.

A function $\Phi: \mathbb{R}^d \to \mathbb{R}^d$ is called an *affine map* if it is of the from $\Phi(x) = \varphi(x) + b$, where $b \in \mathbb{R}^d$ and φ is linear. We note that this decomposition of Φ is unique, and we define det $\Phi = \det \varphi$. In addition, we observe that any k-plane G can be identified with $\mathbb{R}^k$ via some affine bijection. An important example for an affine map is the *translation* by a vector a when the image of x is $x + a$. Another affine map is the *reflection through a point p*; namely, the image of a point x is the point x' such that $p = (x + x')/2$. We say that a compact set S is *centrally symmetric* if its reflected image through some point is S itself. In this case this point is unique and is called the centre of S. We say that S is *o-symmetric* if its centre is the origin.

A self map of the Euclidean space is called *congruence* if it keeps the distance between any two points. Sets that can be mapped bijectively onto each other by a congruence are called *congruent*. Congruences are the affine maps whose linear parts are orthogonal. For example, reflection through a given point is a congruence. Moreover, *reflection through a hyperplane H* is also a congruence, which can be defined as follows: The reflected image of a point x through H is the unique x' such that $(x + x')/2$ lies in H, and $x' - x$ is orthogonal to H.

A.2. Convex Sets in $\mathbb{R}^d$

The properties in this section can be found in T. Bonnesen and W. Fenchel [BoF1987] or R. Schneider [Sch1993b].

Given $x \neq y$ in $\mathbb{R}^d$, the *ray* or *half line xy* is the family of points of the form $x + \lambda(y - x)$ for $\lambda \geq 0$, and x is called the initial point. Moreover, the *segment xy* is the family of points of the form $(1 - \lambda)x + \lambda y$ for $0 \leq \lambda \leq 1$, and x and y are called the endpoints of the segment. Since a set is convex if and only if it contains the segment connecting any two of its points, intersection of convex sets is also convex. Therefore, we may define the *convex hull* conv$X$ of a set X to be the minimal convex set containing X. We observe that conv$X$ is the family of all *convex combinations* of any $x_1, \ldots, x_k \in X$, namely, $\sum_{i=1}^{k} \lambda_i x_i$, where each λ_i is nonnegative, and $\sum_{i=1}^{k} \lambda_i = 1$. According to the *Carathéodory theorem*, it is sufficient to consider subsets of size at most $d + 1$. Thus the convex hull is always contained in the affine hull, and conv$X$ is compact if X is compact.

If C is a compact convex set then $x \in C$ is an *extremal point* if whenever x is contained in a segment $s \subset C$ then x is an endpoint of s. We note that C is the convex hull of its extremal points. Moreover, $x \in C$ is an *exposed point* if there is a hyperplane H such that $H \cap C = x$ and C lies on one side of H. Any exposed point is extremal, and C is the closure of the convex hull of its exposed points. An additional basic property of the convex hull is the *Radon theorem*: If $X \subset \mathbb{R}^d$ has at least $d + 2$ elements then X can be partitioned into two sets X_1 and X_2 satisfying

$$\text{conv} X_1 \cap \text{conv} X_2 \neq \emptyset. \tag{A.2}$$

We note that affine subspaces and half spaces are convex, and any closed convex set is the intersection of the half spaces containing it. Given a compact convex set C, a *supporting hyperplane* at $x \in C$ is a hyperplane H satisfying $x \in H$, and C lies on one side of H. Then a corresponding *normal u* at x is some nonzero vector orthogonal to H that points to the open half space not meeting C. For any point y, there exists a unique closest point $\pi_C(y) = x$ of C to y, and if $y \notin C$ then $y - x$ is a normal at x. The function $\pi_C(\cdot)$ is called the *closest point map* and satisfies

$$\|\pi_C(y_1) - \pi_C(y_2)\| \le \|y_1 - y_2\|. \tag{A.3}$$

Let C be a compact convex set. We observe that C is a convex body in aff$C$, and we call the interior and the boundary of C inside aff$C$ the *relative interior* relint$C$ and the *relative boundary* ∂C of C, respectively. If aff$C$ is k dimensional then C is said to be k dimensional, as well.

A convex *cone* K is a convex set that contains λx for any $x \in K$ and $\lambda \ge 0$. For a set X, the *positive hull* pos$X$ is the smallest convex cone containing X.

If C is a convex set in $\mathbb{R}^d$ then some functions $f, g: C \to \mathbb{R}$ are called *convex* and *concave* if, for any $x, y \in C$ and $0 < \lambda < 1$,

$$f(\lambda x + (1 - \lambda) y) \le \lambda f(x) + (1 - \lambda) f(y),$$
$$g(\lambda x + (1 - \lambda) y) \ge \lambda g(x) + (1 - \lambda) g(y),$$

respectively. We note that convex or concave functions are continuous.

For a compact convex set C, the *support function* is defined to be

$$h_C(u) = \max_{x \in C} \langle x, u \rangle \ \text{ for } \ u \in \mathbb{R}^d.$$

In particular, if $u \ne o$ then the hyperplane with equation $\langle x, u \rangle = h_C(u)$ is a supporting hyperplane with exterior normal u. The support function is convex and positive homogeneous; namely,

$$h_C(\lambda u + \mu v) = \lambda h_C(u) + \mu h_C(v)$$

holds for $u, v \in \mathbb{R}^d$ and nonnegative $\lambda, \mu \in \mathbb{R}$. In turn, any convex and positive homogeneous function is the support function of a unique compact convex set.

Given a compact convex set C and $x \in \mathrm{bd}C$, the family of outward directions

$$N(C, x) = \{y \mid \forall z \in C - x, \ \langle y, z \rangle \le 0\}$$

is called the *normal cone* at x. Naturally, $N(C, x)$ is a convex cone, and x is the closest point of C to a $y \in \mathbb{R}^d$ if and only if $y - x \in N(C, x)$.

The *Minkowski sum* of the compact convex sets C and K is

$$C + K = \{x + y : x \in C \ \text{and} \ y \in K\}.$$

It is a compact convex set, and readily $h_{C+K} = h_C + h_K$. In addition, the *inner parallel set* $C \ominus K$ is the family of points x such that $x + K \subset C$. Let us observe that $C \ominus K$ is compact convex but it might be empty.

We call two convex bodies C and K *homothetic* if $C = \lambda K + x$ for some $\lambda > 0$ and $x \in \mathbb{R}^d$.

Let C be a convex body that contains the origin o in its interior. We define the *distance function* $f_C(\cdot)$ of C by

$$f_C(x) = \min\{\lambda \geq 0 : x \in \lambda C\} \text{ for } x \in \mathbb{R}^d.$$

The convexity of C yields that $f_C(\cdot)$ is convex and positive homogeneous. However, if f is a convex and positive homogeneous function that is positive for $x \neq o$ then there exists a unique convex body C such that $f = f_C$. We note that if C is o-symmetric then f_C is actually a norm, which we denote by $\|\cdot\|_C$.

Next, let C be any convex body. A point $x \in \partial C$ is called smooth if the normal cone $N(C, x)$ is a half line, namely, if there exists only one supporting hyperplane at x. We call the entire convex body *smooth* if each boundary point is smooth. In addition, C is called *strictly convex* if its boundary contains no line segment, or equivalently, if any supporting plane intersects C in a single point.

A.3. The Space of Compact Convex Sets in $\mathbb{R}^d$

The interested reader should consult the surveys P. M. Gruber [Gru1993a,Gru1993c] for more detailed discussion. The distance of a point x and a closed set C in $\mathbb{R}^d$ is

$$d(x, C) = \min_{y \in C}\{d(x, y)\}.$$

In addition, if K and C are compact convex sets then their *Hausdorff distance* is

$$d_H(K, C) = \max\{\max_{x \in C} d(x, K), \max_{y \in K} d(y, C)\}.$$

It is easy to see that the Hausdorff distance is a metric on the space of compact convex sets. We always consider the space of compact convex sets with this topology. In particular, we say that a sequence $\{C_n\}$ of compact convex sets tends to the compact convex set K if $d_H(C_n, K)$ tends to zero. Sometimes this fact is expressed by saying that $\{C_n\}$ *approximates* K. In particular, any compact convex set can be approximated by convex bodies that are both smooth and strictly convex.

We call a family of compact convex sets bounded if the sets are contained in a fixed ball. A fundamental property of the space of compact convex sets is

Theorem A.3.1 (Blaschke selection theorem). *Any bounded sequence of compact convex sets contains a convergent subsequence.*

A subset of the space of compact convex sets is called nowhere dense if its closure does not contain any open set; it is called *meager* if it is covered by a countable family of nowhere dense sets. The Blaschke selection theorem yields (in a nontrivial way) that the family of compact convex sets in $\mathbb{R}^d$ is not meager. Therefore, we say that a property of compact convex sets is *typical* if it holds for all compact convex sets but for a meager family. For example, a typical compact convex set is a smooth and strictly convex body.

A.4. The Spherical Space and the Hyperbolic Space

We discuss models of the spherical and of the hyperbolic d-space that lie in $\mathbb{R}^{d+1}$. The d-dimensional *spherical space*

$$S^d = \left\{(x_0, \ldots, x_d) \in \mathbb{R}^{d+1} : x_0^2 + \ldots + x_d^2 = 1\right\}$$

is the boundary of the unit ball B^{d+1} of $\mathbb{R}^{d+1}$. For $1 \leq k \leq d - 1$, the k-planes in S^d are the intersections with linear $(k + 1)$-subspaces of $\mathbb{R}^{d+1}$. Such a k-plane is a copy of S^k, and if $k = 1$ it is also known as a great circle. The spherical distance of x, $y \in S^d$ is their angle (in radians) as vectors in $\mathbb{R}^{d+1}$. Given nonantipodal x, $y \in S^d$, the spherical *segment* xy is the shorter great circle arc between x and y; hence, its length is the distance of x and y. Moreover, the *ray* xy is the semicircle starting from x and passing through y. In addition, the *angle* of two segments or rays xy and xz is the angle of the two corresponding tangent vectors in $\mathbb{R}^{d+1}$ at x. Finally, volume on S^d coincides with the d-dimensional Hausdorff measure in $\mathbb{R}^{d+1}$ (see Section A.10).

The d-dimensional *hyperbolic space* is represented by the *hyperboloid model*, namely,

$$H^d = \left\{(x_0, \ldots, x_d) \in \mathbb{R}^{d+1} : x_0^2 - x_1^2 - \ldots - x_d^2 = 1 \text{ and } x_0 > 0\right\}.$$

For $0 \leq k \leq d - 1$, the k-planes of H^d are the nonempty intersections with linear $(k + 1)$-subspaces of $\mathbb{R}^{d+1}$ where 1-planes are also known as lines. Any point of a line divides it into two *rays*, and the *segment* xy is the part of the line xy between x and y. It is somewhat harder to define volume, the distance of two points and the angle of two rays with common initial points in H^d than in S^d, and we refer the reader to M. Berger [Ber1987].

Convex sets in S^d and H^d are the intersections with convex cones of $\mathbb{R}^{d+1}$. Important examples are the *ball* of centre x and of radius r (where $r \le \pi/2$ in S^d), which is the family of points whose distance from x is at most r, or the half spaces, which are also called hemispheres in the spherical case. If X is any set then the *convex hull* of X is the intersection of convex sets containing X. We call a compact convex set with nonempty interior a *convex body*, or *convex domain* if $d = 2$.

A.5. Surfaces of Constant Curvature

This section is concerned with the Euclidean plane $\mathbb{R}^2$ (which is of constant curvature 0), the two-sphere S^2 (constant curvature 1), and the hyperbolic plane H^2 (constant curvature -1). The interested reader may consult M. Berger [Ber1987], G. E. Martin [Mar1996] or J. G. Ratcliffe [Rat1994] for detailed treatment. We write $\mathbb{S}$ when discussing properties that hold simultanously in $\mathbb{R}^2$, S^2, and H^2, and we write $d(x, y)$ to denote the distance of points x, y. In addition, the area of a convex domain C is denoted by $A(C)$, and the length of a segment s is frequently denoted by s, as well. The line whose points are of equal distance from the endpoints of a segment s is called the *perpendicular bisector* of s.

We do not describe the differential geometric meaning of the curvature κ of $\mathbb{S}$, whose role is quite apparent say in the following formula: If α, β, and γ denote the angles of a triangle T in $\mathbb{S}$ then $\alpha + \beta + \gamma = \pi + \kappa A(T)$. In the formulae that follow we write a, b, and c to denote the sides of T that are opposite to α, β, and γ, respectively. Because the sum of the angles does not give information about the area in $\mathbb{R}^2$, in this case we use the formula $A(T) = (1/2) ab \sin \gamma$. Next, in Table A.1 we list the *laws of sines and cosines*, which are indispensable tools when distances and angles are calculated. A *polygon* is a convex domain that is bounded by finitely many segments. Maximal segments in the boundary are called *sides*, and the endpoints of the sides are called *vertices*. A polygon with equal sides and equal angles at the vertices is called a *regular polygon*.

The perimeter $P(Q)$ of a polygon Q is defined to be the sum of the lengths of the sides. Since if $C \subset Q$ are polygons then $P(C) \le P(Q)$, we may define the *perimeter* $P(K)$ of any convex domain K to be the supremum of the perimeters of the polygons contained in K. Now if C is a convex domain that is a proper subset of K then $P(C) < P(K)$.

We write $B(r)$ to denote one circular disc of radius r in $\mathbb{S}$, whose area and perimeter are given in Table A.2. The intersection of a half plane and a circular disc is called a *cap*, provided that it has nonempty interior. If C

Table A.1. *Laws of Sines and Cosines*

Euclidean plane, $\mathbb{R}^2$	
Law of sines	$a : b = \sin\alpha : \sin\beta$
Law of cosines	$c^2 = a^2 + b^2 - 2ab\cos\gamma$
Sphere, S^2	
Law of sines	$\sin a : \sin b = \sin\alpha : \sin\beta$
Law of cosines for sides	$\cos c = \cos a \cos b + \sin a \sin b \cos\gamma$
Law of cosines for angles	$\cos\gamma = -\cos\alpha \cos\beta + \sin\alpha \sin\beta \cos c$
Hyperbolic plane, H^2	
Law of sines	$\operatorname{sh} a : \operatorname{sh} b = \sin\alpha : \sin\beta$
Law of cosines for sides	$\operatorname{ch} c = \operatorname{ch} a \operatorname{ch} b - \operatorname{sh} a \operatorname{sh} b \cos\gamma$
Law of cosines for angles	$\cos\gamma = -\cos\alpha \cos\beta + \sin\alpha \sin\beta \operatorname{ch} c$

Table A.2. *Area and Perimeter of B(r)*

	$A(B(r))$	$P(B(r))$
$\mathbb{R}^2$:	πr^2	$2\pi r$
S^2:	$2\pi(1 - \cos r)$	$2\pi \sin r$
H^2:	$2\pi(\operatorname{ch} r - 1)$	$2\pi \operatorname{sh} r$

is a convex domain in $\mathbb{S}$ ($C \neq S^2$) then its inradius $r(C)$ is the radius of the largest circular disc contained in C, and its circumradius $R(C)$ is the radius of the (unique) smallest circular disc containing C. We say that a polygon P is *inscribed* into C if the vertices of P lie on ∂C, and P is *circumscribed* around C if the sides of P touch C.

Given a line l in $\mathbb{S}$ and $\varrho > 0$ (where $\varrho < \pi/2$ on S^2), the points whose distance is ϱ from l form two curves. Let γ be one of these curves. Then γ is called an *equidistance curve* with respect to l, which is a line in the Euclidean case and known as a *hypercycle* in the hyperbolic case and as a *Lexell circle* in the spherical case. Given a segment of length x on l, the area of the points between γ and l whose projection into l lands in the segment is

$$x \cdot \varrho, \qquad x \cdot \operatorname{sh}\varrho, \quad \text{and} \quad x \cdot \sin\varrho \qquad (A.4)$$

in $\mathbb{R}^2$, in H^2, and in S^2, respectively. The last important property of equidistance curves follows by some simple cut and paste argument:

Lemma A.5.1: *Given the triangle pqr, let l be the line passing through the midpoints of the sides pq and pr, and let γ be the equidistance curve with respect to l that contains p. If p' moves along γ then the area of the triangle $p'qr$ stays the same.*

Next we discuss some properties of triangles under deformation. The first one, namely, Lemma A.5.2 (i), follows by the laws of cosine. Moreover, Lemmas A.5.2 (ii) and A.5.3 are consequences of Lemma A.5.1 and of the fact that if an equidistance curve intersects the interior of a circular disc then it intersects the bounding circle in exactly two points.

Lemma A.5.2: *Let the vertices q and r of the triangle pqr be fixed, and let p rotate around q towards r keeping the side length of pq. Writing s to denote the centre of the circle passing through p, q and r, we have*

  (i) the length of the side pr decreases;
  (ii) the area of the triangle increases when q and s are separated by the line pr, the area is maximal when s is the midpoint of the segment pr, and the area of the triangle decreases when q and s lie on the same side of pr.

Lemma A.5.3: *Given a circle and two points p and q of it, let the point r move from p to q along one of the arcs connecting p and q. Then the area of the triangle pqr increases until r reaches the farthest point of the arc from the line pq, and the area decreases after that position.*

Lemma A.5.2 (i) yields that two circles intersect in at most two points:

Lemma A.5.4: *If two circular discs B_1 and B_2 intersect and do not contain each other, then $\partial B_1 \cap \partial B_2$ consists of two points. Moreover, assuming that B_2 is not smaller than B_1, the shorter arc of ∂B_2 cut off by the common secant s is contained in B_1, and the arc of ∂B_1 cut off by s and lying on the other side of s is contained in B_2.*

A.5.1. Convex Arcs

We call a connected arc on the boundary of a convex domain a *convex arc*. The length $|\sigma|$ of a convex arc σ is defined analogously to the perimeter of a convex domain; say $|\partial C| = P(C)$ for a convex domain C.

Let σ be a convex arc in $\mathbb{S}$, and let $x \in \sigma$. Then there exist some (possibly many) tangent lines to σ at x, and σ lies on the same side of any tangent. However, if x is an endpoint of σ then there exists a unique tangent half line to σ at x.

If σ' is a subarc of σ connecting $x_1, x_2 \in \sigma$ such that the tangent half lines to σ' at x_1 and at x_2 intersect then we write $\sigma(x_1, x_2)$ to denote σ'. In this case $\beta(x_1, x_2)$ denotes π minus the angle of the two tangent half lines. Given $x \in \sigma$, the *curvature* of σ at x is defined to be

$$\kappa_\sigma(x) = \lim_{y \in \sigma,\, y \to x} \frac{\beta(y, x)}{|\sigma(y, x)|} \tag{A.5}$$

whenever the limit exists.

We claim that the curvature exists at all points of σ but a set of one-dimensional measure zero. By projecting radially from the model of $\mathbb{S}$ to some plane in $\mathbb{R}^3$ not containing the origin, we may assume that σ is the graph in $\mathbb{R}^2$ of a convex function f defined on an interval. We deduce by the definition of curvature that it exists at $x = (t, f(t))$ if and only if f is twice differentiable at t, and

$$\kappa_\sigma(x) = \frac{f''(t)}{(1 + f'(t)^2)^{\frac{3}{2}}} \tag{A.6}$$

holds in this case. According to the Rademacher theorem (see R. Schneider [Sch1993b]), a convex function on an interval is twice differentiable at all points but a set of measure zero and hence follows the claim.

When speaking about curves of constant curvature in $\mathbb{S}$, we always assume that the curvature exists at each point. These curves are subarcs of the following curves: Lines are of constant curvature zero. In the Euclidean plane circles of radius r are of curvature $1/r$, and on the sphere circles of radius $r < \pi/2$ are of curvature $1/\tan r$. Moreover, in the hyperbolic plane circles of radius r are of curvature $1/\operatorname{th} r$, and equidistant curves of associated distance r are of curvature $\operatorname{th} r$. Finally the so-called horocycles (which we do not discuss in detail) are of constant curvature one.

Given $\kappa \geq 0$, we say that a compact set C is κ-*convex* if any convex curve of constant curvature κ that connects two points of C (the shorter arc in the case of circles) is contained in C (see, for example, E. Gallego and A. Reventós [GaR1999]). It follows that the intersection of κ-convex sets is κ-convex, and we may speak about the κ-*convex hull*. The case $\kappa = 0$ gives back the usual notion of convexity.

Lemma A.5.5: *A convex domain C is κ-convex for $\kappa > 0$ if and only if whenever the curvature exists at some point of ∂C then it is at least κ.*

Remark. If x is a boundary point of a κ-convex domain C then x is contained in some complete convex curve σ of constant curvature κ such that C lies on the convex side of σ.

Proof. If C is κ-convex then let the curvature exist at some $x \in \partial C$. Since, for any y in (A.5), both curves of constant curvature κ connecting x and y lie in C, we deduce $\kappa_{\partial C}(x) \geq \kappa$.

For the reverse direction, we may assume that $\kappa_{\partial C}(x) \geq \kappa'$ at each $x \in \partial C$ whenever it exists where $\kappa' > \kappa$. Let us suppose that C is not κ-convex; hence, there exists a curve σ of constant curvature κ such that its endpoints lie in ∂C, and the rest of the curve lies outside of C. We may also assume that the endpoints are arbitrary close. Projecting to a suitable Euclidean plane, we obtain positive $\kappa_1 > \kappa_2$ and two convex curves σ_1 and σ_2 in $\mathbb{R}^2$ with the following properties: σ_1 and σ_2 connect the same points, the curvature exists at each point of σ_2 and is at most κ_2, the curvature is at least κ_1 at each point of σ_1 whenever it exists, and $\sigma_1 \subset \text{conv}\,\sigma_2$. We may assume that σ_1 and σ_2 are the graphs of the convex functions f_1 and f_2 on some interval $[a, b]$, where $f_2'(a) = 0$.

Writing f_1^- to denote the left-sided derivative of f_1, we claim that

$$f_1^-(t) \geq f_2'(t) \quad \text{for any } t \in (a, b). \tag{A.7}$$

Let $t_0 \in [a, b)$ satisfy $f_1^-(t_0) \geq f_2'(t_0)$, where we set $f_1^-(a) = 0$. We choose some $\varepsilon > 0$ such that $(1 + f_2'(t)^2)^{3/2} > (\kappa_2/\kappa_1)(1 + f_2'(t_0)^2)^{3/2}$ for $t \in (t_0, t_0 + \varepsilon)$. Now if $t \in (t_0, t_0 + \varepsilon)$ and $f_1''(t)$ exists then $f_1^-(t) \geq f_1^-(t_0)$ and (A.6) yields $f_1''(t) > f_2''(t)$; hence,

$$f_1^-(t) \geq f_1^-(t_0) + \int_{t_0}^t f_1''(s)\,ds > f_2'(t_0) + \int_{t_0}^t f_2''(s)\,ds = f_2'(t).$$

In turn, we conclude (A.7). Since $f_1(b) = f_2(b)$, it follows that σ_1 coincides with σ_2, a contradiction, which verifies Lemma A.5.5. $\qquad\square$

Corollary A.5.6. *A circular disc in H^2 is th r-convex for any $r > 0$.*

Given a convex curve σ in the Euclidean plane, the *total curvature $\beta(\sigma)$* is the total change of angle of the tangent line along σ. More precisely, if σ is a closed curve then $\beta(\sigma) = 2\pi$. Otherwise, we subdivide σ into arcs $\sigma(x_{i-1}, x_i)$, $i = 1, \ldots, k$, in a way that x_0 and x_k are the endpoints, and there is a unique tangent line at x_i for $i = 1, \ldots, k - 1$, and we define $\beta(\sigma) = \sum_{i=1}^k \beta(x_{i-1}, x_i)$. It is not hard to see that $\beta(\sigma)$ does not depend on the subdivision and is always at most 2π.

A.5.2. Estimating the Area and the Perimeter

In various cases we need more general sets in $\mathbb{S}$ than convex domains. Let us consider convex arcs $\sigma_1, \ldots, \sigma_k$ in $\mathbb{S}$ whose total length is less than 2π in the spherical case and satisfy one of the following:

(a) $k = 2$ when σ_1 and σ_2 intersect in their two common endpoints or
(b) $k \geq 3$ when σ_i and σ_{i+1} share a common endpoint for $i = 1, \ldots, k-1$; moreover, σ_k and σ_1 share a common endpoint, and no other intersection occurs.

It is not hard to see that $\sigma_1, \ldots, \sigma_k$ are contained in an open hemisphere in the spherical case. If we remove the union of $\sigma_1, \ldots, \sigma_k$, the rest of $\mathbb{S}$ has two connected components according to the *Jordan curve theorem* (see M. A. Armstrong [Arm1983]). We define the *domain bounded by* $\sigma_1, \ldots, \sigma_k$ to be the closure of the bounded component (the one contained in the open hemisphere in the spherical case).

Given points $x_0, \ldots, x_n \in \mathbb{S}$ such that $x_{i-1} \neq x_i$ for $i = 1, \ldots, n$, the union of the segments $x_{i-1}x_i$ for $i = 1, \ldots, n$ is called the *broken line $x_1, \ldots, x_n$*. If the segments $x_{i-1}x_i, i = 1, \ldots, n$, bound a domain D then we also say that the broken line $x_1, \ldots, x_n, x_1$ bounds D.

The following two isoperimetric type statements are discussed in A. Florian [Flo1993]. In the spherical case we always assume that the perimeter is less than 2π.

Theorem A.5.7 (Isoperimetric inequality). *Among domains of given perimeter, the circular disc is the unique domain of maximal area.*

Theorem A.5.8. *Among polygons, which have at most k sides and given perimeter, the unique polygon of maximal area is the regular k-gon.*

Finally, we verify that if C is a convex domain in $\mathbb{R}^2$ then

$$\frac{1}{2} P(C) \cdot r(C) \leq A(C) \leq P(C) \cdot r(C). \tag{A.8}$$

We may assume that C is a polygon. The lower bound follows by dissecting C into triangles such that the centre of a fixed inscribed circular disc is a common vertex. Next we build a rectangle of height $r(C)$ on the inner side of each side of C. Since these rectangles cover C, we conclude the upper bound in (A.8).

A.5.3. Cell Complexes and the Euler Characteristic

We now describe the type of cell complexes in $\mathbb{S}$ that we will encounter and present their main properties. What seems to be strange at first glance is that, two-gons do occur in a natural way in our context. We refer to the book by

M. A. Armstrong [Arm1983] for proofs and for a detailed discussion from a more general prospective.

For us a *cell complex* Σ in $\mathbb{S}$ consists of vertices (which are points), of edges (with each being a convex curve connecting two different vertices), and of cells (which are domains bounded by finitely many edges as in (a) or in (b) from the previous section). We require that the relative interior of an edge never meets a vertex or another edge, and the interior of a cell never meets any other object. In addition, a nonempty intersection of two cells is either a common vertex or a common edge. Finally, if two edges intersect in two common endpoints then they bound a cell (a "two-gon"). In this book we use the term *simplicial complex* in $\mathbb{S}$ only for a cell complex in which each edge is a segment and each cell is a triangle.

We say that Σ is *connected* if, for any two vertices v and w, there exist edges $e_1, \ldots, e_k$ such that $v \in e_1$, $w \in e_k$, and e_i and e_{i+1} intersect. If Σ is finite then we write $f_0(\Sigma)$, $f_1(\Sigma)$, and $f_2(\Sigma)$ to denote the number of vertices, edges, and cells, respectively, and $\chi(\Sigma)$ to denote the *Euler characteristic*

$$\chi(\Sigma) = f_0(\Sigma) - f_1(\Sigma) + f_2(\Sigma). \tag{A.9}$$

We write supp Σ to denote the union of the elements of Σ. We frequently call Σ a *cell decomposition* of supp Σ (or *triangulation* if Σ is simplicial). According to the *Euler formula*,

$$\chi(\Sigma) = 2 \quad \text{if} \quad \text{supp } \Sigma = S^2; \tag{A.10}$$

$$\chi(\Sigma) = 1 \quad \text{if} \quad \text{supp } \Sigma \text{ is a domain.} \tag{A.11}$$

We present a useful application of the second formula:

Lemma A.5.9: *Let Σ be a cell decomposition of a domain C in $\mathbb{S}$ satisfying $f_2(\Sigma) \geq 2$. If $k_1, \ldots, k_{f_2(\Sigma)}$ denote the numbers of edges of the cells, and b denotes the number of edges of Σ that lie in ∂C, then*

$$\sum_{i=1}^{f_2(\Sigma)} (6 - k_i) \geq b + 6.$$

Proof. Each vertex of Σ is of degree at least three; hence, $3f_0(\Sigma) \leq 2f_1(\Sigma)$, which in turn yields $6f_2(\Sigma) \geq 2f_1(\Sigma) + 6$ by the Euler formula (A.11). However, counting the edges of every cell leads to $\sum_{i=1}^{f_2(\Sigma)} k_i = 2f_1(\Sigma) - b$, completing the proof of Lemma A.5.9. $\square$

A.5.4. Compact Surfaces of Constant Curvature 0, 1, or −1

Our main reference for this part is J. G. Ratcliffe [Rat1994]. We say that a compact surface X is of constant curvature 0, 1, or −1 if it can be obtained as a quotient of $\mathbb{S}$, where $\mathbb{S}$ is $\mathbb{R}^2$, S^2, or H^2, respectively, in the following sense: Let Λ be *lattice*, namely, a group of isometries of $\mathbb{S}$ such that the points $\{gx : g \in \Lambda\}$ of the orbit of any given $x \in \mathbb{S}$ are different (Λ acts freely) and form a discrete set (Λ is discrete). Then X is the family $\mathbb{S}/\Lambda$ of orbits of Λ, and the distance between two points of X is the minimal distance between the points of the corresponding orbits. Moreover, the notion of area naturally transfers to X by the quotient map. We call X *orientable* if each element of Λ keeps the orientation of $\mathbb{S}$ (whatever it means). Now the compact surfaces of constant curvature one are S^2 (orientable) and the projective plane P^2 (nonorientable). We note that P^2 can be obtained by identifying the antipodal points of S^2 and is sometimes called the *elliptic plane*. The orientable examples of curvature zero are the tori (see Section A.13), and the nonorientable ones are the so-called Klein bottles. The rest of the surfaces have hyperbolic structure.

We say that a convex domain $C \subset \mathbb{S}$ embeds into X if different points of int$C$ belong to different orbits of Λ. In this case $A(C') = A(C)$ holds for the image C' of C on X. Moreover, a *metric disc* of radius r and of centre p on X consists of the points of X whose distance from p is at most r, and its inverse image is the union of the circular discs of radius r in $\mathbb{S}$ that are centred at the inverse images of p.

We say that a family $\mathcal{F}$ of some objects in $\mathbb{S}$ is *periodic* with respect to Λ if $g\sigma \in \mathcal{F}$ for any $g \in \Lambda$ and $\sigma \in \mathcal{F}$, and Λ defines only finitely many equivalent classes of $\mathcal{F}$. Any cell decomposition of $\mathbb{S}$ periodic with respect to Λ induces a *cell decomposition* Σ of X (which is a so-called CW-complex). If the cells in $\mathbb{S}$ are triangles then Σ is called a *triangulation* of X. We write $f_0(\Sigma)$, $f_1(\Sigma)$, and $f_2(\Sigma)$ to denote the number of equivalence classes of the vertices, the edges, and the cells, respectively, of the cell decomposition of $\mathbb{S}$, which satisfy

$$f_0(\Sigma) - f_1(\Sigma) + f_2(\Sigma) = \chi(X), \tag{A.12}$$

where $\chi(X)$ is the Euler characteristic of X. In particular, $\chi(S^2) = 2$, $\chi(P^2) = 1$, $\chi(X) = 0$ if $\mathbb{S} = \mathbb{R}^2$, and $\chi(X) < 0$ if $\mathbb{S} = H^2$. Since $3f_0(\Sigma) \leq 2f_1(\Sigma)$, (A.12) yields

$$f_1(\Sigma) \leq 3(f_2(\Sigma) - \chi(X)). \tag{A.13}$$

Example A.5.10. If $k = 3, 4, 5,$ or $k = 6$, or $k \geq 7$ then there exists a compact orientable surface X_k of constant curvature 1, or 0, or −1, respectively, such that X_k can be triangulated into embedded images of a regular triangle T_k of angle $2\pi/k$ and metric discs of radius $R(T_k)$ on X_k are embedded circular discs.

If $k = 3, 4, 5$ then $X_k = S^2$, and the triangulation is obtained by projecting radially the faces of the regular tetrahedron, octahedron, and icosahedron (see Section A.7), respectively, onto S^2. If $k = 6$ then X_k is the quotient of $\mathbb{R}^2$ by a suitable hexagonal lattice (see Section A.13).

The hyperbolic case is more subtle. We start with the triangulation of H^2 into regular triangles of angle $2\pi/k$. For a fixed triangle T, let Γ be the group of isometries generated by the three rotations of angle $2\pi/k$ centred at the vertices of the triangle. It follows by the Selberg lemma (see corollary 5 on p. 327 in J. G. Ratcliffe [Rat1994]) that Γ has a torsion-free normal subgroup Λ' of finite index. In particular, Λ' induces a compact surface $\mathbb{S}/\Lambda'$ that is triangulated into embedded images of T_k. To ensure that circular discs of radius $R(T_k)$ in H^2 embed onto metric discs of radius $R(T_k)$ on X_k, we may choose X_k to be $\mathbb{S}/\Lambda$ for some suitable subgroup Λ of Λ' (see G. Fejes Tóth, G. Kuperberg, and W. Kuperberg [FKK1998]).

A.6. Mixed Areas in $\mathbb{R}^2$

This section considers a common generalization of the area and the perimeter (see T. Bonnesen and W. Fenchel [BoF1987], J. R. Sangwine–Yager [SaY1993], or R. Schneider [Sch1993b] for detailed discussion in any dimension). Given a polygon C with edges $e_1, \ldots, e_k$, and a smooth and strictly convex domain K containing o in its interior, let $x_i \in \partial K$ satisfy that the exterior unit normal u_i to e_i is an exterior normal at x_i to K. Then for positive λ and μ, $\lambda C + \mu K$ can be decomposed into three pieces: into λC, into the parallelograms conv$\{\lambda e_i, \lambda e_i + \mu x_i\}$, and into the pieces at the vertices of λC whose suitable translates tile μK. Therefore,

$$A(\lambda C + \varrho K) = A(K)\lambda^2 + 2A(C, K)\lambda\mu + A(K)\mu^2, \qquad (A.14)$$

where the *mixed area* of K and C is

$$A(C, K) = \frac{1}{2} \sum_{i=1}^{k} |e_i| \cdot h_K(u_i). \qquad (A.15)$$

The mixed area can be extended to any pair of compact convex sets using (A.14) and continuity. The formula (A.14) shows that $A(\cdot, \cdot)$ is actually symmetric in its variables; hence, we deduce by (A.15) that $A(\cdot, \cdot)$ is monotonic and linear in both variables. In addition, $A(C, K)$ stays invariant if both C and K are transformed by the same affine map with determinant one. We observe that $A(C, C) = A(C)$, and if K is the unit disc then $A(C, K) = S(C)/2$.

The isoperimetric inequality can be generalized for mixed areas; namely, the Minkowski inequality states for convex domains C and K that

$$A(C, K)^2 \geq A(C) \cdot A(K), \tag{A.16}$$

and equality holds if and only if K and C are homothetic. To state a stability version of the Minkowski inequality, let us introduce the *circumradius* and the *inradius* with respect to a given domain. If K is a convex domain and C is a compact convex set then we write $R_K(C)$ and $r_K(C)$ to denote the smallest R and the largest r, respectively, such that a translate of RK contains C and a translate of rK is contained in C. In particular, if K is the unit disc then $R_K(C) = R(C)$ and $r_K(C) = r(C)$. Now the Bonnesen inequality strengthens the Minkowski inequality in the form (see H. Groemer [Gro1993])

$$A(C, K)^2 - A(C) \cdot A(K) \geq \frac{A(K)^2}{4} (R_K(C) - r_K(C))^2. \tag{A.17}$$

Finally, we verify that

$$A(K, C) \cdot r_K(C) \leq A(C) \leq 2A(K, C) \cdot r_K(C). \tag{A.18}$$

Since a translate of $r_K(C) K$ is contained in C, the lower bound is a consequence of the monotonicity of the mixed areas. For the upper bound, we may assume that $r_K(C) = 1$, C is a polygon, and K is a strictly convex domain containing the origin in its interior. For any side e of C, let u be the exterior unit normal, and let y be the point ∂K with exterior normal u. We write $\Pi(e)$ to denote the parallelogram that is the convex hull of e and $e - y$; hence, the area of $\Pi(e)$ is $|e| \cdot h_C(e)$. Since the parallelograms $\Pi(e)$ cover C as e runs through the sides of C, we conclude (A.18) by (A.15).

A.7. Polyhedral Sets and Polytopes in $\mathbb{R}^d$

See M. M. Bayer and C. W. Lee [BaL1993] or G. Ziegler [Zie1995] for detailed discussion of polyhedral sets and polytopes.

We call a set P a *polyhedral set* in $\mathbb{R}^d$ if it is the intersection of finitely many half spaces. A *face* of P is the intersection of P and of some supporting hyperplane where faces of dimension 0, 1, or $d - 1$ are called *vertices*, *edges*, or *facets*, respectively. We note that a nonempty intersection of some faces of P is a face of P as well. If P is d dimensional then the boundary of P is the union

Table A.3. *Regular Polytopes in Higher Dimensions*

	Dimension	Vertex Figures	Facets
Regular simplex	$d \geq 4$	$d + 1$ simplices	$d + 1$ simplices
Cube	$d \geq 4$	2^d simplices	$2d$ cubes
Crosspolytope	$d \geq 4$	$2d$ crosspolytopes	2^d simplices
24 cells	4	24 cubes	24 octahedra
120–cells	4	600 simplices	120 dodecahedra
600–cells	4	120 icosahedra	600 simplices

of the facets, and any $(d - 2)$-face is contained in exactly two facets. If P is the positive hull of d independent vectors then P is called a *simplicial cone*.

A polyhedral set P is called a *polytope* if it is bounded and d dimensional; hence, P is a convex body. Moreover, P is the convex hull of its vertices, and in turn the convex hull of a finite family of points that are not contained in any hyperplane is always a polytope. The simplest example for a polytope is the *simplex*. It has $d + 1$ vertices in $\mathbb{R}^d$, and the convex hull of any n vertices is a face that is an $(n - 1)$-simplex. A simplex of dimension two is called a *triangle*, and one of dimension three is called a *tetrahedron*. Next we call the part of $\mathbb{R}^d$ between two parallel hyperplanes a *strip*, and we call the intersection of d pairwise nonparallel strips a *parallelepiped* (or a *parallelogram* in the planar case). We say that a polytope P is *inscribed* into a convex body C if the vertices of P lie on ∂C, and P is *circumscribed* around C if the facets of P touch C.

We define a *flag* of a polytope P to be a sequence $F_0 \subset \ldots \subset F_{d-1}$ of faces such that dim $F_i = i$. In addition, we say that P is a *regular polytope* if, for any two flags $\{F_i\}$ and $\{G_i\}$, there exists a congruence that maps P onto P, and F_i onto G_i for $i = 0, \ldots, d - 1$. It follows that each face is regular, and any two faces of the same dimension are congruent. The five three-dimensional regular polytopes are the *Platonic solids*: the regular tetrahedron (four vertices of degree three and four triangular faces), the cube (eight vertices of degree three and six square faces), the octahedron (six vertices of degree four and eight triangular faces), the dodecahedron (twenty vertices of degree three and twelve pentagonal faces), and the icosahedron (twelve vertices of degree five and twenty triangular faces). If v is a vertex of some regular d-polytope then the neighbours of v form a regular $(d - 1)$-polytope that we call the *vertex figure* at v. In Table A.3 we list the regular polytopes in higher dimensions.

We say that a polytope P in $\mathbb{R}^d$ is *simplicial* if all its facets are $(d - 1)$-simplices. Among the regular polytopes, the simplicial ones are the simplex, the crosspolytope, the icosahedron, and the 600-cells.

A.8. Associated Balls and Ellipsoids in $\mathbb{R}^d$

Let C be a compact convex set in $\mathbb{R}^d$. There exists a unique ball of minimal radius containing C, which is called the *circumscribed ball* of C. We call its centre the *circumcentre* of C and its radius the *circumradius* $R(C)$ of C. According to the Jung inequality (see T. Bonnesen and W. Fenchel [BoF1987]), the circumradius and the diameter of C are related in the following way:

Theorem A.8.1 (Jung inequality). *If C is a convex body in $\mathbb{R}^d$ then*

$$\sqrt{\frac{2d}{d+1}} \cdot R(C) \leq \operatorname{diam} C \leq 2R(C).$$

We write $r(C)$ to denote the maximal radius of any ball contained in C, and we call $r(C)$ the *inradius* of C. There may exist more balls with radius $r(C)$ contained in C, and any of these balls is called an *inscribed ball*. Next let u be a unit vector. Then the *width* of C in direction u is $h_C(u) + h_C(-u)$. This is equal to the distance of the supporting hyperplanes normal to u (hence the name), which is, in turn, the length of the projection of C into the line of u. We define the *width* $w(C)$ of C to be the minimum of the widths in all directions. It is not hard to see that there exists a secant s of C whose length is $w(C)$, and the two hyperplanes orthogonal to s at the endpoints are supporting hyperplanes of C. Now the analogue of the Jung inequality is the Steinhagen inequality (see T. Bonnesen and W. Fenchel [BoF1987]):

Theorem A.8.2 (Steinhagen inequality). *If C is a convex body in $\mathbb{R}^d$ then*

$$w(C) \leq \begin{cases} 2\sqrt{d} \cdot r(C) & \text{if d is odd;} \\ \frac{2(d+1)}{\sqrt{d+2}} \cdot r(C) & \text{if d is even.} \end{cases}$$

The ratio of the circumradius and the inradius can be arbitrarily large for convex bodies. Therefore, we sometimes approximate convex bodies by *ellipsoids*, namely, affine images of balls. According to the result of F. John (see K. Ball [Bal1997]), we have

Theorem A.8.3 (John theorem). *For any convex body C in $\mathbb{R}^d$, there exist x and an ellipsoid E centred at the origin such that*

$$E \subset C - x \subset d \cdot E.$$

It follows that any convex body C in $\mathbb{R}^d$ has a translate C' satisfying

$$-C' \subset d \cdot C'. \tag{A.19}$$

Given a compact convex set C in $\mathbb{R}^d$, and $1 \le m \le d$, let $r_m(C)$ be the radius of the largest m-dimensional ball contained in C. In particular, $2r_1(C)$ is the diameter and $r_d(C)$ is the inradius. If $r_m(C)$ is small then C is close to some affine $(m-1)$-plane L, namely (see Puhov [Puh1979]),

$$C \subset L + (m+1) r_m(C) \cdot B^d. \tag{A.20}$$

We quickly sketch a proof for (A.20), where we may assume that C is at least m dimensional. Let T be an m-dimensional simplex of maximal m-content (see Section A.9) contained in C, and let F be an $(m-1)$-dimensional face of T of maximal $(m-1)$-content. Then the opposite vertex of T is of distance at most $(m+1) r_m(T)$ from F; hence, $C \subset \text{aff} F + (m+1) r_m(T) \cdot B^d$.

A.9. Volume, Surface Area, and Lebesgue Measure in $\mathbb{R}^d$

The main references for this section are M. Berger [Ber1987], T. Bonnesen and W. Fenchel [BoF1987], and R. Schneider [Sch1993b].

Borel sets in $\mathbb{R}^d$ form the smallest family of subsets of $\mathbb{R}^d$ such that open sets are Borel sets, the family is closed under countable union and intersection, and the complement of a Borel set is also a Borel set. In particular, closed sets are Borel sets. The d-dimensional *Lebesgue measure* is the unique function on the Borel sets of $\mathbb{R}^d$, which is one for the unit cube, and additive for countable union of disjoint Borel sets. Actually, the Lebesgue measure can be extended to the smallest family $\mathcal{F}$ containing the Borel sets such that $\mathcal{F}$ is closed under countable union and taking a complement, and $\mathcal{F}$ contains all subsets of any Borel set whose Lebesgue measure is zero. The sets in this enlarged family are called (Lebesgue) *measurable sets*. We note that the Lebesgue measure stays invariant under any affine map whose determinant is ± 1, in particular under isometries. In the theory of convex sets, the Lebesgue measure of a measurable set σ is traditionally called *volume* and denoted by $V(\sigma)$. It turns out that if a convex body C is strictly contained in a convex body K then $V(C) < V(K)$. With respect to rescaling by a positive λ, the volume satisfies

$$V(\lambda C) = \lambda^d V(C).$$

A bounded set σ in $\mathbb{R}^d$ is called *Jordan measurable* if the Lebesgue measure of the boundary is zero, or equivalently, if

$$\lim_{\varepsilon \to 0} V(\text{bd} \, \sigma + \varepsilon B^d) = 0. \tag{A.21}$$

In particular any compact convex set C is Jordan measurable.

Let F be a compact convex set whose affine hull is k dimensional. Then the (k-dimensional) volume of F inside aff$F$ is denoted by $|F|$ and is called the *relative content* of F. To simplify notation, we sometimes write $|F|$ even if F is full dimensional. Let dim $F = d - 1$, and let H be a hyperplane not parallel to aff$F$. We write u and v to denote unit normals to H and aff$F$, respectively. Then the orthogonal projection F' of F into H satisfies

$$|F'| = |\langle u, v \rangle| \cdot |F|. \tag{A.22}$$

Now if Q is a polytope then its projection into any hyperplane is twice covered by the projections of the facets (except for the projections of the ($d - 2$)-faces of Q). It follows by (A.22) that if $v_1, \ldots, v_n$ denote the exterior normals to the facets of Q satisfying that $\|v_i\|$ is the content of the corresponding facet for each i then

$$v_1 + \cdots + v_n = 0. \tag{A.23}$$

We define the *surface area* $S(Q)$ of a polytope Q to be the sum of the contents of the facets of Q. For a lower dimensional polytope G, the surface area is defined to be twice the ($d - 1$)-volume. It follows by (A.22) that if $P \subset Q$ holds for the polytopes P and Q then $S(P) \leq S(Q)$. Therefore the surface area can be extended to a continuous function on all compact convex sets. If the compact convex set C is strictly contained in the convex body K then $S(C) < S(K)$, and in addition

$$S(\lambda C) = \lambda^{d-1} S(C)$$

holds for any nonnegative λ.

A real nonnegative function f on $\mathbb{R}^d$ is called *measurable* if the family of x satisfying $f(x) < \alpha$ is measurable for each α. We note that if f is continuous everywhere but at the points of a set of measure zero then f is measurable. In this case the integral

$$\int_{\mathbb{R}^d} f(x) \, dx$$

of f with respect to the Lebesgue measure is well defined (possibly infinite). The value of the integral is not affected by changing f on a set of measure zero. If f is bounded on $\mathbb{R}^d$ and zero outside of a bounded subset then the integral is finite, as, say, for the volume of a convex body C,

$$V(C) = \int_{\mathbb{R}^d} \chi_C(x) \, dx,$$

where χ_C is the characteristic function of C.

Next we recall the *Fubini theorem* in a special case, namely, when the measurable function f is either continuous or continuous in a compact set and zero outside. We write any $x \in \mathbb{R}^d$ in the form $x = (y, z)$, where $y \in \mathbb{R}^n$ and $z \in \mathbb{R}^m$, $d = m + n$. Then

$$\int_{\mathbb{R}^d} f(x)\, dx = \int_{\mathbb{R}^n} \left(\int_{\mathbb{R}^m} f(y, z)\, dz \right) dy.$$

A typical application is that if C is a pyramid whose height is h and whose base is the $(d - 1)$-dimensional compact convex set M then

$$V(C) = \frac{h}{d} |M|. \tag{A.24}$$

Finally, let us discuss the volume and surface area of the unit ball. We write B^d to denote the unit ball centred at o in $\mathbb{R}^d$, and let κ_d denote its volume. Readily $\kappa_0 = 1$ and $\kappa_1 = 2$, and the Fubini theorem yields

$$\kappa_d = 2\kappa_{d-1} \int_0^1 (1 - y^2)^{\frac{d-1}{2}}\, dy. \tag{A.25}$$

Therefore, in principle the value of κ_d can be determined by induction, say $\kappa_2 = \pi$ and $\kappa_3 = 4\pi/3$. We note that approximating the unit ball by circumscribed polytopes and (A.24) yields $S(B^d) = d\kappa_d$.

To have an explicit formula for κ_d, we need Euler's gamma function

$$\Gamma(t) = \int_0^\infty x^{t-1} e^{-x}\, dx$$

for $t > 0$ (see E. Artin [Art1964]). The gamma function satisfies

$$\Gamma\left(\frac{1}{2}\right) = \sqrt{\pi}, \quad \Gamma(1) = 1, \quad \text{and} \quad \Gamma(t + 1) = t\Gamma(t) \tag{A.26}$$

and can be approximated using the celebrated *Stirling formula*

$$\left(\frac{t}{e}\right)^t \sqrt{2\pi t} < \Gamma(t + 1) < \left(\frac{t}{e}\right)^t \sqrt{2\pi(t + 1)}. \tag{A.27}$$

The volume of the unit ball is (see L. Fejes Tóth [FTL1964])

$$\kappa_d = \frac{\pi^{\frac{d}{2}}}{\Gamma\left(\frac{d}{2} + 1\right)}. \tag{A.28}$$

Let us observe that κ_{d-1}/κ_d is increasing as a function of d by (A.25). Since (A.26) yields $2\pi\kappa_d = (d + 2)\kappa_{d+2}$, we deduce the estimates

$$\sqrt{\frac{d}{2\pi}} < \frac{\kappa_{d-1}}{\kappa_d} < \sqrt{\frac{d + 1}{2\pi}}. \tag{A.29}$$

A.10. Hausdorff Measure and Lipschitz Functions in $\mathbb{R}^d$

The material of this section is discussed in depth in the monographs by C. A. Rogers [Rog1998], and K. J. Falconer [Fal1986], and R. Schneider [Sch1993b].

Frequently, we consider not only the total boundary of a convex body but subsets of it as well. To measure these subsets, we review the notion of Hausdorff measure following C. A. Rogers [Rog1998] and K. J. Falconer [Fal1986]). If σ is a subset of $\mathbb{R}^d$ and $k > 0$ then, for any $\delta > 0$, we define

$$\mathcal{H}_\delta^k(\sigma) = \frac{\kappa_k}{2^k} \inf \left\{ \Sigma_{i=1}^\infty (\operatorname{diam} U_i)^k : \{U_i\} \text{ covers } \sigma \text{ and diam } U_i < \delta \right\}.$$

Since $\mathcal{H}_\delta^k(\sigma)$ increases if δ tends to zero, we can define the outer Hausdorff k-measure of σ to be

$$\mathcal{H}^k(\sigma) = \lim_{\delta \to 0} \mathcal{H}_\delta^k(\sigma).$$

We allow $\mathcal{H}^k(\sigma) = \infty$ as well. Hausdorff measurable sets form the smallest family of sets with the following properties: Borel sets and sets of outer Hausdorff k-measure zero are Hausdorff measurable. Moreover the complement of a Hausdorff measurable set and countable union of Hausdorff measurable sets are Hausdorff measurable. On this family $\mathcal{H}^k(\cdot)$ is the *Hausdorff k-measure*; namely, the measure of the countable union of pairwise disjoint sets is the sum of the measures of the sets.

Measurable functions and *integrals* with respect to $\mathcal{H}^k(\cdot)$ are defined analogously as with respect to the Lebesgue measure. We frequently encounter integrals that are easier to evaluate "radially." In this case we use *polar coordinates*; namely, if f is a nonnegative measurable function in $\mathbb{R}^d$ then

$$\int_{\mathbb{R}^d} f(y)\,dy = \int_0^\infty \int_{S^{d-1}} f(rx)\,dx\,dr = \int_{S^{d-1}} \int_0^\infty f(rx)\,dr\,dx. \quad (A.30)$$

We observe that the normalization of $\mathcal{H}^k(\cdot)$ is chosen in a way that, on affine k-planes, the k-dimensional Lebesgue measure and the Hausdorff k-measure coincide; hence, we frequently speak about k-measure instead of the Hausdorff k-measure. In addition, if C is a convex body in $\mathbb{R}^d$ then

$$S(C) = \mathcal{H}^{d-1}(\partial C).$$

We note that the boundary of C inherits a natural topology from the topology of $\mathbb{R}^d$; namely, open sets of ∂C are the intersections of open sets of $\mathbb{R}^d$ with ∂C. We say that a $\sigma \subset \partial C$ is *Jordan measurable* on ∂C if the relative boundary of σ as a subset of ∂C is of $(d-1)$-measure zero. In this case sometimes we write

$$|\sigma| = \mathcal{H}^{d-1}(\sigma).$$

A function f from a subset of $\mathbb{R}^n$ into $\mathbb{R}^n$ is called a *Lipschitz function* if there exists some $c > 0$ such that

$$\|f(x) - f(y)\| \le c \cdot \|x - y\|$$

for any x, y, where f is defined. The definition of the Hausdorff measure yields that if f is defined on σ and $\mathcal{H}^k(\sigma)$ is finite then

$$\mathcal{H}^k(f(\sigma)) \le c^k \cdot \mathcal{H}^k(\sigma). \tag{A.31}$$

Typical examples of Lipschitz functions are the closest point map (see (A.3)), the support function, and the distance function (if it exists). In addition, given convex bodies C and K whose interior contains the origin, the radial projection between the boundaries of C and K is Lipschitz as well.

Finally, we discuss some related properties of the boundary of a convex body C in $\mathbb{R}^d$ (see R. Schneider [Sch1993a,Sch1993b] for references). The boundary points that are not smooth form a set of Hausdorff $(d - 1)$-measure zero. In particular, if g is a continuous function on S^{d-1} and u_x is an exterior unit normal at $x \in \partial C$ then $f(x) = g(u_x)$ is a measurable function on ∂C with respect to $\mathcal{H}^{d-1}(\cdot)$. Therefore, we may write

$$\int_{\partial C} f(x) \, dx$$

to denote the integral with respect to $\mathcal{H}^{d-1}(\cdot)$, and the integral depends only on the values at smooth points.

A.11. Intrinsic Volumes in $\mathbb{R}^d$

Our point of view is rather close to the one presented in the books by H. Hadwiger [Had1957a] and D. Klain and G. C. Rota [KlR1997], but some facts can be found only in R. Schneider [Sch1993b]. In this section $| \cdot |$ stands for the relative content of convex sets.

Let P be a polytope in $\mathbb{R}^d$. If F is an i-face of P then the exterior normal cone $N_P(x)$ is the same for any $x \in \operatorname{relint} F$, and we write $N_P(F)$ to denote this common cone. Since $\dim N_P(F) = d - i$, we define the "relative angle" $\sigma(N(F))$ of the cone to be the quotient of $|B^d \cap N(F)|$ and $\kappa_{d-i} = |B^{d-i}|$. Now $\operatorname{relint} F + N(F)$ is the family of points x such that the unique closest point of P to x lies in the relative interior of F. Thus if $\varrho > 0$ then the Steiner formula

$$V(P + \varrho B^d) = \sum_{i=0}^{d} \kappa_{d-i} V_i(P) \varrho^{d-i} \tag{A.32}$$

holds, where the coefficient $V_i(P)$ is the *intrinsic i-volume*

$$V_i(P) = \sum_{F \ i \ \text{face}} |F| \cdot \sigma(N(F)). \tag{A.33}$$

Here $V_0(P) = 1$, $V_d(P) = V(P)$, and $V_{d-1}(P) = S(P)/2$.

The intrinsic i-volume is readily a continuous function of P. Using this property, the definition of the intrinsic volumes can be extended to any compact convex set M. Then (A.32) still holds, and $V_i(M)$ is continuous. In addition, if λ is nonnegative then $V_i(\lambda M) = \lambda^i \cdot V_i(M)$. The intrinsic volumes of the unit ball in $\mathbb{R}^d$ are

$$V_i(B^d) = \frac{\binom{d}{i} \kappa_d}{\kappa_{d-i}}, \quad i = 0, \ldots, d.$$

A very useful property of the intrinsic volumes is their dimension invariance; namely, if $\mathbb{R}^d$ is embedded into $\mathbb{R}^m$, $m > d$, then, for any compact convex set C in $\mathbb{R}^d$, the i-intrinsic volume is the same in $\mathbb{R}^d$ and in $\mathbb{R}^m$.

Next we consider some integral representations of the intrinsic volumes. If C is a convex body then

$$V_{d-1}(C) = \frac{1}{2\kappa_{d-1}} \int_{S^{d-1}} |C/u| \, du \tag{A.34}$$

holds according to the formula of A. L. Cauchy. This follows for polytopes by (A.22), and approximation yields the result for general convex bodies. Since a polynomial determines its coefficients, we deduce by (A.32) that $V_{d-1}(C + \varrho B^d)$ is a polynomial in ϱ, and the coefficients depend on $V_i(C)$, $i = 0, \ldots, d - 1$. However, the projection of $C + \varrho B^d$ into $u^\perp$ is $(C/u) + \varrho B_u$, where B_u is the unit ball in $u^\perp$, and hence the content of the projection is a polynomial in ϱ. Comparing the coefficients in these two representations yields

$$V_i(C) = \frac{\kappa_{d-1-i}}{\kappa_{d-1}\kappa_{d-i}} \int_{S^{d-1}} V_i(C/u) \, du. \tag{A.35}$$

An immediate corollary is that the intrinsic i-volume is monotonic. In addition, if C and K are convex bodies and C is strictly contained in K then $V_i(C) < V_i(K)$.

Given $1 \le i \le j \le d - 1$, we now generalize (A.35) in a way that C is projected into j-dimensional linear subspaces (see R. Schneider [Sch1993b] for details). We write $\text{Gr}(d, j)$ to denote the family of linear j-spaces in

$\mathbb{R}^d$ (the *Grassmannian*) and $\mu_{d,j}$ to denote the unique invariant measure on $\text{Gr}(d, j)$ satisfying $\mu_{d,j}(\text{Gr}(d, j)) = 1$. Here invariance means that if Σ is any measurable family of j-dimensional linear subspaces and Q is an orthogonal transformation then $\mu_{d,j}(Q\Sigma) = \mu_{d,j}(\Sigma)$. It follows *via* (A.35) that

$$V_i(C) = \frac{V_i(B^d)}{V_i(B^j)} \int_{\text{Gr}(d,j)} V_i(C|L) \, d\mu_{d,j}(L). \qquad (A.36)$$

In particular, $V_i(C)$ is proportional to the *mean i-dimensional projection* of C. If $i = j = 1$ then we obtain

$$V_1(C) = \frac{1}{2\kappa_{d-1}} \int_{S^{d-1}} h_C(u) + h_C(-u) \, du,$$

where $h_C(u) + h_C(-u)$ is the width of C in direction u. Therefore, $V_1(C)$ is proportional to the *mean width* of C, and we also deduce that

$$V_1(\alpha \, C + \beta \, K) = \alpha \cdot V_1(C) + \beta \cdot V_1(K)$$

holds for any nonnegative α and β.

We have already seen that, given the area of convex domains, the perimeter is minimal for the circular disc (see Theorem A.5.7). Also, the isoperimetric inequality generalizes to higher dimensions; namely, if K is a convex body in $\mathbb{R}^d$ and $i = 1, \ldots, d - 1$ then

$$\frac{V_i(K)}{V(K)^{\frac{i}{d}}} \geq \frac{V_i(B^d)}{V(B^d)^{\frac{i}{d}}} \qquad (A.37)$$

with equality if and only if K is a ball (see H. Hadwiger [Had1957a] or K. Leichtweiß [Lei1980]).

The classical way to deform a convex body K into a more symmetric one is the *Steiner symmetrization*. Let H be a hyperplane, and we write K' to denote the reflected image of K through H. Then the Steiner symmetrization produces the convex body K_H, which is the union of $(1/2)(l \cap K + l \cap K')$ as l runs through the lines perpendicular to H. We deduce by the Fubini theorem that $V(K_H) = V(K)$. We note that the Steiner symmetrization actually decreases $V_i(K)$ for $i = 1, \ldots, d - 1$ unless K is symmetric in a hyperplane parallel to H (see H. Hadwiger [Had1957a] or K. Leitweiß [Lei1980]); hence, its subsequent applications lead to the isoperimetric inequality.

A.12. Mixed Volumes in $\mathbb{R}^d$

All the basic facts can be found in R. Schneider [Sch1993b], and see also T. Bonnesen and W. Fenchel [BoF1987] and J. R. Sangwine–Yager [SaY1993] for nice introductions. First we review the definition of mixed volumes that is analogous to the definition of the intrinsic volumes. We note that mixed volumes in $\mathbb{R}^d$ have d variables in general, but in our applications, the ones depending on two convex bodies suffice.

We recall that if C is a convex body and $x \in \partial C$ then $N(C, x)$ is the *normal cone* at x, namely, the family of y such that $\langle y, z - x \rangle \leq 0$ holds for any $z \in C$. Let M be a polytope. We write $\mathcal{F}(M)$ to denote the family of faces of M, and for $F \in \mathcal{F}(M)$, we define the exterior normal cone at F to be $N_M(F) = N(M, x)$ for some $x \in \text{relint} F$. Each $N_M(F)$ is a polyhedral cone with codimension dim F, and the family of all $N_M(F)$ form a cell decomposition of $\mathbb{R}^d$.

Now let K be a smooth and strictly convex body containing o in its interior. If F is a proper face of M then we write $\sigma_K(F)$ to denote the union of the segments conv$\{o, x\}$, where $x \in \partial K$ and $N(K, x) \subset N_M(F)$. Then the sets $F + \sigma_K(F)$ are full dimensional and have disjoint interiors, and

$$M + \varrho K = \bigcup_{F \in \mathcal{F}(M)} F + \varrho \, \sigma_K(F)$$

holds for $\varrho > 0$. It follows by the Fubini Theorem that

$$V(F + \varrho \, \sigma_K(F)) = |F| \cdot |N_M(F) \cap (K/\text{aff} F)| \cdot \varrho^{d - \dim F}.$$

Applying the preceding considerations to λM, we deduce the formula

$$V(\lambda M + \mu K) = \sum_{i=0}^{d} \binom{d}{i} V(M, K; i) \cdot \lambda^{d-i} \mu^i \qquad (A.38)$$

for $\lambda, \mu \geq 0$, where the coefficients satisfy

$$\binom{d}{i} V(M, K; i) = \sum_{\dim F = d-i} |F| \cdot |N_M(F) \cap (K/\text{aff} F)|. \qquad (A.39)$$

This formula is due to H. Minkowski, and $V(M, K; i)$ is called a *mixed volume*. If $i = 1$ then (A.39) is especially simple: Writing $F_1, \ldots, F_k$ to denote the facets of M, and u_j to denote the exterior unit normal at F_j, we have

$$d \, V(M, K; 1) = \sum_{j=1}^{k} |F_j| \cdot h_K(u_j). \qquad (A.40)$$

It is easy to see that the mixed volumes are bounded if M and K are contained in a fixed ball; hence, (A.38) yields that the mixed volumes are continuous. In particular, they can be defined by (A.38) for any compact convex set M and K, and (A.40) holds whenever M is a polytope and K is a compact convex set. If M is a convex body and u_x is an exterior unit normal at $x \in \partial M$ then (A.40) generalize to

$$d\, V(M, K; 1) = \int_{\partial M} h_K(u_x)\, dx. \qquad (A.41)$$

Therefore $V(M, K; 1)$ is linear in K. Changing the role of M and K shows that $V(M, K; i) = V(K, M; d - i)$, and it follows by (A.39) that $V(\cdot, \cdot; i)$ is monotonic in both variables. Finally, if $\lambda \geq 0$ then

$$V(M, \lambda K; i) = \lambda^i\, V(M, K; i).$$

Certain mixed volumes have fundamental geometric meaning, say $V(M) = V(M, K; 0)$, and in general

$$\binom{d}{i} V(B^d, M; i) = \kappa_{d-i}\, V_i(M).$$

We note that (A.39) and continuity yield that if M is a $(d - i)$-dimensional compact convex set then

$$\binom{d}{i} V(M, K; i) = |M| \cdot |K/\mathrm{aff} M|. \qquad (A.42)$$

We now review some related inequalities for given convex bodies C and K in $\mathbb{R}^d$. Probably the most fundamental one is the Minkowski inequality, which states

$$V(C, K; 1)^d \geq V(C)^{d-1} V(K), \qquad (A.43)$$

and equality holds if only if C and K are homothetic. Actually, if K is the unit ball then the Minkowski inequality is equivalent to the classical isoperimetric inequality between surface area and volume. In general the Minkowski inequality is equivalent to the *Brunn–Minkowski inequality*; namely,

$$V(\lambda\, C + (1 - \lambda)\, K)^{\frac{1}{d}} \geq \lambda\, V(C)^{\frac{1}{d}} + (1 - \lambda)\, V(K)^{\frac{1}{d}} \qquad (A.44)$$

holds for $0 < \lambda < 1$, and equality holds if and only if C and K are homothetic. Next, the difference body $(C - C)/2$ satisfies

$$V(C) \leq V\left(\frac{1}{2}(C - C)\right) \leq \frac{1}{2^d} \binom{2d}{d} V(C). \qquad (A.45)$$

Here the lower bound is the consequence of the Brunn–Minkowski inequality, and equality holds if and only if C is centrally symmetric. Moreover, the upper bound is *the Rogers–Shephard inequality*, and equality holds if and only if C

is a simplex. We note that the Stirling formula (A.27) yields

$$\binom{2d}{d} < 4^d. \tag{A.46}$$

The Brunn–Minkowski inequality (A.44) can be generalized in the following form: If C, K, and M are convex bodies then

$$V(\lambda\, C + (1 - \lambda)\, K, M; i)^{\frac{1}{d-i}} \geq \lambda\, V(C, M; i)^{\frac{1}{d-i}} + (1 - \lambda)\, V(K, M; i)^{\frac{1}{d-i}}.$$

Therefore, if C is a convex body and M is a centrally symmetric convex body, then

$$V(C, M; i) \leq V\left((C - C)/2, M; i\right). \tag{A.47}$$

We write $r_K(C)$ to denote the maximal r such that a translate of $r\,K$ is contained in C, and we write $R_K(C)$ to denote the minimal R such that a translate of $R\,K$ contains C. We claim that

$$V(C, K; 1) \cdot r_K(C) \leq V(C) \leq d\, V(C, K; 1) \cdot r_K(C). \tag{A.48}$$

Here the lower bound is a consequence of the fact that a translate of $r_K(C)\, K$ is contained in C. To verify the upper bound, we may assume that C is a polytope and that K is a strictly convex body containing the origin in its interior. We write $F_1, \ldots, F_k$ to denote the facets of C and write u_i to denote the exterior unit normal to F_i. In addition, let y_i be the point of ∂K where the exterior normal is u_i, and let s_i be the segment connecting the origin and $-r_K(C)\, y_i$. Then the cylinders $F_i + s_i$, $i = 1, \ldots, k$, cover C, which, in turn, yields (A.48) by (A.40).

A.13. Lattices in $\mathbb{R}^d$ and Tori

Any property of lattices we need can be found in P. M. Gruber [Gru1993b], P. M. Gruber and C. G. Lekkerkerker [GrL1987], or J. H. Conway and N. J. A. Sloane [CoS1999].

For independent vectors $v_1, \ldots, v_d$ of $\mathbb{R}^d$, the discrete set

$$\Lambda = \left\{ \sum_{i=1}^{d} \alpha_i v_i \,:\, \alpha_i \in \mathbb{Z} \right\}$$

is called a *lattice*. We say that some vectors $u_1, \ldots, u_d \in \Lambda$ form a *basis* of Λ if $u_1, \ldots, u_d$ generate Λ over $\mathbb{Z}$. In this case there exists an integer matrix A such that $\det A = \pm 1$ and $u_i = A\, v_i$ holds for $i = 1, \ldots, n$, and hence $|\det[u_1, \ldots, u_d]|$ is the same for any basis $u_1, \ldots, u_d$. This common value is known as the *determinant* $\det \Lambda$ of Λ.

A lattice Λ' that is a subset of Λ is called a *sublattice*. Any basis of Λ' is of the form $Av_1, \ldots, Av_d$, where A is an integer matrix; therefore, $\det \Lambda' = |\det A| \cdot \det \Lambda$.

Let K be a convex body in $\mathbb{R}^d$. We call the lattice Λ a *packing lattice* or *covering lattice* with respect to K if $\Lambda + K$ form a packing, or if $\Lambda + K$ covers $\mathbb{R}^d$, respectively. According to the Mahler theorem:

> *There exist a packing lattice for K of minimal determinant and a covering lattice for K of maximal determinant.* $\qquad$ (A.49)

We call a family $\mathcal{K}$ of convex bodies *periodic* with respect to the lattice Λ if $x + K \in \mathcal{K}$ holds for any $x \in \Lambda$ and $K \in \mathcal{K}$, and Λ defines only finitely many equivalence classes of $\mathcal{K}$.

Example A.13.1.

- *A typical example for lattices is the* integer lattice $\mathbb{Z}^d$.
- *In a plane the* hexagonal lattice *is formed by the centres of congruent regular hexagons in a tiling. The inscribed circular discs naturally form a packing, and the circumscribed circular discs cover the plane.*
- *In $\mathbb{R}^3$ the* face-centred cubic lattice *(fcc) is generated by side vectors of a regular tetrahedron. The two-dimensional sections of an fcc parallel to the faces of the tetrahedron form layers of hexagonal lattices, and the sections parallel to the two opposite edges of the tetrahedron form layers of square lattices.*

Next we discuss tori (see J. G. Ratcliffe [Rat1994] for details). We fix a lattice Λ in $\mathbb{R}^d$ and identify two points x and y of $\mathbb{R}^d$ if $y - x \in \Lambda$; hence, the resulting d-dimensional compact manifold is the *torus* $T = \mathbb{R}^d/\Lambda$. We note that the torus naturally inherits the additive structure of $\mathbb{R}^d$; hence, we can speak about a translate $\sigma + x$ of a $\sigma \subset T$ by an $x \in T$ and about $x - \sigma$. Unless otherwise stated, the distance between two elements of $\mathbb{R}^d/\Lambda$ is the minimal Euclidean distance between the inverse images in $\mathbb{R}^d$. Moreover, the Lebesgue measure in $\mathbb{R}^d$ induces a measure on T, which satisfies $V(T) = \det \Lambda$, and $V(\sigma + x) = V(x - \sigma) = V(\sigma)$ for any $x \in T$ and measurable $\sigma \subset T$.

We say that a convex body K in $\mathbb{R}^d$ *embeds* into $\mathbb{R}^d/\Lambda$ if the quotient map is injective on $\mathrm{int}K$, which is equivalent saying that $\Lambda + K$ is a packing. Moreover, K *embeds isometrically* if the distance between any two $x, y \in K$ and between their images on T coincides. In particular, if $\Lambda + \varrho B^d$ is a packing then any convex body K in $\mathbb{R}^d$ satisfying $R(K) \leq \varrho$ embeds isometrically into T. Therefore, given a lattice Λ and a convex body K, there exists a positive integer l such that any congruent copy of K in $\mathbb{R}^d$ embeds isometrically into $\mathbb{R}^d / l\Lambda$.

Let us describe how periodic arrangements correspond canonically to finite arrangements on tori. On the one hand, any finite arrangement of embedded convex bodies on $\mathbb{R}^d / \Lambda$ induces a periodic arrangement in $\mathbb{R}^d$ with respect to Λ. On the other hand, let $\mathcal{K}$ be a periodic arrangement with respect to the lattice Λ. Taking a suitable sublattice of Λ, we may assume that the image of any element K of $\mathcal{K}$ embeds into $\mathbb{R}^d / \Lambda$; hence, the embedded images form a finite arrangement on $\mathbb{R}^d / \Lambda$.

Let T be a torus, and let $\sigma \subset T$ be measurable. Given $x_1, \ldots, x_n \in T$, a translate of σ contains $nV(\sigma)/V(T)$ points out of $x_1, \ldots, x_n$ in average because

$$\int_T \#((\sigma + x) \cap \{x_1, \ldots, x_n\}) \, dx = \int_T \sum_{i=1}^n \chi_{\sigma + x}(x_i) \, dx \qquad \text{(A.50)}$$

$$= \sum_{i=1}^n \int_T \chi_{x_i - \sigma}(x) \, dx = n \cdot V(\sigma).$$

A.14. A Little Bit of Probability

The existence of a configuration of N points with a certain property will sometimes be shown by verifying that the required configurations occur with positive probability among all N-tuples. A nice reference on this subject is the book by A. Rényi [Ren1970]. Let X be either a d-dimensional torus or the d-sphere S^d, and let $|\cdot|$ denote the Hausdorff d-measure on X. We define the related *probability measure* μ on the measurable subsets of X by $\mu(Y) = |Y|/|X|$.

We throw points $x_1, \ldots, x_N$ independently and with uniform distribution on X. In particular, the probability that $x_i \in Y$ for a measurable $Y \subset X$ is $\mu(Y)$; hence, the probability that none of $x_1, \ldots, x_N$ is contained in Y is $(1 - \mu(Y))^N$. Next, let C be a convex body embedded into X if X is a torus and a spherical ball of acute radius r if X is the sphere. For any $x \in X$, we write $C(x)$ to denote the translate of C by x if X is a torus and the spherical ball of radius r and centred at x if X is the sphere. Therefore, the probability that a certain $x \in X$ is not contained in $C(x_1), \ldots, C(x_N)$ is $(1 - \mu(C))^N$. In other words, the expected probability measure of the part of X not covered by $C(x_1), \ldots, C(x_N)$ is $(1 - \mu(C))^N$.

Bibliography

[Acz1952]   J. Aczél. Solution to Problem 35 [in Hungarian]. *Mat. Lapok*, 3:94–95, 1952.

[AiZ1999]   M. Aigner, G. Ziegler. *Proofs from the Book*. First edition. Springer-Verlag, 1999.

[AiZ2001]   M. Aigner, G. Ziegler. *Proofs from the Book*. Second edition. Springer-Verlag, 2001.

[And1999]   N. N. Andreev. A spherical code. *Russ. Math. Surv.*, 54:251–253, 1999.

[Ant?]   K. M. Anstreicher. The thirteen spheres: A new proof. *Disc. Comp. Geom.*, accepted.

[Arh1998]   V. Arhelger. *Parameterabhängige dichte Packungen konvexer Körper*. PhD thesis, Universität Siegen, 1998.

[ABB2001]   V. Arhelger, U. Betke, K. Böröczky Jr. Large finite lattice packings. *Geom. Dedicata*, 85:157–182, 2001.

[Arm1983]   M. A. Armstrong. *Basic Topology*. Springer-Verlag, 1983.

[Art1964]   E. Artin. *The Gamma Function*. Holt, Rinehart, and Winston, 1964.

[BHP1997]   M. Baake, J. Hermisson, P. A. B. Pleasants. The torus parametrization of quasiperiodic LI-classes. *J. Phys. A: Math. Gen.*, 30:3029–56, 1997.

[BHR1997]   M. Baake, J. Hermisson, C. Richard. A guide to the symmetry structure of quasiperiodic tiling classes. *J. Phys. I France*, 7:1003–1018, 1997.

[Bal1992]   K. M. Ball. A lower bound for the optimal density of lattice packings. *Int. Math. Res. Not.*, 10:217–221, 1992.

[Bal1997]   K. M. Ball. An elementary introduction to modern convex geometry. In: *Flavors of Geometry*, Silvio Levy (ed.), Cambridge Univ. Press, 1997.

[BaR1952]   R. P. Bambah, C. A. Rogers. Covering the plane with convex sets. *J. London Math. Soc.*, 27:304–314, 1952.

[BRZ1964]   R. P. Bambah, C. A. Rogers, H. Zassenhaus. On coverings with convex domains. *Acta Arith.*, 9:191–207, 1964.

[BaW1971]   R. P. Bambah, A. C. Woods. On plane coverings with convex domains. *Mathematika*, 18:91–97, 1971.

[BaS1981]   E. Bannai, N. J. A. Sloane. Uniqueness of certain spherical codes. *Canad. J. Math.*, 33:437–449, 1981.

[BaE1951]   P. Bateman, P. Erdős. Geometrical extrema suggested by a lemma of Besicovitch. *Am. Math. Mon.*, 58:306–314, 1951.

357

[BaL1993]   M. M. Bayer, C. W. Lee. Combinatorial aspects of convex polytopes. In: *Handbook of Convex Geometry*, P. M. Gruber, J. M. Wills (eds.), North Holland, 485–534, 1993.

[Bav1996]   C. Bavard. Disques extrémaux et surfaces modulaires. *Ann. Fac. Sci. Toulouse Math.*, 5:191–202, 1996.

[BáB2003]   V. Bálint, V. Bálint, Jr. On the number of points at distance at least 1 in the unit cube. *Geombinatorics*, 12:157–166, 2003.

[Bár1994]   I. Bárány. The densest $(n + 2)$-set in $\mathbb{R}^n$. In: *Intuitive geometry*, K. Böröczky, G. Fejes Tóth (eds.), Colloq. Math. Soc. J. Bolyai 63, North Holland, 7–10, 1994.

[Ben1874]   C. Bender. Bestimmung der grössten Anzahl gleicher Kugeln, welche sich auf eine Kugel von demselben Radius, wie die übrigen, auflegen lassen. *Arch. Math. Phys. (Grunert)*, 56:302–306, 1874.

[Ber1994]   M. Berger. *Geometry I*. Springer-Verlag, 1994.

[Ber1987]   M. Berger. *Geometry II*. Springer-Verlag, 1987.

[BeB1999]   U. Betke, K. Böröczky Jr. Asymptotic formulae for the lattice point enumerator. *Canad. J. Math.*, 51:225–249, 1999.

[BeB?]   U. Betke, K. Böröczky Jr. Finite lattice packings and the Wulff-shape. *Mathematika*, accepted.

[BeG1984]   U. Betke, P. Gritzmann. Über L. Fejes Tóths Wurstvermutung in kleinen Dimensionen. *Acta Math. Hungar.*, 43:299–307, 1984.

[BGW1982]   U. Betke, P. Gritzmann, J. M. Wills. Slices of L. Fejes Tóth's sausage conjecture. *Mathematika*, 29:194–201, 1982.

[BeH1998]   U. Betke, M. Henk. Finite packings of spheres. *Disc. Comp. Geom.*, 19:197–227, 1998.

[BeH2000]   U. Betke, M. Henk. Densest lattice packings of 3-polytopes. *Comput. Geom.*, 16:157–186, 2000.

[BHW1994]   U. Betke, M. Henk, J. M. Wills. Finite and infinite packings. *J. reine angew. Math.*, 453:165–191, 1994.

[BHW1995a] U. Betke, M. Henk, J. M. Wills. Sausages are good packings. *Disc. Comp. Geom.*, 13:297–311, 1995.

[BHW1995b] U. Betke, M. Henk, J. M. Wills. A new approach to covering. *Mathematika*, 42:251–263, 1995.

[BHW?]   U. Betke, M. Henk, J. M. Wills. *(In)finite packings. A joint theory for finite and infinite packing problems*. Vieweg, accepted.

[BeS?]   U. Betke, A. Schürmann. Lattices of optimal finite lattice packings. *Monats. Math.*, accepted.

[BeB1984]   A. Bezdek, K. Bezdek. Eine hinreichende Bedingung für die Überdeckung des Einheitswürfels durch homothetische Exemplare im n-dimensionalen euklidischen Raum. *Beitr. Alg. Geom.*, 17:5–21, 1984.

[BeF1999]   A. Bezdek, F. Fodor. Minimal diameter of certain sets in the plane. *J. Combin. Theory Ser. A*, 85:105–111, 1999.

[BKK1995]   A. Bezdek, K. Kuperberg, W. Kuperberg. Mutually contiguous translates of a plane disk. *Duke Math. J.*, 78:19–31, 1995.

[Bez1979]   K. Bezdek. *Optimal coverings of circles* [in Hungarian]. PhD thesis, Roland Eötvös University, Budapest, 1979.

[Bez1982]   K. Bezdek. Ausfüllungen eines Kreises durch kongruente Kreise in der hyperbolische Ebene. *Studia Sci. Math. Hungar.*, 17:353–366, 1982.

[Bez1983]   K. Bezdek. Über einige Kreisüberdeckungen. *Beitr. Alg. Geom.*, 14:7–13, 1983.

[Bez1984]   K. Bezdek. Ausfüllungen in der hyperbolische Ebene durch endliche Anzahl kongruente Kreise. *Ann. Univ. Sci. Budapest*, 27:113–124, 1984.

[Bez1987]   K. Bezdek. Densest packing of small number of congruent spheres in polyhedra. *Ann. Univ. Sci. Budapest*, 30:177–194, 1987.

[Bez1991]   K. Bezdek. The problem of illumination of the boundary of a convex body by affine subspaces. *Mathematika*, 38:362–375, 1991.

[Bez2002]   K. Bezdek. On the maximum number of touching pairs in a finite packing of translates of a convex body. *J. Combin. Theory Ser. A*, 98:192–200, 2002.

[BeB1997]   K. Bezdek, T. Bisztriczky. A proof of Hadwiger's covering conjecture for dual cyclic polytopes. *Geom. Dedicata*, 68:29–41, 1997.

[BBC2001]   K. Bezdek, G. Blekherman, R. Connelly, B. Csikós. The polyhedral Tammes problem. *Arch. Math.*, 76:314–320, 2001.

[BeB2003]   K. Bezdek, P. Brass. On the (k+)-neighbour packings and one-sided Hadwiger configurations. *Contrib. Alg. Geom.*, 44:493–498, 2003.

[BBHP1991] T. Bisztriczky, K. Böröczky Jr., H. Harborth, L. Piepmeyer. On the smallest limited snake of unit disks. *Geom. Dedicata*, 40:319–324, 1991.

[BiH2001]   T. Bisztriczky, H. Harborth. Smallest limited edge-to-edge snakes in Euclidean tessellations. *Congr. Numer.*, 149:155–159, 2001.

[Bla1963]   N. M. Blachman. The closest packing of equal spheres in a larger sphere. *Am. Math. Mon.*, 70:526–529, 1963.

[BlF1963]   N. M. Blachman, L. Few. Multiple packing of spherical caps. *Mathematika*, 10:84–88, 1963.

[Bli1929]   H. F. Blichfeldt. The minimum value of quadratic forms, and the closest packings of spheres. *Math. Ann.*, 101:605–608, 1929.

[Bli1969]   G. Blind. Über Unterdeckungen der Ebene durch Kreise. *J. reine angew. Math.*, 236:145–173, 1969.

[BlB2002]   G. Blind, R. Blind. Packings of unequal circles in a convex set. *Disc. Comp. Geom.*, 28:115–119, 2002.

[Bli1999]   V. Blinovsky. Multiple packing of the Euclidean sphere. *IEEE Trans. Inform. Theory*, 45:1334–1337, 1999.

[Blu1953]   L. M. Blumenthal. *Theory and Applications of Distance Geometry*. Clarendon Press, 1953.

[BoF1993]   V. Boju, L. Funar. Generalized Hadwiger numbers for symmetric ovals. *Proc. Am. Math. Soc.*, 119:931–934, 1993.

[BoS1972]   B. Bollobás, N. Stern. The optimal structure of market areas. *J. Economic Theory*, 4:174–179, 1972.

[Bol1960]   V. G. Boltyanskii. The problem of illuminating the boundary of a convex body [in Russian]. *Izv. Mold. Fil. Akad. Nauk. SSSR*, 76:77–84, 1960.

[BMS1997]   V. G. Boltyanskii, H. Martini, P. Soltan. *Excursions into Combinatorial Geometry*. Springer-Verlag, 1997.

[Bol1832]   F. Bolyai. *Tentamen Juventutem Studiosam in Elementa Matheseos Purae, Elementaris Ac Sublimioris, Methodo Intuituitiva, Evidentiaque Huic Propria, Introducendi*. First edition, 1832–33; Second edition, 1904.

[BoF1987]   T. Bonnesen, W. Fenchel. *Theory of Convex Bodies*. BCS Assoc., Moscow (Idaho), 1987. Translated from German: *Theorie der konvexen Körper*, Springer-Verlag, 1934.

[Bor1933]   K. Borsuk. Drei Sätze über die n-dimensionale euklidische Sphäre. *Fund. Math.*, 20:177–190, 1933.

[Bow2000]   L. Bowen. Circle packing in the hyperbolic plane. *Math. Phys. Electron. J.*, 6:paper no. 6, 2000.

[BoR2003]   L. Bowen, Ch. Radin. Densest packing of equal spheres in hyperbolic space. *Disc. Comp. Geom.*, 29:23–29, 2003.

[BoR?]   L. Bowen, Ch. Radin. Optimally dense packings of hyperbolic space. *Geom. Dedicata*, accepted.

[Bör1971]   K. Böröczky. Über die Newtonsche Zahl regulärer Vielecke. *Period. Math. Hungar.* 1:113–119, 1971.

[Bör1974]   K. Böröczky. Sphere packings in spaces of constant curvature I [in Hungarian]. *Mat. Lapok*, 25:265–306, 1974.

[Bör1978]   K. Böröczky. Packing of spheres in spaces of constant curvature. *Acta Math. Hungar.*, 32:243–261, 1978.

[Bör1983]   K. Böröczky. The problem of Tammes for $n = 11$. *Stud. Sci. Math. Hungar.*, 18:165–171, 1983.

[Bör1987]   K. Böröczky. On an extremum property of the regular simplex in S^d. In: *Intuitive Geometry*, K. Böröczky, G. Fejes Tóth (eds.), Colloq. Math. Soc. János Bolyai 48, Elsevier, 117–121, 1987.

[Bör2003a]   K. Böröczky. The Newton–Gregory problem revisited. In: *Discrete Geometry: In Honor of W. Kuperberg's 60th Birthday*, A. Bezdek (ed.), Marcel Dekker, 103–110, 2003.

[Bör?a]   K. Böröczky. The density of hyperbolic circle packings with respect to Delone triangles, in preparation.

[BöB1996]   K. Böröczky, K. Böröczky Jr. Isoperimetric problems for polytopes with given number of vertices. *Mathematika*, 43:237–254, 1996.

[BöF1964]   K. Böröczky, A. Florian. Über die dichteste Kugelpackung im hyperbolischen Raum. *Acta Math. Hungar.* 15:237–245, 1964.

[BöL?]   K. Böröczky, Zs. Lángi. On the relative distance of six points in a plane convex body, *Stud. Sci. Math. Hungar.*, accepted.

[BöS1996]   K. Böröczky, V. Soltan. Translational and homothetic clouds for a convex body. *Stud. Sci. Math. Hungar.*, 32:93–102, 1996.

[BöS2003a]   K. Böröczky, L. Szabó. Arrangements of thirteen points on a sphere. In: *Discrete Geometry: In Honor of W. Kuperberg's 60th Birthday*, A. Bezdek (ed.), Marcel Dekker, 111–184, 2003.

[BöS2003b]   K. Böröczky, L. Szabó. Fourteen to sixteen points on a sphere. *Stud. Sci. Math. Hungar.*, 40:407–421, 2003.

[Bör1993]   K. Böröczky Jr. About four-ball packings. *Mathematika*, 40:226–232, 1993.

[Bör1994]   K. Böröczky Jr. Mean projections and finite packings of convex bodies. *Monats. Math.*, 118:41–54, 1994.

[Bör1997]   K. Böröczky Jr. Packings of four and five balls. In: *Intuitive Geometry*, I. Bárány, K. Böröczky (eds.), Bolyai Soc. Math Studies 6, *J. Bolyai Math. Soc.*, 265–276, 1997.

[Bör2001]   K. Böröczky Jr. About the error term for best approximation with re-
            spect to the Hausdorff related metrics. *Disc. Comp. Geom.*, 25:293–309,
            2001.

[Bör2002]   K. Böröczky Jr. Discrete point sets in the hyperbolic plane, *Stud. Sci.
            Acad. Hungar.*, 39:21–36, 2002.

[Bör2003b] K. Böröczky Jr. Finite packing and covering by congruent convex do-
            mains. *Disc. Comp. Geom.*, 30:185–193, 2003.

[Bör?b]   K. Böröczky Jr. Finite coverings in the hyperbolic plane, *Disc. Comp.
            Geom.*, accepted.

[BöF2002]   K. Böröczky Jr., G. Fejes Tóth. Stability of some inequalities for three–
            polyhedra. *Rend. Circ. Mat. Palermo,* 70:93–108, 2002.

[BöH1995]   K. Böröczky Jr., M. Henk. Radii and the Sausage Conjecture. *Canad.
            Math. Bull.*, 38:156–166, 1995.

[BLS2000]   K. Böröczky Jr., D. G. Larman, S. Sezgin, C. Zong. On generalized kissing
            numbers and blocking numbers. *Rend. Circ. Mat. Palermo*, 65:39–57,
            2000.

[BöR?]   K. Böröczky Jr., I. Z. Ruzsa. Note on an inequality of Wegner, submitted.

[BöS1998a] K. Böröczky Jr., U. Schnell. Asymptotic shape of finite packings. *Canad.
            J. Math.*, 50:16–28, 1998.

[BöS1998b] K. Böröczky Jr., U. Schnell. Wulff-shape for non-periodic arrangements.
            Lett. Math. Phys., 45:81–94, 1998.

[BöS1999]   K. Böröczky Jr., U. Schnell. Quasi-crystals and Wulff-shape. *Disc. Comp.
            Geom.*, 21:421–436, 1999.

[BSW2000]   K. Böröczky Jr., U. Schnell, J. M. Wills. Quasicrystals, parametric den-
            sity and Wulff-shape. In: *Directions in Mathematical Quasicrystals,*
            M. Baake, R. V. Moody (eds.), CRM Monographs Series 13, Am. Math.
            Soc., 259–276, 2000.

[BöS1995]   K. Böröczky Jr., V. Soltan. Smallest maximal snakes of translates of
            convex domains. *Geom. Dedicata*, 54:31–44, 1995.

[BöT?]   K. Böröczky Jr., G. Tardos. The longest segment in the complement of a
            packing. *Mathematika*, accepted.

[BöW1997]   K. Böröczky Jr., J. M. Wills. Finite sphere packings and critical radii.
            Beitr. Alg. Geom., 38:193–211, 1997.

[BöW2000]   K. Böröczky Jr., G. Wintsche. Sphere packings in the regular crosspoly-
            tope. *Ann. Univ. Sci. Budapest.*, 43:151–157, 2000.

[BöW2003]   K. Böröczky, Jr., G. Wintsche. Covering the sphere by equal spherical
            balls. In: *Discrete and Computational Geometry – The Goodman–Pollack
            Festschrift*, B. Aronov, S. Basú, M. Sharir, J. Pach (eds.), Springer-Verlag,
            237–253, 2003.

[BöW?]   K. Böröczky, Jr., G. Wintsche. Covering the crosspolytope by equal balls,
            in preparation.

[Bra1996]   P. Brass. Erdős distance problems in normed spaces. *Comput. Geom.*,
            6:195–214, 1996.

[BMP?]   P. Brass, W. Moser, J. Pach. *Research Problems in Discrete Geometry,*
            Springer-Verlag, accepted.

[But1972]   G. J. Butler. Simultaneous packing and covering in Euclidean space. *Proc.
            London Math. Soc.*, 25:721–735, 1972.

[CaP1991]   V. Capoyleas, J. Pach. On the perimeter of a point set in the plane. In: *Discrete and computational geometry*, J. E. Goodman, R. Pollack, W. Steiger (eds.), DIMACS Ser. Discrete Math. Theoret. Comput. Sci. 6, Am. Math. Soc., 67–76, 1991.

[CRY1991]   S. A. Chepanov, S. S. Ryshkov, N. N. Yakovlev. On the disjointness of point systems [in Russian]. *Trudy Mat. Inst. Steklov.*, 196:147–155, 1991.

[Cho1995]   T. Y. Chow. Penny-packings with minimal second moments. *Combinatorica*, 15:151–158, 1995.

[CoK?]   H. Cohn, A. Kumar. Optimality and uniqueness of the Leech lattice among lattices. http://research.microsoft.com/~cohn/Leech, in preparation.

[CDH1995]   J. H. Conway, T. D. Duff, R. H. Hardin, N. J. A. Sloane. Minimal-energy clusters of hard spheres. *Disc. Comp. Geom.*, 14:237–259, 1995.

[CoS1999]   J. H. Conway, N. J. A. Sloane. *Sphere Packings, Lattices and Groups*. Third edition. Springer-Verlag, 1999.

[Cox1963]   H. S. M. Coxeter. An upper bound for the number of equal nonoverlapping shperes that can touch another of the same size. *Convexity*, V. L. Klee (ed.), Proc. Sympos. Pure Math. 7, Am. Math. Soc., 53–71, 1963.

[CFR1959]   H. S. M. Coxeter, L. Few, C. A. Rogers. Covering space with equal spheres. *Mathematika*, 6:147–157, 1959.

[Csó1977]   G. Csóka. The number of congruent spheres that cover a given sphere of three-dimensional space is not less than 30 [in Russian]. *Stud. Sci. Math. Hungar.*, 12:323–334, 1977.

[DLM2000]   L. Dalla, D. G. Larman, P. Mani–Levitska, Ch. Zong. The blocking numbers of convex bodies. *Disc. Comp. Geom.*, 24:267–277, 2000.

[Dan1960]   L. Danzer. Drei Beispiele zu Lagerungsproblemen. *Arch. Math.*, 11:159–165, 1960.

[Dan1986]   L. Danzer. Finite point-sets on S^2 with minimum distance as large as possible. *Discrete Math.*, 60:3–66, 1986.

[DaH1951]   H. Davenport, Gy. Hajós. Problem 35 [in Hungarian]. *Mat. Lapok*, 2:68, 1951.

[Del1934]   B. N. Delone. Sur la sphère vide. *Bull. Acad. Sci. URSS, VII. Ser.*, 793–800, 1934.

[Del1972]   P. Delsarte. Bounds for unrestricted codes by linear programming. *Philips Res. Rep.*, 27:272–289, 1972.

[Dic1939]   L. E. Dickson. *Modern Elementary Theory of Numbers*. Univ. Chicago Press, 1939.

[Din1943]   A. Dinghas. Über einen geometrischen Satz von Wulff über die Gleichgewichtsform von Kristallen. *Z. Kristallographie*, 105:304–314, 1943.

[Dir1850]   G. L. Dirichlet. Über die Reduction der positiven quadratischen Formen mit drei unbestimmten ganzen Zahlen. *J. reine angew. Math.*, 40:216–219, 1850.

[DoL1995]   K. Doliwka, M. Lassak. On relatively short and long sides of convex pentagons. *Geom. Dedicata*, 56:221–224, 1995.

[DLR1992]   P. G. Doyle, J. C. Lagarias, D. Randall. Self-packing of centrally symmetric convex bodies in $\mathbb{R}^2$. *Disc. Comp. Geom.*, 8:171–189, 1992.

[Dow1944]   C. H. Dowker. On minimum circumscribed polygons. *Bull. Am. Math. Soc.*, 50:120–122, 1944.

[ErF1983]   P. Erdős, Z. Füredi. The greatest angle among n points in the d-dimensional Euclidean space. In: *Combinatorial mathematics*, C. Berge, D. Bresson, P. Camion, J.-F. Maurras, F. Sterboul (eds.), North-Holland Math. Stud. 75, North-Holland, 275–283, 1983.

[ErG1975]   P. Erdős, R. L. Graham. On packing squares with equal squares. *J. Combin. Theory Ser. A*, 19:119–123, 1975.

[ErR1962]   P. Erdős, C. A. Rogers: Covering space with convex bodies. *Acta Arith.*, 7:281–285, 1962.

[ErZ2001]   T. Ericson, V. Zinoviev. *Codes on Euclidean Spheres*. North–Holland, 2001.

[Fal1986]   K. J. Falconer. *The Geometry of Fractal Sets*. Cambridge Univ. Press, 1986.

[Fár1950]   I. Fáry. Sur la densité des réseaux de domaines convexes. *Bull. Soc. Math. France*, 178:152–161, 1950.

[FLM1991]   Z. Füredi, J. C. Lagarias, F. Morgan. Singularities of minimal surfaces and networks and related extremal problems in Minkowski space. In: *Discrete and computational geometry*, J. E. Goodman, R. Pollack, W. Steiger (eds.), DIMACS Ser. Discrete Math. Theoret. Comput. Sci. 6, Am. Math. Soc, 95–109, 1991.

[FTG1969]   G. Fejes Tóth. Kreisüberdeckungen der Sphäre. *Stud. Sci. Math. Hungar.*, 4:225–247, 1969.

[FTG1970]   G. Fejes Tóth. *The loosest covering of the sphere by circles* [in Hungarian]. MSc. thesis, Roland Eötvös University, Budapest, 1970.

[FTG1972]   G. Fejes Tóth. Covering the plane by convex discs. *Acta Math. Hungar.*, 23:263–270, 1972.

[FTG1974]   G. Fejes Tóth. Solid sets of circles. *Stud. Sci. Math. Hungar.*, 9:101–109, 1974.

[FTG1976]   G. Fejes Tóth. Multiple packing and covering of the plane with circles. *Acta Math. Hungar.*, 27:135–140, 1976.

[FTG1977]   G. Fejes Tóth. On a Dowker-type theorem of Eggleston. *Acta Math. Hungar.*, 29:131–148, 1977.

[FTG1981]   G. Fejes Tóth. Ten-neighbour packing of equal balls. *Period. Math. Hungar.*, 12:125–127, 1981.

[FTG1987]   G. Fejes Tóth. Finite coverings by translates of centrally symmetric convex domains. *Disc. Comp. Geom.*, 2:353–363, 1987.

[FTG1988]   G. Fejes Tóth. Note to a paper of Bambah, Rogers and Zassenhaus. *Acta Arith.*, 50:119–122, 1988.

[FTG1995a] G. Fejes Tóth. Densest packings of typical convex sets are not lattice-like. *Disc. Comp. Geom.*, 14:1–8, 1995.

[FTG1995b] G. Fejes Tóth. Covering the plane with two kinds of circles. *Disc. Comp. Geom.*, 13:445–457, 1995.

[FTG1997]   G. Fejes Tóth. Packing and covering. In: *Handbook of Discrete and Computational Geometry*, J. E. Goodman, J. O'Rourke (eds.), CRC Press Ser. Discrete Math. Appl., CRC Press, 19–41, 1997.

[FTG2001]   G. Fejes Tóth. A stability criterion to the moment theorem. *Stud. Sci. Acad. Hungar.*, 34:209–224.

[FTG?]   G. Fejes Tóth. Thinnest covering a circle by eight, nine or ten congruent circles, in preparation.

[FTG?a]   G. Fejes Tóth. Thinnest covering a circle by eight, nine or ten congruent circles. submitted.

[FTG?b]   G. Fejes Tóth. Covering with fat convex discs. submitted.

[FTF1973]   G. Fejes Tóth, L. Fejes Tóth. Remark on a paper of C. H. Dowker. *Period. Math. Hungar.*, 3:271–274, 1973.

[FTF1973a] G. Fejes Tóth, L. Fejes Tóth. Remark on a paper of C. H. Dowker. *Period. Math. Hungar.*, 3:271–274, 1973.

[FTF1973b] G. Fejes Tóth, L. Fejes Tóth. On totally separable domains. *Acta Math. Hungar.*, 24:229–232, 1973.

[FTF1989]   G. Fejes Tóth, L. Fejes Tóth. A geometrical analogue of the phase transformation of crystals. *Comput. Math. Appl.*, 17:251–254, 1989.

[FGW1984]   G. Fejes Tóth, P. Gritzmann, J. M. Wills. Sausage-skin problems for finite coverings. *Mathematika*, 31:117–136, 1984.

[FGW1990]   G. Fejes Tóth, P. Gritzmann, J. M. Wills. On finite multiple packings. *Arch. Math.*, 55:407–411, 1990.

[FKK1998]   G. Fejes Tóth, G. Kuperberg, W. Kuperberg. Highly saturated packings and reduced coverings. *Monats. Math.*, 125:127–145, 1998.

[FTK1993]   G. Fejes Tóth, W. Kuperberg. Packing and covering. In: *Handbook of Convex Geometry*, P. M. Gruber, J. M. Wills (eds.), North Holland, 799–860, 1993.

[FTK1995]   G. Fejes Tóth, W. Kuperberg. Thin non-lattice covering with an affine image of a strictly convex body. *Mathematika*, 42:239–250, 1995.

[FTZ1994]   G. Fejes Tóth, T. Zamfirescu. For most convex discs thinnest covering is not lattice-like. In: *Intuitive Geometry*, K. Böröczky, G. Fejes Tóth (eds.), Colloq. Math. Soc. J. Bolyai 63, North Holland, 105–108, 1994.

[FTL1940]   L. Fejes Tóth. Über einen geometrischen Satz. *Math. Z.*, 46:83–85, 1940.

[FTL1949]   L. Fejes Tóth. Über dichteste Kreislagerung und dünnste Kreisüberdeckung. *Comment. Math. Helv.*, 23:342–349, 1949.

[FTL1950]   L. Fejes Tóth. Some packing and covering theorems. *Acta Sci. Math. Szeged*, 12:62–67, 1950.

[FTL1959]   L. Fejes Tóth. Verdeckung einer Kugel durch Kugeln. *Publ. Math. Debrecen*, 6:234–240, 1959.

[FTL1964]   L. Fejes Tóth. *Regular Figures*. Pergamon Press, 1964.

[FTL1965]   L. Fejes Tóth. Minkowskian distribution of discs. *Proc. Am. Math. Soc.*, 16:999–1004, 1965.

[FTL1967a] L. Fejes Tóth. Minkowskian circle-aggregates. *Math. Ann.*, 171:97–103, 1967.

[FTL1967b] L. Fejes Tóth. On the number of equal discs that can touch another of the same kind. *Stud. Sci. Math. Hungar.*, 2:363–367, 1967.

[FTL1968]   L. Fejes Tóth. Solid circle-packings and circle-coverings. *Stud. Sci. Math. Hungar.*, 3:401–409, 1968.

[FTL1970]   L. Fejes Tóth. Über eine affininvariante Masszahl bei Eipolyedern. *Stud. Sci. Math. Hungar.*, 5:173–180, 1970.

[FTL1971]   L. Fejes Tóth. Perfect distribution of points on a sphere. *Period. Math. Hungar.*, 1:25–33, 1971.

[FTL1972]   L. Fejes Tóth. *Lagerungen in der Ebene, auf der Kugel und im Raum*. Springer-Verlag, 1972.

[FTL1975a] L. Fejes Tóth. Research Problems. *Period. Math. Hungar.*, 6:197–199, 1975.

[FTL1975b] L. Fejes Tóth. On Hadwiger numbers and Newton numbers of a convex body. *Stud. Sci. Math. Hungar.*, 10:111–115, 1975.

[FTL1999] L. Fejes Tóth. Minkowski circle packings on the sphere. *Disc. Comp. Geom.*, 22:161–166, 1999.

[Flo1993] A. Florian. Extremum problems for convex discs and polyhedra. In: *Handbook of Convex Geometry*, P. M. Gruber, J. M. Wills (eds.), North Holland, 177–221, 1993.

[Flo2001] A. Florian. Packing of incongruent circles on the sphere. *Monats. Math.*, 133:111–129, 2001.

[FlH2003] A. Florian, A. Heppes. On the non-solidity of some packings and coverings with circles. In: *Discrete Geometry: In Honor of W. Kuperberg's 60th Birthday*, A. Bezdek (ed.), Marcel Dekker, 279–290, 2003.

[FlH1999] A. Florian, A. Heppes. Packing circles of two different sizes on the sphere, II. *Period. Math. Hungar.*, 39:125–127, 1999.

[Fod1999] F. Fodor. The densest packing of 19 congruent circles in a circle. *Geom. Dedicata*, 74:139–145, 1999.

[Fod2000] F. Fodor. The densest packing of 12 congruent circles in a circle. *Beitr. Alg. Geom.*, 41:401–409, 2000.

[Fod2003a] F. Fodor. The densest packing of 13 congruent circles in a circle. *Beitr. Alg. Geom.*, 44:431–440, 2003.

[Fod2003b] F. Fodor. The densest packing of 14 congruent circles in a circle, *Studies Univ. Zilina*, 16:25–34, 2003.

[FoG1969] J. H. Folkman, R. L. Graham. A packing inequality for compact convex subsets of the plane. *Canad. Math. Bull.*, 12:745–752, 1969.

[FoT1996] P. W. Fowler, T. Tarnai. Transition from spherical circle packing to covering: Geometrical analogues of chemical isomerization. *Roy. Soc. London Proc. Ser. A*, 452:2043–2064, 1996.

[FoT1999] P. W. Fowler, T. Tarnai. Transition from circle packing to covering on a sphere: The odd case of 13 circles. *Roy. Soc. London Proc. Ser. A*, 455:4131–4143, 1999.

[FPR1984] P. Frankl, J. Pach, V. Röld. How to build a barricade. *Monats. Math.*, 98:93–98, 1984.

[Fri2003] E. Friedman. Packing of unit squares in squares: a survey, and new results. *Electron. J. Combin.*, 7:Dynamic Survey 7, 2000.

[Für2003] Z. Füredi. Covering a triangle with homothetic copies. In: *Discrete Geometry: In Honor of W. Kuperberg's 60th Birthday*, A. Bezdek (ed.), Marcel Dekker, 435–445, 2003.

[FLM1991] Z. Füredi, J. C. Lagarias, F. Morgan. Singularities of minimal surfaces and networks and related extremal problems in Minkowski space. In: *Discrete and Computational Geometry*, J. E. Goodman, R. Pollack, W. Steiger (eds.), DIMACS Ser. Discrete Math. Theoret. Comput. Sci. 6:95–109, 1991.

[FüL1994] Z. Füredi, P. A. Loeb. On the best constant for the Besicovitch covering theorem. *Proc. Am. Math. Soc.*, 121:1063–1073, 1994.

[Gal1996]   Sh. I. Galiev. Multiple packings and coverings of a sphere. *Disc. Math. Appl.*, 6:413–426, 1996.

[GaR1999]   E. Gallego, A. Reventós. Asymptotic behaviour of λ-convex sets in the hyperbolic plane. *Geom. Dedicata*, 76:275–289, 1999.

[GaW1992]   P. M. Gandini, J. M. Wills. On finite sphere-packings. *Math. Pannon.*, 3:19–29, 1992.

[GaZ1992]   P. M. Gandini, A. Zucco. On the sausage catastrophe in 4-space. *Mathematika*, 39:274–278, 1992.

[Gar1995]   R. Gardner. *Geometric Tomography*. Cambridge Univ. Press, 1995.

[Gau1840]   C. F. Gauß. Untersuchungen über die Eigenschaften der positiven ternären quadratischen formen von Ludwig August Seeber. *J. reine angew. Math.*, 20:312–320, 1840.

[Gib1878]   J. W. Gibbs. *Trans. Conn. Academy*, 3:479, 1878, see also *The Collected Works of J. Willard Gibbs*, I., p. 219, Yale Univ. Press, 1948.

[GoM1960]   I. Z. Gohberg, A. S. Markus. One problem on covering convex figures by similar figures [in Russian]. *Izv. Mold. Fil. Akad. Nauk. SSSR*, 76:87–90, 1960.

[Gol1971]   M. Goldberg. On the densest packing of equal spheres in a cube. *Math. Mag.*, 44:199–208, 1971.

[Gol1977]   G. Golser. Dichteste Kugelpackungen im Oktaeder. *Stud. Sci. Math. Hungar.*, 12:337–343, 1977.

[GGM1998]   Y. Gordon, O. Guédon, M. Meyer. An isomorphic Dvoretzky's theorem for convex bodies. *Stud. Math.*, 127:191–200, 1998.

[GrL1995]   R. L. Graham, B. D. Lubachevsky. Dense packings of equal disks in an equilateral triangle: From 22 to 34 and beyond. *Electron. J. Combin.*, 2:article 1, 1995.

[GLN1998]   R. L. Graham, B. D. Lubachevsky, K. J. Nurmela, P. R. J. Östegård. Dense packings of congruent circles in a circle. *Discrete Math.*, 181:139–154, 1998.

[GLS1997]   R. L. Graham, B. D. Lubachevsky, F. H. Stillinger: Patterns and structures in disk packings. *Period. Math. Hungar.*, 34:123–142, 1997.

[GrS1990]   R. L. Graham, N. J. A. Sloane. Penny-packing and two-dimensional codes. *Disc. Comp. Geom.*, 5:1–11, 1990.

[GWZ1972]   R. L. Graham, H. S. Witsenhausen, H. J. Zassenhaus. On tightest packings in the Minkowski plane. *Pac. J. Math.*, 41:699–715, 1972.

[Gri1984]   P. Gritzmann. *Finite Packungen und Überdeckungen*. Habilitationsschrift, Universität Siegen, 1984.

[GrW1993a]   P. Gritzmann, J. M. Wills. Lattice points. In: *Handbook of Convex Geometry*, P. M. Gruber, J. M. Wills (eds.), North Holland, 765–797, 1993.

[GrW1993b]   P. Gritzmann, J. M. Wills. Finite packing and covering. In: *Handbook of Convex Geometry*, P. M. Gruber, J. M. Wills (eds.), North Holland, 861–897, 1993.

[Gro1960]   H. Groemer. Über die Einlagerung von Kreisen in einen konvexen Bereich. *Math. Z.*, 73:285–294, 1960.

[Gro1961]   H. Groemer. Abschätzungen für die Anzahl der konvexen Körper, die einen konvexen Körper berühren. *Monats. Math.*, 65:74–81, 1961.

[Gro1963]   H. Groemer. Existenzsätze für Lagerungen im Euklidischen Raum. *Math. Z.*, 81:260–278, 1963.

[Gro1982]   H. Groemer. Covering and packing properties of bounded sequences of convex sets. *Mathematika*, 29:18–31, 1982.

[Gro1985]   H. Groemer. Coverings and packings by sequences of convex sets. In: *Discrete geometry and convexity*, Proc. Conf. New York 1982, J. E. Goodman, E. Lutwak, J. Malkevitch, R. Pollack (eds.), Ann. N. Y. Acad. Sci., 440:262–278, 1985.

[Gro1993]   H. Groemer. Stability of geometric inequalities. In: *Handbook of Convex Geometry*, P. M. Gruber, J. M. Wills (eds.), North Holland, 125–150, 1993.

[Gru1986]   P. M. Gruber. Typical convex bodies have surprisingly few neighbours in densest lattice packings. *Stud. Sci. Math. Hungar.*, 21:163–173, 1986.

[Gru1993a]   P. M. Gruber. The space of convex bodies. In: *Handbook of Convex Geometry*, P. M. Gruber, J. M. Wills (eds.), North Holland, 301–318, 1993.

[Gru1993b]   P. M. Gruber. Geometry of numbers. In: *Handbook of Convex Geometry*, P. M. Gruber, J. M. Wills (eds.), North Holland, 739–763, 1993.

[Gru1993c]   P. M. Gruber. Baire categories in convexity. In: *Handbook of Convex Geometry*, P. M. Gruber, J. M. Wills (eds.), North Holland, 1327–1346, 1993.

[Gru2001]   P. M. Gruber. Optimal configurations of finite point sets in Riemannian 2-manifolds. *Geom. Dedicata*, 84:271–320, 2001.

[Gru2002]   P. M. Gruber. Optimale Quantisierung. *Math. Semesterber.*, 49:227–251, 2002.

[Gru?a]   P. M. Gruber. Optimum quantization and its applications. *Adv. Math.*, accepted.

[Gru?b]   P. M. Gruber. *Discrete and convex geometry*, book in preparation.

[GrL1987]   P. M. Gruber, C. G. Lekkerkerker. *Geometry of Numbers*. North Holland, 1987.

[Grü1961]   B. Grünbaum. On a conjecture of H. Hadwiger. *Pac. J. Math.*, 11:215–219, 1961.

[Grü1963]   B. Grünbaum. Strictly antipodal sets. *Israel J. Math.*, 1:5–10, 1963.

[GrS1987]   B. Grünbaum, G. C. Shephard. *Tilings and Patterns*. Freeman, 1987.

[Gün1875]   S. Günther. Ein stereometrisches Problem. *Arch. Math. Phys. (Grunert)*, 57:209–215, 1875.

[Had1955]   H. Hadwiger. Volumschätzung für die einen Eikörper überdeckenden und unterdeckenden Parallelotope. *Elem. Math.*, 10:122–124, 1955.

[Had1957a]   H. Hadwiger. *Vorlesungen über Inhalt, Oberfläche und Isoperimetrie*. Springer-Verlag, 1957.

[Had1957b]   H. Hadwiger. Über Treffanzahlen bei translationsgleichen Eikörpern. *Arch. Math.*, 8:212–213, 1957.

[Had1960]   H. Hadwiger. Ungelöste Probleme, Nr. 38. *Elem. Math.*, 15:121, 1960.

[HLS1959]   C. J. A. Halberg, Jr., E. Levin, E. G. Straus. On contiguous congruent sets in Euclidean space. *Proc. Amer. Math. Soc.*, 10:335–344, 1959.

[Hal2000]   T. C. Hales. Cannonballs and honeycombs. *Not. Am. Math. Soc.*, 47:440–449, 2000.

[Hal?]   T. C. Hales. An abridged proof of the Kepler Conjecture, submitted.

[Har1974]   H. Harborth. Lösung zu Problem 664A. *Elem. Math.*, 29:14–15, 1974.

[Hár1986]   L. Hárs. The Tammes problem for $n = 10$. *Stud. Sci. Math. Hungar.*, 21:439–451, 1986.

[HeR1980]   R. C. Heitmann, Ch. Radin. The ground state for sticky disks. *J. Statist. Phys.*, 22:281–287, 1980.

[Hen1995]   M. Henk. *Finite and Infinite Packings*. Habilitationsschrift, Universität Siegen, 1995.

[Hen?]   M. Henk. Free planes in lattice sphere packings, Adv. Geometry, accepted.

[HZZ2002]   M. Henk, G. Ziegler, Ch. Zong. On free planes in lattice ball packings. *Bull. London Math. Soc.*, 34:284–290, 2002.

[HeZ2000]   M. Henk, Ch. Zong. Segments in ball packings. *Mathematika*, 47:31–38, 2000.

[HeK1997]   A. Heppes, G. Kertész. Packing circles of two different sizes on the sphere. In: *Intuitive Geometry*, I. Bárány, K. Böröczky (eds.), Bolyai Soc. Math Studies 6, J. Bolyai Math. Soc., 357–365, 1997.

[Hep1955]   A. Heppes. Über mehrfache Kreislagerungen. *Elem. Math.*, 10:125–127, 1955.

[Hep1960]   A. Heppes. Ein Satz über gitterförmige Kugelpackungen. *Ann. Univ. Sci. Budapest*, 3–4:89–90, 1960/61.

[Hep1967]   A. Heppes. On the number of spheres which can hide a given sphere. *Canad. J. Math.*, 19:413–418, 1967.

[Hep1998]   A. Heppes. Coverings of the elliptic plane by four congruent circles [in Hungarian]. *Mat. Lapok*, 8–9:4–6, 1998–99.

[Hep1999]   A. Heppes. Densest circle packing on the flat torus. *Period. Math. Hungar.*, 39:129–134, 1999.

[Hep2001]   A. Heppes. Packing of bounded domains on a sphere of constant curvature. *Acta Math. Hungar.*, 91:245–252, 2001.

[Hep2003]   A. Heppes. Covering the plane with fat ellipses without non-crossing assumption. *Disc. Comp. Geom.*, 29:477–481, 2003.

[HeM1997]   A. Heppes, J. B. M. Melissen. Covering a rectangle with equal circles. *Period. Math. Hungar.*, 34:65–81, 1997.

[Hop1874]   R. Hoppe. Bemerkung der Redaction. *Arch. Math. Phys. (Grunert)*, 56:307–312, 1874.

[Hor1971]   I. Hortobágyi. Durchleuchtung gitterförmiger Kugelpackungen mit Lichtbündeln. *Stud. Sci. Math. Hungar.*, 6:147–150, 1971.

[HoR1975]   J. Horváth, S. S. Ryskov. Estimation of the radius of a cylinder that can be imbedded in every lattice packing of n-dimensional unit balls. *Math. Notes*, 17:72–75, 1975.

[Hsi2001]   W. Y. Hsiang. *Least Action Principle of Crystal Formation of Dense Packing Type and Kepler's Conjecture*. World Scientific, 2001.

[Hux1996]   M. N. Huxley. *Area, Lattice Points, and Exponential Sums*. Oxford Univ. Press, 1996.

[Imr1964]   M. Imre. Kreislagerungen auf Flächen konstanter Krümmung. *Acta Math. Hungar.*, 15:115–121, 1964.

[Ism1998]   D. Ismailescu. Covering the plane with copies of a convex disk. *Disc. Comp. Geom.*, 20:251–263, 1998.

[Jan1992]   C. Janot. *Quasicrystals: A Primer*. Oxford Univ. Press, 1992.

[Jan1998]   J. Januszewski. Covering a triangle with sequences of its homothetic copies. In: *Proceedings of the International Scientific Conference on Mathematics*, Univ. Žilina, 29–34, 1998.

[Jan2003]   J. Januszewski. Translative covering a convex body by its homothetic copies. *Studia Sci. Math. Hungar.*, 40:341–348, 2003.

[JaL1994]   J. Januszewski, M. Lassak. On-line covering by boxes and by convex bodies. *Bull. Polish Acad. Sci. Math.*, 42:69–76, 1994.

[JaL1995a]   J. Januszewski, M. Lassak. On-line covering the unit square by squares and the three-dimensional unit cube by cubes. *Demonstratio Math.*, 28:143–149, 1995.

[JaL1995b]   J. Januszewski, M. Lassak. Efficient on-line covering of large cubes by convex bodies of diameters at most one. *Bull. Polish Acad. Sci. Math.*, 43:305–315, 1995.

[JaL1997]   J. Januszewski, M. Lassak. On-line packing sequences of cubes in the unit cube. *Geom. Dedicata*, 67:285–293, 1997.

[Jun1901]   H. W. E. Jung. Über die kleinste Kugel die eine räumliche Figur einschliesst. *J. reine angew. Math.*, 123:241–257, 1901.

[KaL1978]   G. A. Kabatjanskii, V. I. Levenšteĭn. Bounds for packings on a sphere and in a space. *Problems Inform. Transmission*, 14:1–17, 1978.

[KaK1993]   J. Kahn, G. Kalai. A counterexample to Borsuk's conjecture. *Bull. Am. Math. Soc.*, 29:60–62, 1993.

[KeS2002]   M. Kearney, P. Shiu. Efficient packing of unit squares in a square. *Electron. J. Combin.*, 9:Article 14, 2002.

[Ker1939]   R. Kershner. The number of circles covering a set. *Am. J. Math.*, 61:665–671, 1939.

[Ker1988]   G. Kertész. On totally separable packings of equal balls. *Acta Math. Hungar.*, 51:363–364, 1988.

[KiW1987]   K. Kirchner, G. Wengerodt. Die dichteste Packung von 36 Kreisen in einem Quadrat. *Beitr. Alg. Geom.*, 25:147–159, 1987.

[KlR1997]   D. Klain, G. C. Rota. *Introduction to Geometric Probability*. Cambridge Univ. Press, 1997.

[KLS2000]   J. Koolen, M. Laurent, A. Schrijver. Equilateral dimension of the rectilinear space. *Designs, Codes Cryp.*, 21:149–164, 2000.

[KuK1990]   G. Kuperberg, W. Kuperberg. Double lattice packings of convex bodies in the plane. *Disc. Comp. Geom.*, 5:389–397, 1990.

[KuK?]   G. Kuperberg, W. Kuperberg. Ball packings and coverings with respect to the unit cube, in preparation.

[Kup1987]   W. Kuperberg. An inequality linking packing and covering densities of plane convex bodies. *Geom. Dedicata*, 23:59–66, 1987.

[Kup?]   W. Kuperberg. An extremum property of the cross-polytope. http://front.math.ucdavis.edu/math.MG/0112290.

[LaP2003]   J. C. Lagarias, P. A. B. Pleasants. Repetitive Delone sets and perfect quasicrystals, *Ergodic Theory Dynam. Sys.*, 23:831–867, 2003.

[Lar1966a]   D. G. Larman. An asymptotic bound for the residual area of a packing of discs. *Proc. Cambridge Philos. Soc.*, 62:699–704, 1966.

[Lar1966b]   D. G. Larman. A note on the Besicovitch dimension of the closest packing of spheres in $\mathbb{R}^n$. *Proc. Cambridge Philos. Soc.*, 62:193–195, 1966.

[LaR1970]   D. G. Larman, C. A. Rogers. Paths in the one-skeleton of a convex body. *Mathematika*, 17:293–314, 1970.

[LaR1973]   D. G. Larman, C. A. Rogers. The finite dimensional skeletons of a compact convex set. *Bull. London Math. Soc.*, 5:145–153, 1973.

[LaT1984]   D. G. Larman, N. K. Tamvakis. The decomposition of the n-sphere and the boundaries of plane convex domains. In: *Convexity and Graph Theory*, M. Rosenfeld, J. Zaks (eds.), North-Holland Math. Stud. 87, North-Holland, 209–214, 1984.

[LaZ1999]   D. G. Larman, Ch. Zong. On the kissing numbers of some special convex bodies. *Disc. Comp. Geom.*, 21:233–242, 1999.

[Las1993]   M. Lassak. Estimation of the volume of parallelotopes contained in convex bodies. *Bull. Polish Acad. Sci. Math.*, 41:349–353, 1993.

[Las1997]   M. Lassak. On-line potato-sack algorithm efficient for packing into small boxes. *Period. Math. Hungar.*, 34:105–110, 1997.

[Las2002]   M. Lassak. On-line algorithms for the q-adic covering of the unit interval and for covering a cube by cubes. *Beitr. Alg. Geom.*, 43:537–549, 2002.

[LaZ1991]   M. Lassak, J. Zhang. An on-line potato-sack theorem. *Disc. Comp. Geom.*, 6:1–7, 1991.

[Lau1943]   M. von Laue. Der Wulffsche Satz für die Gleichgewichtsform von Kristallen. *Z. Kristallographie A*, 105:124–133, 1943.

[Lán2003]   Zs. Lángi. On the relative lengths of sides of convex polygons. *Stud. Sci. Math. Hungar.*, 40:115–120, 2003.

[Lee1956]   J. Leech. The problem of the thirteen spheres. *Math. Gazette*, 40:22–23. 1953.

[Lei1980]   K. Leichtweiß. *Konvexe Mengen*. Springer–Verlag, 1980.

[Lev1979]   V. I. Levenšteĭn. On bounds for packings in n-dimensional Euclidean space. *Sov. Math. Dokl.*, 20:417–421, 1979.

[Lev1955]   F. W. Levi. Überdeckung eines Eibereiches durch Parallelverschiebung seines offenen Kerns. *Arch. Math.*, 6:369–370, 1955.

[Lin1973]   J. Linhart. Die Newtonsche Zahl von regelmässigen Fünfecken. *Period. Math. Hungar.*, 4:315–328, 1973.

[Mah1947]   K. Mahler. On the minimum determinant and the circumscribed hexagons of a convex domain. *Indagationes Math.*, 9:326–337, 1947.

[MaM1991]   E. Makai Jr., H. Martini. On the number of antipodal or strictly antipodal pairs of points in finite subsets of $\mathbb{R}^d$. In: *Applied geometry and discrete mathematics*, P. Gritzmann, B. Sturmfels (eds.), DIMACS Ser. Discrete Math. Theoret. Comput. Sci. 4, Am. Math. Soc., 457–470, 1991.

[MaM1993]   E. Makai Jr., H. Martini. On the number of antipodal or strictly antipodal pairs of points in finite subsets of $\mathbb{R}^d$. II. *Period. Math. Hungar.*, 27:185–198, 1993.

[MaP1983]   E. Makai Jr., J. Pach. Controlling function classes and covering Euclidean space. *Stud. Sci. Math. Hungar.*, 18:435–459, 1983.

[MaP?]   P. Mani-Levitska, J. Pach. Decomposition problems for multiple coverings of unit balls. manuscript.

[MEP2003] V. N. Manoharan, M. T. Elsesser, D. J. Pine. Dense packing and symmetry in small clusters of microspheres. *Science*, 301:483–487, 2003.

[Mar?] M. Cs. Markót. Optimal packing of 28 equal circles in a unit square–the first reliable solution. *SIAM J. Optimization*, accepted.

[Mar1996] G. E. Martin. *The Foundations of Geometry and the Non-Euclidean Plane.* Springer-Verlag, 1996.

[Mar1999] T. H. Marshall. Asymptotic volume formulae and hyperbolic ball packing. *Ann. Acad. Sci. Fenn. Math.*, 24:31–43, 1999.

[Mat2002] J. Matoušek. *Lectures on Discrete Geometry.* Springer-Verlag, 2002.

[McM1980] P. McMullen. Convex bodies which tile space by translation. *Mathematika*, 27:113–121, 1980.

[MeM1968] J. Meir, L. Moser. On packing of squares and cubes. *J. Combin. Theory*, 5:126–134, 1968.

[MeS1965] A. Meir, J. Schaer. On a geometric extremum problem. *Canad. Math. Bull.*, 8:21–27, 1965.

[Meg1996] A. Megeney. *The Besicovitch and the Hausdorff dimension of the residual set of packings of convex bodies in $\mathbb{R}^n$.* Master Thesis, University College London, 1996.

[Mel1993] J. B. M. Melissen. Densest packing of eleven congruent circles in an equilateral triangle. *Am. Math. Mon.*, 100:916–925, 1993.

[Mel1994a] J. B. M. Melissen. Densest packing of eleven congruent circles in a circle. *Geom. Dedicata*, 50:15–25, 1994.

[Mel1994b] J. B. M. Melissen. Densest packing of six equal circles in a square. *Elem. Math.*, 49:27–31, 1994.

[Mel1994c] J. B. M. Melissen. Optimal packings of eleven equal circles in an equilateral triangle. *Acta Math. Hungar.*, 65:389–393, 1994.

[Mel1997] J. B. M. Melissen. *Packing and covering with circles.* PhD dissertation, University of Utrecht, 1997.

[MeS2000] J. B. M. Melissen, P. C. Schuur. Covering a rectangle with six and seven circles. *Discrete Appl. Math.*, 99:149–156, 2000.

[MeS2001] M. Meyer and U. Schnell. On finite lattice coverings. *J. Geometry*, 72:132–150, 2001.

[Mil1987] R. Milano. *Configurations optimales de disques dans un polygone régulier.* Mémoire de Licence Université Libre de Bruxelles, 1987.

[Mil1985] V. D. Milman. Almost Euclidean quotient spaces of subspaces of a finite-dimensional normed space. *Proc. Am. Math. Soc.*, 94:445–449, 1985.

[Mol1952] J. Molnár. Ausfüllung und Überdeckung eines konvexen sphärischen Gebieten durch Kreise, I. *Publ. Math. Debrecen*, 2:266–275, 1952.

[Mol1953] J. Molnár. Ausfüllung und Überdeckung eines konvexen sphärischen Gebieten durch Kreise, II. *Publ. Math. Debrecen*, 3:150–157, 1953.

[Mol1955] J. Molnár. On inscribed and circumscribed polygons of convex regions [in Hungarian]. *Mat. Lapok*, 6:210–218, 1955.

[Mol1959] J. Molnár. Unterdeckung und Überdeckung der Ebene durch Kreise. *Ann. Univ. Sci. Budapest*, 2:33–40, 1959.

[Mol1960] J. Molnár. Über ϱ-konvexe Gebiete II. *Magyar Mat. Kongr. Budapest*, 51–53, 1960.

[Mol1966]   J. Molnár. Aggregati di cerchi di Minkowski [in Italian]. *Ann. Mat. Pura Appl.*, (4) 71:101–108, 1966.

[MoM1967]   J. Moon, L. Moser. Some packing and covering theorems. *Colloq. Math.*, 17:103–110, 1967.

[Mos1960]   L. Moser. Problem 24. *Canad. Math. Bull.*, 3:78, 1960.

[Mus?]   O. R. Musin. The kissing number in four dimensions. arXiv:math.MG/ 0309430, submitted.

[Nur1999]   K. J. Nurmela. *Circle Coverings in the Plane*. TUCS Gen. Publ., 15, Turku, 71–78, 1999.

[Nur2000]   K. J. Nurmela. Conjecturally optimal coverings of an equilateral triangle with up to 36 equal circles. *Exp. Math.*, 9:241–250, 2000.

[NuÖ1997]   K. J. Nurmela, P. R. J. Östegård. Packing up to 50 equal circles in a square. *Disc. Comp. Geom.*, 18:111–120, 1997.

[NuÖ1999]   K. J. Nurmela, P. R. J. Östegård. More optimal packings of equal circles in a square. *Disc. Comp. Geom.*, 22:439–457, 1999.

[NÖS1999]   K. J. Nurmela, P. R. J. Östegård, R. aus dem Spring. Asymptotic behavior of optimal circle packings in a square. *Canad. Math. Bull.*, 42:380–385, 1999.

[OdS1979]   A. M. Odlyzko, N. J. A. Sloane. New bounds on the number of unit spheres that can touch a unit sphere in n dimensions. *J. Combin. Theory Ser. A*, 26:210–214, 1979.

[Ole1961a]   N. Oler. An inequality in the geometry of numbers. *Acta Math.*, 105:19–48, 1961.

[Ole1961b]   N. Oler. A finite packing problem. *Canad. Math. Bull.*, 4:153–155, 1961.

[ONF?]   B. Orvos–Nagyné Farkas. The hyperbolic version of the moment theorem. in preparation.

[Pac1986]   J. Pach. Covering the plane with convex polygons. *Disc. Comp. Geom.*, 1:73–81, 1986.

[PaA1995]   J. Pach, P. K. Agarwal. *Combinatorial Geometry*. Wiley-Interscience, 1995.

[Pay1997]   Ch. Payan. Empilement de cercles égaux dans un triangle équilatéral. À propos d'une conjecture d'Erdős-Oler. *Discrete Math.*, 165/166:555–565, 1997.

[PWM1992]   R. Peikert, D. Würtz, M. Monagan, C. de Groot. Packing circles in a square: A review and new results. In: *Lect. Notes Control Inf. Sci.*, 180:45–54, 1992.

[Pen1974]   R. Penrose. The role of aesthetics in pure and applied mathematical research. *Bull. Inst. Math. Appl.*, 10:266–271, 1974.

[Pet1971]   C. M. Petty. Equilateral sets in Minkowski space. *Proc. Am. Math. Soc.*, 29:369–374, 1971.

[Pic1899]   G. Pick. Geometrisches zur Zahlenlehre. *Lotus Prag*, 311–319, 1899.

[Pir1969]   U. Pirl. Der Mindestanstand von n in der Einheitskreisscheibe gelegenen Punkten. *Math. Nachr.*, 40:111–124, 1969.

[Pór?a]   A. Pór. On e-antipodal polytopes. submitted.

[Pór?b]   A. Pór. On the Besicovitch dimension of the residual set of a sphere packing. submitted.

[PóT?]   A. Pór, I. Talata. Gaps of kissing number. in preparation.

[Puh1979] S. V. Puhov. Inequalities for the Kolmogorov and Bernšteĭn widths in
           Hilbert space. *Math. Notes*, 25:320–326, 1979.

[Rad1981] Ch. Radin. The ground state for soft disks. *J. Statist. Phys.*, 26:365–373,
           1981.

[Ran1955] R. A. Rankin. The closest packing of spherical caps in *n* dimensions.
           Proc. Glasgow Math. Assoc., 2:139–144, 1955.

[Rat1994] J. G. Ratcliffe. *Foundations of Hyperbolic Manifolds.* Springer-Verlag,
           1994.

[Rei1922] K. Reinhardt. Extremale Polygone gegebenen Durchmesser. *Jahres-
           bericht Deut. Math. Ver.*, 31:251–270, 1922.

[Rei1934] K. Reinhardt. Über die dichteste gitterförmige Lagerung kongruenter
           Bereiche in der Ebene und eine besondere Art konvexer Kurven. *Abh.
           Math. Sem. Hansischer Univ.*, 10:216–230, 1934.

[Ren1970] A. Rényi. *Probability Theory.* North-Holland, 1970.

[Rob1961] R. M. Robinson. Arrangements of 24 points on a sphere. *Math. Ann.*,
           144:17–48, 1961.

[Rog1950] C. A. Rogers. A note on coverings and packings. *J. London Math. Soc.*,
           25:327–331, 1950.

[Rog1951] C. A. Rogers. The closest packing of convex two-dimensional do-
           mains. *Acta Math.*, 86:309–321, 1951. Correction. ibid. 104:305–306,
           1960.

[Rog1957] C. A. Rogers. A note on coverings. *Mathematika*, 4:1–6, 1957.

[Rog1958] C. A. Rogers. The packing of equal spheres. *Proc. London Math. Soc.*,
           8:609–620, 1958.

[Rog1963] C. A. Rogers. Covering a sphere with spheres. *Mathematika*, 10:157–164,
           1963.

[Rog1964] C. A. Rogers. *Packing and Covering*, Cambridge Univ. Press, 1964.

[Rog1998] C. A. Rogers. *Hausdorff Measures.* Cambridge Univ. Press, 1998.

[RoS1957] C. A. Rogers, G. C. Shephard. The difference body of a convex body.
           Arch. Math., 8:220–233, 1957.

[RoV1978] K. F. Roth, R. C. Vaughan. Inefficiency in packing squares with unit
           squares. *J. Combin. Theory Ser. A*, 24:170–186, 1978.

[Rud1969] M. Ruda. Packing of circles in rectangles [in Hungarian]. *Magyar Tud.
           Akad. Mat. Fiz. Oszt. Közl.*, 19:73–87, 1969.

[Rud2000] M. Rudelson. Distances between non-symmetric convex bodies and the
           MM^* estimate. *Positivity*, 4:161–178, 2000.

[SaY1993] J. R. Sangwine–Yager. Mixed Volumes. In: *Handbook of Convex
           Geometry*, P. M. Gruber, J. M. Wills (eds.), North Holland, 43–71, 1993.

[San1976] L. A. Santaló. *Integral Geometry and Geometric Probability.* Addison-
           Wesley, 1976.

[Sch1965] J. Schaer. The densest packing of nine circles in a square. *Canad. Math.
           Bull.*, 8:273–277, 1965.

[Sch1966a] J. Schaer. On the densest packing of spheres in a cube. *Canad. Math.
           Bull.*, 9:265–270, 1966.

[Sch1966b] J. Schaer. The densest packing of five spheres in a cube. *Canad. Math.
           Bull.*, 9:271–274, 1966.

[Sch1966c] J. Schaer. The densest packing of six spheres in a cube. *Canad. Math. Bull.*, 9:275–280, 1966.

[Sch1994] J. Schaer. The densest packing of ten congruent spheres in a cube. In: *Intuitive Geometry*, K. Böröczky, G. Fejes Tóth (eds.), Colloq. Math. Soc. J. Bolyai 63, North-Holland, 403–424, 1994.

[Sch1961] W. M. Schmidt. Zur Lagerung kongruenter Körper in Raum. *Monats. Math.*, 65:154–158, 1961.

[Sch1963a] W. Schmidt. On the Minkowski-Hlawka theorem. *Illinois J. Math.*, 7:18–23, 1963. Correction, ibid. 7:714, 1963.

[Sch1993a] R. Schneider. Convex surfaces, curvature and surface area measures. In: *Handbook of Convex Geometry*, P. M. Gruber, J. M. Wills (eds.), North Holland, 273–299, 1993.

[Sch1993b] R. Schneider. *Convex Bodies – The Brunn–Minkowski Theory*, Cambridge Univ. Press, 1993.

[Sch1999] U. Schnell. Periodic sphere packings and the Wulff-Shape. *Beitr. Alg. Geom.*, 40:125–140, 1999.

[Sch2000a] U. Schnell. Wulff-shape and density deviation. *Geom. Dedicata*, 79:51–63, 2000.

[ScW2000] U. Schnell, P. M. Wills. Densest packings of more than three d-spheres are nonplanar. *Disc. Comp. Geom.*, 24:539–549, 2000.

[Sch1942] I. J. Schoenberg. Positive definite functions on spheres. *Duke Math. J.*, 9:96–108, 1942.

[Sch2000b] P. Scholl. *Finite Kugelpackungen*. Diploma thesis, Universität Siegen, 2000.

[SSW2003] P. Scholl, A. Schürmann, J. M. Wills. A discrete isoperimetric inequality and its application to sphere packings. In: *Discrete and Computational Geometry – The Goodman–Pollack Festschrift*, B. Aronov, S. Basú, M. Sharir, J. Pach (eds.), Springer-Verlag, 751–765, 2003.

[Sch2000c] A. Schürmann. On parametric density of finite circle packings. *Beitr. Alg. Geom.*, 41:329–334, 2000.

[Sch2000d] A. Schürmann. *Plane finite packings*. PhD thesis, Siegen, 2000.

[Sch2002] A. Schürmann. On extremal finite packings. *Disc. Comp. Geom.*, 28:389–403, 2002.

[Sch?] A. Schürmann. The spherical conjecture in Minkowski geometry. *Arch. Math.*, accepted.

[Sch1955] K. Schütte. Überdeckungen der Kugel mit höchstens acht Kreisen. *Math. Ann.*, 129:181–186, 1955.

[Sch1963b] K. Schütte. Minimale Durchmesser endlicher Punktmenge mit vorgeschriebenem Mindestabstand. *Math. Ann.*, 150:91–98, 1963.

[ScW1951] K. Schütte, B. L. van der Waerden. Auf welcher Kugel haben 5, 6, 7, 8 oder 9 Punkte mit Mindestabstand Eins Platz? *Math. Ann.*, 123:96–124, 1951.

[ScW1953] K. Schütte, B. L. van der Waerden. Das Problem der dreizehn Kugeln. *Math. Ann.*, 53:325–334, 1953.

[Sch1970] B. L. Schwartz. Separating points in a square. *J. Recr. Math.*, 3:195–204, 1970.

[SeM1944] B. Segre, K. Mahler. On the densest packing of circles. *Am. Math. Mon.*, 51:261–270, 1944.

[Sen1990]   M. Senechal. Brief history of geometrical crystallography, In: *Historical Atlas of Crystallography*, J. Lima-de-Faria (ed.), Kluwer, 1990.

[Sen1995]   M. Senechal. *Quasicrystals and Geometry.* Cambridge Univ. Press, 1995.

[Sha1959]   C. E. Shannon. Probability of error for optimal codes in a Gaussian channel. *Bell System Tech. J.*, 38:611–656, 1959.

[Slo]   N. J. A. Sloane. personal homepage. http://www.research.att.com/~njas

[Sol1997]   B. Solomyak. Dynamics of self similar tilings. *Ergodic Theory Dynam. Sys.*, 17:695–738, 1997.

[Swi1953]   H. P. F. Swinnerton-Dyer. Extremal lattices of convex bodies. *Proc. Cambridge Philos. Soc.*, 49:161–162, 1953.

[SzU2001]   L. Szabó, Z. Ujváry–Menyhárt. Maximal facet-to-facet snakes of unit cubes. *Beitr. Alg. Geom.*, 42:203–217, 2001.

[SzU2002]   L. Szabó, Z. Ujváry–Menyhárt. Clouds for planar convex bodies. *Aequationes Math.*, 63:292–302, 2002.

[SzS?]   P. G. Szabó, E. Specht. Packing up to 200 equal circles in a square, submitted.

[Sze1952]   T. Szele. Solution to Problem 35 [in Hungarian]. *Mat. Lapok*, 3:95, 1952.

[Tal1998a] I. Talata. Exponential lower bound for the translative kissing numbers of d-dimensional convex bodies. *Disc. Comp. Geom.*, 19:447–455, 1998.

[Tal1998b] I. Talata. On a lemma of Minkowski. *Period. Math. Hungar.*, 36:199–207, 1998.

[Tal1999a] I. Talata. Solution of Hadwiger–Levi's covering problem for duals of cyclic $2k$-polytopes. *Geom. Dedicata*, 74:61–71, 1999.

[Tal1999b] I. Talata. The translative kissing number of tetrahedra is 18. *Disc. Comp. Geom.*, 22:231–248, 1999.

[Tal1999c] I. Talata. On extensive subsets of convex bodies. *Period. Math. Hungar.*, 38:231–246, 1999.

[Tal2000a] I. Talata. On translational clouds for a convex body. *Geom. Dedicata*, 80:319–329, 2000.

[Tal2000b] I. Talata. A lower bound for the translative kissing numbers of simplices. *Combinatorica*, 20:281–293, 2000.

[Tal?a]   I. Talata. Even translative kissing numbers in dimension three, in preparation.

[Tal?b]   I. Talata. An example for a convex body whose translative kissing number is odd, in preparation.

[Tal?c]   I. Talata. On the translative kissing number of direct products of convex bodies, in preparation.

[Tal?d]   I. Talata. On equilateral dimensions of strictly convex bodies. in preparation.

[Tam1970]   P. Tammela. An estimate of the critical determinant of a two dimensional convex symmetric domain. *Izv. Vyss. Ucebn. Zaved. Mat.*, 1970:103–107, 1970.

[Tam1930]   P. M. L. Tammes. On the origin of number and arrangement of the places of exit on the surface of pollen grains. *Rec. Trav. Bot. Neerl.*, 27:1–84, 1930.

[Tar1984]   T. Tarnai. Spherical circle-packing in nature, practice and theory. *Structural Topology*, 9:39–58, 1984.

[Tar2001]　　T. Tarnai. Mechanical model of the pattern formation of lotus receptacles. *Int. J. Solids Struc.*, 38:2161–2170, 2001.

[Tar?]　　　T. Tarnai. Packing on a sphere. *Roy. Soc. London Proc. Ser. A*, in preparation.

[TaG1991]　T. Tarnai, Zs. Gáspár. Arrangement of 23 points on a sphere (on a conjecture of R. M. Robinson). *Roy. Soc. London Proc. Ser. A*, 433:257–267, 1991.

[TaG1995]　T. Tarnai, Zs. Gáspár. Covering a square. *Elem. Math.*, 50:167–170, 1995.

[Tho1996]　A. C. Thompson. *Minkowski Geometry*. Cambridge Univ. Press, 1996.

[Thu1892]　A. Thue. On new geometric and number theoretic theorems [in Norwegian]. *Forhdl. Skand. Naturforskermode*, 352–353, 1892.

[Thu1910]　A. Thue. Über die dichteste Zusammenstellung von kongruenten Kreisen in einer Ebene. *Christiana Vid. Selsk. Skr.*, 1:3–9, 1910.

[Thu1977]　A. Thue. *Selected Papers*, Universitetsforlaget, Oslo, 1977.

[Tót?]　　　G. Tóth. Decomposing multiple coverings, in preparation.

[VaW1994]　S. Vassallo and J. M. Wills. On mixed sphere packings. *Beitr. Alg. Geom.*, 35:67–71, 1994.

[Ven1954]　B. A. Venkov. On a class of Euclidean polyhedra [in Russian]. *Vestnik Leningrad. Univ. Ser. Mat. Fiz. Him.*, 9:11–31, 1954.

[Ver?]　　　J-L. Verger-Gaugry. Covering a ball with smaller equal balls in $\mathbb{R}^n$. *Disc. Comp. Geom.*, accepted.

[Vin1950]　S. Vincze. On a geometrical extremum problem. *Acta Sci. Math. Szeged*, 12:136–142, 1950.

[Vor1908]　G. F. Voronoi. Deuxième Mémoire. Recherches sur les parallélloèdres primitifs. *J. reine angew. Math.*, 134:198–287, 1908.

[Wae1952]　B. L. van der Waerden. Punkte auf der Kugel. Drei Zusatze. *Math. Ann.*, 125:213–222, 1952.

[Weg1980]　G. Wegner. Zur einem ebenen Überdeckungsproblem. *Stud. Sci. Math. Hungar.*, 15:287–297, 1980.

[Weg1986]　G. Wegner. Über endliche Kreispackungen in der Ebene. *Stud. Sci. Math. Hungar.*, 21:1–28, 1986.

[Weg1992]　G. Wegner. Relative Newton numbers. *Monats. Math.*, 114:149–160, 1992.

[Wil1983]　J. M. Wills. Reseach problem 33. *Period. Math. Hungar.*, 14:189–191, 1983.

[Wil1993]　J. M. Wills. Finite sphere packings and sphere coverings. *Rend. Sem. Math. Messina*, Serie II/17, Supp. No. 2:99–103, 1993.

[Wil1996]　J. M. Wills. Lattice packings of spheres and the Wulff-shape. *Mathematika*, 43:229–236, 1996.

[Wil1997]　J. M. Wills. On large lattice packings of spheres. *Geom. Dedicata*, 65:117–126, 1997.

[Wil2000]　J. M. Wills. The Wulff-shape of large periodic sphere packings. In: *Discrete Mathematical Chemistry*, P. Hansen, P. Fowler, M. Zheng (eds.), DIMACS Ser. Discrete Math. Theoret. Comput. Sci. 51, Am. Math. Soc., 367–375, 2000.

[Win2003]　G. Wintsche. *Finite sphere packings and coverings* [in Hungarian]. PhD thesis, Technical University Budapest, 2003.

[Wul1901] G. Wulff. Zur Frage der Geschwindigkeit des Wachstums und der Auflösung der Krystalflächen. *Z. Kristallographie Mineral*, 34:499, 1901.

[Wyn1965] J. M. Wyner. Capabilities of bounded discrepancy decoding. *Bell System Tech. J.*, 44:1061–1122, 1965.

[Zie1995] G. Ziegler. *Lectures on Polytopes*. Springer-Verlag, 1995.

[Zon1994] Ch. Zong. An example concerning the translative kissing number of a convex body. *Disc. Comp. Geom.*, 12:183–188, 1994.

[Zon1995a] Ch. Zong. On a conjecture of Croft, Falconer and Guy on finite packings. *Arch. Math.*, 64:269–272, 1995.

[Zon1995b] Ch. Zong. Some remarks concerning kissing numbers, blocking numbers and covering numbers. *Period. Math. Hungar.*, 30:233–238, 1995.

[Zon1996a] Ch. Zong. *Strange Phenomena in Convex and Discrete Geometry*. Springer-Verlag, 1996.

[Zon1996b] Ch. Zong. The kissing numbers of tetrahedra. *Disc. Comp. Geom.*, 15:239–252, 1996.

[Zon1997a] Ch. Zong. The translative kissing number of the Cartesian product of two convex bodies, one of which is two-dimensional. *Geom. Dedicata*, 65:135–145, 1997.

[Zon1997b] Ch. Zong. A problem of blocking light rays. *Geom. Dedicata*, 67:117–128, 1997.

[Zon1999] Ch. Zong. *Sphere packings*. Springer-Verlag, 1999.

Index